PERFORMANCE OF PROTECTIVE CLOTHING

A symposium
sponsored by
ASTM Committee F-23
on Protective Clothing
Raleigh, NC, 16–20 July 1984

ASTM SPECIAL TECHNICAL PUBLICATION 900
Roger L. Barker, North Carolina State
University, and Gerard C. Coletta,
Risk Control Services, editors

ASTM Publication Code Number (PCN)
04-900000-55

1916 Race Street, Philadelphia, PA 19103

Library of Congress Cataloging-in-Publication Data

Performance of protective clothing.

(ASTM special technical publication; 900)
"ASTM publication code number (PCN) 04-900000-55."
Papers presented at the First International
Symposium on the Performance of Protective Clothing.
Includes bibliographies and index.
1. Clothing, Protective—Testing—Congresses.
I. Barker, Roger L. II. Coletta, Gerard C.
III. ASTM Committee F-23 on Protective Clothing.
IV. International Symposium on the Performance of
Protective Clothing (1st : 1984 : Raleigh, N.C.)
V. Series
HD7395.C5P47 1986 687'.16 86-10706
ISBN 0-8031-0461-8

Library of Congress Catalog Card Number: 86-10706

NOTE

The Society is not responsible, as a body,
for the statements and opinions
advanced in this publication.

Printed in Ann Arbor, MI
June 1986

Foreword

The papers in this publication, *Performance of Protective Clothing*, were presented at the International Symposium on the Performance of Protective Clothing, which was held in Raleigh, North Carolina, on 16–20 July 1984. The meeting was sponsored by ASTM Committee F-23 on Protective Clothing. This symposium was the first to bring together all areas of interest in the field of protective clothing for occupational exposures, and other such symposia are planned.

The symposium chairmen were Roger L. Barker, North Carolina State University School of Textiles, and Gerard C. Coletta, Risk Control Services. Both men also served as editors of this publication.

Related ASTM Publications

ASTM Performance Standards for Textile Fabrics, 1983, 03-413083-18

ASTM Standards for Electrical Protective Equipment for Workers, 1985, 03-618085-21

A Note of Appreciation to Reviewers

The quality of the papers that appear in this publication reflects not only the obvious efforts of the authors but also the unheralded, though essential, work of the reviewers. On behalf of ASTM we acknowledge with appreciation their dedication to high professional standards and their sacrifice of time and effort.

ASTM Committee on Publications

ASTM Editorial Staff

Contents

RESISTANCE TO PESTICIDES—FIELD PERFORMANCE AND CLEANING PROCEDURES

RISK ASSESSMENT OF CHEMICAL EXPOSURE HAZARDS IN SELECTING PROTECTIVE CLOTHING

Heat Stress, Fit Testing, and Other Performance Requirements for Protective Clothing

Summary

Indexes

Introduction

The papers contained in this Special Technical Publication address a broad range of topics relating to the performance of clothing used for protection against chemical and thermal occupational hazards. These papers were presented at the first International Symposium on the Performance of Protective Clothing, held in Raleigh, North Carolina, on 16–20 July 1984. The meeting was sponsored by ASTM Committee F-23 on Protective Clothing.

Topics in this volume under the general subject of chemical protection include test methodology for evaluating the permeation resistance of protective clothing materials, field evaluation methods for end-use items of protective clothing, decontamination techniques, and risk assessment in the selection and use of chemical protective clothing. Also, a number of papers deal with the performance of protective clothing used against pesticides, a chemical hazard that has recently begun to attract a high level of interest.

Topics under the general subject of thermal protective clothing include research studies on laboratory test methods for measuring the insulative effects of materials, as well as the protective performance of materials exposed to hazardous levels of heat. The specific role of new, heat-resistant fabrics in insulating against flame and molten metals is discussed, as are specialized clothing systems for industrial and fire-fighting applications. Subjects important to the design and selection of protective garments are covered, including garment fit, sizing requirements, and the impact of protective clothing on human comfort in hot environments.

Background

In the mid-1970s, the Industrial Safety Equipment Association (ISEA), a protective clothing and equipment manufacturers' trade organization, voiced an opinion that the lack of standards for industrial protective clothing had created problems that have impact manufacturers and users alike. At the request of ISEA, and with agreement from protective clothing users, research organizations, and several government groups, ASTM sponsored the formation of its Committee F-23 on Protective Clothing in June 1977.

Initially, the focus of Committee F-23 was on chemical protective clothing. However, in 1980 a group sponsored by the Aluminum Association, and primarily concerned with protection against thermal hazards like molten metal splash, was invited to join Committee F-23 as a separate subcommittee. Because of this action, the scope of Committee F-23 was expanded to include all aspects of base materials and protective clothing end-use items used for protection against occupational exposures to chemical and thermal hazards.

Besides the paramount need for uniform standards for industrial protective apparel, there was a recognized need for available, reliable performance data for protective clothing. The purpose of the Raleigh symposium was to bring together, for the first time, all those interested in protective clothing for occupational exposures, to present the findings of relevant research and, perhaps most important, to stimulate discussion and further development efforts.

Response to the Symposium

By every measure, the response to this symposium was overwhelming. More than 50 technical papers were presented by experts from the United States and abroad. The authors came from industrial, academic, and government laboratories; from technical organizations and corporate safety programs; and as private consultants. Many of these authors are widely recognized as current experts in diverse aspects of laboratory testing and materials development, in clothing design, and in occupational risk assessment procedures. Many have first-hand experience and knowledge of the needs and requirements of protective clothing in the workplace.

Truly international in scope, the 250 attendees brought a diversity of interests and perspectives. They included research scientists, engineers, apparel and materials manufacturers, industrial safety professionals, and fire fighters.

Overview of the Subject Matter

The papers in this volume are separated into eight sections, each of which represents an important topic in the areas of testing, performance, selection, or use of protective clothing. The sections are arranged to distinguish between chemical protection, thermal protection, and other related topics that are generally important to the performance of protective clothing.

The first four sections are devoted to chemical protection. The first of these addresses the permeation resistance of base materials. The second focuses on field performance and cleaning processes for clothing systems used in protection against pesticides. Discussions on risk assessment as a part of the selection and use of chemical protection clothing are featured in the third. The testing of seams and closures, fully encapsulated ensembles, and procedures for maintenance and decontamination of protective clothing are presented in the fourth section.

The next three sections are concerned with thermally protective materials and clothing. Laboratory test methods and the evaluation of the thermal performance of clothing materials used against radiant heat or flame are discussed in the fifth section. New developments in the technology of flame-resistant fibers and finishes for fabrics are the subjects of papers in the sixth. The next section is concerned with clothing systems for industrial and fire-fighting applications.

The eighth section differs from the previous seven in that the papers address topics which, although distinct from thermal or chemical resistance, are critical nonetheless to the performance of protective clothing. These topics include sizing, fit testing, and heat stress.

Significance of This Special Technical Publication

This volume represents the first Special Technical Publication (STP) by ASTM in the field of industrial protective clothing. These papers embody the understanding and diverse perspectives of many of the world's top experts in the field of protective clothing. This volume should be a valuable source of information for anyone interested in chemical- and heat-resistant materials and clothing, as well as in principles and applications of laboratory testing methodologies.

Further, this STP is intended to stimulate broader interest in the ongoing activities of ASTM Committee F-23 in the areas of chemical and thermal resistance and in the related activities of anthropometric sizing and comfort of protective clothing. ASTM will continue its efforts in this field by publishing papers from the second International Symposium on the Performance of Protective Clothing, scheduled to be held in Tampa, Florida, in January 1987.

Dr. J. Donald Millar, Director of the U.S. National Institute for Occupational Safety and Health (NIOSH), was the keynote speaker at this first symposium. Dr. Millar pointed out that "the ultimate goal [of NIOSH] is the prevention of occupational disease and injury" and that "protective clothing can help prevent some of the leading work-related illnesses and injuries." The goal of achieving superior industrial protective clothing is brought ever closer, not only by vigorous research efforts but also by forums that make results available to scientists, to manufacturers, and to users of protective apparel.

The research discussed herein contributes to the body of authoritative literature on protective clothing. The editors hope that these findings will focus additional efforts in this important area, an area that has impact on the safety and health of hundreds of thousands of workers throughout the world.

Roger L. Barker

North Carolina State University School of Textiles, Raleigh, NC 27650; symposium chairman and editor.

Gerard C. Coletta

Risk Control Services, Inc., Tiburon, CA 94920; symposium chairman and editor.

Permeation Resistance of Chemical Protective Clothing Materials

Stephen L. Davis,[1] *Charles E. Feigley,*[2] *and*
George A. Dwiggins[2]

Comparison of Two Methods Used to Measure Permeation of Glove Materials by a Complex Organic Mixture

REFERENCE: Davis, S. L., Feigley, C. E., and Dwiggins, G. A., **"Comparison of Two Methods Used to Measure Permeation of Glove Materials by a Complex Organic Mixture,"** *Performance of Protective Clothing, ASTM STP 900,* R. L. Barker and G. C. Coletta, Eds., American Society for Testing and Materials, Philadelphia, 1986, pp. 7–21.

ABSTRACT: Previous studies measuring permeation of liquid chemicals through glove materials by measurement of radiolabeled phenol in a liquid receiving medium were replicated with a method using a flowing gaseous receiving medium in which permeant was measured by a flame ionization detector (FID). The breakthrough times and permeation rates for liquefied coal and toluene permeating polyvinyl chloride (PVC), nitrile rubber, and natural rubber glove specimens were compared. For liquefied coal, the FID method indicated that all glove types exhibited much greater resistance to steady-state permeation (although earlier breakthrough was detected for PVC and natural rubber). For toluene permeation through natural rubber and the two PVC materials, the radiolabeled tracer method indicated greater resistance. Many of the marked differences in observed results are not attributable to instrument sensitivity. When a challenging substance is a complex mixture, large differences in permeation parameters measured by different techniques can be the result of various constituents permeating at different rates. Methods that do not quantify specifically the compounds of toxicological significance may provide misleading information on the degree of protection provided.

KEY WORDS: protective clothing, industrial hygiene, liquefied coal, permeation rate, occupational health, breakthrough time

Various methods to measure the extent of permeation through protective glove materials have been reported. These include the use of scintillation

[1]Industrial hygienist, IT Corp., Knoxville, TN 37923. Formerly graduate student, School of Public Health, University of South Carolina, Columbia, SC 29208.

[2]Associate professor, and assistant professor, respectively, School of Public Health, University of South Carolina, Columbia, SC 29208.

counters [1], spectrophotometric detectors [2,3], and flame ionization detectors (FIDs) [4,5]. No single method delivers satisfactory results for all permeants and glove materials. This is particularly true in measuring the permeation of complex mixtures.

One such mixture, which poses acute and chronic hazards to occupationally exposed workers, is liquefied coal. Liquids derived from coal have been associated with numerous adverse health effects, including increases in the incidence of human skin cancers [6,7]. This possibility of adverse health effects makes the reduction of occupational exposures a matter of concern to health professionals. Gloves have been used at some facilities to reduce contact with potentially hazardous liquids, but the degree of protection conferred by a particular glove varies with the nature of the contaminant.

The permeation of liquefied coal through glove materials used in liquefaction pilot plants has been studied by Bennett et al [1]. They used radiolabeled phenol as a tracer in the challenge material and a scintillation counter to determine the amount of tracer in grab samples of saline receiving medium. Although it was inferred from previous work [2] that minor constituents may sometimes serve as indicators of the permeation of bulk materials, the degree to which the breakthrough time and permeation rate of phenol approximate those of the other constituents was not known. The study reported here was undertaken to define better the protective capabilities of the previously tested glove materials and to obtain a better understanding of permeation results obtained by means of the radiolabeled tracer and saline receiving medium. The specific objective was to compare the results of permeation experiments using a flowing gas receiving medium and an FID with the results previously obtained using the radiolabeled tracer method [1] for the same types of protective gloves and portions of the same liquefied coal sample. To facilitate this comparison, experiments using toluene as a challenge material were performed before those using liquefied coal.

Experimental Method

Circular sections (7.25 cm in diameter) were cut from the palms of test gloves and weighed (to the nearest 0.0001 g), and their thicknesses were measured with a micrometer at five points (to the nearest 0.0025 mm). The specimens were placed between two polytetrafluoroethylene (PTFE) gaskets and tightened between the two chambers of a permeation cell to a torque of 2×10^8 dyne · cm (15 ft · lbf) on the retaining bolts. A flow of 55 mL/min of dry nitrogen at 26°C passed through the receiving chamber and subsequently into the FID of a Tracor 222Q gas chromatograph (GC). The glove material remained in the cell for a minimum of 1.5 h prior to addition of the challenge material to allow the vaporization of volatile compounds from the glove specimens and, thus, to establish a steady FID response baseline.

The FID was then calibrated by injecting known quantities of toluene

[Fisher, American Chemical Society (ACS) grade] into the nitrogen stream at the modified GC injection port. Therefore, permeation rates for liquefied coal are reported as the mass of toluene that would produce an equivalent FID response. The FID response per carbon atom varies with the molecular species and could not be determined since the composition of permeant vapors from liquefied coal was not known. In the toluene experiments, breakthrough was defined as an increase of 0.18 $mg/m^2/s^1$ over background, the lowest detectable increase at the attenuation settings used (input, 10^4; output, 64). In experiments involving liquefied coal, this value was 0.00032 $mg/m^2/s^1$, as toluene (attenuation: input, 10; output, 32).

Upon completion of calibration, 70 mL of challenge material was added to the challenge chamber, thus initiating the experiment. The toluene experiments continued for a minimum of 2 h and until a steady-state permeation rate was attained. The liquefied coal experiments lasted 24 h.

Initial experiments were performed using a glass permeation cell (Pesce Laboratories), as specified in the ASTM Test for Resistance of Protective Clothing Materials to Permeation by Hazardous Liquid Chemicals (F 739-81). Because of frequent breakage of the glass cell, a stainless steel permeation cell with similar internal structure and dimensions, shown in Fig. 1, was

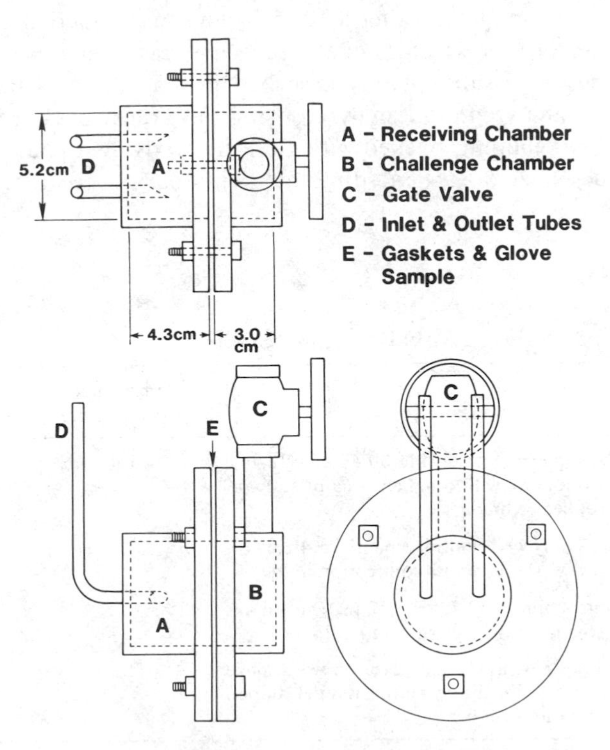

FIG. 1—*Stainless steel permeation cell.*

fabricated and used for the remainder of the study. Experiments employing the same challenge and glove materials revealed no difference in results between the two cells.

The four glove types described in Table 1 were selected for study because they were used in three coal liquefaction pilot plants and because they were tested by Bennett et al by the radiolabeled tracer method. The liquefied coal challenge material was a heavy fuel oil product stream from the pilot plant operated by Ashland Synthetic Fuels, Inc., in Catlettsburg, Kentucky, utilizing the H-Coal process (Hydrocarbon Research, Inc.).

Experiments to determine the efficacy of glove decontamination following exposure to liquefied coal using the tracer method were also replicated with gas phase receiving medium and FID measurement. Following the 24-h exposure experiments, two polyvinyl chloride (PVC)-1 glove specimens were air dried for 48 h and then washed using an identical procedure and agents used to wash gloves by one coal liquefaction pilot plant. This consists of alkali and alkali plus detergent cycles followed by seven rinse cycles and a dry cycle. The glove specimens were then placed in the cell for a postwash exposure experiment.

To investigate the possibility that the composition of the permeating materials changed during the course of the experiments, a separate permeation experiment was conducted using a PVC-1 glove and liquefied coal. Six sequential charcoal tube samples of the receiving gas stream were collected. These samples were subsequently desorbed in 0.5 mL of carbon disulfide (Fisher, Spectranalyzed) and analyzed using the Tracor gas chromatograph [183 by 0.32 cm column, packed with 10% free fatty acid phase (FFAP) on Chromasorb W, acid-washed, dimethyldichlorosilane-treated, 80 to 100 mesh].

TABLE 1—*Glove types studied.*

Glove	Description	Approximate Thickness, mm	Approximate Density, g/cm^3
PVC-1	black polyvinyl chloride 30.48-cm (12-in.) work glove with roughened palm and cotton jersey lining	1.4	0.77
PVC-2	black polyvinyl chloride smooth 25.4-cm (10-in.) work glove with knit cotton lining	1.1	0.83
Milled nitrile	white unlined 30.48-cm (12-in.) milled nitrile laboratory glove with no lining	0.29	1.1
Natural rubber	yellow 25.4-cm (10-in.) textured work glove composed of heavy cotton flannel dipped in natural rubber	2.2	0.69

Comparison of Permeation Measurement Methods

The two permeation measurement techniques of interest here differ in several fundamental respects. The method used by Bennett et al involves the addition of radiolabeled phenol to the challenging substance, the permeation of labeled challenging substance through the glove material and into the saline receiving medium, and the measurement of labeled phenol by liquid scintillation counting. It should be noted that only the amount of permeation by the radiolabeled component is measured and specimens are discrete, that is, the total permeation of one component over a time interval is assessed. Consequently, the exact instant of breakthrough cannot be determined. In contrast, the method employed in this study utilizes a flowing gas as a receiving medium and an FID as a detector of permeation. Permeation measurement therefore is quasi-instantaneous, and the detection method responds to virtually every constituent in the challenge material, although the sensitivity differs for different compounds.

A direct comparison of data generated by experiments using the two methods is possible only if two assumptions are made. One must assume that all components of the liquefied coal/radiolabeled phenol mixture permeate the glove material and enter the receiving liquid at the same rate and that the response of the FID to toluene is comparable to its average response to the components of liquefied coal. These assumptions are considered in the following discussion.

Experimental Results and Discussion—Toluene

The results for the eight toluene experiments (two runs with each glove type) are presented in Fig. 2. All four glove types exhibited breakthrough followed by a very rapid increase in permeation rate until a nearly constant, steady-state rate was attained. Steady-state permeation rates varied by less than a factor of two from the least to the greatest. Replicate experiments with the natural rubber glove specimens showed the largest variation in permeation rate, an observation consistent with the greater variation in thickness of the natural rubber gloves.

The glove density and mean thickness (prior to exposure), the observed breakthrough times, and the steady-state permeation rates for each experiment are presented in Table 2*a*. Breakthrough occurred for all experiments in less than 20 min. The natural rubber glove material, the thickest of those tested, gave the longest time to breakthrough and the lowest permeation rate. Of the two PVC glove types, the thicker material showed longer breakthrough times and lower permeation rates. However, nitrile rubber, the thinnest material tested, showed breakthrough times between those of the two PVC materials, although its steady-state permeation rate was higher than that of either

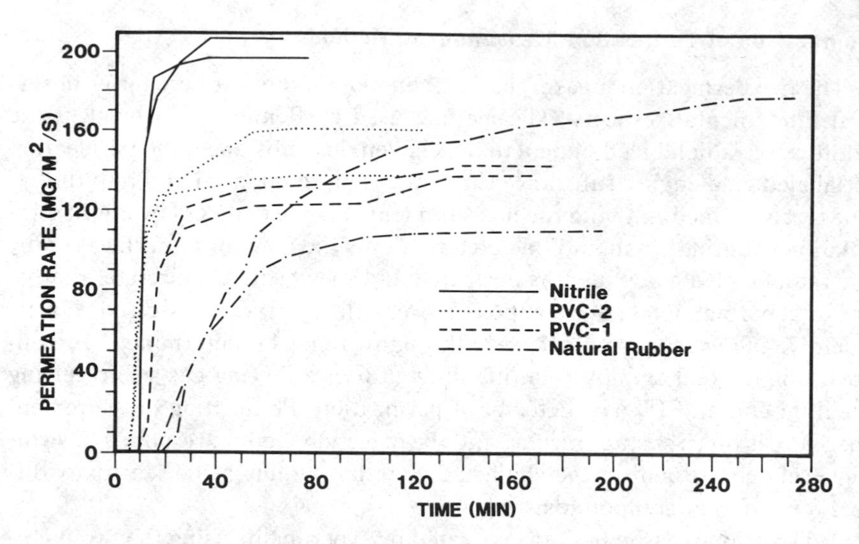

FIG. 2—*Permeation by toluene from the time of application.*

PVC material. Thus, both glove material and thickness influenced resistance to permeation.

To examine only the influence of the glove material, the measures of performance were normalized for thickness. The breakthrough times were divided by thickness squared, and the steady-state permeation rates were multiplied by thickness. Such adjustments for glove thickness, which have both theoretical [*8*] and experimental [*9*] support, permit the comparison of the intrinsic resistances to permeation displayed by various materials. Table 2*a* shows that

TABLE 2*a*—*Mean glove thicknesses and densities, toluene breakthrough times, and steady-state permeation rates.*

Glove	Average Thickness, mm	Average Density, g/cm³	Breakthrough Time, min	Permeation Rate, mg/m²/s	Normalized Breakthrough Time, min/mm²	Normalized Permeation Rate, mg·mm/m²/s
PVC-1	1.349	0.798	9.0	138	4.9	186
PVC-1	1.360	0.774	9.0	143	4.9	194
PVC-2	1.062	0.820	5.5	162	4.9	172
PVC-2	1.100	0.979	4.7	138	3.9	152
Nitrile rubber	0.328	0.987	7.5	197	69.7	65
Nitrile rubber	0.306	1.084	7.5	207	80.1	63
Natural rubber	1.911	0.733	16.5	108	4.5	206
Natural rubber	2.401	0.686	18.0	152	3.1	365

TABLE 2b—*Mean glove thicknesses and densities, toluene breakthrough times and steady-state permeation rates, reported by Bennett et al [1].*

Glove	Average Thickness, mm	Average Density, g/cm^3	Break-through Time, min	Permeation Rate, $mg/m^2/s$[a]	Normalized Breakthrough Time, min/mm^2	Normalized Permeation Rate, $mg \cdot mm/m^2/s$[a]
PVC-1	1.40	0.84	30 to 60	36	15 to 31	50
PVC-1	1.38	0.84	30 to 45		16 to 24	
PVC-2	1.00	0.84	15 to 30	47	15 to 30	51
PVC-2	1.17	0.84	15 to 30		11 to 22	
Nitrile rubber	0.39	1.04	0 to 15	640	0 to 99	250
Nitrile rubber	0.38	1.04	0 to 15		0 to 104	
Natural rubber	1.67	0.80	30 to 45	41	11 to 16	75
Natural rubber	2.01	0.80	30 to 45		7 to 11	

[a]Averages are presented. It is assumed that the ratio of phenol to toluene is the same on both sides of the glove material; that is, if a given fraction of the activity permeated over a time interval, then it is assumed that the same fraction of the total mass of challenge material also permeated over the time interval.

the normalized breakthrough times for nitrile rubber are the greatest and that those for natural rubber are the lowest, in contrast with the actual breakthrough times. Nitrile rubber appears to offer the greatest resistance to breakthrough and to subsequent permeation, when thickness is considered in this way.

The radiolabeled tracer study [*1*] reported that breakthrough occurred at between 30 and 60 min for PVC-1, between 15 and 30 min for PVC-2, prior to 15 min for milled nitrile, and between 30 and 45 min for natural rubber. Comparison with the data in Table 2*a* shows that the FID method detected breakthrough much earlier for natural rubber and both PVC materials, although nearly identical glove materials were tested. The nitrile breakthrough times determined by the FID were within the range observed by the tracer method, although the exact time of breakthrough could not be determined by the tracer method.

The steady-state permeation rates measured by means of the radiolabeled tracer technique were comparable for PVC-1, PVC-2, and natural rubber (like the rates reported in Table 2*a* for these same glove materials). The assumption of identical composition on both sides of the glove specimen permits the calculation of total permeation rates at steady-state from the radiolabeled tracer data. For PVC-1, PVC-2, and natural rubber, these estimates are approximately one-fourth to one-third of the steady-state permeation rates observed in the present study, as Table 2*b* shows. Apparently, when steady-state permeation rates had been attained, the ratio of phenol to toluene was lower on the receiving side of these specimens than in the challenge material, indicating that toluene had a greater permeability than phenol. The opposite relationship exists for nitrile rubber. The aforementioned assumption leads to the conclusion that Bennett et al measured a steady-state permeation rate three times higher than that reported in Table 2*a*, a result which suggests that the ratio of phenol to toluene was higher on the receiving side of this glove specimen. However, Bennett et al reported the deterioration of the nitrile rubber glove specimens. No such deterioration occurred under the experimental regimen employed in this study.

Differences in instrument sensitivity may contribute to the differences in breakthrough times reported in these studies. The collection of grab samples in the radiolabeled tracer technique allows the determination of breakthrough only within a period determined by the sampling frequency. In addition, the radiolabeled phenol and the toluene may not reach the receiving side of the glove material in the proportions existing on the challenge side. If it is assumed that phenol and toluene appear simultaneously on the receiving side of the glove specimen, in the same ratio as that on the challenge side, then the average permeation rate required to detect breakthrough with the radiolabeled tracer method over the 15-min sampling intervals would be 3.7 $mg/m^2/s^1$ (of the total mass of challenge material). The results from the present study (Fig. 2) indicate that the permeation rate of toluene increases very rapidly after initial detection in the receiving medium. Thus, the scintillation counting method should detect breakthrough no more than about 5 min later than the FID used in the present study, if the assumption is valid. Therefore, differential permeation rates, not sensitivity differences, probably account for most of the differences in breakthrough times determined by the two methods. The ratio of the toluene to phenol permeation rates probably exceeded the ratio of toluene to phenol in the challenge material for PVC-1, PVC-2, and natural rubber.

The normalized breakthrough times determined by these experiments and by the radiolabeled tracer experiments were compared with the results of two other studies [*3,5*] in Table 3. Nelson et al [*3*] used a MIRAN spectrometer to measure the organic vapor concentration in a flowing gas stream receiving medium. Schoch et al [*5*] also employed a flowing gas stream receiving medium, but they determined the outlet vapor concentration with an FID. The normalized breakthrough times given by the studies employing FID measurement (Table 2*a* and the last four lines of Table 3) agree extremely well, even though the gloves tested may have had different manufacturers. The normalized breakthrough times observed by FID were much shorter than those determined by either radiolabeled tracer or spectrophotometric detection for both PVC and natural rubber. Breakthrough times for pure phenol, when normalized for thickness (Table 3), were much longer than normalized break-

TABLE 3—*Breakthrough times reported by Nelson et al [3] and by Schoch et al [5].*

Glove	Thickness, mm	Challenge Substance	Break-through Time, min	Normalized Breakthrough Time, min/mm²	Reference
PVC	0.12	toluene	0.2	14	[3]
PVC	0.13	toluene	0.2	12	[3]
PVC	0.31	toluene	2.9	30	[3]
PVC	0.20	toluene	3.3	82	[3]
Natural rubber	0.45	toluene	3.0	15	[3]
Natural rubber	0.59	toluene	4.1	12	[3]
PVC	0.12	phenol	3.2	220	[3]
PVC	0.13	phenol	8.0	470	[3]
PVC	0.31	phenol	32.0	330	[3]
PVC	0.20	phenol	15.0	380	[3]
PVC	1.60	toluene	12	4.7	[5]
Nitrile rubber[a]	0.45	toluene	21	100	[5]
Nitrile rubber[b]	0.45	toluene	[c]	[c]	[5]
Natural rubber	0.42	toluene	0.5	2.8	[5]

[a]Smooth palm area.
[b]Coarse palm area.
[c]No breakthrough detected in 65 min.

through times for phenol in toluene (Table 2*b*). Thus, it is possible that the presence of toluene greatly increased the early permeation rate of phenol in the tracer experiments. (However, permeability apparently remains less than that of toluene, as discussed earlier.)

Experimental Results and Discussion—Liquefied Coal

In contrast with the characteristics of toluene permeation, liquefied coal exhibited a much slower rise in permeation rate after longer breakthrough times and a failure to achieve a steady-state permeation rate within 24 h, as Fig. 3 shows. No permeation through nitrile glove specimens was detected. The permeation rates were not dramatically different for the other glove materials; the most rapid rate detected was only 2.1 times the slowest rate.

Table 4*a* presents experimental data on tested glove materials, breakthrough times, and permeation rates after 24 h. The normalization of breakthrough times, which adjusts for glove thickness, results in close agreement among the natural rubber and the PVC materials. Normalized permeation rates show that the natural rubber offered less resistance to permeation than the PVC materials. (It is assumed that the flowing gaseous receiving medium contained all of the liquefied coal permeant, that is, none adhered to the re-

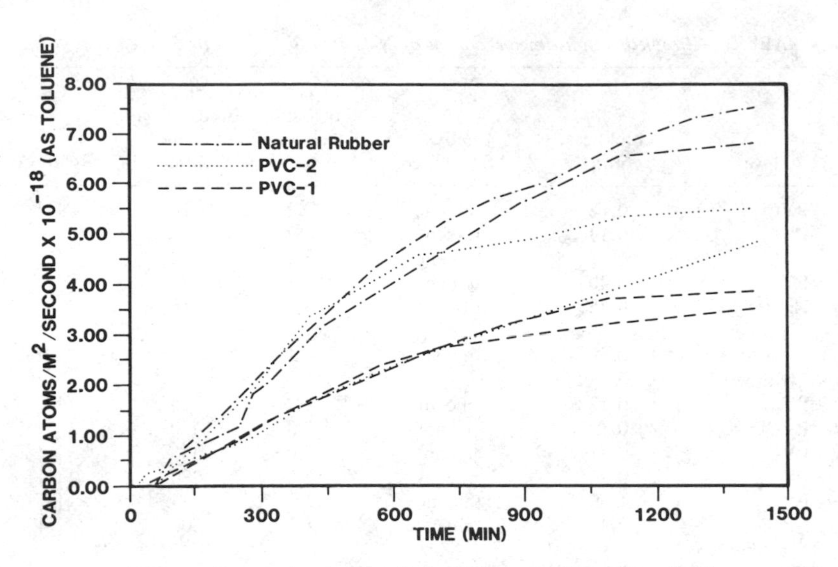

FIG. 3—*Permeation by liquefied coal from the time of application.*

ceiving side of the glove specimens. Although the many components having low vapor pressures could present such a problem, such compounds probably also have relatively low permeation rates [*10,11*].

Permeation rates reported by Bennett et al [*1*] (Table 4*b*) can be compared with those in Table 4*a* if the aforementioned assumption concerning invariant composition across the glove material is employed. This assumption leads to the conclusion that Bennett et al recorded permeation rates which were about two orders of magnitude higher than those reported in Table 4*a* for PVC-1, PVC-2, and natural rubber. (Note that the values in Table 4*a* are expressed in micrograms per metre squared per second and that those in Table 4*b* are given in milligrams per metre squared per second.) The normalized permeation rate implied by this treatment of their nitrile rubber data actually exceeds any value reported in Table 4*a* for any glove material. This provides convincing evidence that the assumption of invariant composition is not valid for these materials, even after 24 h. It appears clear that phenol permeated the glove preferentially, relative to most other volatile constituents of liquefied coal. Thus, the liquefied coal permeation rates suggested by radiolabeled phenol permeation are much too large.

Breakthrough was not detected earlier in the experiments of Bennett et al, however, as Tables 4*a* and 4*b* show. This apparently conflicting result is explained, at least in large measure, by the sensitivity differences in the detection techniques employed in the two studies. If one assumes invariant composition across the glove material, then a permeation rate of 3.7 $mg/m^2/s^1$ (of the total challenge liquid) was required to permit detection of breakthrough during the first 6 h of the experiments of Bennett et al, when 15-min sampling

TABLE 4*a*—*Mean glove thicknesses and densities, breakthrough times, and permeation rates after 24 h of exposure to liquefied coal.*

Glove	Average Thickness, mm	Average Density, g/cm^3	Break-through Time, min	Permeation Rate at 24 h, μg/m^2/s[a]	Normalized Breakthrough Time, min/mm^2	Normalized Permeation Rate at 24 h, μg·mm/m^2/s[a]
PVC-1	1.33	0.798	36	77	20	100
PVC-1	1.53	0.698	68	83	29	130
PVC-2	1.21	0.760	24	100	16	130
PVC-2	1.17	0.773	34	120	25	140
Nitrile rubber	0.23	1.090	[b]	[b]	[b]	[b]
Nitrile rubber	0.295	1.128	[b]	[b]	[b]	[b]
Natural rubber	2.33	0.653	72	150	13	350
Natural rubber	2.11	0.689	64	160	14	350

[a]Since the FID was calibrated with toluene, these rates are reported as a flow of toluene yielding an equivalent FID response.
[b]No breakthrough was detected.

TABLE 4*b*—*Mean glove thicknesses and densities, breakthrough times, and estimated permeation rates after 24 h of exposure to liquefied coal, reported by Bennett et al [1].*

Glove	Average Thickness, mm	Average Density, g/cm^3	Break-through Time, min[a]	Estimated Rate of Permeation at 24 h, mg/m^2/s	Normalized Breakthrough Time, min/mm^2[a]	Normalized Permeation Rate at 24 h, mg·mm/m^2/s
PVC-1	1.20	0.84	375 to 390	13	260 to 270	16
PVC-2	1.11	0.84	165 to 180	20	130 to 150	22
PVC-2	1.06	0.84	210 to 225		190 to 200	
Nitrile rubber	0.36	1.04	720 to 1410	1.3	5 600 to 10 900	0.5
Natural rubber	1.88	0.80	285 to 300	16	81 to 85	30

[a]The range given represents the time interval of the discrete specimen in which permeation was first detectable in the liquid receiving medium.

intervals were used. This rate is much greater than any observed in this study (Table 4*a*). The observation of breakthrough at any time during the first 6 h of the experiments by Bennett et al offers further evidence that phenol moved more rapidly through these glove materials than most other volatile constituents.

Bennett et al used a single 12-h sampling interval during the final half of their 24-h experiments. During this longer interval, breakthrough was detected for nitrile rubber. The assumption employed here gives a minimum detectable 12-h permeation rate of 0.077 $mg/m^2/s^1$, which could have been detected in the present study if it actually occurred. This suggests preferential passage of phenol through the nitrile rubber glove as well.

It is interesting to compare the values for breakthrough time in Tables 4*a* and 4*b* with those reported by Nelson et al [*4*] for shale oil and coal tar extract. One might expect that their data, shown in Table 5, would agree more closely with those reported in Table 4*a*, because of the similarity of the two methods. The fact that their data agree more closely with radiolabeled tracer results suggests that a lower sensitivity of the detection method used by Nelson et al was possibly responsible, since the increase in permeation rate following breakthrough is quite gradual. An equally plausible reason for this result is that the composition of the challenge materials differed.

Experimental Results and Discussion—Wash Experiments

The glove cleaning regimen described did not effect the complete removal of liquefied coal contaminants from exposed glove specimens. Vapor from residual liquefied coal was detected immediately after the two washed PVC-1 specimens were placed in the cell for the second test (before the addition of challenge material). Vapor emanation declined rapidly; within 1 h it approached the baseline rate of evolution for unexposed glove specimens. When washed specimens were exposed to a second liquefied coal challenge, the rate of vapor emanation remained low during an interval comparable to the previously measured breakthrough time. Comparable permeation rates (after breakthrough) were observed for the washed and unwashed specimens of the PVC-1 material.

TABLE 5—*Breakthrough times and normalized breakthrough times for shale oil and coal tar extract reported by Nelson et al [4].*

		Shale Oil		Coal Tar Extract	
Material	Thickness, mm	Time, min	Normalized Time, min/mm^2	Time, min	Normalized Time, min/mm^2
PVC	0.13	9.0	530	4.6	270
	0.16	8.4	330	25.2	980
	0.31	27.0	280	47.4	490
	0.23	10.8	200	39.6	750
Natural rubber	0.48	28.8	120	19.8	86
	0.55	28.2	93	60	200
Nitrile rubber	0.31	60	620	60	620

Bennett et al detected the presence of residual radiolabeled phenol on the receiving side of the PVC-1 specimens within 15 min after the addition of challenge material in similar experiments testing washed gloves. No phenol was found in specimens taken before the addition of challenge material. The absence of detectable quantities in these earlier specimens probably was due to the depletion of this contaminant species at locations near the glove surface and to the higher limit of detection of the tracer method. Washing had no apparent effect on permeation rates following breakthrough.

Conclusion

The various experimental techniques by which the permeation of challenge liquids through specimens of glove material is studied may provide confusing and contradictory data, particularly when the challenging substance is a mixture. Salient factors which may lead to such confusion are differences in the receiving medium, the species detected on the receiving side, and the sensitivity of the method. The various constituents of a mixture may permeate at different rates, and the relative composition of the permeant may vary with time. This is true of liquefied coal, as the chromatographs presented in Fig. 4 show. It is clear that the composition of the vapor detected on the receiving side of the PVC-1 glove changed dramatically from the 2 to 4-h time interval to the 22 to 24-h interval. If only one chemical species is detected as a measure of permeation, then an erroneous estimate of total permeation may be obtained. The data presented here demonstrate the contradictions which may arise from a comparison between a method detecting only one component of a mixture and one responding to all volatile components of a mixture. Ideally, investigations of permeation by mixtures should utilize measurement techniques capable of quantifying individual permeant compounds. The measurement of one species might be appropriate, however, if one were concerned about only that particular chemical within the mixture.

This study, using a flowing gas receiving medium and a flame ionization detector, permits one to rank the four glove types tested according to dermal protection against liquefied coal. Normalized breakthrough times and permeation rates (Table 4*a*) show that nitrile rubber is the best and natural rubber is the worst of the glove materials. However, normalization of these descriptors accounts for variations in thickness, and therefore describes an intrinsic property of the glove material—not the glove itself. The greater thickness of the natural rubber glove causes it to confer protection for a somewhat longer period than either of the PVC gloves. It is important to note that the nitrile rubber glove was the thinnest of the four and that no permeation through it was detected.

Normalized breakthrough times and normalized permeation rates give approximately the same ordinal relationship among the glove materials for pro-

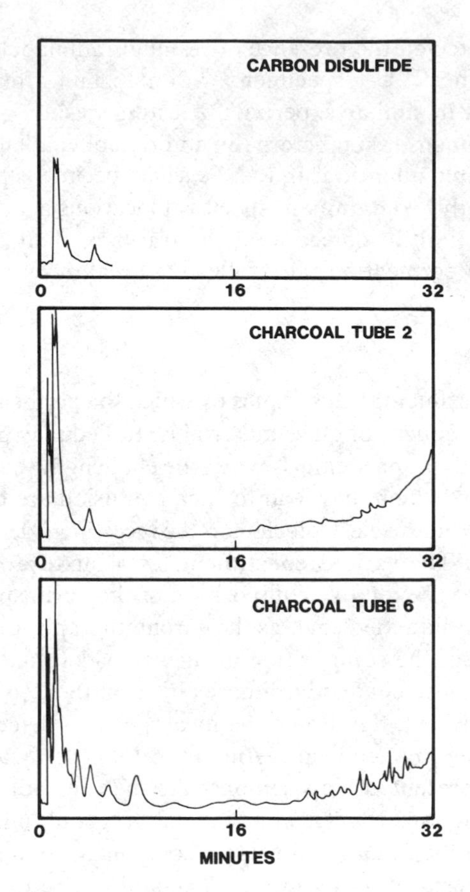

FIG. 4—*Chromatographs of carbon disulfide and of vapor samples collected on the receiving side of the cell during an early (Charcoal Tube 2) and a later (Tube 6) time interval. The input attenuation for Charcoal Tube 2 was half that used during the generation of the other two presented chromatographs.*

tection against toluene (Table 2*a*). However, permeation through the thin nitrile gloves did occur. Thus, the nitrile glove exhibited a resistance to permeation similar to that of the two PVC gloves. The natural rubber glove had the longest breakthrough time, as a consequence of its thickness.

The washing procedure utilized in this study and in that of Bennett et al [*1*] did not remove all traces of contaminant. After washing by this procedure, a subsequent exposure apparently is not a requirement for the emanation of constituents of liquefied coal from the glove. The potential for adverse health effects may not be assessed because the identities of the permeating species are unknown.

References

[1] Bennett, R. D., Feigley, C. E., Oswald, E. O., and Hill, R. H., *American Industrial Hygiene Association Journal*, Vol. 44, No. 6, June 1983, pp. 447-452.
[2] Weeks, R. W. and Dean, B. J., *American Industrial Hygiene Association Journal*, Vol. 38, No. 12, Dec. 1977, pp. 721-725.
[3] Nelson, G. O., Lum, B. Y., Carlson, G. J., Wong, C. M., and Johnson, J. S., *American Industrial Hygiene Association Journal*, Vol. 42, No. 3, March 1981, pp. 217-225.
[4] Nelson, G. O., Carlson, G. J., and Buerer, A. L., "Glove Permeation by Shale Oil and Coal Tar Extract," Report No. UCRL-52893, Lawrence Livermore Laboratory, University of California, Livermore, CA, Feb. 1980.
[5] Schoch, D. H., Tersegno, L. K., Winter, J. E., Bush, D. G., and James, R. L., "Testing of 'Impervious' Gloves for Permeation by Organic Solvents," American Industrial Hygiene Conference, Cincinnati, OH, 6-10 June 1982.
[6] Sexton, R. J., *Archives of Environmental Health*, Vol. 1, Sept. 1960, pp. 15/181-26/192 and 42/208-65/231.
[7] Buhl, P. H., "Coal Liquefaction, a Case Study of the Role of Health Effects Research for an Emerging Technology," American Industrial Hygiene Conference, Cincinnati, OH, 6-10 June 1982.
[8] Fujita, H. in *Diffusion in Polymers*, J. Crank and G. S. Park, Eds., Academic Press, London, 1968, pp. 76-79.
[9] Sansone, E. B. and Jonas, L. A., *Environmental Research*, Vol. 26, 1981, pp. 340-346.
[10] Van Krevelan, D. W., *Properties of Polymers: Correlations with Chemical Structure*, Elsevier, New York, 1972, p. 292.
[11] Kumins, C. A. and Kwei, T. K. in *Diffusion in Polymers*, J. Crank and G. S. Park, Eds., Academic Press, New York, 1968, Chapter 4, pp. 107-140.

Jimmy L. Perkins[1] *and Michael C. Ridge*[1]

Use of Infrared Spectroscopy in Permeation Tests

REFERENCE: Perkins, J. L. and Ridge, M. C., **"Use of Infrared Spectroscopy in Permeation Tests,"** *Performance of Protective Clothing, ASTM STP 900*, R. L. Barker and G. C. Coletta, Eds., American Society for Testing and Materials, Philadelphia, 1986, pp. 22-31.

ABSTRACT: The use of infrared (IR) analysis for permeation work has both advantages and disadvantages. Most advantageous is its ability to monitor a permeation test constantly without taking samples, as is necessary for most gas chromatography (GC) systems. Infrared detection also requires little attendance, and breakthrough time is very easy to discern.

Disadvantages include a longer delay in detected breakthrough time (because of the volume of the system) than with GC systems. However, the relative error introduced is quite small. Another problem is the necessity to purge the IR gas cell during calibration or before adding a different sample. This makes simultaneous testing of several cells difficult if not impossible. Last, a high static pressure [12.5 to 25 cm (5 to 10 in.) of H_2O] on the polymer membrane in the ASTM cell is created by the usual high flow rate (5 to 10 L/min) in commonly used IR systems. Although this pressure does not seem to affect permeation, these high flow rates are not necessary, and indeed, lower flow rates will reduce the error in detection of breakthrough time. The calculations and data to support these statements are reported.

KEY WORDS: permeation, infrared spectroscopy, Viton, ASTM permeation cell, protective clothing

Within the last five to ten years, industrial hygienists in industry, government, and academia have begun conducting a wide array of tests which help in selecting appropriate protective clothing and glove material for a given chemical exposure. ASTM, realizing that standard test methods would allow more meaningful results, established ASTM Committee F-23 on Protective Clothing. This committee then published the ASTM Test for Resistance of Protective Clothing Materials to Permeation by Hazardous Liquid Chemicals (F 739-81).

[1]Associate professor and graduate assistant, respectively, University of Alabama in Birmingham, Center for Occupational Health and Safety, Birmingham, AL 35294.

ASTM Test F 739-81 does not require a specific analytical method, but leaves that decision to the investigator. Presumably, selection of an analytical method would be based on factors such as the equipment on hand, the nature of the analyte, potential interferences, the need to conduct duplicate tests simultaneously, automation, and so forth. To date, gas chromatography (GC) appears to be the analytical method of choice for volatile organics. Using automated, microprocessor-based GC systems, the triplicate tests recommended in the standard ASTM method can be run simultaneously [*1*].

Infrared spectroscopy (IR) is also used and has several advantages over GC. Unfortunately, several disadvantages also exist. Single-beam IR instruments currently exist which are portable, easy to calibrate, and quite sensitive for many gases and vapors. While GC has also become portable, some portable models (the Michrometer from Microsensor Technology, Inc.) are not as sensitive as IR. Others (HNU and Photovac) are as sensitive as or more sensitive than IR.

Infrared Spectroscopy

Single-beam IR instruments have been applied to industrial hygiene and other field problems for several years. Two popular models are the Foxboro-Wilks Miran 1A and the Miran 80. Both are single-beam instruments with adjustable 0.75 to 20.25-m cell path length. Their slit widths are similarly adjustable, and analytical wavelength ranges are identical; however, their similarities end at this point. The physical similarities, including the gas cell, can be seen in Fig. 1.

The Miran 1A is a manual machine. Wavelength setting and zero adjusts are accomplished by manual control. Gain is not compensated for on the Miran 1A. Absorption or transmission can be read from a meter or a strip chart recorder. Drift is specified at less than 0.006 absorption units per 8 h. The simplicity of the Miran 1A yields virtually maintenance-free operation and allows one to master the machine quite easily.

The Miran 80 is a microprocessor-based instrument. Simplicity has been exchanged for significantly greater accuracy and the ability to account for interference and to analyze several components simultaneously. Wavelengths are electronically set, accurate to three decimal places (in micrometres). Zeros and gains are set electronically. This allows for highly precise repeatability.

The Miran 80 can analyze up to 18 wavelengths in a 2-min period. Consequently, more than one vapor can be analyzed, and a reference wavelength set in a nonabsorbing region can be used to compensate for drift or interferences in much the same manner as with a double-beam instrument. Unfortunately, the increased sophistication of the Miran 80 requires considerably more training than is required for the Miran 1A; this can result in more operator errors and equipment malfunctions. There is also the continued uneasiness (regard-

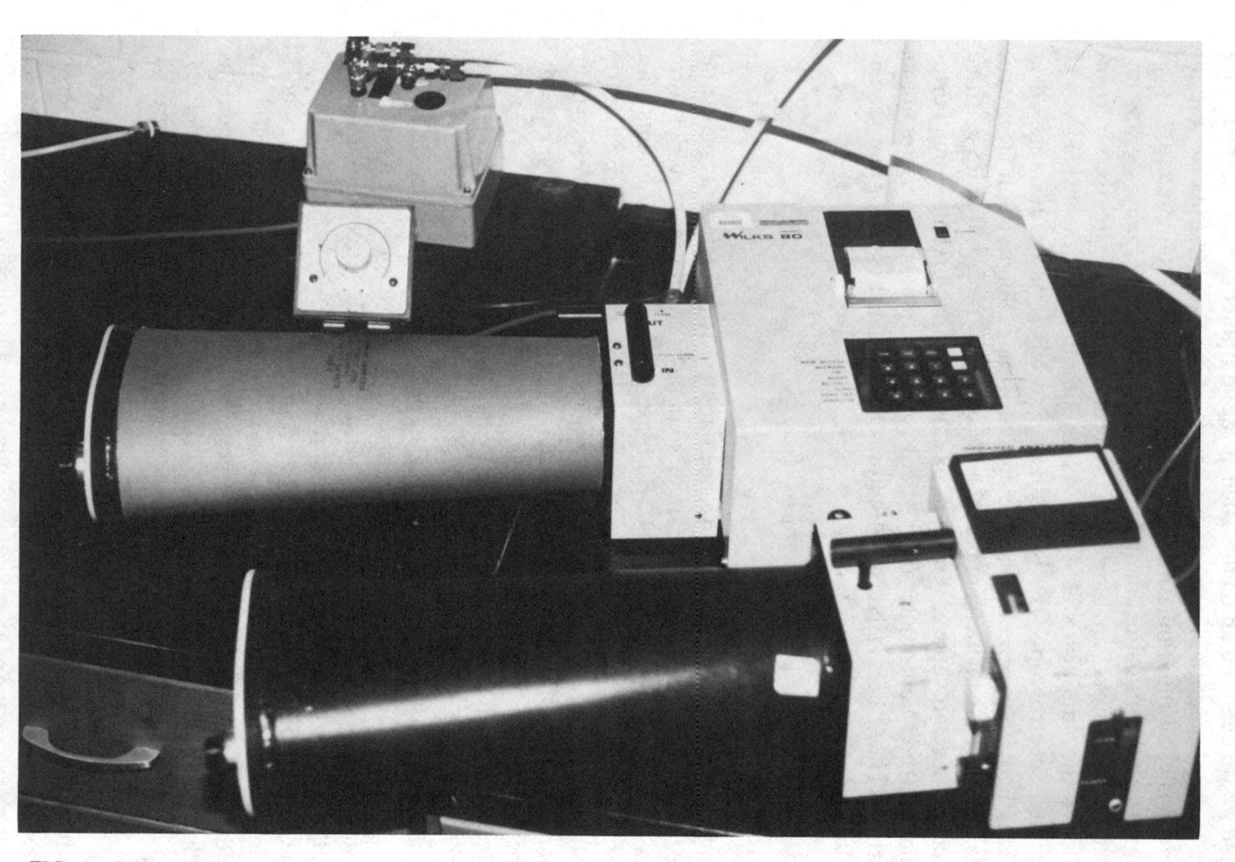

FIG. 1—*Miran 1A* (bottom) *and Miran 80* (top). *Note that the only difference is the microprocessor in the Miran 80. The total length of the Miran 1A is 68.5 cm (27 in.).*

less of the operator's level of knowledge) that the numbers entered to the keyboard, their order, and so on, are not exactly correct. Although, this can be checked, it requires more time, and still does not dispell the fear that the "garbage in-garbage out" principle is at work. These problems have been the authors' experience.

Calibration for either instrument is relatively easy and can be accomplished in less than 30 min from start to finish, assuming that the appropriate wavelength or wavelengths are known in advance. Finding the correct wavelength or wavelengths requires one or two extra "purge" steps. When a vapor is introduced to the Miran 1A or 80 for the purpose of finding the best absorbing wavelength, the gas cell must be purged or decontaminated prior to calibration, so that gains and zeros can be reset. In other words, scanning across all wavelengths (2.5 to 14.5 μm) to find a "good" wavelength and monitoring one or more wavelengths (for calibration purposes) are two different operations and require different gain and zero settings. Failure to obtain different gain and zero settings will introduce random errors of unknown magnitude. With the Miran 80, a second purge step is recommended, since after calibration, it is advisable to introduce known vapor concentrations to check that the microprocessor yields the correct result.

Purge Time

The purge time is an unknown quantity in IR gas cell work. The Miran gas cell is about 5.6 L in volume. The calibration pump flow rate is around 10 L/min for an open system, that is, the system is continuously exhausted rather than recirculated. Consequently, even under poor mixing conditions, 1 h should be sufficient to purge contaminants based on the following equation [2]

$$\ln C_p - \ln C_0 = -\frac{Q\Delta T}{VK} \tag{1}$$

where C_p is the concentration after purging, C_0 is the initial concentration, Q is the flow rate, ΔT is the purge time, V is the volume, and K is a mixing constant ($K = 10$ would represent very incomplete mixing) [2]. If C_0 is set at 1000 ppm, and K is set at 3, then C_p is approximately 2.5 ppm, after 10 min of purging ($\Delta T = 10$). For $\Delta T = 20$ min, C_p decreases to 0.007 ppm. Each additional minute of purging decreases C_p by a factor of 0.55.

However, a further potential problem is the Teflon-lined gas cell. While molecular interactions with Teflon are rare, Teflon does have pores which may fill with certain molecules. This phenomenon is apparently dependent on molecular size. We have found that water vapor, methanol, and carbon disulfide are three molecules that appear to fill these pores and require longer purge times. A heater jacket which will heat the gas cell to 110°C will ensure decontamination, but one must wait for the cell to return to room temperature, as IR work is temperature dependent.

A heater jacket is also a must for avoiding condensation in the gas cell. If, for example, the ASTM permeation cell is kept at 23°C and the laboratory temperature (and consequently the Miran temperature) is 18°C, condensation of water vapor and particularly the permeant chemical will occur in the Miran gas cell. This condensation will damage the optics in the gas cell. A heater jacket will solve this problem. It is also wise to use dried air when purging the cell at room temperature, followed by a purge and fill with nitrogen as required by ASTM Test F 739-81 procedures.

In-Series or Multiple Tests

Certainly any analytical method that saves time is an attractive approach to permeation testing. Since the tests are done in triplicate, the Radian GC system [*1*] is attractive because it performs three tests and a blank simultaneously. The IR system cannot always be used in this manner since the time needed to purge the Miran gas cell between samplings of each ASTM cell could become quite long, particularly with respect to breakthrough time. As indicated in Eq 1, a purge time of 10 min would reduce the concentration by a factor of 400, not to infinity. The initial purge, in a series of purges, may reduce the concentration to below the detection limit in a specific period of time. However, after each successive purge and refilling, the remaining concentration will build and probably become detectable. If three different ASTM cells are being sampled, and the steady-state rate is being sought at a relatively high concentration, the length of purge time could necessarily become long. In addition, even if the test start times for the three ASTM cells are staggered, the necessary purge time will cause considerable variability in the three detected breakthrough times, particularly if they are less than 30 min.

On the other hand, since the Miran 80 can analyze up to eleven gases at a time, another option is to run two or three tests in series, using different permeants. The configuration in Fig. 4 shows how such a system would look. Over the long run, the same end results would be accomplished as with the Radian GC system.

One important consideration in series tests is the possibility of Permeant A (Fig. 2) back permeating into the vapor side of Membrane B, or vice versa. This could have an effect on the polymer properties and permeation rates. However, we have determined in our research that there are more immediate problems in attempting in-series testing with the Miran.

In a typical permeation test, the steady-state rate may not be reached until some high concentration is reached—say, 1000 ppm in a closed-loop system. In order to detect breakthrough at about 1 ppm, another calibration curve (or

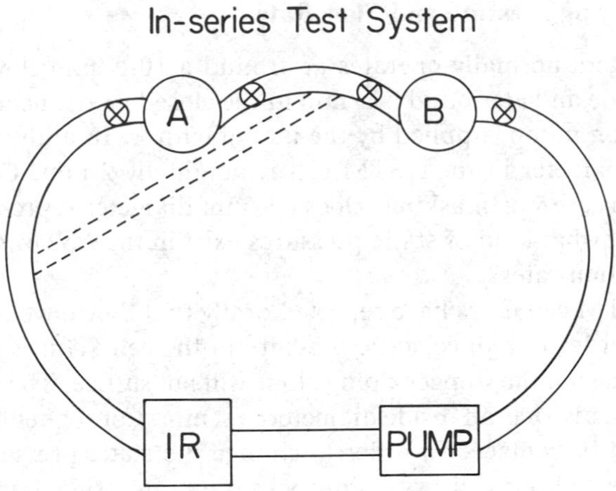

FIG. 2—*Schematic of proposed in-series testing system using two different challenge materials. The dashed line represents a normal one-component system.*

perhaps two) must be obtained. Usually a more sensitive wavelength must be used for the lower concentration and a less sensitive wavelength for the 1000-ppm concentration. Each calibration requires one analysis program in the microprocessor. However, as one adds programs, memory is not sufficient for eleven components. For two vapors or components and one reference wavelength, five programs are available. If, as in the preceding example, three programs are needed for each vapor, the worst-case situation is that three squared, or nine programs, would be needed to cover all possible combinations of the two components—that is, Components A and B both having low concentrations, A and B both having high concentrations, A with high and B with low concentration, and so on. Nine programs are not possible on the Miran 80.

If one has an idea of the magnitude of breakthrough times and permeation rates, the number of analyses can be reduced, that is, some combinations can be ruled out, and perhaps the maximum five analyses available will suffice. However, calibration becomes more complex, for obvious reasons, and one is still not certain that back permeation is not affecting the results.

In effect, we have found any potential short cuts using closed-loop IR to be more time-consuming in the long run. Another alternative is to use an open-loop system so that concentrations at steady-state rate are lower. While this solves the calibration problem, the possible problem of extended detection of breakthrough time may follow. Further research in this area is necessary in order to make IR a more efficient tool in permeation work. Consequently, in-series or parallel testing is not recommended with the Miran.

Effects of Varying Pressure and Flow Rates

The IR system normally operates at around a 10-L/min flow rate in the open-loop mode and at around 7 L/min in the closed mode, using the closed-loop calibration pump supplied by the manufacturer. In addition, the stopcock orifice diameter in the ASTM cell is nominally 2 mm. Consequently, three questions are worth asking. Does a 2-mm diameter represent a critical orifice? If not, what kind of static pressures exist in the ASTM cell? Do they affect permeation rates?

Several ASTM cell users have reported orally that they have bored out the stopcocks in order to reduce static pressure in the cell. This is a good idea, however, boring out the stopcock plug alone will not suffice. The glass portion of the stopcock also has an inside diameter of 2 mm. Consequently, the entire stopcock must be replaced in order to change the static pressure. We have modified several of our cells with stopcocks 8 mm in inside diameter.

This modification does make a substantial difference. Figure 3 shows how static pressure varies with the flow rate and stopcock orifice size. At a flow rate of 9 L/min, an 8-mm stopcock results in a 6.5-cm (2.7-in.) H_2O pressure

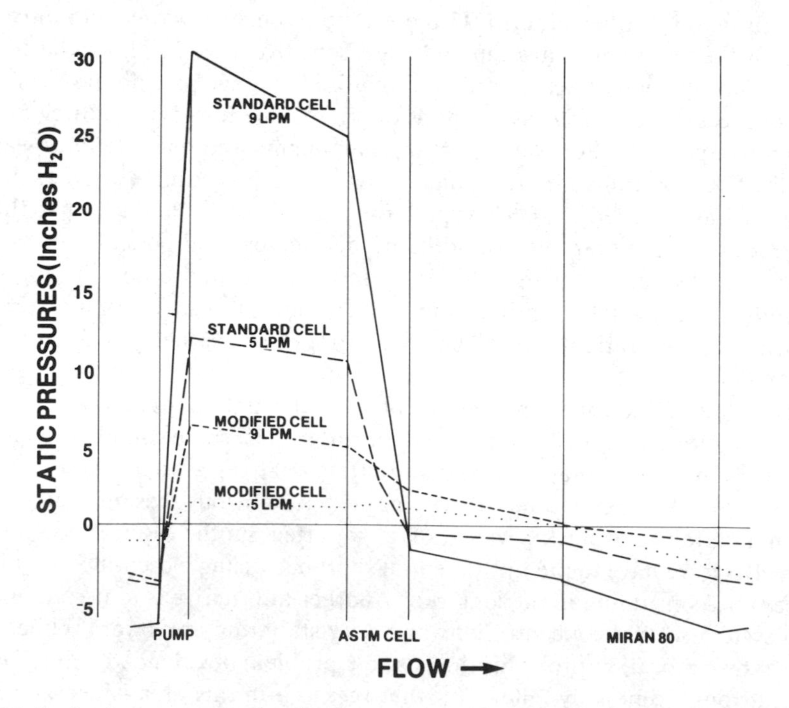

FIG. 3—*Static pressure in a closed system containing the Miran 80, ASTM permeation cell or ASTM modified cell (8-mm in internal diameter stopcocks), a bellows pump, and 37.5 m (13 ft) of 0.8 cm (5/16-in.) tubing versus various flow rates.*

drop across the ASTM cell, while a 66-cm (26.3-in.) H_2O pressure drop results for the 2-mm stopcock. Average static pressure in the standard cell is 30 cm (12 in.), while in the modified cell it is 9 cm (3.7 in.).

At 5 L/min, the flow rate average pressure is about 2.5 cm (1 in.) by water gage in the modified cell and 12.5 cm (5 in.) by water gage in the standard cell. These data indicate that pressure loss does vary substantially with stopcock diameter and flow rate. Does this affect the evaporation or diffusion step in the permeation process? Will a large [37.5 to 62.5 cm (15 to 25 in.) of H_2O] pressure drop in the ASTM cell cause a leak in the system, which may be further enhanced by the pulsing of the pump?

We have detected extremely small leaks in the closed system by virtue of a small drift in the Miran readout caused by an influx of ambient water vapor and carbon dioxide. The leak or leaks cannot be detected with a soap solution. In addition, the ASTM cell leaks when under pressure. The amount of leakage varies from one cell to the next and depending on the pressure. However, some cells, given a 25-cm (10-in.) static pressure load, will depressurize overnight. Again, the solution to this problem is to use an open-loop system, provided the detection time is not appreciably affected. On the other hand, the drift caused in the IR detection is only on the order of 1 ppm/h. Nevertheless, the test is not being conducted according to ASTM specification if inward leaks occur.

As for pressure effects on permeation rate, Table 1 shows the results of eleven tests at varying pressure. Viton (nominally, 0.25 mm) was challenged with methylene chloride. The flow rates and stopcocks have been varied to attain the pressures. While there appears to be no relation between time to steady state or permeation rate and pressure, there does appear to be a relation between detection of breakthrough and pressure.

Figure 4 shows the plot of pressure versus detected breakthrough time. The

TABLE 1—*ASTM cell pressure* (P_{cell}) *(inches of* H_2O*), detection of breakthrough time* (T_D)*, time to first steady state* (T_{SS1})*, and first steady-state permeation rate* (J) *for eleven tests of Viton and methylene chloride.*

P_{cell}, in. of H_2O	T_D, min	T_{SS1}, min	J, mg/m^2/s
1.3	81	17	2.69
1.4	83	16	2.74
1.5	83	22	2.41
1.6	78	17	2.74
2.0	65	20	2.81
2.0	73	19	1.92
3.3	78	24	1.93
3.8	80	20	2.41
6.8	71	16	2.74
7.8	70	>14	>2.41
8.3	74	23	2.83

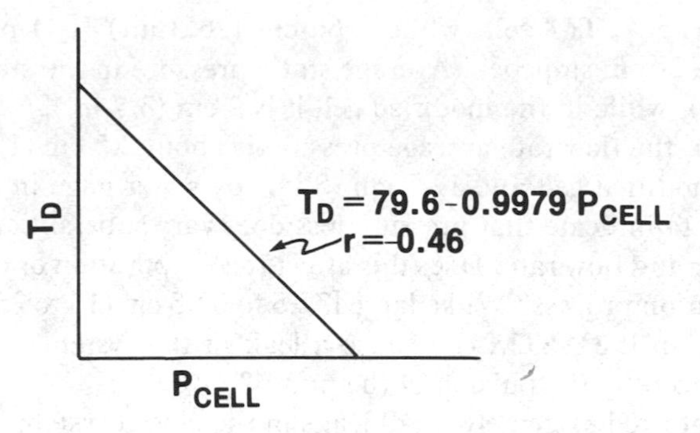

FIG. 4—*Detection of breakthrough time versus static pressure in the ASTM cell using Viton and methylene chloride.*

slope appears to be very near −1. Presumably, as pressure increases, time to detect breakthrough, T_D, decreases.

There are obviously other sources of variation in the T_D values. The variance, r^2, value (0.2) indicates that only 20% of the variation of T_D is explained by pressure. Indeed a *t*-test reveals that the slope and correlation coefficient, *r*, are not different from zero (*P*, 0.05 level). There is obviously a tremendous amount of scatter in the data, and consequently the value for the slope, appearing to be near −1, is probably an artifact of this particular data set. Had one or two points been different, a slope near zero could have occurred. The conclusion is that these data do not reveal any relationship between permeation parameters and static pressure at flow rates greater than 3 L/min.

Discussion and Conclusion

Infrared analysis provides constant monitoring of any atmosphere. This allows easy detection of breakthrough time and does not require constant sampling or attention during a permeation test. The major drawbacks appear to be associated with attempts to perform more than one test simultaneously, and a probable longer detection time for breakthrough than in GC systems.

A potential solution to the first problem is the use of the Miran 80, an open-loop system, an in-series testing of different permeants. However, this is not recommended without further research. The possibility of testing the same permeant in three separate ASTM cells, while purging the Miran cell between samplings of each ASTM cell, is also attractive. However, purge times that will ensure nondetectable levels are minimally on the order of 4 to 5 min. Hence, this would make accurate detection of breakthrough time for all three cells impossible, unless a staggered start were performed and breakthrough times were greater than about 30 min.

Acknowledgments

The authors wish to thank Vernon Rose for his review. This work was partially funded by a grant from the National Institute for Occupational Safety and Health, 1-RO1-OHO1932.

References

[*1*] Harless, J. M., Garcia, D. B., and Sorensen, B. A., "The Radian/National Toxicology Program Glove Testing Project," Paper presented at the American Industrial Hygiene Conference, Philadelphia, PA, May 1983.

[*2*] Mutchler, J. E. in *The Industrial Environment, Its Evaluation and Control, National Institute for Occupational Safety and Health*, Cincinnati, OH, 1973, p. 580.

Mark W. Spence[1]

A Proposed Basis for Characterizing and Comparing the Permeation Resistance of Chemical Protective Clothing Materials

REFERENCE: Spence, M. W., "**A Proposed Basis for Characterizing and Comparing the Permeation Resistance of Chemical Protective Clothing Materials,**" *Performance of Protective Clothing, ASTM STP 900*, R. L. Barker and G. C. Coletta, Eds., American Society for Testing and Materials, Philadelphia, 1986, pp. 32–38.

ABSTRACT: A common problem among users of chemical protective clothing is the lack of a practical, consistent basis for comparing and characterizing the permeation barrier effectiveness of different protective clothing materials. This becomes a particular problem for generic materials (for example, neoprene) for which differences in raw materials, additives, and manufacturing techniques can result in differences in permeation resistance for the same material made by different manufacturers.

An approach to resolving this problem is described which uses permeation testing with a solubility-parameter-based set of test chemicals as a basis for comparing clothing materials. A proposed test battery is presented, and the rationale behind it, as well as its potential usefulness, is discussed.

KEY WORDS: chemical protective clothing, permeation, permeation testing, chemical test battery, permeation resistance, solubility parameters, protective clothing

Today's user of chemical protective clothing is faced with a formidable task when selecting appropriate clothing. Among the many factors that must be considered are cost, construction style, availability, and mode of use (disposable versus reusable). However, the most important factor is the effectiveness of the clothing as a barrier to the chemicals of interest. Barrier effectiveness is best determined by permeation testing, but until recently very little permeation test data have been available. The advent of a standardized method for conducting such testing, ASTM Test for Resistance of Protective Clothing

[1]Senior research chemist, Health and Environmental Sciences, Dow Chemical U.S.A., Midland, MI 48460.

Materials to Permeation by Hazardous Liquid Chemicals (F-739-81), has helped the situation considerably. However, there are still many items of chemical protective clothing for which no permeation test data are available.

Even if data are available, rarely have two items of interest been tested against the same set of chemicals, which makes comparisons difficult. The situation becomes a particular problem when one tries to compare generic clothing materials from different manufacturers. Differences in raw materials, additives, and manufacturing techniques can result in differences in permeation resistance for items of the same material (for example, neoprene) made by different manufacturers. An example of such a situation is demonstrated by the permeation data shown in Table 1. Even though both the gloves tested were identical in thickness and generic composition, the breakthrough time for carbon tetrachloride is twice as long for Glove A as for Glove B, while for propylene glycol monomethyl ether, the breakthrough order is reversed.

Because of the resources involved, every protective clothing item cannot be tested against every chemical or mixture of chemicals. What is needed, then, is a practical, consistent basis for making general comparisons of the permeation barrier characteristics of protective clothing materials. This paper presents a proposed approach to providing such a basis through permeation testing with a specific battery of test chemicals.

Discussion

Solubility Parameter

Many manufacturers of protective clothing use test batteries consisting of various common commodity chemicals, but these provide little basis for generalizations concerning the type of chemicals that will rapidly permeate a particular material. To be most useful, a test battery should consist of chemicals chosen to characterize the permeation barrier properties of protective clothing materials.

The most promising chemical property for characterizing permeation resistance appears to be the solubility of the chemical in the protective clothing material. Crank and Park [*1*] have stated that the permeation of chemicals

TABLE 1—*Variation in permeation resistance for two nitrile gloves.*

		Breakthrough Time, min	
Manufacturer	Description	Carbon Tetrachloride	Propylene Glycol Monomethyl Ether
A	0.43 mm, flock lined	127	59
B	0.46 mm, flock lined	63	96

through polymer membranes is primarily a function of the solubility of the chemical in the polymer.

To describe the relative solubility of chemicals and polymers in each other quantitatively, polymer and paint chemists have long used *solubility parameters*. Basically, the solubility parameter is a quantification of the old rule "like dissolves like." The solubility parameter is a number assigned to a chemical or polymer that describes its solvency characteristics—essentially, its polarity. Solubility parameters for liquid chemicals can be calculated from the energy of vaporization and molar volume of the chemicals using the expression

$$SP = \left(\frac{\Delta E_v}{V}\right)^{1/2}$$

where

SP = solubility parameter, $MPa^{1/2}$,
ΔE_v = energy of vaporization, J/mol, and
V = molar volume of the liquid, cm^3/mol.

There are several types of solubility parameter scales [*2*], but historically the most widely used system has been the Hildebrand solubility parameter. On the Hildebrand scale, solubility parameter values for liquid chemicals range from a low of about 15 for a nonpolar compound such as hexane to a high of 47 for a polar compound such as water. Solubility parameter values for polymers used in chemical protective clothing range from 16 for butyl rubber (nonpolar) to 27 for polyvinyl alcohol (polar). More detailed descriptions and tabulations of solubility parameters are available elsewhere [*2–4*].

The utility of solubility parameters lies in the ease with which relative solubility comparisons can be made. One simply compares the numerical value of the solubility parameters for two different chemicals (or a chemical and a polymer); the smaller the difference, the greater the solubility of the two chemicals in each other. To take a simple example, consider a comparison of the solubilities of hexane (SP = 14.9) and methanol (SP = 30.7) in water (SP = 47.0). The differences in solubility parameter between water and methanol and between water and hexane are 16.3 and 32.1, respectively. Thus, methanol would be expected to have a greater solubility in water than hexane. It should be noted that the solubility parameter approach is comparative, not absolute; it gives information on which solute of a series will be most soluble in a given solvent but makes no prediction of the absolute magnitude of the solubility of a given solute.

Extending this concept to permeation through polymeric protective clothing material yields the conclusion that: the greater the solubility of a chemical in a polymer material, the lower the permeation resistance of the material for that chemical. This has been demonstrated experimentally [5], as is shown by the data in Table 2.

Proposed Test Battery

The proposed solubility-parameter-based battery of test chemicals is actually a modification of the method described by Burrell [3] for determining the solubility parameters for polymers. While solubility parameter values for liquid chemicals can be calculated from the energy of vaporization and the molar volume of the chemical, as described earlier, solubility parameters for polymers must be determined experimentally. The specimens of the polymer are placed into a series of test liquids, which cover the range of solubility parameters. The degree of polymer swelling or the observed solubility of the polymer is noted for each liquid. The polymer's solubility parameter, then, is taken to be the same as that of the liquid in which the polymer swells the most or is the most completely soluble.

Since the classic Hildebrand solubility parameters take into account only the dispersion forces between molecules, it was noted that they were increasingly inaccurate as the hydrogen bonding between molecules increased. To attempt to alleviate this problem, Burrell divided solvents into three hydrogen bonding classes: poorly hydrogen bonded, moderately hydrogen bonded, and strongly hydrogen bonded. For each hydrogen bonding class, a "solvent spectrum" was chosen consisting of a series of solvents spanning the range of solubility parameters. Using these three "solvent spectra," three solubility parameters would then be determined for each polymer, as described earlier, one for each class of solvents. Subsequent solubility parameter comparisons could be made by using the appropriate polymer solubility parameter for the chemical in question.

A logical extension of Burrell's solvent spectra approach is to apply it to selecting a battery of chemicals for permeation testing. The advantage of permeation testing of a material with such a solubility-parameter-based test battery is that it provides information on the chemical solubility parameter ranges for which the material would act best (and worst) as a permeation barrier. If the permeation data using this same battery of test chemicals were available for many materials from many manufacturers, the task of comparing the permeation resistance characteristics of the various materials would become much easier.

The utility of this solubility parameter approach in comparing generic materials from different manufacturers is illustrated by returning to the example given earlier for nitrile gloves. As shown by the data in Table 1, the breakthrough time order for Gloves A and B when tested with carbon tetrachloride was reversed when the gloves were tested against propylene glycol monomethyl ether. An interpretation of these data becomes clear when we note that

TABLE 2—*Correlation of permeation resistance with the difference in solubility parameter (DSP).*

	Chemical (Solubility Parameter)					
	Hexane (SP = 14.7)		Tetrachloroethylene (SP = 20.3)		Allyl Alcohol (SP = 25.8)	
Material (Solubility Parameter)	DSP	BT[a] (PR)[b]	DSP	BT (PR)	DSP	BT (PR)
Butyl rubber (SP = 16.4)	1.7	2.1 (44)	3.9	80 (149)	9.4	>480 (. . .)
Neoprene (SP = 18.4)	3.7	4.2 (5.2)	1.9	12 (163)	7.4	141 (0.24)
Polyvinyl alcohol (SP = 26.6)	11.9	>480 (. . .)	6.3	>480 (. . .)	0.8	26 (5.1)

[a]BT = breakthrough time, in minutes.
[b]PR = steady-state permeation rate, in milligrams per square metre per second.

the solubility parameters for carbon tetrachloride and propylene glycol monomethyl ether are 17.8 and 20.9, respectively. This information, taken with the permeation data, leads one to the conclusion that the nitrile material used by Manufacturer A has a higher solubility parameter (that is, is more polar) than that used by Manufacturer B. This general information about the permeation barrier characteristics of the two nitrile materials would prove useful in the future when deciding which one would offer the best permeation resistance to a particular chemical; if the chemical had a low solubility parameter, Glove A would be the better choice, while Glove B would perform better with chemicals of high solubility parameter.

Using the concept of testing with a spectrum of chemicals covering the range of solubility parameters and hydrogen bonding classes, the proposed permeation test chemical battery shown in Table 3 was selected. Additional criteria that were considered in selecting chemicals for the proposed test battery were the following:

1. Functional group variety—It is desirable to have as many different chemical functional groups as possible represented to screen materials for any unusual interactions.
2. Maximized permeation potential—So that protective clothing comparisons are made on the basis of "worst-case" data, it is desirable to use chemicals with the highest potential to permeate polymer materials. According to permeation theory [1], permeation would be maximized for small molecules with high vapor pressures.
3. Availability and ease of handling—The chemicals in a test battery should be readily available and not unusually difficult to handle.
4. Amenability to current testing techniques—The majority of current permeation testing is conducted using a nitrogen or air collection medium and analysis by gas chromatography with flame ionization detection. The chemicals in the proposed test battery could all be tested by this technique.

TABLE 3—*Proposed test chemical battery.*

Chemical	SP	Hydrogen Bonding Class
Hexane	14.7	poor
Diethyl ether	16.0	moderate
Diethyl amine	16.4	strong
Toluene	18.2	poor
Methyl ethyl ketone	19.0	moderate
n-Pentanol	21.1	strong
Acetonitrile	24.3	poor
Dimethylformamide	24.8	moderate
Methanol	29.7	strong

Summary

A solubility-parameter-based battery of permeation test chemicals has been proposed as a practical, consistent basis for comparing and characterizing the permeation resistance of chemical protective clothing materials. Widespread adoption of such a test battery by protective clothing manufacturers would provide protective clothing users with the information necessary to (1) compare the same generic material from different manufacturers, (2) select the best candidate materials for more definitive, case-specific permeation testing, (3) characterize the types of chemicals for which a particular material offers particularly good (or poor) permeation resistance, and (4) demonstrate equivalence or superiority of particular items of chemical protective clothing in order to facilitate substitution without extensive testing. A protective clothing manufacturer might also use such a test battery as the basis for a quality assurance program. Ultimately, a more extensive test chemical battery based on the same scheme could even be used to determine the solubility parameter of novel chemical protective clothing materials.

In short, information resulting from widespread use of such a test battery would increase users' confidence in protective clothing selection and use.

References

[1] Crank, J. and Park, G. S., Eds., *Diffusion in Polymers*, Academic Press, New York, 1968.
[2] Barton, A. F., *Handbook of Solubility Parameters and Other Cohesion Parameters*, CRC Press, Boca Raton, FL, 1983.
[3] Burrell, H., "Solubility Parameter Values" in *Polymer Handbook*, J. Brandrup and E. H. Immergut, Eds., Interscience, New York, 1966.
[4] Synder, L. R., "Solutions to Solution Problems-1," *Chemtech*, Vol. 9, No. 12, 1979, pp. 750–755.
[5] Spence, M. W., "Chemical Permeation Through Protective Clothing Materials: An Evaluation of Four Critical Variables," Paper presented at the American Industrial Hygiene Conference, Portland, OR, 1981.

Dirk M. Baars,[1] *Dana B. Eagles,*[1] *and Jeffrey A. Emond*[1]

Test Method for Evaluating Adsorptive Fabrics

REFERENCE: Baars, D. M., Eagles, D. B., and Emond, J. A., **"Test Method for Evaluating Adsorptive Fabrics,"** *Performance of Protective Clothing, ASTM STP 900*, R. L. Barker and G. C. Coletta, Eds., American Society for Testing and Materials, Philadelphia, 1986, pp. 39–50.

ABSTRACT: Albany International Research Co. has developed a test apparatus and method for evaluating the protection afforded by various textile materials against low levels of toxic vapors. The test apparatus consists of a flow system that provides a constant challenge gas flow rate and concentration to be passed through the test specimen and a detection system that continuously monitors the concentration of the toxic component in the effluent gas stream. In the present apparatus, carbon tetrachloride is used as the challenge material at a concentration of 5.0×10^{-6} g/cm^3. The traces of effluent concentration versus time obtained from the tests can be used to quantify total sorption capacity and sorption rate for the sample fabrics. Reproducibility studies have yielded coefficients of variation of 3.4% for total capacity and 5.5% for the sorption rate constant determined by this test method.

KEY WORDS: adsorption tests, adsorption test equipment, protective clothing, gas mixtures, adsorption capacity, adsorption kinetics, chemical agent detection

Concern has been growing in the military, as well as in industry, about provision of clothing that will protect the wearer against toxic chemical vapors. This has created the need for a test that will accurately measure the effectiveness of various textile materials in toxic component removal. This testing is of great importance in both development and quality control of candidate products.

An apparatus and method for performing such a test have been developed at Albany International Research Co. The test involves passing a gas stream containing a constant level of a toxic component through a sorptive fabric specimen and continuously monitoring the concentration of this component in the effluent. Carbon tetrachloride (CCl_4) has been the challenge vapor of

[1] Senior research engineer, assistant director, and research chemist, respectively, Albany International Research Co., Dedham, MA 02026.

choice since it is a standard in testing of sorptive materials [see ANSI/ASTM Test for Carbon Tetrachloride Activity of Activated Carbon, D 3467-76 (1983)]. The effluent concentration data obtained can be analyzed for sorption rate and capacity, which allows one to predict the protection afforded by the fabric in use.

Test Apparatus

A schematic of the sorption test apparatus is shown in Fig. 1. The apparatus can be broken down into three major sections, as follows: (1) a challenge gas mixture generator to provide a constant challenge gas flow rate and concentration, (2) a sample cell to house the test specimen, and (3) a detection system to continuously measure and record the concentration of the challenge vapor in the effluent stream. Each of these sections is described in detail in this paper.

Challenge Stream Generator

A variety of challenge materials and concentrations can be used in this test. In the present apparatus, the challenge stream generator provides a 16.7-cm^3/s (1-L/min) flow of nitrogen containing 5.0×10^{-6} g/cm^3 CCl_4. This is accomplished as described in the following paragraph.

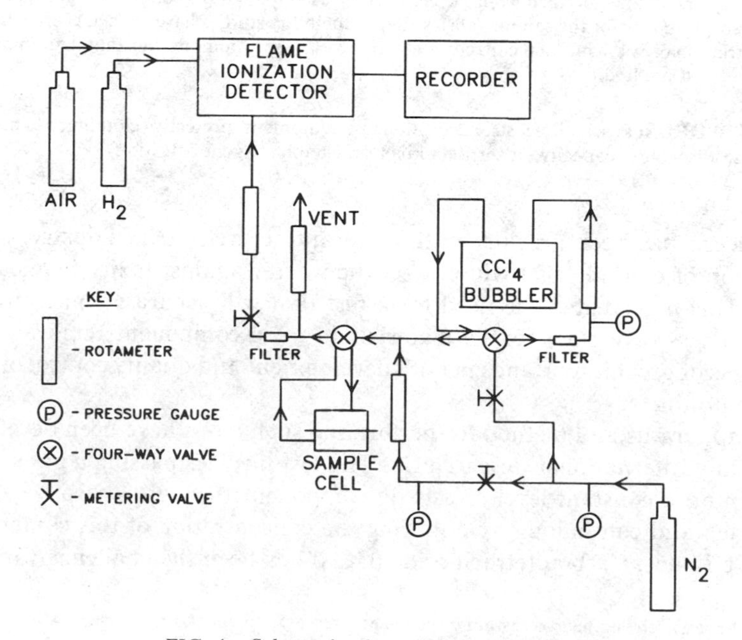

FIG. 1—*Schematic of sorption test apparatus.*

A gas cylinder of nitrogen, equipped with a two-stage regulator, is set to provide a feed pressure of 69 kPa to the flow system. The nitrogen flow is split and regulated by precision metering valves so that 0.32 cm^3/s passes through a glass bubbler containing liquid CCl_4 (reagent grade) and 16.4 cm^3/s bypasses this bubbler. The liquid CCl_4 is maintained at a constant temperature of 0°C by immersing the bubbler in an ice bath contained in a Dewar flask. The nitrogen stream passing through the bubbler becomes saturated with CCl_4 vapor. The CCl_4 concentration in this gas mixture can be approximated from its known vapor pressure at 0°C. The bubbler and bypass streams are combined downstream to yield the desired challenge gas mixture flow rate and CCl_4 concentration at ambient conditions.

Sample Cell

The sample cell design follows that of the Dawson cup given in the military specification for cloth, laminated, nylon tricot knit, polyurethane foam laminate for chemical protection (MIL-C-43858). Photographs of this cup are shown in Fig. 2. During a test, the inlet gas stream enters the top half of the cup, is dispersed across the specimen by a diffuser plate, and exits from the bottom half of the cup. Specimens are cut out of the fabric to be tested in the shape of a circular disk with a diameter of 13.6 cm. The specimen is mounted in the sample cell with a copper screen backing for support. The two halves of the cup, with the specimen sandwiched in between, are sealed around the edge with butyl rubber so that a circular 100-cm^2 cross-sectional area of specimen is left exposed to the gas stream. This mounting procedure has the flexibility to test specimens of different thicknesses as well as to test multiple layers of specimens. The sealed sample cell is immersed in a water bath to allow temperature control and leak detection.

Detection System

A small fraction of the effluent stream from the sample cell (0.50 cm^3/s) is passed through a flame ionization detector (FID) to measure the CCl_4 concentration, while the remainder is vented to a fume hood. An FID was chosen as the detector because of its large linear range (10^8) and good detectability (10^{-12} g/s). A strip chart recorder is used to record the data as a trace of effluent CCl_4 concentration versus elapsed time from the start of the test. Computer data acquisition has also been used effectively to provide easier data handling as well as improved overall accuracy and sensitivity.

Plumbing and Instrumentation

All of the transfer lines used in the test setup are fabricated from 6.35-mm (1/8-in.) stainless steel tubing. This results in minimal residence time of the

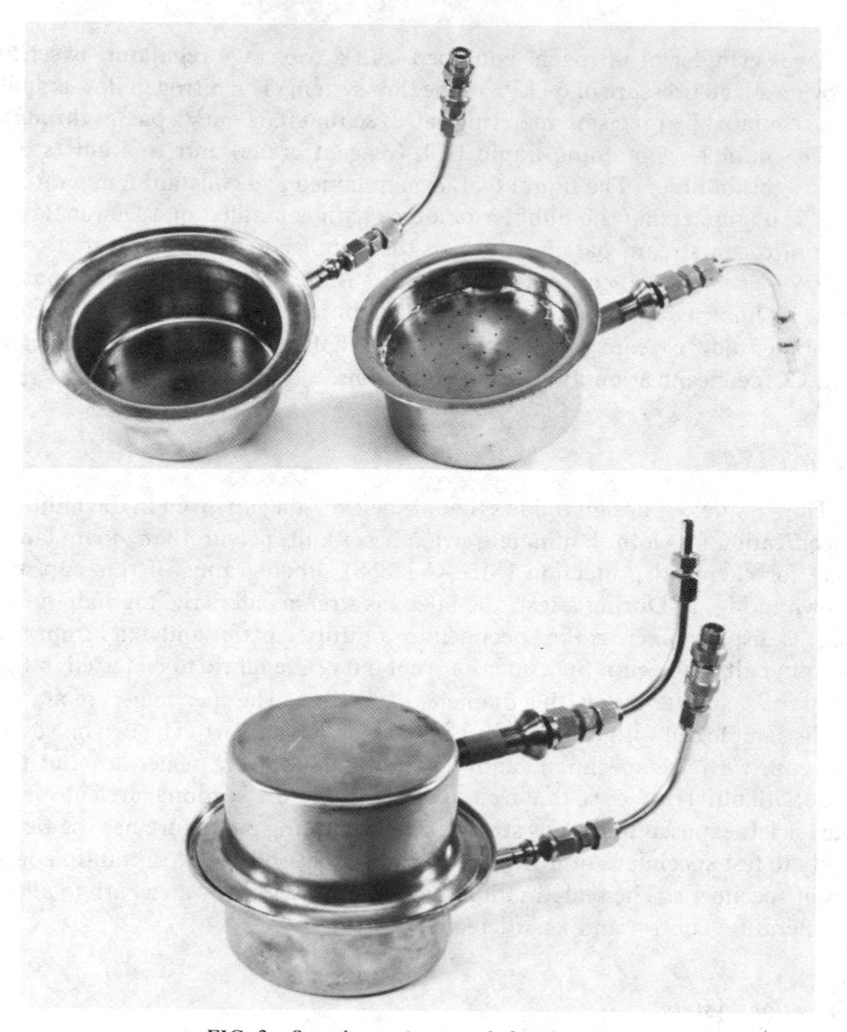

FIG. 2—*Sample cup (open and closed views).*

gas in the flow system while good structural durability is retained. All permanent metal-to-metal connections are made with Swagelok compression fittings. Connections at the glass bubbler and at the sample cell are made with Cajon O-ring seal fittings. Four calibrated rotameters are used to monitor flow rates through the bubbler, bypassing the bubbler, through the detector, and out the vent. Precision metering valves are used to control these flows at the proper levels. Two four-way valves are included to allow for bypassing the bubbler or the sample cell, or both. Borosilicate glass microfiber particulate filters (Balston Type 9922-05) are positioned in line before the bubbler and before the detector to avoid contamination in these areas. Three pressure gages are used to measure nitrogen feed pressure, CCl_4 bubbler head pressure, and fabric specimen pressure drop.

Test Procedure

The standard test procedure involves operating the test apparatus sequentially in four different modes. Changes in the operating mode are made simply by adjusting the set positions of the two four-way valves. Each of the four modes is described in the following sections, in the order they are used during the test procedure.

Baseline Mode

To operate in baseline mode, the four-way valves are set so that both the CCl_4 bubbler and the sample cell are bypassed, and pure nitrogen is fed to the detector. This allows the operator to zero the detector and recorder properly.

Challenge Stream Mode

In the challenge stream mode, the sample cell is still bypassed, but the gas is allowed to pass through the bubbler. A trace of the inlet concentration wave is obtained, which is characterized by a steep rise in CCl_4 concentration from 0 to 5.0×10^{-6} g/cm^3. In this way, proper operation of the flow and detection systems can be checked before the specimen is committed to be tested.

Degassing Mode

After a brief period of operation in baseline mode to remove all CCl_4 from the lines, the system is operated in "degassing" mode. In this mode, the bubbler is bypassed while allowing pure nitrogen to pass through the test specimen. The specimen is checked for any residual volatiles that might be given off and cause an errant detector response. All the flows can now also be adjusted to compensate for the additional pressure drop of the test specimen in line.

Test Mode

The actual test is begun at this point by switching the four-way valve that allows nitrogen to pass through the CCl_4 bubbler. Initially, there is no detectable amount of CCl_4 in the effluent stream from the sample cell, which is seen on the strip chart recorder as a continuation of the baseline. The length of this section of the effluent concentration curve from the start of the test to the first detectable sign of CCl_4 is defined as the breakthrough time for that test specimen. The breakthrough time value will, of course, depend on the sensitivity settings of the detector and chart recorder. In the present test protocol, CCl_4

concentrations as low as 1×10^{-10} g/cm^3 can be detected. After breakthrough, the CCl_4 concentration rises at an increasing rate, goes through an inflection point, and asymptotically approaches the inlet concentration corresponding to complete saturation of the test fabric. A typical test curve is shown in Fig. 3. The inlet CCl_4 concentration is rechecked at the end of the test by operating, once again, in the challenge stream mode.

Results Analysis

Extrapolation of Curve to Saturation

In the interest of time, a test is normally terminated when the effluent CCl_4 concentration reaches a level greater than 90% of the inlet concentration. The remainder of the curve is then extrapolated from the collected data in order to calculate the total equilibrium sorption capacity of the test specimen. Two methods of extrapolation have been used—linear and exponential.

Linear extrapolation is performed by drawing a straight line out from the last experimental data point at a slope equal to the average slope observed in the final portion of the experimental curve. The time at which the effluent concentration becomes equivalent to the inlet concentration corresponding to complete saturation is then calculated from the equation for this line, as follows

$$t_0 = t_f + \frac{C_0 - C_f}{M} \tag{1}$$

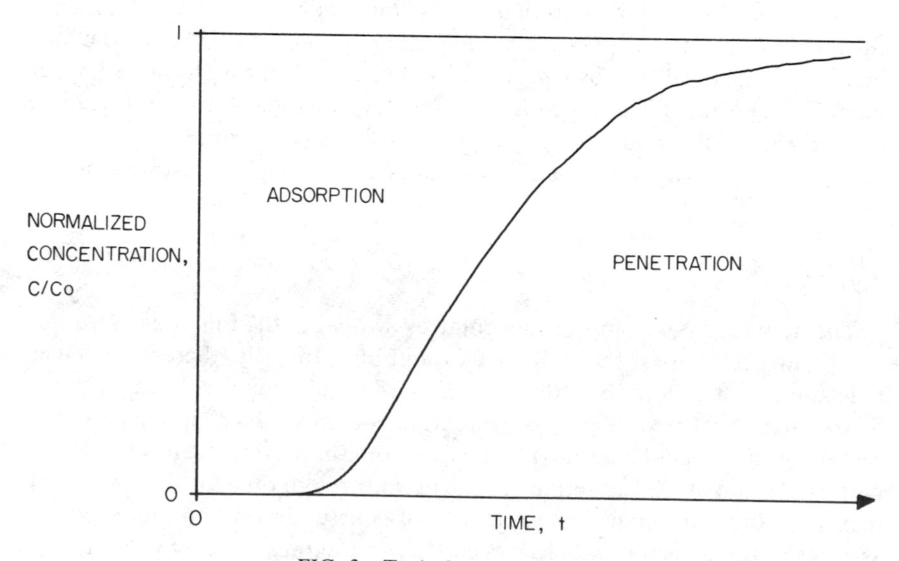

FIG. 3—*Typical test result curve.*

where

t_0 = time at which effluent concentration equals inlet concentration,
t_f, C_f = time and effluent CCl_4 concentration from the last experimental data point,
C_0 = inlet CCl_4 concentration, and
M = average slope in the final portion of the experimental curve.

The calculated data point, (t_0,C_0), is then taken to be the final point on the curve.

Exponential extrapolation is a preferred method in that it mimics the observed asymptotic behavior. This extrapolation is performed by curve fitting the last section of the experimental curve to the following function

$$\frac{C}{C_0} = 1 - Ae^{-bt} \tag{2}$$

where

t = time from start of test, s,
C = effluent CCl_4 concentration, g/cm^3,
C_0 = inlet CCl_4 concentration, g/cm^3, and
A, b = constants.

The "best fit" values of constants A and b are evaluated from a linear regression on a plot of $\ell n\ (1 - C/C_0)$ versus t. The derived exponential equation is then used to describe the remainder of the curve from the time of the last data point to infinite time.

Equilibrium Sorption Capacity

From the acquired data of effluent CCl_4 concentration versus time, a calculation of equilibrium sorption capacity at a vapor concentration of 5.0×10^{-6} g/cm^3 can be made. Graphically, the method for doing this is as follows. The area under the effluent CCl_4 concentration curve is proportional to the total mass of CCl_4 that is *not* adsorbed by the fabric specimen. The area between the curve and a horizontal line at the inlet concentration level, C_0, is proportional to the total mass of CCl_4 that *is* adsorbed by the fabric specimen (see Fig. 3). Thus, if the size of this adsorption area is evaluated and multiplied by the proper proportionality factor, the result will be the total mass of CCl_4 adsorbed.

In the case of a linear extrapolation, the mass of CCl_4 adsorbed is calculated as

$$W_0 = C_0 Q \int_0^{t_0} \left(1 - \frac{C}{C_0}\right) dt \tag{3}$$

where

W_0 = mass of CCl_4 adsorbed at equilibrium,
Q = flow rate of gas through the specimen, and
t_0 = calculated time of fabric saturation (from Eq 1).

The integral is approximated by an appropriate numerical method, such as the trapezoidal rule or Simpson's rule [1]. The calculated mass of CCl_4 adsorbed will be a conservative estimate because of the linear extrapolation.

If an exponential extrapolation is used, the mass of CCl_4 adsorbed is calculated as

$$W_0 = C_0 Q\left[\int_0^{t_f}\left(1 - \frac{C}{C_0}\right)dt + \int_{t_f}^{\infty}\left(1 - \frac{C}{C_0}\right)dt\right] \quad (4)$$

The first integral term in Eq 4 is, again, estimated by a numerical method using the experimental data. The second integral term is the estimated adsorption taking place in the extrapolated portion of the curve. Using the obtained exponential curve fit, this indefinite integral can be expressed as a finite quantity as follows

$$\int_{t_f}^{\infty}\left(1 - \frac{C}{C_0}\right)dt = \int_{t_f}^{\infty} Ae^{-bt}\,dt = \frac{A}{b}e^{-bt_f} \quad (5)$$

The value of this quantity is easily calculated from the known values of A, b, and t_f.

The calculated equilibrium sorption capacity in each case assumes that the inlet wave front of CCl_4 is a perfect step function from a concentration of 0 to 5.0×10^{-6} g/cm^3. This is not actually true and must be corrected. A correction factor is obtained by running a test with an empty sample cell and measuring apparent adsorption. This value, which has been found to be approximately 0.1 g CCl_4 in the present test setup, is then subtracted from the sample test results to give corrected total sorption capacities. These sorption capacities are then generally normalized by dividing by the weight of the test fabric or active sorptive ingredient exposed to the gas stream.

Sorption Kinetics

Sorption rate information is also available from the effluent CCl_4 concentration curve. This is important because it is usually desirable to know not only the total sorptive capacity of a test specimen but also how fast it utilizes this capacity.

The simplest rate information obtained is breakthrough time. The magnitude of the breakthrough time can be used as a quality control measure for trial products if the test conditions are set equivalent to the actual challenge conditions the fabric will face in use. It is not a true indication of sorption rate, however, since this breakthrough time value is a function of total CCl_4 sorption capacity as well as rate, and thus depends on specimen size.

Correlations exist in the literature [*2-5*], which predict sorption rate equations based on assumed mechanisms and rate-limiting steps. One or more rate constants can be calculated from these correlations which reflect the sorption rate and are usually a function only of temperature. The major drawback of these correlations is that they are not general but apply only to certain test specimens and in most cases only to certain parts of the test result curves. One such correlation is that developed by Wheeler [*5*] for adsorption in a fixed-bed catalytic reactor. A modified form of Wheeler's equation has been used effectively by Jonas and Svirbely [*6*] to model the early regions of CCl_4 sorption curves obtained from packed beds of activated carbon. The equation they used was as follows

$$t_b = \frac{W_e}{C_0 Q}\left[W - \frac{\rho_B Q \ln\left(\frac{C_0}{C_x}\right)}{K_v}\right] \tag{7}$$

where

C_0 = inlet concentration of gas, g CCl_4/cm^3,
Q = gas flow rate, cm^3/s,
ρ_B = bulk density of activated carbon, g carbon/cm^3,
C_x = breakthrough concentration, g CCl_4/cm^3,
W_e = normalized breakthrough capacity, g CCl_4/g carbon,
W = mass of activated carbon, g carbon,
t_b = breakthrough time, s, and
K_v = pseudo-first-order rate constant, s^{-1}.

The values for all the parameters in the equation except the rate constant, K_v, and breakthrough capacity, W_e, are known from the test conditions and results. The unknowns can be calculated from a linear regression on a plot of breakthrough time versus weight of sorbent. The slope and *y*-intercept of this line from Eq 7 will be

$$\text{slope} = \frac{W_e}{C_0 Q} \tag{8}$$

and

$$Y = \text{intercept} = -\frac{W_e\, \rho_B \ln\left(\frac{C_0}{C_x}\right)}{C_0 K_v} \tag{9}$$

From Eqs 8 and 9, the values of W_e and K_v can be calculated. This procedure has been effectively used to describe the sorption rate in a variety of activated carbon-containing fabrics.

Test Reproducibility

Reproducibility of the results obtained from the sorption test has been measured by running replicate samples from a single piece of sorptive fabric. Thirty test specimens were cut from a sample fabric containing 20% by weight of activated carbon as a sorptive component. Three sets of tests were conducted using one, two, three, and four layers of sample fabric. The last four-layer result was discarded because operator error resulted in the collection of insufficient data. The remaining results were analyzed for total sorptive capacity at 5.0×10^{-6} g/cm^3 CCl_4 and sorptive rate using the procedures described previously.

Total sorptive capacity was calculated from each test curve using Eq 3. The results are summarized in Table 1. A coefficient of variation of 3.4% was calculated for the eleven data points, indicating excellent reproducibility.

The three sets of test results were also used to calculate values for breakthrough capacity, W_e, and rate constant, K_v, in Eq 7. Best-fit straight lines were obtained on plots of breakthrough time versus weight of activated carbon for each of the data sets. The breakthrough concentration was set at 1% of the inlet concentration. The bulk density of activated carbon was estimated

TABLE 1—*Reproducibility test: total adsorption capacity of 5 mg/L CCl_4.*

Test Set No.	No. of Layers	Carbon Loading, g/100 cm^2	Capacity, g CCl_4/g Carbon
1	1	0.311	0.460
	2	0.622	0.477
	3	0.934	0.495
	4	1.245	0.469
2	1	0.311	0.493
	2	0.622	0.505
	3	0.934	0.474
	4	1.245	0.452
3	1	0.311	0.467
	2	0.622	0.484
	3	0.934	0.466

Average capacity = 0.477 g CCl_4/g carbon
Standard deviation = 0.016 g CCl_4/g carbon
Coefficient of variation = 3.4%

TABLE 2—*Reproducibility test: linear regression fits of breakthrough time versus weight of carbon.*

Test Set No.	W, g Carbon	t_b, s	Linear Regression Fit
1	0.311	2.0×10^2	
	0.622	9.00×10^2	$t_b = 2.21 \times 10^3 W - 4.85 \times 10^2$
	0.934	1.57×10^3	
	1.245	2.27×10^3	$r^2 = 1.0000$
2	0.311	2.4×10^2	
	0.622	9.90×10^2	$t_b = 2.42 \times 10^3 W - 5.12 \times 10^2$
	0.934	1.76×10^3	
	1.245	2.50×10^3	$r^2 = 1.0000$
3	0.311	1.9×10^2	$t_b = 2.55 \times 10^3 W - 5.31 \times 10^2$
	0.622	8.40×10^2	
	0.934	1.59×10^3	$r^2 = 0.9993$

at 0.093 g/cm^3 for the sample fabric. The results of the curve fitting are given in Table 2. All three sets of data yielded fairly good straight-line fits with coefficients of determination, r^2, ranging from 0.9993 to 1.0000.

The values of breakthrough capacity, W_e, and rate constant, K_v, were calculated from the curve fits using Eqs 8 and 9. Results of these calculations are given in Table 3. Again, good reproducibility is observed with coefficients of variation of about 5% for both quantities (W_e and K_v).

Acknowledgment

The authors are grateful to Drs. R. B. Davis and C. E. Kramer for their guidance and support. We also acknowledge the efforts of C. H. Park, S. R. Pinkham, K. A. Greenwood, and Dr. S. C. Stern, who have helped maintain and operate the test equipment.

TABLE 3—*Reproducibility test: values of breakthrough capacity and rate constant.*

Test Set No.	W_e, g CCl_4/g Carbon	K_v, s^{-1}
1	0.185	32.7
2	0.202	33.8
3	0.188	30.3
Average	0.192	32.3
Standard deviation	0.009	1.8
% coefficient of variance	4.7	5.5

References

[*1*] Davis, P. J. and Polonsky, I. in *Handbook of Mathematical Functions*, Dover Publications, New York, 1972, Chapter 25, pp. 885-887.
[*2*] Basmadjian, D., *Industrial and Engineering Chemistry Process Design and Development*, Vol. 19, No. 1, 1980, pp. 129-144.
[*3*] Masamune, S. and Smith, J. M., *American Institute of Chemical Engineers Journal*, Vol. 10, 1964, pp. 246-252.
[*4*] Rosen, J. B., *Industrial and Engineering Chemistry*, Vol. 46, No. 8, Aug. 1954, pp. 1590-1594.
[*5*] Wheeler, A. and Robell, A. J., *Journal of Catalysis*, Vol. 13, 1969, pp. 299-305.
[*6*] Jonas, L. A. and Svirbely, W. J., *Journal of Catalysis*, Vol. 24, 1972, pp. 446-459.

Norman W. Henry III[1]

How Protective Is Protective Clothing?

REFERENCE: Henry, N. W. III, **"How Protective Is Protective Clothing?"** *Performance of Protective Clothing, ASTM STP 900*, R. L. Barker and G. C. Coletta, Eds., American Society for Testing and Materials, Philadelphia, 1986, pp. 51–58.

ABSTRACT: A standard method to evaluate the permeation resistance of protective clothing has been developed. This method has generated numerous data on the resistance of various clothing materials to specific chemicals. Three chemicals for which little data were available are formaldehyde, chlorine, and hydrogen cyanide. Experience with the method in evaluating materials for their resistance to these chemicals is presented as well as a review of current methods for determining protective clothing performance. Results from the tests show that nitrile-butadiene, neoprene, and butyl rubber offer the best resistance to formaldehyde, chlorine, and hydrogen cyanide, respectively, and that both breakthrough time and permeation rate data are needed to determine performance.

KEY WORDS: protective clothing, butyl rubber, nitrile-butadiene rubber (NBR), neoprene, polyethylene-(PE) coated Tyvek, latex rubber, polyvinyl chloride (PVC), breakthrough time, permeation rate

All clothing is protective to some extent. It is the degree of protection that is important, particularly in the chemical industry, in which protective clothing is worn to prevent exposure to chemicals during production, distribution, storage, and use. These chemicals may be gases, liquids, or solids, but whatever their state, working with them safely means knowing how protective the clothing is or, more specifically, how resistant it is to chemical permeation.

Three chemicals for which relatively little permeation data exist are chlorine, hydrogen cyanide, and formaldehyde. Chlorine is a severely irritating gas. Hydrogen cyanide is a liquid below 26°C which, if absorbed in the skin, may rapidly become lethal. Formaldehyde, a gas usually supplied as an aqueous solution containing 37% (by weight) of formaldehyde, has a pronounced irritant effect on the skin. Gloves and suits are worn to prevent exposure to

[1]Research chemist, E. I. du Pont de Nemours & Co., Inc., Haskell Laboratory for Toxicology and Industrial Medicine, Newark, DE 19711.

these chemicals, but no permeation data exist to show which type of protective clothing material is the most resistant.

Procedure

To determine the permeation resistance of glove and suit materials to each of these chemicals, the newly adopted ASTM Test for Resistance of Protective Clothing Materials to Permeation by Hazardous Liquid Chemicals (F 739-81) was used. Although this method currently specifies liquids, it is being modified to include gases. Chlorine was tested following a modified protocol that consisted of continuously passing the gas over the outside surface of the protective clothing material during the test run. Liquid hydrogen cyanide was tested below 26°C, and formaldehyde was tested as a 37% solution.

Materials

All three chemicals were tested against glove or suit materials that are currently being used or are potential candidate materials. Chlorine was tested with natural rubber latex gloves, neoprene, and polyethylene-(PE) coated Tyvek (spun-bonded olefin) suit materials. Hydrogen cyanide (HCN) was tested against suit materials made of butyl rubber, PE-coated Tyvek, and polyvinyl chloride (PVC). Formaldehyde was tested with natural latex and nitrile-butadiene rubber (NBR) glove materials.

The outside surface of three specimens of glove and suit materials was exposed to each of the three chemicals in standard glass ASTM permeation cells. The inside surface of the materials was monitored for breakthrough time and steady-state permeation rate using analytical methods sensitive to and specific for each chemical.

Equipment

Chlorine was monitored continuously for breakthrough using an Interscan chlorine meter sensitive to 0.5 ppm (volume/volume). Hydrogen cyanide was monitored using an Orion conductivity meter equipped with a cyanide selective electrode. Cyanide concentrations as low as 0.3 μg/mL were detected in discrete water samples removed from the collection medium side. Formaldehyde was also analyzed using an aqueous collection medium from which aliquots were withdrawn at selected time intervals and analyzed by the chromotropic acid method.[2] The minimum detection limit for formaldehyde using this method was 1 μg/mL.

[2] *Formaldehyde in Air: NIOSH Manual of Analytical Methods*, DHEW Publication No. (NIOSH) 75-121, National Institute of Occupational Safety and Health, Cincinnati, OH, 1974, pp. 125-1–125-9.

Breakthrough is defined as the time at which a detectable amount of chemical is found permeating the test specimen. The steady-state permeation rate is determined by plotting the concentration of the chemical found against time. At steady state, a straight line is produced. The permeation rate in units of weight of chemical permeating per area per time (milligram per square metre per minute) is calculated by substituting the slope of the steady-state permeation into the permeation equation (ASTM Test F 739-81).

Results

Natural rubber latex gloves and neoprene suit materials were tested against 0.1% chlorine gas in nitrogen supplied from a gas cylinder. Suit material made of PE-coated Tyvek was also tested against chlorine, but at a lower concentration—0.002%. After continuous exposure for 8 h, no detectable amounts of chlorine were found, and no physical effects other than brown discoloration of the natural rubber latex specimens occurred. However, on close examination of the neoprene test specimens, the surface seemed brittle compared with that of an unexposed control specimen. Examination by electron microscopy of the chlorine-exposed specimen revealed that the surface had linear raised cracks with irregularly shaped platelike structures, shown in Fig. 1. These were in contrast to the intact smooth surface observed on an unexposed control specimen. Therefore, even though no breakthrough was observed, structural deterioration had occurred on the surface of the neoprene. The PE-coated Tyvek suit material also showed no visible effects after exposure and was resistant to 0.002% chlorine for 8 h. The PE-coated Tyvek specimens were not examined by electron microscopy.

All three of the test materials exposed to 100% HCN had measurable breakthrough times and permeation rates. Butyl rubber was the most resistant material, followed by PE-coated Tyvek, then PVC. The PVC material, previously used for protection, had the shortest breakthrough time (30 min) compared with 60 min for both PE-coated Tyvek and butyl rubber. A plot of the concentration of HCN, in micrograms per millilitre, found permeating these materials, versus the time in minutes, is shown in Fig. 2. The figure shows that the butyl material tested had the lowest permeation rate. Although no visual effects were noted on the outside surface of either the butyl or PE-coated Tyvek test specimens, a definite discoloration of the "hot pink" PVC test specimen was observed.

Tests with 37% formaldehyde using latex and nitrile-butadiene glove materials demonstrated that the nitrile gloves were more resistant. No breakthrough time was observed after 6 h of continuous exposure. The thinner, more flexible latex gloves, commonly used for protection, had a breakthrough time of 8 min and a relatively high permeation rate. One of the unusual visible effects noted on the latex glove specimens after exposure was the formation of pronounced nodules protruding inward toward the inside surface of the glove

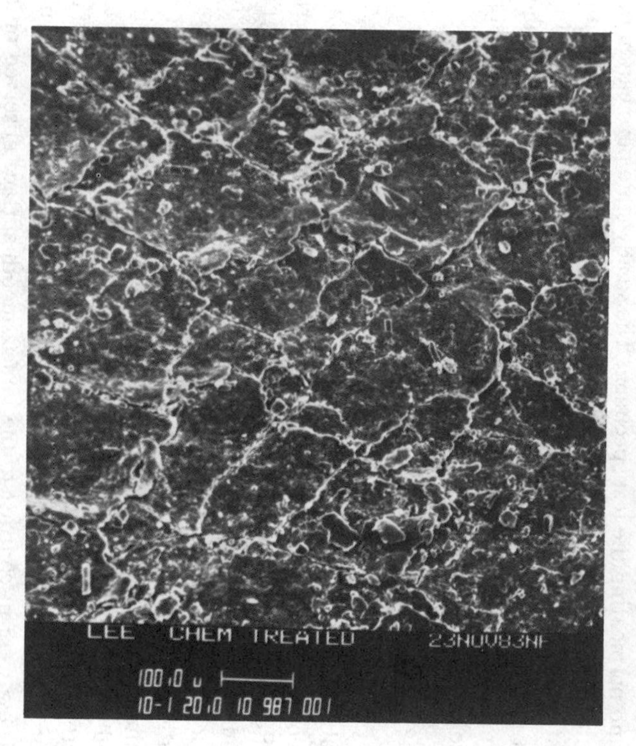

FIG. 1—*Electron photomicrographs of control* (left) *and chlorine-exposed* (right) *neoprene test material (× 100 magnification).*

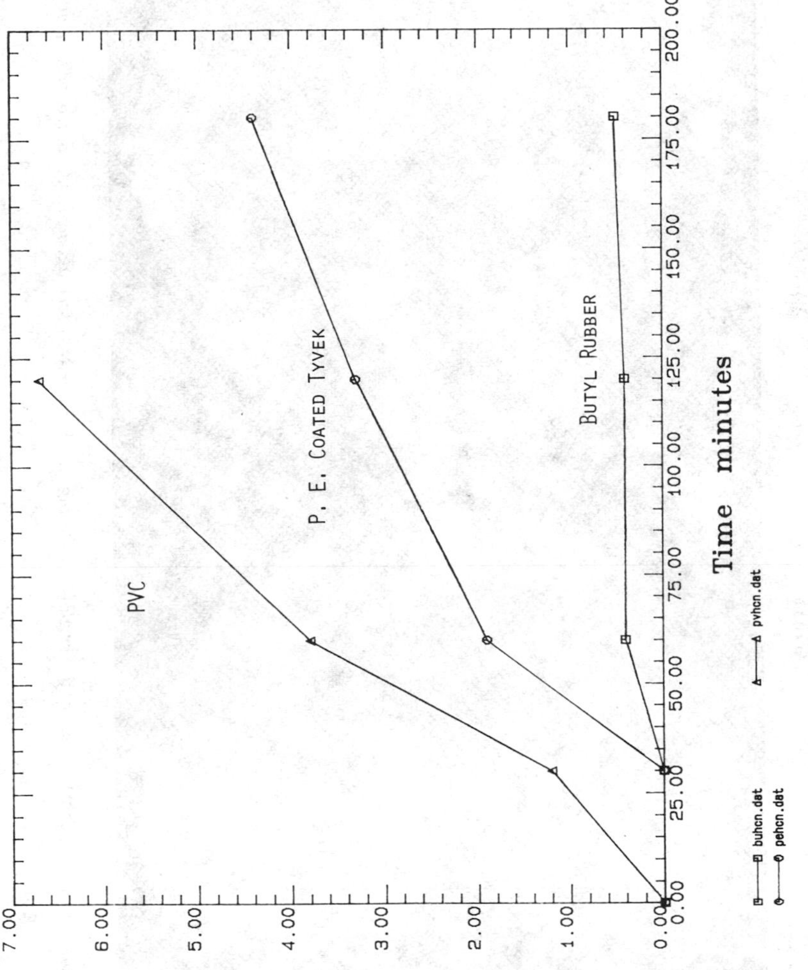

FIG. 2—*Plot of HCN concentration found permeating the test materials versus time in minutes.*

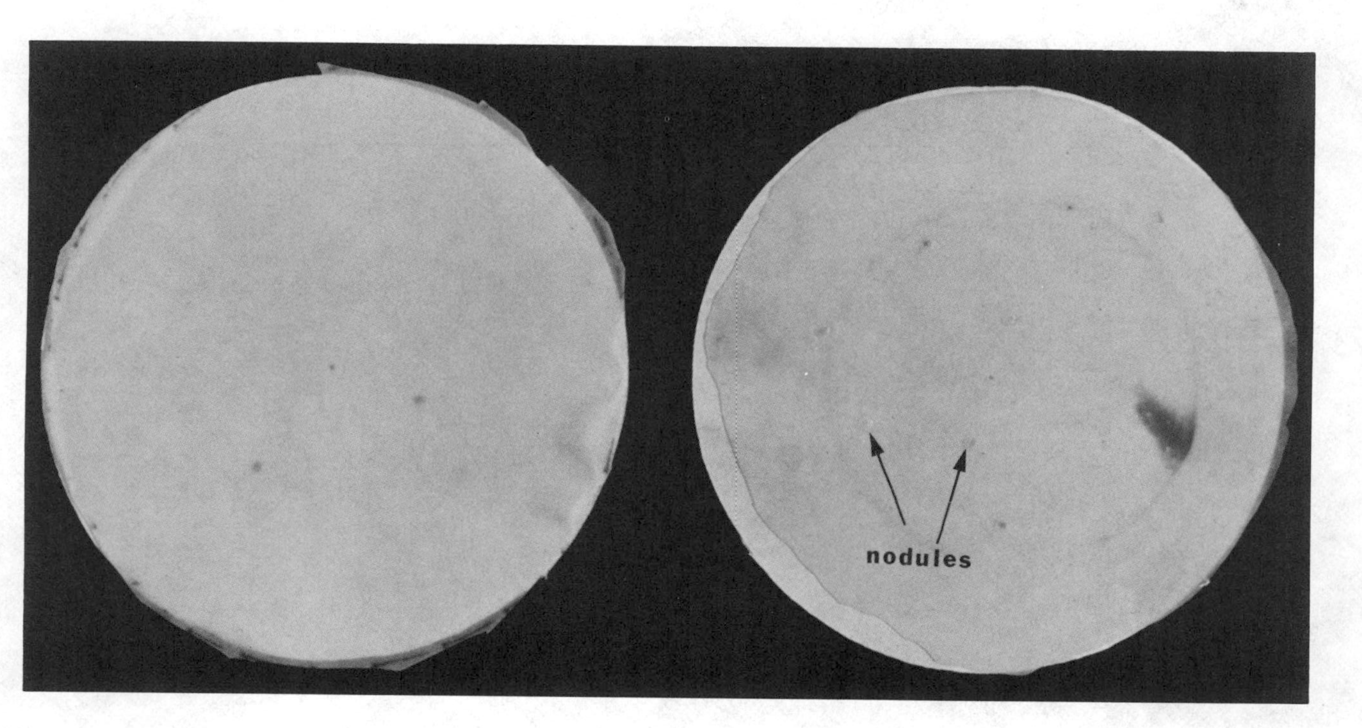

FIG. 3—*Nodules observed* (right) *on latex rubber gloves following exposure to 37% formaldehyde. A control specimen is on the left.*

specimens, as shown in Fig. 3. The results of all the permeation tests, including the thickness of the materials, are shown in Table 1.

Discussion

Tests performed with glove and suit materials against chlorine gas, liquid hydrogen cyanide, and 37% formaldehyde following ASTM Test F 739-81 show that permeation data are needed to determine how protective clothing is. By measuring both breakthrough times and permeation rates, performance data are obtained from which more resistant clothing materials can be selected for protection. Even though these materials are resistant to chemical permeation, consideration should also be given to physical requirements of clothing materials, since materials that rip or tear easily on the job will not provide adequate protection. Tests for measuring physical performance exist and should be used in conjunction with the permeation test method. Consideration should also be given to the fact that the test method evaluates a worst-case situation of continuous liquid or gas exposure, and, therefore, careful evaluation of the results is necessary, depending on the intended use. Except for this limitation, the test method proved to be versatile for gases as well as liquids. Although no breakthrough or permeation was observed with chlorine, the method worked well, since no difficulties were encountered using the modified protocol for gases. This standardized test method for either gases or

TABLE 1—*Summary of permeation data.*[a]

Chemical	Clothing Material	Thickness, mm	Breakthrough Time, h[b]	Permeation Rate, mg/min/m²[c]
Cl_2 (gas)				
0.1%	natural rubber latex	0.46	>8.0	ND[d]
	neoprene	0.41	>8.0	ND
0.002%	PE Tyvek	0.15	>8.0	ND
HCN (liquid), 100%	PVC	0.79	0.5	2.9
	butyl	0.38	1.0	0.15
	PE Tyvek	0.15	1.0	1.1
CH_2O (solution), 37%	natural latex	0.15	0.1	33.3
	nitrile	0.28	>6.0	ND

[a]The values are mean values of triplicate tests.
[b]To convert to seconds, multiply by 3600.
[c]To convert to milligrams per second per square metre, divide by 60.
[d]ND = no detectable permeation during the test run.

liquids allows comparison of different types of clothing materials to provide the best protection. Eventually the test method will stimulate renewed development of better and more resistant clothing materials for protection.

Finally, as more testing is done, my original question, "How protective is protective clothing?," will be answered from data obtained by methods such as the permeation test method used in this study and other methods developed that demonstrate the capability of a material to withstand the challenge to test chemicals under controlled conditions in the laboratory, rather than accidentally in the workplace, where worker exposure can occur.

Krister Forsberg,[1] *Knut G. Olsson,*[1] *and Björn Carlmark*[2]

Testing of Candidate Glove Materials Against Metal Cutting Fluids

REFERENCE: Forsberg, K., Olsson, K. G., and Carlmark, B., **"Testing of Candidate Glove Materials Against Metal Cutting Fluids,"** *Performance of Protective Clothing, ASTM STP 900*, R. L. Barker and G. C. Coletta, Eds., American Society for Testing and Materials, Philadelphia, 1986, pp. 59-66.

ABSTRACT: This materials study demonstrates a testing program based on the requirements of protective glove material against metal cutting fluids in workshops and steel industries.

Permeation testing using various cutting fluids, puncture testing using a standardized penetrometer needle, and tensile strength testing indicated that nitrile rubber is the most protective material. By X-ray fluorescence analysis it was demonstrated that chromium and nickel, which act as allergens, can permeate glove materials when the metals are present in cutting fluids. The results of the permeation test showed a breakthrough time of less than 2 to 3 h for natural rubber and neoprene. Nitrile without additives showed a breakthrough time of >5 h and resistance to puncturing by a force of more than 25 N. The tensile strength test showed a strength of approximately 6 N/mm^2 for nitrile and 12 N/mm^2 for neoprene, measured at the breakthrough limit.

The primary requirements were resistance to cutting fluids and to puncture from work pieces. The secondary requirements were that the glove material should be able to break apart when drawn into moving machine parts and that the glove should be thin and elastic from the functional point of view (grip comfort and tactility).

The authors conclude that nitrile rubber latex best performs the requirements.

KEY WORDS: permeation testing, puncture testing, tensile strength testing, polymer glove materials, metal cutting fluids, chemical protective clothing

Skin contact with cutting fluids is a significant large occupational hygiene problem in workshops and steel industries. All kinds of wet work, regardless of the type of cutting, contain a risk that should be preventable, for example, by the use of protective gloves.

[1]Researcher, Department of Work Science, and polymer engineer, Department of Polymer-technology, respectively, The Royal Institute of Technology, S-100 44 Stockholm, Sweden.

[2]Physicist, Scandlab, S-191 07 Sollentuna, Sweden.

Protective gloves are used today in several work processes, but rotating machine parts, for example, create a risk of accidents with them. Another problem is that metallic pollutants appear in the cutting fluid, for example, chromium and nickel.

Skin contact with various cutting fluids requires chemically resistant protective gloves. Work pieces, shavings, and other fragments with sharp edges require protective gloves with physical resistance. Many work tasks require a good fit in the protective glove as well. Functional requirements vary from light to very hard grip and control of tactility. From a protective and functional point of view, the material thickness must be optimized. Work with metal cutting fluids requires a number of criteria in the selection of protective glove materials.

This materials study was designed to use different test methods with the aim of selecting the best protective glove material. With regard to chemical resistance, the permeation of two industrial cutting fluids through specimens from candidate glove materials was measured. The physical resistance was measured by puncture testing using a standardized penetrometer needle. The tensile strength of the candidate materials was also tested to evaluate the risk of gloves catching in rotating parts.

Procedure

Permeation Testing

A two-chambered permeation cell with an inner diameter of 30 mm was used to perform testing. A sketch of the test cell is shown in Fig. 1. The collecting medium was nitrogen or water. A so-called open-loop system was used for gas collection. Certain tested fluids have a very low vapor pressure at room temperature, thus, gaseous collection can be inappropriate. The test cell, therefore, is equipped with stop nuts at the inlet and the outlet. Every half hour after application of the test fluid, the nuts are loosened, so that 5 mL of water can be sprayed in the lower chamber. The nuts are then tightened, and the test cell is turned up and down for approximately 1 min. If the test liquid permeates, the permeant is "cleaned away" in this manner, and, after the collecting medium (water or isopropyl alcohol), is tapped, it can be analyzed with a gas chromatograph or other appropriate analytical instrument. When nitrogen was used as the collecting medium, the flow rate was three to four exchanges per minute. The analytical method was gas chromatography/flame ionization detection (GC/FID), when nitrogen was used as a collecting medium, and ultraviolet (UV), respectively, when water was used. The procedure is described in more detail in the ASTM Test for Resistance of Protective Clothing Materials to Permeation by Hazardous Liquid Chemicals (F 739-81). In connection with the determination of breakthrough time (in minutes)

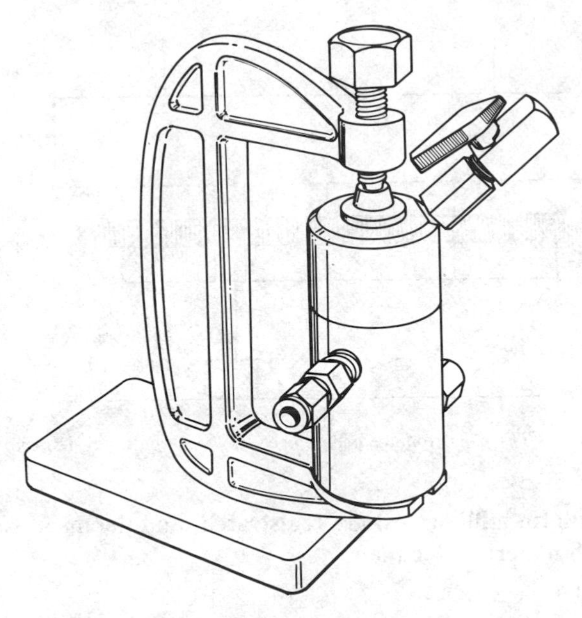

FIG. 1—*Two-chamber chemical permeation test cell.*

and permeation rate (in milligrams per square centimetre per minute), the metal analysis was made (three tests) by X-ray fluorescence analysis.

Challenge Chemicals

Two different metal cutting fluids were included in the study. The fluids were taken from industrial production. Liquid A (Esso Somentor 33) was a cutting oil and Liquid B (Blasocut Blaser) was an emulsifiable cutting fluid. Liquid A was collected by gas and liquid modes. Liquid B was only collected by liquid mode.

Puncture Testing

The test method has been described earlier by Coletta et al [*1*]. The method measures the force required to cause a standardized penetrometer needle to puncture a material specimen. The needle is 2.03 mm in diameter and has a tip radius of 0.25 mm. The tip angle is 26°. Figure 2 shows a side projection of the penetrometer and how the material specimen is fastened. The puncture equipment is connected to a tension testing machine, such as an Instron. The rate of travel is 500 mm/min. Three tests from one material specimen are registered as curves by a recorder. The force (in Newtons) and the break-

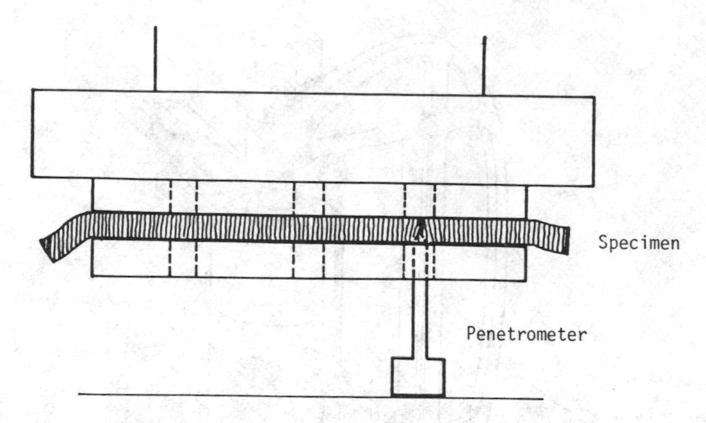

FIG. 2—*The penetrometer needle breaking through the test specimen.*

through length (in millimetres) are registrated, and the mean value is calculated for each material specimen.

Tensile Strength Testing

The tensile strength was tested with an Instron in accordance with Swedish Standard (SS 16 22 02). This method agrees with the ASTM Test for Rubber Properties in Tension (D 412-83). Three variables were determined: the limit of breaking (Newtons per millimetre squared), that is, the draw tension at breaking; elongation, that is, by how many percentage points more than its length the material stretches; and modulus, which is the ratio of the tension and the deformation in the direction of the force (at 300% of stretching).

Candidate Materials

Eight different candidate materials were selected from the three most common rubber materials. The materials are described in Table 1. Seven of the

TABLE 1—*Description of candidate materials.*

No.	Barrier Material	Description
1	natural rubber	natural rubber without additives, latex dipped
2	neoprene	chloroprene rubber without additives, latex dipped
3	neoprene	chloroprene rubber with carbon, latex dipped
4	neoprene	chloroprene rubber with carbon and chlorination, latex dipped
5	nitrile	nitrile rubber without additives, latex dipped
6	nitrile	nitrile rubber with carbon, latex dipped
7	nitrile	nitrile rubber with carbon and chlorination, latex dipped
8	nitrile	nitrile rubber without additives, solvent dipped

materials were latex dipped and one was solvent dipped. Carbon black was added to two of the materials.

Results

Permeation Testing

Five of the candidate glove materials were tested against two different cutting fluids. Permeation data are presented in Table 2. The results indicated that latex-dipped nitrile was the most resistant material. The number of permeation tests was limited; however, it was judged that relevant information could be obtained from this small number of tests. The results from the puncture testing and tensile strength testing had significant influence on the selection of the permeation test. Natural rubber, for example, has a low puncture resistance and was therefore limited to one test. In addition, permeated metals were studied by X-ray fluorescence analysis. The amount of chromium and nickel permeated through three different materials is listed in Table 3. The results of one analysis are shown in Fig. 3.

TABLE 2—*Permeation data from permeation testing of liquids A and B.*[a]

		Liquid A		Liquid B
Material and No.	Thickness, mm	Breakthrough Time, min	Permeation Rate, mg/cm²/min	Breakthrough Time, min
Natural rubber, 1	0.66	...	...	120 to 150
Neoprene, 2	0.40	112	2.6	>150
Nitrile, 5	0.32	>300	...	>150
Nitrile, 7	0.38	>150	...	>180
Nitrile, 8	0.37	>120	...	150 to 180

[a]Liquid A = Esso Somentor 33; Liquid B = Blasocut Blaser.

TABLE 3—*Metal data from permeation testing of Liquid B.*[a]

Material and No.	Time of Liquid Contact, min	Chromium, ng/cm²	Nickel, ng/cm²
Natural rubber, 1	150	58	107
Neoprene, 2	210	35	23
Nitrile, 8	180	117	140

[a]The concentrations of chromium and nickel were approximately 30 and 15 ppm, respectively.

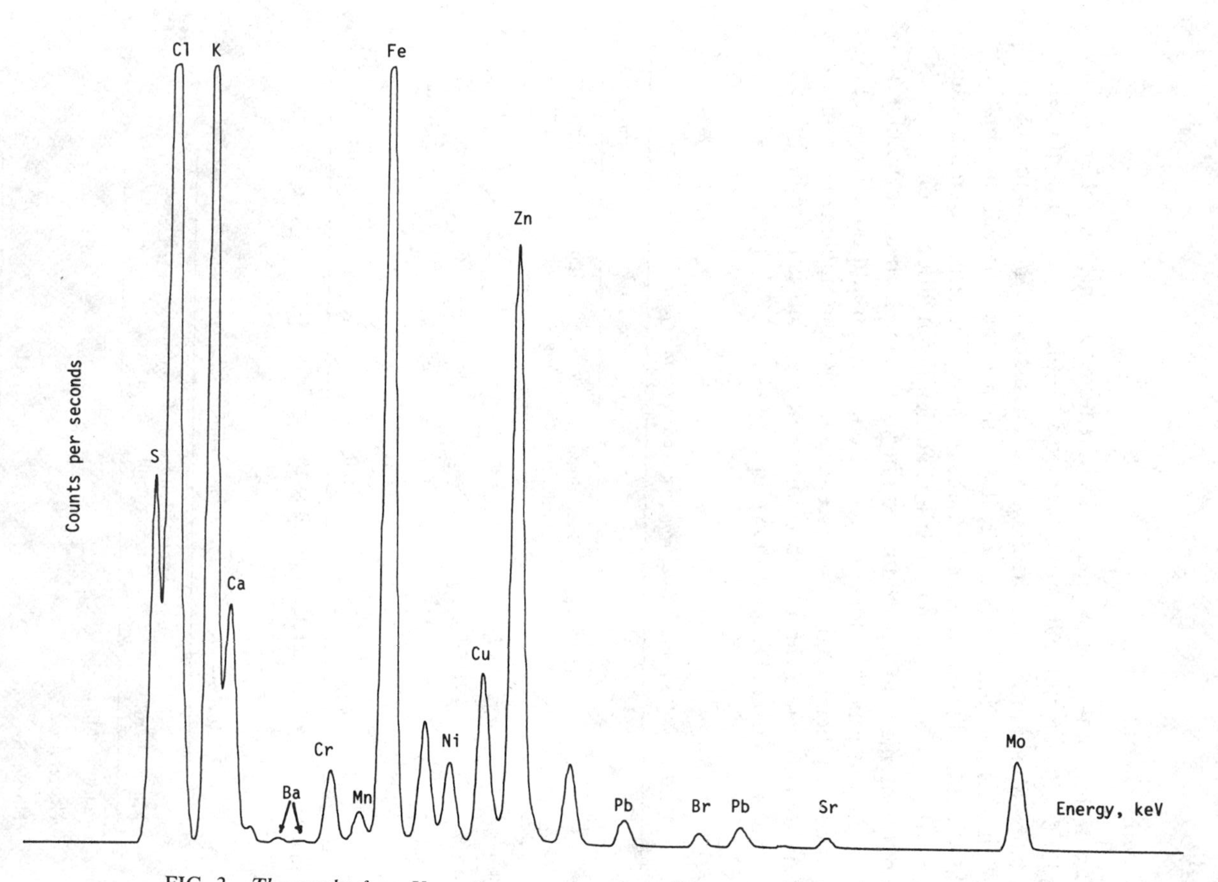

FIG. 3—*The results from X-ray fluorescence analysis. Note the peaks for chromium and nickel.*

Puncture Testing

The physical resistance was tested in all eight candidate glove materials. The puncture data are presented in Table 4. Again, the results indicated that latex-dipped nitrile was the most resistant material. Neoprene material had no punctures until a breakthrough length of 23 mm. The material shows far too high an elasticity to measure the force of breakthrough. A more complete picture of the behavior of the rubber is obtained by tensile stress tests.

Tensile Strength Testing

Six of the candidate glove materials were tested for tensile strength. The aim of the study was to compare nitrile and neoprene with or without additives. Tension tests were made on three punched-out sample specimens, and the mean value was calculated. The results show differences in limit of breaking and elongation. The modulus of elasticity was somewhat higher for nitrile. This study indicated that nitrile rubber was the best material with regard to the risk of being caught in rotating parts. The tensile strength data are presented in Table 5.

TABLE 4—*Puncture data.*

Material and No.	Thickness, mm	Force, N	Breakthrough Length, mm
Natural rubber, 1	0.64 to 0.67	11	11
Neoprene, 2	0.37 to 0.38	no puncture	
Neoprene, 3	0.34 to 0.36	no puncture	
Neoprene, 4	0.37 to 0.38	no puncture	
Nitrile, 5	0.28 to 0.30	26	17
Nitrile, 6	0.34 to 0.36	14	17
Nitrile, 7	0.39 to 0.40	12	15
Nitrile, 8	0.48 to 0.50	15	14

[a]No puncture means >23 mm of breakthrough length. The puncture data are the mean values calculated from three puncture tests.

TABLE 5—*Tensile strength data.*

Material and No.	Thickness, mm	Limit of Breakthrough, N/mm^2	Elongation, %	Modulus, N/mm^2
Neoprene, 2	0.38	12.0	1200	1.4
Neoprene, 3	0.34	10.8	1200	1.3
Neoprene, 4	0.36	14.5	1200	1.4
Nitrile, 5	0.26	5.1	900	1.5
Nitrile, 6	0.36	6.0	900	1.7
Nitrile, 7	0.35	6.4	800	2.1

Discussion

This material study demonstrates a testing program based on the requirements of protective glove material against metal cutting fluids in workshops and steel industries. The number of tests and candidate materials is small, but further study should be focused on nitrile rubber. The authors propose that the manufacturer should try to change the vulcanizing system. One can, *inter alia*, increase the degree of cross-linking by adding a vulcanizing agent. This reduces the elasticity and increases the material resistance. The next step will be to study the possibilities of increasing the acrylonitrile content, which further increases the physical and chemical resistance. To this should be added a study on chlorination.

One cannot expect to find a glove material that provides protection for a large number of tasks and chemicals. A material that meets the requirements for work with metal cutting fluids should also be tested against other chemicals in the same industry. This reduces the number of glove material types to a minimum.

Two of three test methods have been accepted as standard. It was not possible earlier to test glove protection against puncture. A measure of the material resistance to perforation is achieved by the new method of measuring the glove material resistance to puncture by a standardized penetrometer needle. The method is simple and easy to run. The authors propose that ASTM Committee F-23 on Protective Clothing should focus on a puncture testing method.

Acknowledgments

This study has been supported by the Swedish Work Environment Fund under Contract ASF 81-1187.

Reference

[1] Coletta, G. C., Schwope, A. D., Arons, I., King, J., and Sivak, A., "Development of Performance Criteria for Protective Clothing Used Against Carcinogenic Liquids," Report to NIOSH under Contract 210-76-0130, Arthur D. Little, Cambridge, MA, Oct. 1978.

Gunh Mellström[1]

Experiences from Developing a Data Base on the Protective Effects Against Chemicals of Gloves Made from Rubber and Plastic Materials

REFERENCE: Mellström, G., **"Experiences from Developing a Data Base on the Protective Effects Against Chemicals of Gloves Made from Rubber and Plastic Materials,"** *Performance of Protective Clothing, ASTM STP 900*, R. L. Barker and G. C. Coletta, Eds., American Society for Testing and Materials, Philadelphia, 1986, pp. 67–74.

ABSTRACT: A data base has been developed containing test data on protective effects of safety gloves against chemicals. Permeation data and clinical data as well as experimental data are collected and, after dermatological assessment, are put into different files in the data base. The main purpose of this data base is to give up-to-date and relevant information to those who have to choose and recommend suitable gloves to workers handling or working with hazardous chemicals. The information from the data base is presented to the users in the form of printed documents which can be updated regularly.

Today test results are not always comparable because so many different investigation methods are used. To make it possible to compare test results, still more work must be done to establish internationally accepted standards or reference methods (*in vitro* and *in vivo*) and internationally accepted rules for presenting test data. This would make it easier to evaluate the protective effect against chemicals of protective glove materials.

KEY WORDS: data base, protective effect, rubber gloves, plastic gloves, hazardous chemicals, dermatological assessment, protective clothing

In recent years, increased interest in protecting workers against chemicals by using safety gloves made of different polymeric materials has led to an increasing number of investigation reports. To get a general view of what has been achieved and what methods have been used, a data base containing this kind of information has been developed over the past two years at the Occupational Dermatology Unit, Research Department, Swedish National Board of Occupational Safety and Health. At present, the data base is available on a

[1]Research pharmacist, Department of Occupational Dermatology, National Board of Occupational Safety and Health, S-171 84 Solna, Sweden.

microcomputer. The hardware used consists of keyboard, display unit, flexible disk unit, and printer. The software consists of a data base program suitable for general file management.

Data Input

Information on safety gloves is collected from several sources, examined, and put into the data base (Fig. 1). The information comes from manufacturers, investigation reports, scientific periodicals, and the Selective Dissemination of Information (SDI) service from international literature data bases in the fields of chemistry, medicine, and occupational safety and health.

The data base consists of five files, each with a different kind of information (Fig. 2).

Product File

One of the files, the product file, contains the kind of information one usually gets in catalogs from manufacturers and distributor firms, for example, the name of the glove, the material, size, style, thickness, color, and recommended application. This file contains information about 250 different gloves for occupational use on the Swedish market.

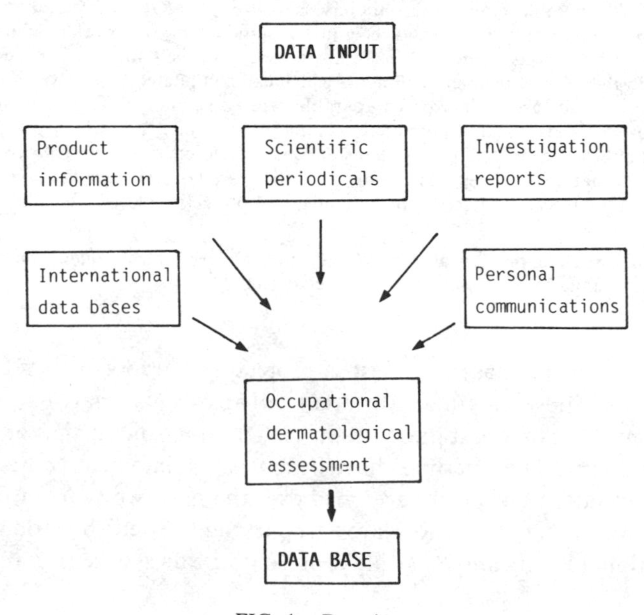

FIG. 1—*Data input.*

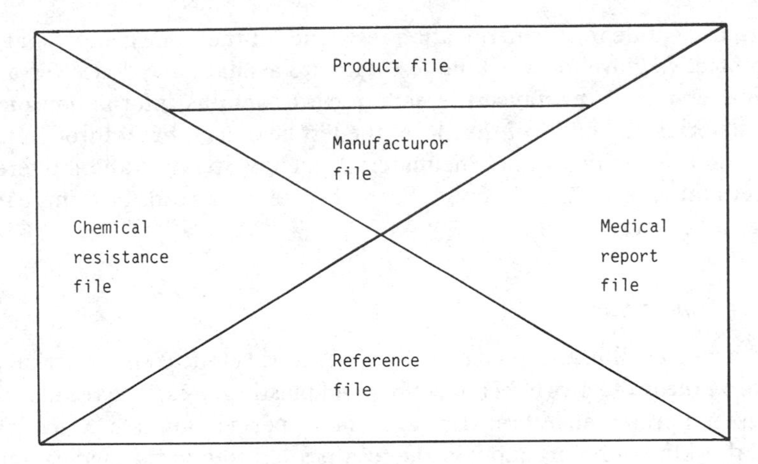

FIG. 2—*Data base structure.*

Manufacturer File

There is also a manufacturer file containing the names and addresses of the different companies in Sweden that currently market safety gloves for occupational use.

Chemical Resistance File

In the chemical resistance file, results from technical measurements of permeation resistance for different glove materials are collected. Most of the permeation resistance measurements are performed by using a permeation cell with two chambers separated by a glove membrane. Today, the only standard method for this kind of measurements is the ASTM Test for Resistance of Protective Clothing Materials to Permeation by Hazardous Liquid Chemicals (F 739-81). This standard method can serve as a reference for the ongoing research in this field. When testing the permeation resistance against highly toxic chemicals in small amounts and chemicals with low vapor pressure or low water solubility, microcells and specified analyzing techniques must be used. Therefore, a test equivalent to ASTM Test F 739-81 would be useful, for assessment of results performed with different testing equipment but an identical testing procedure.

More work is also needed to establish an internationally accepted identification or classification system of different protective material. Today very few hazardous chemical/protective material combinations are tested in relation to the total number of possibilities. Therefore, it would be of great help to know if the permeation resistance data for a protective material in a suit could be used as a guide for the permeation resistance of the same generic material in a protective glove for which no test data are available, and vice versa.

In the chemical resistance file each record contains information on the glove material, the name of the glove, the test chemical, breakthrough time, permeation rate, thickness of the material, the investigator, and a reference number and other data of interest. Each of these field words or terms can be used as a key word.

Medical Report File

In the medical literature you can also find several clinical and experimental reports on the protective effect of rubber and plastic gloves. The results, however, are not presented as breakthrough time or permeation rate. Some examples of this kind of report and how the results are stored in the medical report file will be presented.

Clinical Studies—A common clinical method used is patch testing. Pieces of the glove material are placed between the test chemical and the skin, usually on the back, and are fixed with non allergenic surgical tape for 48 h. After another 24 h, the subject's reactions are assessed for redness, edema, papules, and vesicles. The results of the test are read as the difference in reactivity between protected and unprotected skin. In this way both allergenic and irritating chemicals can be tested, and it is the local effect that is studied. With this method, one can also investigate contact allergy to rubber—a possible side effect when using rubber gloves [*1–8*].

Experimental Studies—Experimental investigations are used to demonstrate effects on healthy human beings or animals and are usually accomplished in a standard manner in a laboratory. The protective effect of the glove material can be determined by measuring its capacity to reduce the percutaneous absorption of a chemical in humans or animals.

For example, solvents can be placed into a small open-bottomed glass reservoir and attached to the skin of a guinea pig, and the blood concentration can then be measured repeatedly over time. The exposed area is then protected with a glove membrane before the solvent is placed in the reservoir. The difference between the blood concentration for normal, unprotected skin and that for protected skin during the same period of time gives a measurement of the protective effect of the glove material [*9,10*].

Workplace-Related Studies—Field studies on the significance of skin protection in exposed workers can be of great value if skin absorption is an important route of occupational exposure. The concentration of the chemical or its metabolites is measured in blood or urine after normal occupational exposure, with and without protective gloves [*11–13*].

In this kind of investigation, the results are presented in quite a different way from those investigations using permeation cells. This fact makes it difficult to find a suitable general data input system. At present, results other than breakthrough time and permeation rate are stored in the medical report file. There, the field headings are the author, title, publication, reference number, test chemicals, glove materials, test method, and results. Each of these records contains at least two big text sections, method and result.

Reference File

Finally, there is the reference file with the title, author, publication, key word, and reference number as field words.

Output Data

The main purpose of developing this data base was to make it easier to give up-to-date and relevant information to the staff working in the Occupational Health Service and to those responsible for corporate health services who have to recommend suitable safety gloves to workers handling or working with hazardous chemicals.

The data base is developed in the research department, and there is no possibility for on-line searching from the outside. The output from the different files in the data base is therefore arranged and presented in three printed documents in a way that makes regular up dating possible [*14–16*]. Part 1 consists of a survey of documentation concerning protective effects and side effects of plastic and rubber protective gloves. It is mainly the information in the chemical resistance file and the medical report file, which has been structured and then presented in investigation reports with information on the title, authors, and publication, and a short description of the tested materials, chemicals, methods used, and results. There is also an index, which makes it easy to find information on a chemical or protective glove material.

The second part is a market survey on the plastic and rubber protective gloves for occupational use on the Swedish market. It consists of data printouts, in which the name of the glove and glove material are the key words.

The third part consists of a compilation of test results from the chemical resistance file, to be used as a guide when choosing plastic and rubber protective gloves. The data printouts have been arranged so that test results, such as breakthrough time, are easy to find and update regularly. Three different lists are included, two of which have glove materials as key words. In one list the breakthrough time is 10 min or less for the tested chemicals. The second list contains information on breakthrough time for 60 min or more in one column, for less than 60 min but more than 30 min in another, and for less than 30 min but more than 10 min in the last column. The third list is like the others but the chemical name is used as a key word. At present, there is no information on permeation rate in these lists because this parameter is not always measured; it is more dependent on the test procedure and the analyzing technique used than the breakthrough time is.

If someone would like to have a data printout with information, for example, on permeation rate, breakthrough time, method, and temperature, it is possible to get one just by asking for it by letter or by making a telephone call. For each special search key there will be a special data printout. The only "standard" printouts are those that have been revised and are available in Part 3 of the printed document.

Because the external access to the data base is very limited today, work is in progress to convert the information to another data base system to which it will be possible in the future to have on-line contact. The data base will also be enlarged, with more files.

The information from the data base is thus distributed to the Occupational Health Service partly in the form of the printed documents for information, education and training and also to some extent by telephone consulting service [*14–16*]. Information from the data base is also available to organizations working in the occupational safety field (Fig. 3).

Recommendations for Improvement

From two years of work developing a data base containing information based on test results from different kinds of studies, the author's experiences can be summarized as follows:

1. An internationally accepted *standard of measuring and presenting* chemical permeation data for protective gloves is highly desirable. This would

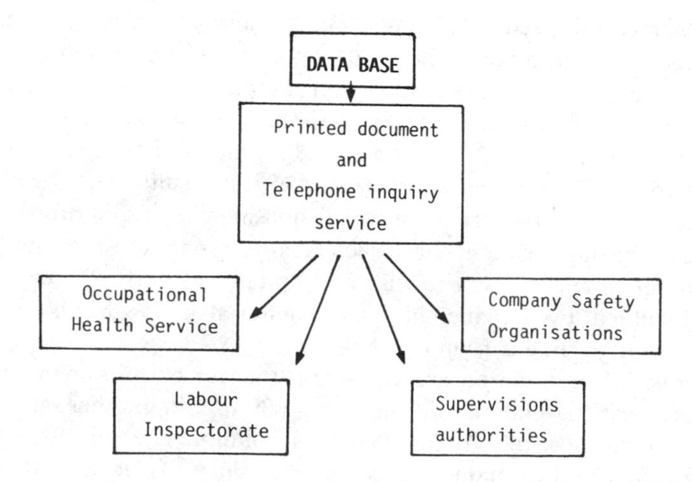

FIG. 3—*Data base output.*

make it possible to perform a much more meaningful evaluation of test results, and these results would be more suitable for data processing.

2. An internationally accepted *coding system* for well-defined protective glove materials [such as Chemical Abstract Service (CAS) numbers for chemicals] would make it easier to identify the glove or the material. It should be used by scientists as well as by manufacturers and salesmen.

3. To get a more *complete view* of the protective effect of gloves against chemicals, one should include the following factors:

(*a*) permeation and degradation data,
(*b*) results from clinical and experimental investigations,
(*c*) dermatological assessment of both the protective and the adverse effects, that is, the sensitizing capacity,
(*d*) the properties of the chemical, and the application area, and
(*e*) how the gloves are to be used.

References

[*1*] Dekker, J. C., Pel, H. J., and Sieswerda, D., "Schijnzekerheden bij het werken met dimetylformamide," *Tijdschrift voor sociale Geneeskunde*, Vol. 52, 1974, pp. 860–862.

[*2*] Lidén, C., "Occupational Dermatoses at a Film Laboratory," *Contact Dermatitis*, Vol. 10, 1984, pp. 77–78.

[*3*] Moursiden, H. T. and Faber, O. "Penetration of Protective Gloves by Allergens and Irritants," *Transaction of St. John's Hospital, Dermatology Society*, Vol. 59, 1973, pp. 230–234.

[*4*] Pegum, J. S., "Penetration of Protective Gloves by Epoxy Resin," *Contact Dermatitis*, Vol. 5, 1979, pp. 281–283.

[*5*] Pegum, J. S. and Medhurst, F. A., "Contact Dermatitis from Penetration of Rubber Gloves by Acrylic Monomer," *British Medical Journal*, Vol. 2, 1971, pp. 141–143.

[*6*] Thestrup-Pedersen, K., Christiansen, J. V., and Zachariae, H., "Precautions for Personnel Applying Topical Nitrogen Mustard to Patients with Mycosis Fungoides, *Dermatologica*, Vol. 165, 1982, pp. 108–113.

[*7*] Thomsen, K. and Ingeman Mikkelsen, H., "Protective Capacity of Gloves Used for Handling of Nitrogen Mustard," *Contact Dermatitis*, Vol. 1, 1975, p. 268.

[*8*] Wall, L. M., "Nickel Penetration Through Rubber Gloves," *Contact Dermatitis*, Vol. 6, 1980, pp. 461–463.

[*9*] Boman, A., Wahlberg, J. E., and Johansson, G., "A Method for the Study of the Effects of Barrier Creams and Protective Gloves on the Percutaneous Absorption of Solvents, *Dermatologica*, Vol. 164, 1982, pp. 157–160.

[*10*] Wahlberg, J. E., Boman, A., and Johansson, G., "En metod att studera effekten av skyddshandskar och barriärkrämer på den percutana resorption en lösningsmedel," Investigation Report, National Board of Occupational Safety and Health, Occupational Dermatology Unit, Research Department, Solna, Sweden, 1980.

[*11*] Brooks, S. M., Anderson, M. S., Emmet, E., Carson, A., Tsay, J-Y., Elia, V., Buncher, R., Karbowsky, R., "The Effects of Protective Equipment on Styrene Exposure in Workers in Reinforced Plastics Industry," *Archives of Environmental Health*, Vol. 35, 1980, pp. 287–294.

[*12*] Hogstedt, C. and Ståhl, R., "Skin Absorption and Protective Gloves in Dynamite Work," *American Industrial Hygiene Association Journal*, Vol. 41, 1980, pp. 367–372.

[*13*] Lauwerys, R. R., Kivits, A., Lhoir, M., Rigolet, P., Houbeau, D., Buchet, J. P., Roels, H. A., "Biological Surveillance of Workers Exposed to DMF and the Influence of Skin Protection on Its Percutaneous Absorption," *International Archives of Occupational Environmental Health*, Vol. 95, 1980, pp. 189–203.

[14] Mellström, G., "Protective Gloves and Barrier Creams, 1983:28—Part I: A Survey of Documentation Concerning Protective Effects and Side Effects of Plastic and Rubber Protective Gloves and Barrier Creams, "Investigation Reports, National Board of Occupational Safety and Health Occupational Dermatology Unit, Research Department, Solna, Sweden, 1983.
[15] Mellström, G., "Protective Gloves and Barrier Creams, 1983:29—Part II: Market Survey—Plastic and Rubber Protective Gloves—Barrier Creams," Investigation Reports, National Board of Occupational Safety and Health Occupational Dermatology Unit, Research Department, Solna, Sweden, 1983.
[16] Mellström, G., "Protective Gloves and Barrier Creams 1983:30—Part III: Compilation of Test Results to Be Used as Guidance When Choosing Plastic and Rubber Protective Gloves," Investigation Reports, National Board of Occupational Safety and Health, Occupational Dermatology Unit, Research Department, Solna, Sweden, 1983.

C. Nelson Schlatter[1] *and Drew J. Miller*[1]

Influence of Film Thickness on the Permeation Resistance Properties of Unsupported Glove Films

REFERENCE: Schlatter, C. N. and Miller, D. J., **"Influence of Film Thickness on the Permeation Resistance Properties of Unsupported Glove Films,"** *Performance of Protective Clothing, ASTM STP 900*, R. L. Barker and G. C. Coletta, Eds., American Society for Testing and Materials, Philadelphia, 1986, pp. 75-81.

ABSTRACT: Permeation data, where available to date, cover only one style in any product line. Most unsupported glove product lines are made in several styles, generally differing in thickness. This is a report on an investigation of the effects of these differences in thickness on measured permeation properties.

Five typical industrial chemicals were tested: aniline, dimethylformamide (DMF), hexane, methyl isobutyl ketone (MIBK), and toluene. Film specimens were made in thicknesses of about 15 mils and about 25 mils from each of five unsupported glove formulations. Two of these formulations are available commercially as Edmont 29-series neoprene and Sol-Vex nitrile gloves; the other three were experimental butyl latex systems. All the tests were run in duplicate using apparatus and procedures from the ASTM Test for Resistance of Protective Clothing Materials to Permeation by Hazardous Liquid Chemicals (F 739-81). The collecting medium was a continuous stream of dry nitrogen, sampled at intervals with a gas sampling valve. The samples were passed through the analytical column of a Gow-Mac gas chromatograph and detected by flame ionization detection (FID).

For 3 of the possible 25 combinations of chemicals and protective films, no breakthroughs were observed. In 5 others, the chemical did not break through the film soon enough or in sufficient quantity to provide complete data on the effects of film thickness. For 10 combinations, the change in thickness had a much greater effect on breakthrough time than on permeation rate. In only 6 cases did an increase in thickness reduce the measured steady-state permeation rate, and the breakthrough time was increased for 5 of these. One case was an anomaly, since the thicker specimen had a longer breakthrough time but a higher rate measurement. Repeat tests confirmed this anomaly.

We conclude that a change in thickness is more likely to affect breakthrough time than steady-state permeation rate.

KEY WORDS: personal protective equipment, permeation testing, gloves, butyl rubber, neoprene, nitrile rubber, film thickness, protective clothing

[1]Chemist and senior development chemist, Edmont Division, Becton, Dickinson and Co., Coshocton, OH 43812.

Most reported studies on the permeation properties of gloves describe the effects of a small number of chemicals on a limited number of glove styles chosen as a broad representation of the entire field of available products [*1-4*]. More extensive test programs [*5-7*] commonly add additional chemicals without greatly increasing the number of glove styles. To cover the entire field with a limited number of styles, one must select styles from different manufacturers and product lines. Only infrequently have two or more gloves from the same product line and manufacturer, differing only in thickness, been tested. Therefore, most reported permeation studies do not include thickness as a significant variable.

One previous report [*8*] stated that permeation rate is inversely proportional to glove thickness. Breakthrough time was not reported in this paper for the materials that were tested at different thicknesses.

At the Edmont Division of Becton, Dickinson and Co. permeation tests are carried out on experimental formulations as part of the company's ongoing program of developing new and improved glove styles. The authors are now reporting the observed effects of glove thickness on both breakthrough time and permeation rate in some of this experimental work. Since the company makes its own gloves, it is in a position to make and test specimens that vary only in thickness from the exact same batch of rubber compound. It should therefore be possible to eliminate any extraneous formulation and processing variables that might influence the results.

Procedure

Unsupported glove specimens were made in two thicknesses from each of five formulations. The nominal gages were 0.38 mm (15 mil)[2] and 0.64 mm (25 mil). For the three experimental butyl latex systems, the films were built up by multiple dipping and drying steps on porcelain forms before the final drying and curing. The total thickness was controlled by varying the number of dip and dry steps. The neoprene and nitrile rubber gloves were made from formulations used for the company's normal commercial production by dipping the forms first in a coagulant bath, then in the formulated rubber latex. The film thickness was controlled by varying the immersion time in the latex to allow the coagulant picked up on the forms in the first dip to gel an appropriate amount of rubber. These gloves were completed by leaching, drying, and curing. In all cases, the same dip baths, the same ovens, and the same curing times were used for both the thick and the thin specimens from each formulation. Further details of the glove manufacturing processes are proprietary.

Five chemicals were selected for a permeation screening of these specimens: aniline, dimethylformamide (DMF), *n*-hexane, methyl isobutyl ketone

[2]The original thickness measurements were in English units.

(MIBK), and toluene. These represent five different chemical classes with a wide range of polarity and aromatic character. The ASTM Test for Resistance of Protective Clothing Materials to Permeation by Hazardous Liquid Chemicals (F 739-81) was run with the following options and modifications:

1. Two, and not three, replicate specimens were tested. Some of the experimental laboratory gloves were available only in limited quantities.
2. The permeated chemicals were collected in a continuous stream of flowing nitrogen. A gas sampling valve with a 2-mL sample loop was used to divert samples from this nitrogen stream to a Gow-Mac Model 750 gas chromatograph with a flame ionization detector for analysis.
3. The nominal duration of some tests was 6 h. Some tests were interrupted sooner if it became clear that a permeation rate maximum had been achieved; others were allowed to continue longer as a matter of experimental convenience.

Results

Experimental data on all the specimens are listed in Table 1. The 25 possible combinations of glove formulation and test chemical may be separated into the following groups:

1. *No breakthrough was observed (three cases):* DMF/butyl A, butyl B; hexane/nitrile.
2. *Insufficient material permeated to reach valid conclusions (five cases):* aniline/butyl A, butyl B, butyl C; DMF/butyl C and neoprene.
3. *The thinner specimen had a quicker breakthrough with little effect on the permeation rate (ten cases):* aniline/neoprene, nitrile; DMF/nitrile; MIBK/neoprene, nitrile; toluene/butyl A, butyl B, butyl C, neoprene, and nitrile.
4. *The thinner specimen had a quicker breakthrough with a faster permeation rate (five cases):* hexane/butyl A, butyl B, butyl C; MIBK/butyl B and butyl C.
5. *The thinner specimen had an equal breakthrough time and a faster permeation rate (one case):* hexane/neoprene.
6. *The thinner specimen had a quicker breakthrough time and a slower permeation rate (one anomalous case):* MIBK/butyl A. The thicker specimen was retested, and its higher permeation rate was confirmed.

Discussion

Of the 25 combinations of glove formulations and chemicals tested, only 6 showed a higher permeation rate for the thinner gloves. However, 15 showed a faster breakthrough time for the thinner gloves, including 5 of the 6 that had higher permeation rates. Our first conclusion, therefore, is that break-

TABLE 1—*Permeation data versus film thickness (gage) for five glove rubber formulations (test results are in duplicate).*

Glove Material	Thickness (gage), mm (mil)		Chemical	Breakthrough Time, min		Time to Maximum Rate, min		Maximum Rate, mg/m²/s	
	Low Gage	High Gage		Low Gage	High Gage	Low Gage	High Gage	Low Gage	High Gage
Butyl A	0.38 to 0.41	0.61 to 0.66	aniline	>390	>1320	4260[a]	...	0.33	ND[b]
	(15 to 16)	(24 to 26)		>390	>1320	4260	...	0.33	ND[b]
			DMF	1320	>4260[a]	...	...	ND[b]	ND[b]
				1380	>4260[a]	...	...	ND[b]	ND[b]
			hexane	8	20	25	60	>390[c]	304
				10	25	25	100	320	206
			MIBK	90	300	250[d]	480	3.2	8.7[d]
				160	270	220[d]	480	3.7	6.5[d]
			toluene	10	30	40	150	45.5	46.2
				20	30	40	240	42.3	46.0
Butyl B	0.38 to 0.43	0.61 to 0.66	aniline	>420	>1380	1320	...	tr[e]	ND[b]
	(15 to 17)	(24 to 16)		>420	>1380	1320	...	tr[e]	ND[b]
			DMF	>1440	>1440	...	...	ND[b]	ND[b]
				>1440	>1440	...	...	ND[b]	ND[b]
			hexane	2	30	60	110	>390[c]	228
				10	30	140	190	385	204
			MIBK	90	300	360	360	6	0.8
				130	340	360	360	6.8	0.2
			toluene	10	30	35	140	44.5	46.8
				15	40	25	120	50.7	41.8
Butyl C	0.36 to 0.41	0.58 to 0.66	aniline	420	>480	450	...	0.2	ND[b]
	(14 to 16)	(23 to 26)		420	>480	450	...	tr[e]	ND[b]
			DMF	>360	>420	4260[a]	1500	0.2	tr[e]
				>360	>420	4260[a]	...	0.2	ND[b]

			hexane	7	20	60	110	>390[c]	195
				10	20	15	240	>390[c]	248
			MIBK	10	180	20	360	5	2.8
				50	225	120	360	11.7	1.3
			toluene	10	20	30	160	40.8	50
				10	30	90	200	42.2	46.8
Neoprene	0.36 to 0.41 (14 to 16)	0.61 to 0.66 (24 to 26)	aniline	30	120	210	480	1	0.5
				60	150	240	290	1	1.5
			DMF	180	>330	330	...	0.2	ND[b]
				210	>360	270	...	0.2	ND[b]
			hexane	60	60	300	140	20.2	15.2
				60	60	300	180	12.5	8.8
			MIBK	20	30	90	170	50.5	49.7
				20	32	90	150	47.3	46.2
			toluene	5	15	40	40	45.7	45.7
				5	20	90	55	40	39.2
Nitrile	0.36 to 0.41 (14 to 16)	0.61 to 0.66 (24 to 26)	aniline	90	150	210	380	0.5	0.5
				90	175	240	360	0.5	0.5
			DMF	210	>300	1200	1320	1.8	1.8
				210	>300	1320	1320	2.0	1.7
			hexane	>420	>1020	...	...	ND[b]	ND[b]
				>1080	>1020	...	...	ND[b]	ND[b]
			MIBK	30	70	120	200	50.7	48.3
				30	70	120	200	48.3	42.7
			toluene	20	35	120	130	43.3	35.2
				20	40	210	190	33.5	39.7

[a]Test ran over one weekend.
[b]None detected.
[c]Permeation rate exceeded the calibrated range of the detector.
[d]Repeat tests, to confirm the anomaly, gave breakthrough times of 255 and 290 min with permeation rates of 2.7 and 11.2 $mg/m^2/s$.
[e]Trace; an insufficient amount was detected to quantitate.

through time is more likely than permeation rate to be affected by a change in thickness.

Because of low or zero permeation rates, no valid comparisons could be made for 8 of the 25 possible combinations. Of the other 17, the literature [*5, 9*] lists 11 as Not Recommended or Poor. As might have been expected, lower-rated combinations are the ones that are most likely to permeate fast enough to provide complete data in this type of experiment. The Edmont Division rates all four combinations tested in the previous study on permeation rate versus thickness [*8*] as Not Recommended.

If we consider only those cases in which complete data were obtained for glove-chemical combinations that are rated Fair or better, the first conclusion is still valid. Of the six such cases, five show quicker breakthrough time for the thinner specimen, but only three show a higher permeation rate.

According to the study mentioned previously [*8*], the final permeation rate was found to be inversely proportional to glove thickness. Breakthrough time was not studied as a function of thickness. No such simple mathematical relationship can be seen in our data. There are several possible reasons for this:

1. The previous workers tested specimens as much as 0.127 mm (50 mils) thick. Some of the data that they obtained on thinner specimens do not fit their trend lines as accurately. Nonuniformity, leading to more variable results, should be more noticeable on specimens that are thin enough to represent actual unsupported gloves.
2. The previous workers tested for only 1 h. We normally ran each test for 6 h to better represent the actual time spent working in an 8-h shift.
3. The previous workers varied thickness by testing single or multiple layers of rubber, all cut from the same glove. Each of our specimens was a single piece of rubber with its thickness adjusted by changing the conditions under which it was made. Permeation is normally a three-step process: the chemical dissolves in the glove, diffuses through, and desorbs. Extra desorption and resolution steps are added when multilayer specimens are tested, and this may affect the results.
4. The choice of gloves and solvents may affect the consistency of the inverse-proportional relationship. Our own data on hexane fit to within 25%, whereas data on MIBK are off by as much as 95%.

So, what is the significance of this work for actual glove selection? Those materials that completely resist permeation at higher thickness are also highly or totally resistant at lower thicknesses. Therefore, if you can find a glove that offers excellent protection against a hazardous chemical, the thickness of that glove is not a critical variable.

For other materials, thickness normally has a greater than proportional effect on breakthrough time. Increasing thickness by a factor of 1.67 increased the quicker of the two measured breakthrough times by an average factor of 4.2 for the 16 cases in this experiment for which these numbers can be calculated. In only two cases did the ratio of these breakthrough times increase less than the relative thickness, and in two other cases breakthrough time was 15 or 18 times longer for the thicker specimen. These high values are due more to short breakthrough times for the thinner specimen than to unusually long ones for the thick specimen, which implies that thicker films pose less risk of a rapid breakthrough because of nonuniformity of the film. So the choice of a thicker glove can improve worker safety by a greater amount than might be expected.

References

[1] Sansone, E. B. and Tewari, Y. B., *American Industrial Hygiene Association Journal*, Vol. 41, March 1980, pp. 170-174.
[2] Stampfer, J. F., McLeod, M. J., Betts, M. R., Martinez, A. M., and Berardinelli, S. P., *American Industrial Hygiene Association Journal*, Vol. 45, Sept. 1984, pp. 634-641.
[3] Nelson, G. O., Carlson, G. J., and Buerer, A. L., "Glove Permeation by Shale Oil and Coal Tar Extract," Technical Report No. UCRL-52893, Lawrence Livermore Laboratory, Livermore, CA, Feb. 1980.
[4] Spence, M. W., Meyers, O. O., Merteno, J. A., and Hawkins, A. R., "Glove Materials for Chlorinated Solvents: Permeation Resistance Comparison for Four Solvents," Paper No. 155 presented at the American Industrial Hygiene Conference, Detroit, MI, 1984.
[5] "Edmont Chemical Resistance Guide," product literature from the Edmont Division of Becton, Dickinson and Co., Coshocton, OH, Aug. 1983.
[6] "Siebe Norton Permeation Resistance Guide," product literature from Siebe North, Inc., Charleston, NC, 1983.
[7] National Toxicological Program Reports, Radian Corp., Austin, TX, 1984.
[8] Nelson, G. O., Lum, B. Y., Carlson, G. J., Wong, C. M., and Johnson, J. S., *American Industrial Hygiene Association Journal*, Vol. 42, March 1981, pp. 217-225.
[9] Schwope, A. D., Costas, P. P., Jackson, J. O., and Weitzman, D. J., *Guidelines for the Selection of Chemical Protective Clothing*, Arthur D. Little, Cambridge, MA, March 1983.

Joseph E. Winter[1]

Protective Clothing Permeation Testing: Calculations and Presentation of Data

REFERENCE: Winter, J. E., **"Protective Clothing Permeation Testing: Calculations and Presentation of Data,"** *Performance of Protective Clothing, ASTM STP 900,* R. L. Barker and G. C. Coletta, Eds., American Society for Testing and Materials, Philadelphia, 1986, pp. 82–92.

ABSTRACT: Data from protective clothing permeation testing are used to determine which specific protective clothing material will protect a worker better than other similar clothing materials. Currently, data are presented in a graphical form, showing the concentration of challenge chemical in a collection medium versus the elapsed time of chemical contact. The results obtained when using a closed or recirculated collection medium cannot be compared directly with the results when using an open or dynamic collection medium. The relationship between these results and methods for converting to a uniform system of permeation rate versus chemical contact time are presented.

Once in a uniform system, the permeation data are entered into a computer data base with graphical output, thus providing the health professional rapid access to this information.

KEY WORDS: gloves, protective clothing, chemical permeation

ASTM Committee F-23 on Protective Clothing develops standard tests to determine the limitations of protective clothing (gloves are considered to be protective clothing). One of the tests developed by the committee is the ASTM Test for Resistance of Protective Clothing Materials to Permeation by Hazardous Liquid Chemicals (F 739-81). Data obtained using this test can be found in the literature [*1–4*] and in protective clothing manufacturers' advertisements [*5–8*]. Users of protective clothing also use the test to obtain data for their own specific applications.

Some problems with calculations have arisen when using the ASTM method. These problems arise when different laboratory experimental setups are used or when the permeant in the collection medium is continuously mon-

[1]Industrial hygiene chemist, Quality Services Organization, Kodak Park, Eastman Kodak Co., Rochester, NY 14650.

itored instead of taken in discrete samples. A simpler method of calculating and presenting the permeation data is proposed.

Permeation data from this test are one of the many resources used by the health professional when prescribing protective clothing for a particular work environment. The breakthrough time and steady-state permeation rate, as reported when using the ASTM permeation method, are useful to the health professional for calculating the maximum possible skin exposure and relating this to the toxicity of the chemical in question. The graphical representation of the permeation data from the ASTM method cannot be expressed in terms of dose without more intimate knowledge of the test procedure. Therefore, an alternative graphical representation of the data is presented.

The system that Eastman Kodak Co. has developed for archiving these data in a computer data base is described. The system provides the health professional with rapid access to the information in graphical form and the ability to compare different clothing materials or chemicals simultaneously.

Procedure

The procedure used for testing the permeability of protective clothing is that of ASTM Test F 739-81. This method consists of exposing a protective clothing material's outside surface to a challenge chemical, while the inside surface is in contact with a collection medium. The collection medium is then monitored for challenge chemical content.

A test cell is used to contain the challenge chemical and the collection medium. The test cell consists of a two-chambered cell for placing a known amount of surface area of the protective clothing material in contact with the challenge chemical on the clothing's normal outside surface and in contact with the collecting medium on the clothing's normal inside surface. ASTM describes a test cell for this purpose in ASTM Test F 739-81; similar test cells are described in the literature [*1, 9*].

The ASTM procedure does not prescribe an analytical technique to measure the concentration of the permeant in the collection medium. This is because the analytical technique will vary with the challenge chemical, collection medium, laboratory setup, and the equipment available in the laboratory performing the testing. The calculations that follow are applicable to all quantitative analytical techniques after the results are converted into the appropriate units of concentration (for example, milligrams per litre).

Calculations

Nomenclature

The following symbols are used in the equations:

A Area of the protective clothing material specimen contacted with the chemical under test, cm^2

C_i, C Concentration of chemical in the collection medium, where C_i is the concentration at time T_i, mg/L

F Flow rate of fresh collecting medium through the cell, L/min

i An indexing number assigned to each discrete sample, starting with $i = 1$ for the first sample

P Permeation rate, mg/cm^2 · min

T_i, T Time elapsed beginning with the initial chemical contact, where T_i is the time at which the discrete sample, i, was removed, min

V_t Total volume of the collection medium, L

V_s Volume of the discrete sample removed from the collection medium, L

The calculations for obtaining the permeation rate from analytical data vary with the experimental setup and are conveniently separated into three groups. The first is a dynamic system or one that uses a continuous flow of fresh collecting medium. The second is a closed system which does not lose significant collection medium volume when sampling is done. And the third is a closed system which experiences significant volume loss from sampling.

Dynamic System

The experimental setup in which a continuous flow of fresh collecting medium transports the permeant from the collection side of the test cell to the analyzer is shown in Fig. 1. This is often referred to as the "open" system or "open loop" system. The collection medium can be either a liquid or a gas. The analytical technique can involve discrete sampling, as is the case when using a gas chromatograph with gas sampling loops. A continuous monitoring instrument can be used to get a continuous profile of the permeant concentration in the collection medium. Total hydrocarbon analyzers, infrared

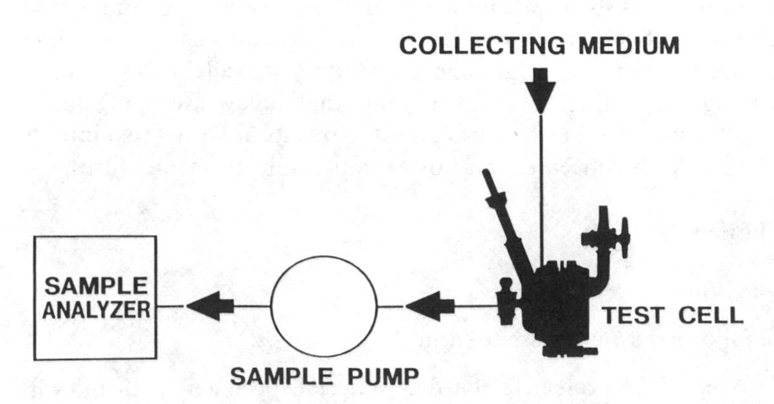

FIG. 1—*Example setup for continuous flow of fresh collecting medium.*

analyzers, and monitoring instruments designed for a specific compound have been used [9,10].

To calculate the permeation rate at any time during the test from the analytical data obtained at that time, the following equation is used

$$P = C\left[\frac{F}{A}\right]$$

The concentration of the permeant in the collection medium, C, at any time, T, is directly proportional to the permeation rate, P, by the factor $[F/A]$. The output of many analytical devices can incorporate this factor and give results directly in permeation rate. For example, the recorder output from a continuous monitoring device can be converted to a permeation rate versus time profile. This is done by converting the concentration axis into permeation rate using the $[F/A]$ factor.

An assumption made when using the calculation for the dynamic system is that the time it takes for the permeant to be transported from the test cell to the analytical instrument is insignificant. Steps should be taken to minimize this source of error, such as increasing the collection medium flow rate.

Closed Systems

Two experimental setups have been popular when using closed systems. The first involves circulating the collection medium through an external pump, as in Fig. 2. In the second, all sampling or monitoring is done within the collection side of the test cell. There are no differences in the calculations between these two setups. However, the flow rate of the collection medium must be optimized to ensure prompt delivery of the permeant to the analytical

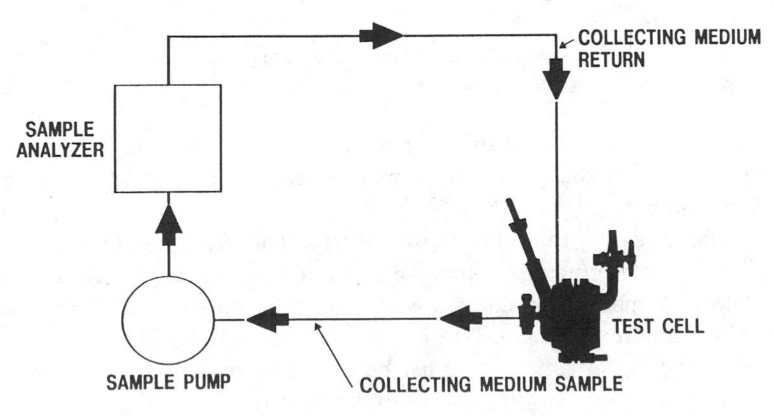

FIG. 2—*Example setup for continuous collecting medium withdrawal, analysis, and return.*

device. The same assumption is made as with the dynamic system, that the permeant transport time is insignificant.

Closed Systems (Without Volume Loss from Sampling)

This calculation is applicable when discrete samples are withdrawn, analyzed by a nondestructive technique, and replaced prior to further sampling, or when the volume of the discrete samples is insignificant in relation to the total collection medium volume (for example, microlitre aliquots). This calculation is also used when a continuous monitoring instrument is used in conjunction with a closed system. When using a continuous monitoring instrument with a closed system, the results are recorded at time intervals as if discrete samples were being taken and analyzed at those times. These times are chosen to best define the permeation rate versus time curve, this is analogous to choosing discrete sampling times that best define this curve.

$$P = \frac{(C_i - C_{i-1})V_t}{(T_i - T_{i-1})A}$$

In this equation, $C_i - C_{i-1}$ represents the increase in concentration since the last discrete sample; multiplying by total volume, V_t, represents the total amount of chemical permeating the protective clothing material since the last discrete sample was taken. Because the last sample was taken at time T_{i-1}, this calculated permeation rate is the average permeation rate over the time period T_{i-1} to T_i.

Closed System (Significant Volume Loss from Sampling)

The following equation is applicable when discrete samples of significant volume are removed and not replaced.

$$P = \frac{(C_i - C_{i-1})[V_t - (i - 1)V_s]}{(T_i - T_{i-1})A}$$

The only difference between this and the previous equation is the correction for the decrease in the total collection medium volume after each discrete sample is taken.

It may be necessary to add a volume of fresh collection medium to the cell after each discrete sample is taken. This should be done only to keep the protective clothing material in contact with the collection medium. In this case, the total volume remains constant but the concentration of permeant in the collection medium at time T_{i-1} must be corrected by the amount it was diluted, $[V_t - V_s]/V_t$, as in the following equation.

$$P = \frac{\left[\dfrac{C_i - C_{i-1}(V_t - V_s)}{V_t}\right] V_t}{(T_i - T_{i-1})A}$$

In all equations for discrete sampling it is important to note that the calculated permeation rate is the average permeation rate over the time period since the last sample was taken, T_{i-1} to T_i, and not the permeation rate at the time the discrete sample was taken.

Graphical Presentation of Data

The ASTM permeation method requires that a plot of the concentration of the challenge chemical in the collection medium versus time be included in each report. One of the reasons that this plot was originally included in the report is that the slope of the line is proportional to the permeation rate (when using a closed system without significant sample loss). The slope from this plot was then used to calculate permeation rate. When using a closed system with significant sample loss, a "corrected concentration" was plotted versus time. This calculation method cannot be used with a dynamic system.

When using the ASTM permeation method, the concentration of challenge chemical in the collection medium is proportional to the total volume of collection medium. Therefore, the data on the plot cannot be compared between laboratories or converted into units that may be meaningful to the health professional because the report does not include the total volume of the collection medium. An example of this type of plot is provided in Fig. 3.

An improved method for providing a graphical representation is to plot

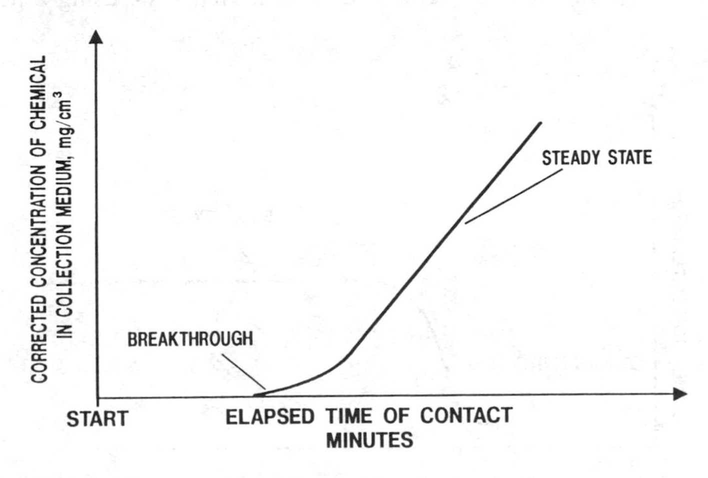

FIG. 3—*Concentration of chemical in collection medium versus time.*

permeation rate versus time. This plot has the advantage of being in units relevant to the users of the information and directly comparable to plots from other laboratories, using any experimental setup.

In the case of the dynamic system, the presentation rate calculated is plotted versus time. When using a continuous monitoring instrument, one can use a strip chart recorder to plot the permeation rate (concentration times the F/A factor) versus time directly.

With closed systems, the average permeation rate over the period T_{i-1} to T_i is plotted against the midpoint of that time period, as calculated by the following equation

$$T = \frac{T_i + T_{i-1}}{2}$$

An example of this type of plot using the same data as were used in generating Fig. 3 is provided in Fig. 4. The area of the plot that represents "steady state" or no change in permeation rate with respect to time is the region of constant slope in Fig. 3. The permeation rate must be calculated from this slope. In Fig. 4, steady state is represented by the region where slope is equal to zero, and the permeation rate can be read directly from the plot.

Computer Data Base

The system that Eastman Kodak Co. has set up for archiving and retrieving the plots uses the Statistical Analysis System (SAS) software package that resides in a time-sharing network on our mainframe computer. Access is through a conveniently located terminal.

A similar system could be developed on a limited scale using a personal

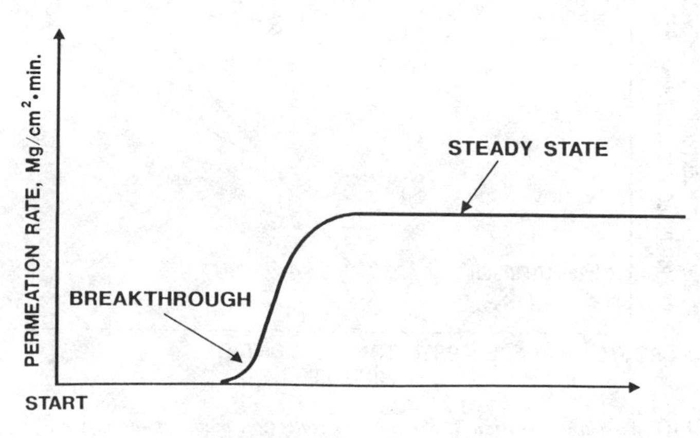

FIG. 4—*Permeation rate versus time.*

computer and one of the spreadsheet packages with graphics capabilities. The following steps outline the data flow sequence, as illustrated in Fig. 5:

1. The appropriate time and corresponding concentration pairs are selected that best define the permeation curve. Only the worst-case test is entered from triplicate test data.
2. These values, together with concentration to permeation rate conversion variables and codes for the chemical, material type, model, and seam type, are entered into a temporary SAS data set using the facilities of SAS/FSP (a full-screen data entry and edit package).
3. The user then invokes a software module which converts concentration to permeation rates. The corresponding time and permeation rate pairs as well as information on the chemical, material type, model, and seam type are then stored into a permanent SAS data base.
4. A programmed SAS/GRAPH routine generates a full series of plots (that is, a single chemical versus all tested materials; a single material versus all tested chemicals). These are printed on a Gould electrostatic plotter located with the mainframe computer. Examples are included in Figs. 6 and 7.
5. Publication-quality plots can be reproduced locally, one at a time, on a Hewlett-Packard 7470A two-pen plotter under the control of an interactive SAS/GRAPH session.

Conclusion

Most investigators who have published reports or have shared data on their permeation studies present tables of data pertaining to the breakthrough times and the steady-state permeation rate. Most of us choose a protective

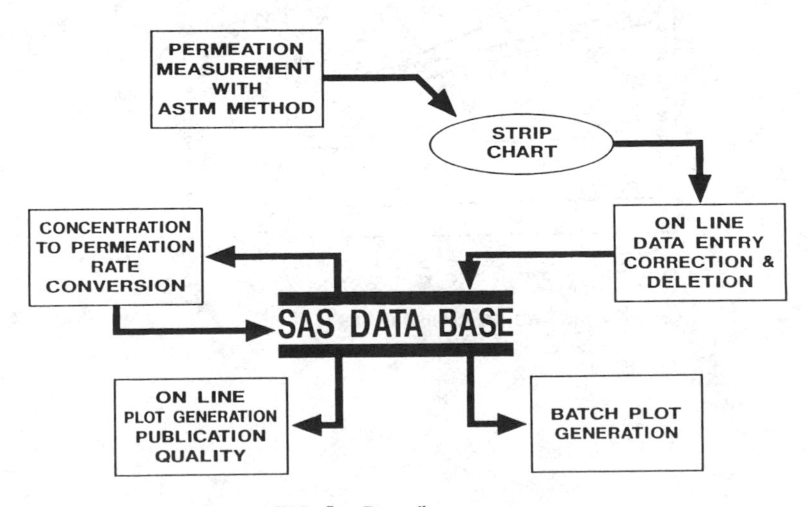

FIG. 5—*Data flow sequence.*

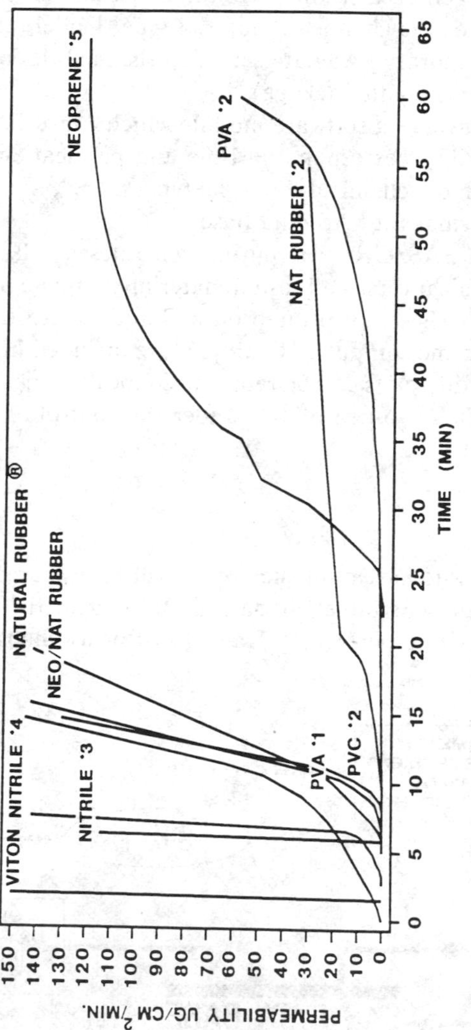

FIG. 6—*Permeation rate versus time plot; a single chemical versus all the materials tested. The solvent is acetone.*

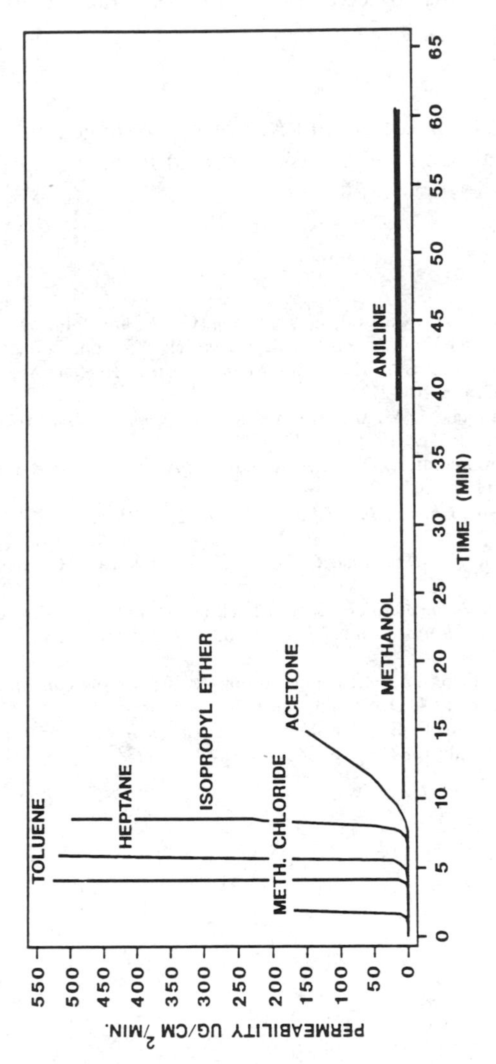

FIG. 7—*Permeation rate versus time plots; a single material versus all the chemicals tested. The material is neoprene/natural rubber.*

clothing material with the highest breakthrough time and lowest permeation rate tempered with other considerations, such as availability, flexibility, durability, and cost. This can be a time-consuming effort if one has many chemicals to consider for a protective clothing application. With permeation rates calculated directly and presented in a graphical format that can be compared between laboratories, this choice can be more easily facilitated.

Acknowledgments

The author wishes to thank Donald A. Trevoy, Management Services Division, Eastman Kodak Co., for his assistance with setting up the computer data base.

References

[1] Nelson, G. O., Lum, B. Y., Carlson, G. J., Wong, C. M., and Johnson, J. S., *American Industrial Hygiene Association Journal*, Vol. 42, March 1981, pp. 217-225.
[2] Weeks, R. W., Jr. and McLeod, M. J., *American Industrial Hygiene Association Journal*, Vol. 43, March 1982, pp. 201-211.
[3] Schwope, A. D., Randel, M. A., and Broome, M. G., *American Industrial Hygiene Association Journal*, Vol. 42, Oct. 1981, pp. 722-725.
[4] Henry, N. W. III and Schlatter, C. N., *American Industrial Hygiene Association Journal*, Vol. 42, March 1981, pp. 202-207.
[5] *Technical Data Report: Chemical Resistance of Gore-Tex™ Industrial Fabric*, W. L. Gore & Associates, Inc.
[6] *Siebe Norton Permeation Resistance Guide*, Siebe Norton Inc., Charleston, SC, April 1983.
[7] *Edmont Chemical Resistance Guide*, Becton Dickinson and Co., Coshocton, OH, 1983.
[8] *Tyvek™ Protective Clothing Guide*, E.I. du Pont de Nemours & Co., Wilmington, DE, 1982.
[9] Berardinelli, S. P., Mickelsen, R. L., and Roder, M. M., *American Industrial Hygiene Association Journal*, Vol. 44, Dec. 1983, pp. 886-889.
[10] Abernathy, R. N., Cohen, R. B., and Shirtz, J. J., *American Industrial Hygiene Association Journal*, Vol. 44, July 1983, pp. 505-513.

Resistance to Pesticides—Field Performance and Cleaning Procedures

Alan P. Nielsen[1] *and Richard V. Moraski*[2]

Protective Clothing and the Agricultural Worker

REFERENCE: Nielsen, A. P. and Moraski, R. V., **"Protective Clothing and the Agricultural Worker,"** *Performance of Protective Clothing, ASTM STP 900*, R. L. Barker and G. C. Coletta, Eds., American Society for Testing and Materials, Philadelphia, 1986, pp. 95-102.

ABSTRACT: The purpose of this paper is to provide an overview of the role of protective clothing in hazard reduction for pesticide applicators and mixer/loaders. Research and other activities of the Office of Pesticide Programs, U.S. Environmental Protection Agency, relating to worker exposure and protective clothing are discussed, including field and laboratory studies. The goal of the program, to identify personal protective equipment that will be effective from performance and worker acceptability standpoints, is also discussed.

KEY WORDS: agricultural worker, applicators, carbaryl, dermal exposure, ground boom sprayer, mancozeb, mixer/loaders, protective clothing, protective equipment, respiratory exposure, pesticide spraying, thermistor, 2,4-dichlorophenoxyacetic acid, (2,4-D)

Significant research into the dermal and respiratory pesticide exposure of agricultural workers involved in spray operations began with Durham and Wolfe in the early 1960s [*1*]. The primary emphasis of their work has been applicators and mixer/loaders involved in airblast pesticide spraying of orchards. Their work [*1-5*] and that of Davis [*6*], who continued these efforts in Wenatchee, Washington, established not only the initial data base on worker exposure, but also field monitoring techniques for quantifying exposure that are still used today.

In the 1970s, exposure monitoring efforts expanded to other types of spray operations, including types of ground rig applications, aerial applications,

[1]Biologist, U.S. Environmental Protection Agency, Office of Pesticide Programs, Hazard Evaluation Div., Exposure Assessment Branch, Washington, DC 20460.

[2]Chemist, U.S. Environmental Protection Agency, Office of Health and Environmental Assessment, Office of Research and Development, Exposure Assessment Group, Washington, DC 20460.

and other types of pesticide use practices. In the last few years, emphasis has been placed not only on the exposure of applicators and mixer/loaders, but also on the value of protective clothing and equipment and on engineering controls such as closed systems in reducing exposure to workers directly involved in pesticide use. Recently, monitoring efforts have been expanded to include agricultural workers who are not involved in the application of pesticides. These workers, while conducting their activities in treated areas, may be exposed to soil, foliar, and airborne pesticide residues [*7,8*]. Farm workers involved in hand harvesting, cultivation, irrigation, scouting, fruit thinning, and many other types of tasks can be exposed to significant amounts of pesticide residue given the right conditions [*8*]. The purpose of this paper, however, will be to discuss pesticide exposure to applicators and mixer/loaders, with emphasis on the value of protective clothing and equipment.

This paper will present an overview of worker exposure monitoring activities and pesticide-related protective clothing research being conducted by the Office of Pesticide Programs (OPP) of the U.S. Environmental Protection Agency (EPA). An assessment of the research needs in this area will also be discussed.

Field Monitoring

In 1980 and 1981, the Health Effects Branch of the Hazard Evaluation Division, OPP, sponsored a series of field monitoring studies at six pesticide hazard assessment projects located throughout the United States [*9*]. The studies were designed to characterize applicator and mixer/loader exposure to pesticides during ground boom spray applications. At that time the majority of research had been done on air blast applications, and it was clear that not enough was known about other types of applications. A recent review of the published exposure data by Hackatorn and Eberhart [*10*] demonstrates that this situation still exists today.

The field monitoring techniques for the applicators and mixer/loaders were basically those outlined by Davis [*6*]. This includes dermal pad dosimeters constructed of alpha-cellulose or gauze, or both (depending on job function and formulation type), attached to disposable Tyvek jackets to assess potential skin exposure. Cotton gloves and alcohol hand rinses were used to assess hand exposure. Personal air samplers were worn by the participants to assess potential respiratory exposure. Altogether, 6 dependent or exposure variables were looked at and 22 independent variables, including crop type, boom length, wind direction, pounds of active ingredient of pesticide applied per hour, spray pressure, nozzle type, job function, formulation type, and many others.

Cycles of mixing/loading and application of three pesticides, 2,4-dichlorophenoxyacetic acid (2,4-D), carbaryl, and mancozeb—were monitored separately. The study is undergoing (EPA) peer review and will be published in its entirety, but some of the data are presented here to demonstrate exposure levels and the potential importance of personal protective equipment.

The average exposure of the ground rig applicators, studywide, was approximately 25 mg/h (range of 1 to 130 mg/h). For the applicators monitored in these studies, dermal exposure excluding the hands, accounted for 26% of the total exposure. Hand exposure represented 74% of the total exposure. The data clearly support previous work by other investigators showing the dermal route to be the main source of exposure, particularly the hands [2,3]. Air blast applicators, as compared with ground boom applicators, are generally thought to receive the highest exposure. A review of the literature [11] shows a mean of 30 mg/h (range of 3 to 300 mg/h), which bears this theory out.

The same trends in worker exposure hold true for the mixer/loaders, but they are even more extreme. Mixer/loaders, in our study, had a total exposure mean of 2900 mg/h (range of 27 to 32 000 mg/h). Hand exposure alone accounted for 99.7% of the total exposure. Dermal exposure of the rest of the body accounted for only 0.2% of the total exposure.

For both applicators and mixer/loaders, respiratory exposure was less than 0.1% of the total. This is usually the case in ground applications except where fumigants (gases) or highly volatile formulations are involved [6]. It should be noted that respiratory exposure should not always be overlooked, since most, if not all, materials are absorbed more rapidly and completely through the lungs than through the skin [1].

Dermal exposure, however, does not necessarily mean that dermal absorption has occurred. Absorption is chemical specific and is mainly dependent on the active ingredient of a pesticide, the formulation type, and the location of the exposed dermal area [12].

Another route of exposure is oral ingestion, although it is not usually considered a significant route of entry during spray operations [13]. As with other routes of exposure, it could be important depending on worker habits, such as wiping the face with contaminated hands or gloves. This, of course, is not a recommended practice.

It is well established that, without adequate safety precautions, workers mixing/loading or applying pesticides, as well as any individuals involved in fixing, adjusting, changing, or even cleaning equipment, could face potentially dangerous levels of exposure to pesticides. This exposure potential is especially real for individuals who perform all these tasks.

Protection

Since dermal absorption is the significant route of pesticide entry [9], any barrier that can be placed between the worker and the chemical to reduce dermal contact can reduce exposure. With the exception of properly filtered air systems in enclosed cabs [14], the only significant type of barrier available to applicators is protective clothing. For mixer/loader safety, closed systems and technological advances in container design and packaging have protective potential, but much more work needs to be done [*15*]. Protective clothing then becomes the only viable alternative.

In 1981, the director of the Office of Pesticide Programs issued a directive to the Hazard Evaluation Division to look into the issues and problems involving protective clothing. The concept of establishing a working group was subsequently approved. The purposes of the working group would include the following:

(*a*) to review and assess the available literature and information on personal protective equipment;
(*b*) to assess the role of protective articles in reducing exposure;
(*c*) to identify monitoring and research areas in which information may need to be generated;
(*d*) to establish lines of communication with researchers and authorities in the area of personal protective equipment, including other federal agencies such as the National Institute of Occupational Safety and Health (NIOSH) and the U.S. Department of Agriculture (USDA), and many others; and
(*e*) to make recommendations to OPP on the use of such information regarding the regulatory process, such as in product labeling.

Since the working group was established, we have compiled and summarized information on protective clothing as it relates to agricultural pesticide uses. The summary also includes abstracts of numerous telephone contacts with investigators involved in personal protective equipment research. A copy of this report can be obtained from the EPA [*16*].

Based on the report, the working group then identified monitoring and research needs pertaining to personal protective equipment. The conclusions were as follows:

1. Additional evaluations need to be conducted in the laboratory on the permeability and penetrability of various fabrics by different classes of pesticides and formulation types. These fabrics could include cotton or other "normal" clothing materials treated with repellent finishes or untreated, as well as commercially available materials, both woven and nonwoven, for example, Tyvek and Gore-Tex. These evaluations will be realized by funding research at the Department of Textiles, Merchandising and Design, University of Tennessee, Knoxville. The research will proceed using the DeJonge penetration test method based on a contained spray apparatus for sample testing [*17*], which was modified from an earlier version [*18*].

2. More field testing of the performance of protective clothing in reducing exposure and maintaining worker comfort needs to be conducted. As in other cases wherein workers are required to wear protective garments, the heat stress factor can be considerable. This could be why wearing protective gar-

ments has not become an accepted practice among pesticide users [*19*]. Currently, a study is being completed by Dr. Nigg of the University of Florida, Lake Alfred, on the efficacy of Tyvek coveralls. The study will focus on reducing the pesticide exposure of applicators that spray citrus groves. Additional field work is being planned to incorporate portable thermistors to be worn on and under clothing to measure the effects of materials on skin and clothing surface temperatures.

When evaluating worker comfort and protective clothing, however, numerous factors must be considered. A recent publication by Goldman [*20*] identifies six key parameters to be considered for a heat balance equation for determining whether or not a given combination of factors results in heat stress to the worker. These parameters include air temperature, ambient air motion, ambient vapor pressure, and the mean radiant heat temperature. The other two factors, which are subject to behavioral temperature regulation, are the task work load and the type of clothing being worn by the worker.

Goldman also discusses body temperatures, heart rate, and skin dampness as other important criteria to be considered when addressing worker comfort and protective clothing. As one would imagine, higher levels of relative humidity at the skin are associated with increasing heat discomfort. The specifics of how to monitor worker comfort and pesticide exposure under field conditions constitute a complex question, which we are currently addressing. Field studies will be coordinated with the laboratory testing being done by the University of Tennessee.

3. Emphasis should be on conducting laboratory evaluations of various types of glove materials for permeation resistance to pesticides. Using the ASTM Test for Resistance of Protective Clothing Materials to Permeation by Hazardous Liquid Chemicals (F 739-85), we are planning to test various types of glove materials. These materials will include various synthetic and natural rubber, viton, and even a disposable shoulder-length polyethylene glove. Consideration was given not only to material type, but also to thickness, since without adequate glove dexterity and flexibility, we do not believe the gloves would be purchased by pesticide users. Factors such as the length of the cuff are also important because the greater the area of wrist and forearm covered, the less the dermal exposure. Also, we want to test gloves that represent a range of costs and consequent disposability.

In regard to specific types of pesticides, we will be concentrating on Category I pesticides [those having an acute dermal median lethal dose (LD_{50}) of ≤ 200 mg/kg of body weight, based on laboratory animal studies]. The largest group of pesticides meeting this criterion are organophosphorous compounds, but other classes will be investigated as well. The ASTM method involves a continuous liquid/fabric interphase and represents an extreme exposure scenario, but the method should still provide a very good benchmark system for rating glove materials.

4. Developing standardized and practical methods should be encouraged. In this instance, besides ASTM Test F 739-85 and the DeJonge penetration test methods [*17*], we are currently funding an effort through the California Pesticide Hazard Assessment Project with Dr. Fenske. The aim of this project is to develop a computer video-image processing technique involving fluorescent tracers. With this method we would be capable of examining dermal exposure of pesticide users qualitatively and also quantitatively. All of the results have not been received concerning this approach, but the technique would appear to have a very practical application for clothing design [*21*].

5. Consolidated, state-of-the-art information should be disseminated to pesticide users. A system for maintaining and updating this information should also be developed. We have established an *ad hoc* committee with the National Agricultural Chemicals Association (NACA), the pesticide manufacturing industry, the USDA Extension Service, and academe to examine these issues. One important and recommended document now available is the USDA/EPA publication, *Applying Pesticides Correctly: A Guide for Private and Commercial Applicators* [*22*]. The guide has recently been revised and published. It contains valuable information on applying pesticides correctly.

Other areas discussed in the summary [*16*] include research to do the following:

(*a*) determine the effects of decontamination procedures, such as laundering, on pesticide retention and the protective capability of various fabrics;
(*b*) investigate material durability and degradation (wear) on the efficacy of protective materials; and
(*c*) determine the effects of clothing design, for example, seam leakage, on protective capability.

Accurate data from these types of studies would enable OPP and the pesticide manufacturers to make the most accurate and consistent protective clothing recommendations possible. It would also provide the pesticide community with guidance on the necessary protection to minimize exposure to hazardous chemicals.

It must be pointed out that even though protective clothing remains the most important means of minimizing overall exposure to pesticides, the protective clothing may itself become a hazard. For example, gloves with elastic wrist bands are not recommended for use because pesticides trapped in the wrist area will have prolonged direct contact with the skin. Increased sweating will then increase the absorption. Gloves that are not waterproof or chemically resistant, that readily trap residues, such as woven fabrics and leather, and that are not protective of the wrist area are not recommended for use. It has also been demonstrated that significant hand exposures can still occur, even when wearing impermeable gloves, if they are not decontaminated both inside and outside or are not properly maintained [*23*].

The OPP is mandated under the Federal Insecticide, Fungicide, and Rodenticide Act (FIFRA), as amended, and the Code of Federal Regulations, Title 40, Section 162.10, Labeling Requirements of Pesticide Products, to ascertain that all chemicals have appropriate precautionary use information on the label. This information also includes instructions to the worker on the proper personal protective equipment for the safe use of pesticides.

Day-to-day evaluation of potential pesticide hazards through the various EPA registration review processes clearly indicate the need for developing information on personal protective equipment. Protective clothing requirements will become more important, not only for dealing with acutely toxic pesticides, but also for dealing with some products identified as potentially causing chronic health problems. The EPA mandate for appropriate precautionary labeling applies in both cases.

Our overall opinion at this time is that a very extensive, inclusive, and well-coordinated effort is needed to resolve these pertinent issues. We believe the cost to accomplish the stated objectives should be shared by the federal and state governments and private industry, including chemical and protective clothing manufacturers.

We would urge any scientists or organizations that have laboratory or field performance data on pesticides and personal protective equipment to contact the EPA as soon as possible.

References

[1] Durham, W. F. and Wolfe, H. R., *Bulletin WHO*, Vol. 26, 1962, pp. 75–91.

[2] Wolfe, H. R., Durham, W. F., and Armstrong, J. F., *Archives of Environmental Health*, Vol. 14, 1967, pp. 622–633.

[3] Wolfe, H. R., *Weeds, Trees and Turf*, Vol. 12, No. 4, 1973, p. 12.

[4] Durham, W. F., Wolfe, J. H., and Elliot, J. W., *Archives of Environmental Health*, Vol. 24, 1972, pp. 381–387.

[5] Wolfe, H. R., Staiff, D. C., Armstrong, J. F., and Comer, S. W., *Archives of Environmental Health*, Vol. 25, 1972, pp. 29–31.

[6] Davis, J. E., *Residue Reviews*, Vol. 75, 1980, pp. 33–50.

[7] Nigg, H. N., "Worker Reentry in Florida Citrus: Pesticides in the Agricultural Environment," Government Report, National Technical Information Service (NTIS)/PB 80-177728, Issue 17, Springfield, VA, 1980.

[8] Zwieg, G., Gao, R., and Popendorf, W. J., *Journal of Agricultural and Food Chemistry*, Vol. 31, 1983, pp. 1109–1113.

[9] U.S. Environmental Protection Agency, "Applicator and Mixer/Loader Exposure to Pesticides During Ground Boom Spraying Operations," Draft Report Contract No. 68-01-6271, Task No. 48, Versar, Springfield, VA, 1984.

[10] Hackathorn, D. R. and Eberhart, D. C., "Data Base Proposal for Use in Predicting Mixer/Loader/Applicator Exposure," in *American Chemical Society Symposium 273, Dermal Exposure Related to Pesticide Use*, American Chemical Society, Washington, DC, 1985, pp. 341–356.

[11] Reinert, J. C. and Severn, D. J., "Dermal Exposure to Pesticides: The Environment Protection Agency's Viewpoint," in *American Chemical Society Symposium 273, Dermal Exposure Related to Pesticide Use*, American Chemical Society, Washington, DC, 1985, pp. 357–368.

[12] Maibach, H. I., Feldman, R. J., Milby, T. H., and Serat, W. F., *Archives of Environmental Health*, Vol. 23, 1971, pp. 208–211.

[13] Bohmont, B. L., *The New Pesticide User's Guide*, Reston Publishing Company, Reston, VA, 1983.
[14] Taschenberg, E. F. and Bourke, J. B., *Bulletin of Environmental Contamination and Toxicology*, Vol. 13, No. 3, 1975, pp. 263-268.
[15] Jacobs, W. W., "Closed Systems for Mixing and Loading," Paper presented at the Determination and Assessment of Pesticide Exposure Workshop, U.S. Department of Agriculture, New Jersey Agricultural Experiment Station, U.S Environmental Protection Agency, Hershey, PA, 1980.
[16] Bodden, M., Cioffi, J., Fong, V., McLaughlin, M., and Russell, S., "Summary of Research on Protective Materials for Agricultural Pesticide Uses," Contract No. 68-01-5944, Mitre Corporation, McLean, VA, 1983.
[17] DeJonge, J. O., Easter, E. P., Leonas, K. K., and King, R. M., "Protective Apparel Research," in *American Chemical Society Symposium 273, Dermal Exposure Related to Pesticide Use*, American Chemical Society, Washington, DC, 1985, pp. 403-411.
[18] Orlando, J., Branson, D., Ayers, G., and Leavitt, R., *Journal of Environmental Science and Health*, Vol. 16, No. 5, 1981, pp. 617-628.
[19] DeJonge, J. O., "Clothing as a Barrier to Pesticide Exposure," Paper presented at the American Chemical Society Meeting, Seattle, WA, 1983.
[20] Goldman, R. F., "Heat Stress in Industrial Protective Encapsulating Garments," in *Protecting Personnel at Hazardous Waste Sites*, Butterworth, Stoneham, MA, 1985, pp. 215-266.
[21] Fenske, R. A., Leffingwell, J. T., and Spear, R. C., "Evaluation of Flourescent Tracer Methodology for Dermal Exposure Assessment," in *American Chemical Society Symposium 273, Dermal Exposure Related to Pesticide Use,* American Chemical Society, Washington, DC, 1985, pp. 377-394.
[22] U.S. Department of Agriculture/Environmental Protection Agency, *Applying Pesticides Correctly, A Guide for Private and Commercial Applicators*, North Carolina Agricultural Extension Service, Raleigh, NC, Revised, 1984.
[23] Maddy, K. T., Wang, R. G., and Winter, C. K., "Dermal Exposure Monitoring of Mixers, Loaders, and Applicators of Pesticides in California," Worker Health and Safety Unit Report, HS 1069, California Department of Food and Agriculture, Sacramento, CA, 1983.

Margaret H. Rucker,[1] *Kaye M. McGee,*[2] *and Tatyana Chordas*[3]

California Pesticide Applicators' Attitudes and Practices Regarding the Use and Care of Protective Clothing

REFERENCE: Rucker, M. H., McGee, K. M., and Chordas, T., **"California Pesticide Applicators' Attitudes and Practices Regarding the Use and Care of Protective Clothing,"** *Performance of Protective Clothing, ASTM STP 900*, R. L. Barker and G. C. Coletta, Eds., American Society for Testing and Materials, Philadelphia, 1986, pp. 103–113.

ABSTRACT: The objectives of this study were to (1) obtain data on attitudes and practices of California pesticide applicators and household members responsible for the care of pesticide-soiled clothing, (2) identify relationships between attitudes and practices, and (3) determine if any group might be especially at risk. A survey of growers and commercial applicators suggested that the growers were the higher-risk group; not only did they use more hazardous chemicals, but they also were less likely to agree with the need for certain precautions. In general, users of more toxic chemicals indicated they recognized that fact but appeared to downplay the potential danger. With respect to storage and cleaning of pesticide-soiled clothing, the data indicate that a sizable proportion of the sample stored and washed such clothing with other household garments. This finding is cause for some concern since health problems can develop from secondary exposure to pesticides.

KEY WORDS: pesticide application, protective clothing, secondary pesticide exposure

A number of studies have indicated that dermal exposure to pesticides is a major health hazard and that this hazard could be reduced by appropriate use and care of protective clothing. One of the earliest reports of pesticide poisonings, summarizing eleven episodes of poisoning from parathion residues involving more than 70 workers, suggested that percutaneous absorption was the primary route of entry [*1*]. The authors noted that removal of protective clothing and continued wearing of contaminated clothing favored

[1]Associate professor, Division of Textiles and Clothing, University of California, Davis, CA 95616.
[2]Owner of a custom knitwear company, Kaye's Babes, Vallejo, CA 94591.
[3]Retail store buyer, Pumpkins & Monkeys, Davis, CA 95616.

dermal absorption in several of these cases. More recently, Lamb conducted a series of monitoring studies with azinphos methyl to determine the reentry time for citrus pickers and found dermal exposure to be more serious than inhalation exposure [*2*]. The author concluded that protection from dermal exposure by use of gloves, a daily change of work clothes, and bathing would considerably reduce the exposure hazard. Other studies have shown the skin to be the main route of absorption during spraying or aerosol operations [*3,4*]. Data gathered by the Western Pesticides Laboratory in Wenatchee, Washington, provide further support for the importance of protecting the skin [*5*]. In a study of 31 different work activities involving ten different pesticides, the potential dermal exposure to each pesticide in every work situation was much greater than the potential respiratory exposure.

Workers' use of protective clothing as it relates to frequency and seriousness of pesticide exposure has received scant attention from other researchers. One exception is a study reported by DeJonge, Vredevoogd, and Henry [*6*] in which fruit growers, in comparison with other farmers, were found to have the highest beliefs in protection and to take the most protective action. The authors suggest that this is so because fruit growers use the air blast method of pesticide application and therefore are at greater risk than other farmers. Their higher perceived need for protection leads to greater adoption of protective practices. The farmers in this study were also asked for preferences regarding protective clothing. Responses to these items indicated that shirts and pants made of either cotton or cotton/polyester were the most popular choices. Other articles indicate that a variety of everyday items are actually being worn during pesticide application [*7,8*].

In addition to the direct effects on workers, unsafe practices with respect to protective clothing can also pose a threat to other members of the household. Work by Finley et al [*9*] and Easley et al [*10*] indicates that when pesticide-contaminated clothing is washed with other family clothing, secondary contamination will occur. Other studies suggest that families of exposed workers may develop health problems after secondary exposure through laundering of work clothing at home [*11,12*]. To date, however, little information is available on practices of family members regarding cleaning and storage procedures for contaminated clothing.

Several studies have suggested that attitudes toward the risks and benefits involved in pesticide application differ and may affect current practices as well as receptivity to educational efforts. Vlek and Stallen [*13*] found that when their subjects were grouped by occupation, agricultural personnel tended to rate pesticide application as a relatively acceptable risk. Acceptability in this case was apparently related to perceptions of large-scale benefits, with judgments of riskiness being less important. Other investigators have noted that attitudes and practices regarding protective clothing vary substantially within the category of agricultural worker [*5,14,15*].

In summary, a review of the literature indicated that more complete information is needed on both applicators' and household members' attitudes and practices regarding clothing worn when applying pesticides. Therefore, an exploratory study was designed to meet the following objectives:

(*a*) to provide data on attitudes and practices of California pesticide applicators and household members who are responsible for the care of pesticide-soiled clothing;
(*b*) to identify relationships between attitudes and practices; and
(*c*) to identify any groups who may be especially at risk because of factors such as inaccurate information or attitudes related to questionable practices.

Procedure

A two-part questionnaire was developed for use in this study. The first part was intended for the applicator, and the second part was designed to be completed by the person in the household who was responsible for the care of the clothing. Items in the applicator section included questions about pesticides used, sources of information about pesticide application, clothing in relation to pesticide exposure, use of safety equipment, and attitudes toward pesticide use and clothing worn for pesticide application. Items in the section for people responsible for care of the household's clothing included questions on attitudes and practices regarding storage and cleaning of garments worn for pesticide application.

A sample of 100 commercial pesticide applicators was obtained from a list provided by the California Department of Food and Agriculture, and a sample of 100 growers was selected from lists supplied by County Agricultural Commissioners. The systematic sampling procedure recommended by Babbie [*16*] was used to choose each group of 100 people. These 200 people were sent a copy of the questionnaire with a cover letter asking them to complete and return the questionnaire or, if no one in the household applied pesticides, to send back the blank questionnaire.

Twenty-three of the applicators and 15 of the growers returned completed questionnaires. An additional 17 applicators and 24 growers returned blank forms. Five of the applicators' forms and 2 of the growers' forms were returned as undeliverable. Frequency distributions were calculated for all items. Chi-square tests were used to assess homogeneity of distribution of responses for four comparisons of particular interest in this study. These comparisons were of commercial applicators versus growers, users of restricted versus nonrestricted pesticides, users of Categories 1 and 2 pesticides (highly to moderately toxic compounds) versus Category 3 pesticides (slightly toxic to relatively nontoxic compounds), and users of air blast spraying versus users of other application methods.

Results and Discussion

The data in Table 1 indicate the respondents' use of and satisfaction with various sources of information on pesticide application. As might be expected, the most frequently used sources were personal experience and labels on pesticide containers. It should be noted that not only are labels convenient sources of information closest to the point of application, they also represent legal documents specifying instructions that the applicators are required to follow. However, not all respondents were satisfied with the information on the labels. Causes of dissatisfaction included problematic specification of pesticide categories, insufficient information on toxicity, and lack of specific information regarding some points such as compatability with other pesticides, target pests, and effects of various environmental conditions.

The next most frequently used sources were chemical companies and pesticide dealers; factors such as convenience and perceived expertise may be related to high use of these sources. Pesticide dealers were also found to be a primary source of information in a survey of farmers in Iowa and Illinois [*17*]. As was pointed out in a report of that survey, the number of industry representatives and dealers in relation to other sources, such as university extension specialists and farm advisers, contributes to their success in communicating with farmers. Furthermore, 27% of the Midwestern sample felt that dealers were a "*highly* qualified source of information on *all* aspects of agricultural chemicals and their use," and another 55% considered them to be a "qualified source of information on *some* aspects of agricultural chemicals and their use."

Sources that were relatively low in terms of usage include friends and neighbors and medical personnel. The respondents commented that most of their friends and neighbors were not well informed and consequently were apt to offer confused and conflicting advice. With respect to medical personnel, respondents had questions about how to go about obtaining such advice. Use of

TABLE 1—*Responses to questions about sources of information on pesticide application.*

Source	Number Using Each Source	Mean Satisfaction Score[a]
My own experience	37	1.03
Package label or insert	37	1.08
Chemical company	36	1.06
Pesticide dealer	29	1.03
Magazine or newspaper articles	26	1.12
USDA Extension Service	25	1.00
Friends and neighbors	18	1.11
Medical personnel	12	1.08
Employer	10[b]	1.00

[a]Values are 1 (satisfied) and 2 (dissatisfied).
[b]Reflects the high number (25) of self-employed respondents.

employers was also low, but this was because the majority of the respondents were self-employed.

The responses to a series of questions about use of protective equipment are presented in Table 2. Face shield, goggles, and face mask usage fell, on the average, between "sometimes" and "seldom," respirators were used a little more frequently. Next to "not necessary," "not comfortable" was the reason most often given for not wearing an item. Goggles represented special problems for some of the respondents; three noted that they interfered with their eyeglasses, and another complained that goggles "fog up." One respondent commented that sunglasses were worn instead of goggles for protection against accidental splashes. These data suggest the need for improved design of goggles to make them more generally acceptable for use in application of pesticides.

Table 3 contains the data compiled from responses to selected clothing-

TABLE 2—*Frequency of use of safety equipment with reasons for low use.*

	Face Shield	Goggles	Face Mask	Respirator
Means[a]	3.82	3.55	3.43	2.81
Reasons for low use[b]				
Not available	2	0	0	2
Too expensive	0	0	0	1
Not comfortable	4	4	4	6
Not necessary	15	8	17	4
Other	1	7	3	2

[a]Scale ranges from 1 (always) to 5 (never).
[b]Multiple responses were allowed.

TABLE 3—*Responses to selected clothing-related questions.*

Question	Mean	Number
Do you wear disposable protective garments for applying pesticides?		8
Do you wear other special protective garments for applying pesticides?		5
How often would you say pesticide gets on your clothes?[a]	4.19	
When pesticide gets on your clothing, how often does it get through the clothing to the skin?[b]	2.06	
For the clothes you usually wear when applying pesticides, how effective do you feel they are in protecting you from pesticides?[c]	2.37	
After your clothes become soiled with pesticide, when do you usually change clothes?[d]	3.17	

[a]Scale ranges from 1 (never) to 6 (constant contact during application season).
[b]Scale ranges from 1 (never) to 5 (always).
[c]Scale ranges from 1 (very effective) to 5 (very ineffective).
[d]Scale ranges from 1 (immediately) to 4 (at the end of the work day).

related questions. As found in previous studies, most respondents wore a variety of everyday items when applying pesticides. Only eight respondents reported wearing special disposable protective garments. One of these eight and an additional four said they wore other special protective garments. Most of the respondents felt that pesticides were often deposited on clothing; the mean response fell between "once a week during application season" and "nearly every day during application season." However, they were much less likely to feel that pesticides got through their clothing to the skin. The mean response to this item was close to "seldom." Moreover, responses to a question specifically asking about perceived effectiveness of clothing in protecting from pesticides have an average value somewhat above the midpoint of the scale toward the "very effective" end. Although field studies such as those conducted by Finley and her colleagues and Fenske et al suggest that these views are apt to be overly optimistic perceptions of the protection provided by typical work garments [*18–20*], other research, such as the work by Gold et al, indicates that work clothing can provide an effective barrier to some pesticides under certain conditions [*21*].

With respect to changing clothes that become soiled with pesticide, the majority of respondents indicated that they waited until the end of the day to do so, although seven said "immediately." It should be noted that actions are apt to vary with the nature of the soiling. One respondent specifically stated that he changed immediately "if spill," but at the end of the work day "if normal drift."

The responses to the cost and risk versus benefit items are presented in Table 4. These data are in keeping with the suggestion by Vlek and Stallen [*13*] that for activities such as pesticide application, in which there is the potential for high benefits but also high risks, agricultural personnel are apt to emphasize the benefits and attach less importance to the risks. In general, the respondents in the present study felt the benefits of pesticide application were appreciably higher than the risks.

TABLE 4—*Responses to cost and risk versus benefit questions.*

Variable	Mean	Median, $
Cost of pesticide application		2000
Savings from pesticide application		8000
Overall risks associated with pesticide application[a]	2.17	
Overall benefits associated with pesticide application[a]	4.51	
Likelihood of immediate harm[b]	2.12	
Seriousness of immediate harm[c]	2.26	
Likelihood of long-term harm[b]	2.13	
Seriousness of long-term harm[c]	2.31	

[a]Scale ranges from 1 (very low) to 5 (very high).
[b]Scale ranges from 1 (very unlikely) to 5 (very likely).
[c]Scale ranges from 1 (very mild) to 5 (very serious).

Growers' responses and commercial applicators' responses are compared in Table 5. These comparisons indicated that commercial applicators were less likely to use restricted pesticides. Similarly, they were less likely to use pesticides classified in the more toxic categories, although the latter difference did not reach statistical significance. Consistent with these differences in types of pesticides used, the commercial applicators were significantly less likely to wear a face mask or face shield. They were also less likely to wear a respirator or goggles, although these differences were not significant. On the other hand, they were more likely to express uniform agreement regarding the importance of certain precautions, such as washing the hands after handling pesticides and keeping children out of the area when working with pesticides. This finding suggests the possibility that some psychological phenomenon such as cognitive dissonance or a defensive-avoidance reaction might be operating among people who apply pesticides (see Ref *22*). People at relatively high risk may place less importance on some precautions to convince themselves that a high risk cannot be that risky or they would not expose themselves to it (dissonance reduction). Alternatively, they may find that placing importance on precautions reminds them of risks, which they find less stressful if ignored (defensive-avoidance reaction).

A comparison of users of restricted versus nonrestricted pesticides, shown in Table 6, produced supportive findings regarding an inverse relationship between use of potentially more dangerous chemicals and minimization of those dangers. Users of restricted chemicals were more likely to feel that pesticide frequently gets on clothes but less likely to believe that it gets through to

TABLE 5—*Significant differences between commercial applicators and growers.*

Variable	Applicator Mean	Grower Mean	χ^2
Use of restricted pesticides[a]	0.56	1.60	9.36**
How often do you wear a face mask?[b]	4.25	2.57	6.39*
How often do you wear a face shield?[b]	4.60	3.15	5.17*
It is important to wash your hands after handling pesticides.[c]	1.06	1.40	3.89*
It isn't safe to let children hang around when working with pesticides.[c]	1.00	1.47	6.64**

[a]Scale values are 0 (nonrestricted), 1 (restricted with qualifications), and 2 (restricted without qualifications).

[b]Scale ranges from 1 (always) to 5 (never).

[c]Scale ranges from 1 (strongly agree) to 5 (strongly disagree).

*$P < 0.05$.

**$P < 0.01$.

TABLE 6—*Differences between users of restricted pesticides and users of nonrestricted pesticides.*

Variable	Restricted Mean	Nonrestricted Mean	χ^2
How often would you say pesticide gets on your clothes?[a]	4.56	3.00	3.88*
How often does pesticide get through clothing to the skin?[b]	1.95	2.75	2.59
The risks involved in getting pesticide on clothes is nothing compared to breathing pollution in the air.[c]	2.74	3.40	3.20**

[a]Scale ranges from 1 (never) to 6 (constant contact).
[b]Scale ranges from 1 (never) to 5 (always).
[c]Scale ranges from 1 (strongly agree) to 5 (strongly disagree).
$*P < 0.05$.
$**P < 0.10$.

the skin. Furthermore, they were more likely to express agreement with the statement that the risk involved in getting pesticide on clothes is nothing compared to that of breathing pollution in the air. Similarly, as indicated in Table 7, when users of different categories of pesticides were compared, it was found that users of the more toxic chemicals acknowledged the fact by showing significantly higher disagreement with the statement that "the pesticides I use are not very toxic." However, a concomitant deemphasis of the risks is suggested by significantly more agreement with the statement that "you can't do a good job of applying pesticides if you are always worried about the risks."

TABLE 7—*Significant differences between users of Categories 1 and 2 pesticides and users of Category 3 pesticides.*

Variable	Categories 1 and 2 Means	Category 3 Mean	χ^2
The pesticides I use are not very toxic.[a]	3.68	2.17	9.07**
You can't do a good job of applying pesticides if you are always worried about the risks.[b]	2.96	4.17	3.92*

[a]Category 1 consists of highly toxic compounds, Category 2 of moderately toxic compounds, and Category 3 of slightly toxic to relatively nontoxic compounds.
[b]Scale ranges from 1 (strongly agree) to 5 (strongly disagree).
$*P < 0.05$.
$**P < 0.01$.

When applicators were classified by method of pesticide application, air blast versus other methods (Table 8), once again there appeared to be an inverse relationship between actual risk and opinions about risk. Although the users of the air blast method reported wearing face masks more often, they showed less agreement with a statement about keeping equipment in good repair. In addition, they indicated more agreement with statements on being tough enough to take exposure to pesticides without harm and on there being little chance of pesticide getting on the skin if clothes are washed after every pesticide application. This latter finding could be interpreted as either less concern about pesticide getting on the skin or more concern about frequent washing of protective clothing. The fact that the air blast group also tended to show more agreement with the statement that, when applying pesticide, it is safe to wear the same clothes day after day without washing them suggests that the former interpretation is more likely.

The data on storage and cleaning of clothing worn when applying pesticides are presented in Table 9. These data indicate that 43% of the launderer sample stored such clothing with the family laundry at least sometimes, and 36% of the sample washed the clothing with the family laundry at least sometimes. These findings are cause for some concern since, as was previously indicated, health problems can develop through secondary exposure to pesticides when work clothing comes in contact with other household garments.

TABLE 8—*Significant differences between applicators using air blast and applicators using other methods of pesticide application.*

Variable	Air Blast Mean	Other Methods Mean	χ^2
How often do you wear a face mask?[a]	2.73	4.44	7.58**
I feel I am tough enough to take exposure to pesticides without harm.[b]	4.00	4.72	4.91*
Pesticide application equipment should be kept in tip-top repair to avoid accidents.[b]	1.45	1.10	4.69*
If clothes are washed after every pesticide application, there is not much chance of pesticide getting on the skin through the clothes.[b]	2.18	2.89	5.22*

[a]Scale ranges from 1 (always) to 5 (never).
[b]Scale ranges from 1 (strongly agree) to 5 (strongly disagree).
*$P < 0.05$.
**$P < 0.01$.

TABLE 9—*Responses to storage and cleaning items.*

Question	Always	Usually	Sometimes	Seldom	Never
Where do you usually store clothing worn during pesticide application?					
With the family laundry	5	2	3	1	12
In the closet with other clothes	7	1	2	0	13
In a separate container	11	2	1	2	7
In a plastic bag	2	1	2	0	17
On a special hook or hanger away from other clothes	8	3	1	0	13
How do you wash the clothing that was worn when applying pesticides?					
With the family laundry	2	4	2	2	12
In a separate load	15	7	4	1	1
Prerinsed before put in washer	3	2	5	2	11
Washed out by hand	0	0	1	1	21
Washed two or more times in the washer	1	2	6	5	9

Conclusions

In conclusion, these findings suggest that applicators of the more toxic pesticides may be in double jeopardy; not only are they dealing with more hazardous substances, but they also perceive some precautions to be less important. The same trend appears when applicators are classified by method of application. Moreover, providing additional information to the applicators may not be of much help; applicators seemed to be aware of the objective hazards, but this knowledge did not affect their appreciation for certain safety measures. In addition, there appears to be some cause for concern about household laundry practices involving contact between pesticide-soiled clothing and other garments. Indications are that such practices may lead to health problems through secondary contamination of the family laundry.

Limitations

These findings, while strongly suggestive of several problem areas, are based on a limited sample of California growers and commercial applicators. Additional work is now in progress as part of the NC-170 Regional Research Project to expand the data base and provide additional information on some of the issues raised in the present study.

Acknowledgments

We would like to thank the Northern California Occupational Health Center, Davis Component, University of California Department of Internal Medicine, for providing a grant to support this project.

References

[1] Quinby, G. E. and Lemmon, A. B., *Journal of the American Medical Association*, Vol. 166, No. 7, 15 Feb. 1958, pp. 740-746.
[2] Lamb, D. W. in *Field Worker Exposure During Pesticide Application*, W. F. Tordoir and E. A. H. van Heemstra-Lequin, Eds., Elsevier, Amsterdam, 1980, pp. 121-127.
[3] Batchelor, G. S. and Walker, K. C., *A.M.A. Archives of Industrial Hygiene and Occupational Medicine*, Vol. 10, Dec. 1954, pp. 522-529.
[4] Culver, D., Caplan, P., and Batchelor, G. S., *A.M.A. Archives of Industrial Health*, Vol. 13, Jan. 1956, pp. 37-50.
[5] Wolfe, R. H., Durham, F., and Armstrong, J. F., *Archives of Environmental Health*, Vol. 14, April 1967, pp. 622-633.
[6] DeJonge, J. O., Vredevoogd, J., and Henry, M. S., *Clothing and Textiles Research Journal*, Vol. 2, No. 1, Fall/Winter 1983-1984, pp. 9-14.
[7] Davies, J. E., Freed, V. H., Enos, H. F., Barquet, A., Morgade, C., and Danauskas, J. X., *Residue Reviews*, Vol. 75, 1980, pp. 7-20.
[8] Southwick, J. W., Mecham, H. D., Cannon, P. M., and Gortatowski, M. J., "Pesticide Residues in Laundered Clothing," *Proceedings*, Third Conference of Environmental Chemicals and Human and Animal Health, Colorado State University, Fort Collins, CO, 1974.
[9] Finley, E. L., Metcalfe, G. I., McDermott, F. G., Graves, J. B., Schilling, P. E., and Bonner, F. L., *Bulletin of Environmental Contamination and Toxicology*, Vol. 12, No. 3, Sept. 1974, pp. 268-274.
[10] Easley, C. B., Laughlin, J. M., Gold, R. E., and Tupy, D. R., *Archives of Environmental Contamination and Toxicology*, Vol. 12, No. 1, 1983, pp. 71-76.
[11] Bellin, J. S., *Occupational Health and Safety*, Vol. 50, No. 6, June 1981, pp. 39-42.
[12] Li, F. P., Lokich, J., Lapey, J., Neptune, W. B., and Wilkins, E. W., Jr., *Journal of the American Medical Association*, Vol. 240, No. 5, 4 Aug. 1978, p. 467.
[13] Vlek, C. and Stallen, P. J., *Organizational Behavior and Human Performance*, Vol. 28, No. 2, Oct. 1981, pp. 235-271.
[14] Church, J., "An Investigation into the Differences in Attitudes and Practices Regarding Protective Clothing for Pesticide Users as Affected by Age, Educational Level, and Type of Farm," Unpublished paper, Department of Human Environment and Design, Michigan State University, East Lansing, MI, Summer 1980.
[15] Henry, M. S., "Users' Perceptions of Attributes of Functional Apparel," unpublished masters thesis, Michigan State University, East Lansing, MI, 1980.
[16] Babbie, E., *The Practice of Social Research*, Wadsworth, Belmont, CA, 1983.
[17] Turim, J., Reese, C. D., Kempter, J., and Muir, W., *Farmers Pesticide Use Decisions and Attitudes on Alternate Crop Protection Methods*, U.S. Environmental Protection Agency and Council on Environmental Quality, Washington, DC, 1974.
[18] Finley, E. L., Bellon, J. M., Graves, J. B., and Koonce, K. L., *Louisiana Agriculture*, Vol. 20, No. 3, Spring 1977, pp. 8-9.
[19] Finley, E. L., Graves, J. B., Hewitt, F. C., Morris, M. F., Harmon, C. W., Iddings, F. A., Schilling, P. E., and Koonce, K. L., *Bulletin of Environmental Contamination and Toxicology*, Vol. 22, Nos. 4/5, July 1979, pp. 590-597.
[20] Fenske, R. A., Leffingwell, J. T., and Spear, R. C., "Field Testing of Fluorescent Tracer Methodology for Dermal Exposure Assessment," Paper presented at the American Chemical Society Meeting, St. Louis, MO, April 1984.
[21] Gold, R. E., Leavitt, J. R. C., Holcslaw, T., and Tupy, D., *Archives of Environmental Contamination and Toxicology*, Vol. 11, No. 1, 1982, pp. 63-67.
[22] Harrison, A. A., *Individuals and Groups*, Brooks/Cole, Monterey, CA, 1976.

Donna H. Branson,[1] *George S. Ayers,*[2] *and Maureen S. Henry*[3]

Effectiveness of Selected Work Fabrics as Barriers to Pesticide Penetration

REFERENCE: Branson, D. H., Ayers, G. S., and Henry, M. S., **"Effectiveness of Selected Work Fabrics as Barriers to Pesticide Penetration,"** *Performance of Protective Clothing, ASTM STP 900,* R. L. Barker and G. C. Coletta, Eds., American Society for Testing and Materials, Philadelphia, 1986, pp. 114-120.

ABSTRACT: The effectiveness of seven work fabrics as barriers to three volumes of four carbon-14-labeled pesticides was determined in a controlled laboratory study. Significant differences were found for fabric, pesticide, and volume. Three laminate fabrics offered significantly better protection against the pesticides tested than fabrics now worn for pesticide application. The test fabrics were least effective barriers against parathion and Guthion and most effective barriers for paraquat. In general, as the volume increased, the effectiveness of the test fabrics decreased.

KEY WORDS: pesticide application, protective clothing, pesticide penetration, pesticide permeation, Guthion, paraquat, dinoseb, parathion

Dermal exposure is a serious concern for many types of pesticide application and field procedures. The use of protective clothing as a barrier is considered vital for minimizing dermal exposure for those who work with and around pesticides. Yet data documenting the protection afforded by various fabrics to pesticides are limited and fragmentary [*1-3*]. Also, different methodologies have been used in these studies. The effectiveness of protective clothing may be influenced by a complex set of garment characteristics, environmental conditions, and characteristics of pesticides.

[1]Associate professor, Department of Clothing, Textiles and Merchandising, Oklahoma State University, Stillwater, OK 74078.

[2]Associate professor, Department of Entomology, Michigan State University, East Lansing, MI 48824.

[3]Instructor, Department of Home Economics, Family Life and Consumer Education, Central Michigan University, Mount Pleasant, MI 48859.

The overall purpose of this research was to contribute to the knowledge base in this area in order to support fabric recommendations for the development of improved protective clothing that would afford pesticide applicators dermal protection and an acceptable level of thermal comfort. The effectiveness of seven work fabrics as barriers to three volumes of four pesticides was determined. The results presented here focus on the evaluation of the seven fabrics as new fabrics. However, the larger study also examined the effectiveness of the test fabrics after laboratory-controlled abrasion and repeated launderings were performed.

Procedure

A controlled laboratory study was conducted using a test methodology previously developed by the authors. A split-plot experimental design, which was completely randomized within the main plot and factorial within the split, was used with three replications.

Test Fabrics

Two fabrics commonly worn by the agricultural work force and five potential protective fabrics served as the test fabrics. A 100% cotton 500.17-g/m^2 (14.75-oz/yd^2) denim with a Sanforset finish and a 100% cotton woven chambray shirting-weight fabric were the two fabrics selected to represent typical clothing now used for pesticide application. The five potential protective fabrics included three laminate variations, an uncoated spun-bonded 100% olefin, and a 65% polyester/35% cotton 193.3-g/m^2 (5.7-oz/yd^2) coated poplin. Laminate 1 consisted of a microporous membrane of polytetrafluoroethylene laminated between a face fabric of 100% nylon ripstop and an inner layer of nylon tricot. Laminate 2 consisted of a microporous membrane of polytetrafluoroethylene laminated between an outer fabric of 100% polyester taffeta and an inner fabric of polyester tricot. Laminate 3 was a two-layer laminate with the microporous membrane serving as the face fabric and laminated to a polyester/cotton woven fabric backing.

Pesticides

The choice of pesticides was based on an assessment of their potential health hazard, which included toxicity, the method and frequency of application, and the extent of usage in the United States. The following four pesticides were selected: parathion, Guthion, paraquat, and dinoseb.

Solutions for pesticide penetration testing were made from mixtures of carbon-14 (^{14}C) isotopically labeled pesticides, commercial formulations, and water so that the final solutions were representative of field sprays used in agricultural pesticide application. A concentration of 2.995 kg/m^3

(0.025 lb/gal), or 0.3% on a weight per weight basis, was used for the four pesticides. The radioactivity per unit volume was also held constant.

The mixtures were proportioned as follows: the parathion solution was 6 μCi of ^{14}C-parathion, 18.75 μL parathion 8E formulation, and 4.98 mL of water. The Guthion solution was 6 μCi of ^{14}C-Guthion, 75 μL of Guthion 2S formulation, and 5.92 mL of water. The paraquat solution was made as follows: 6 μCi of ^{14}C-paraquat, 75 μL Paraquat-Plus formulation, and 5.32 mL of water. The dinoseb solution was 6 μCi of ^{14}C-dinoseb, 50 μL of Dinitro-5 formulation, and 5.95 mL of water.

Volume

Each pesticide was tested at the following three volumes: 25, 50, and 75 μL.

Protocol

Fabric specimens were cut to ensure a representative sampling of the total fabric, and the identical warp and filling yarns were not included. Preparation of the fabric specimens for pesticide testing was begun by applying a narrow bead of seam sealer to the edges of the 38.1-mm (1.5-in.) square fabric specimens. This was done to prevent the pesticide from migrating from the original contact location to the outer fabric edge and around the edge to the gauze layers beneath the test fabric. The focus of the study was on determining whether pesticide went through the test fabrics rather than around them.

The fabric assembly, which included a prepared test fabric specimen, multiple layers of cotton gauze, and an aluminum foil backing, was placed between two 44.45-mm (1.75-in.) square aluminum plates and secured with metal clamps. The top plate featured a 19.05-mm (0.75-in.) circular opening positioned in the center of the plate. The holders were designed to eliminate the need for fabric cutting prior to pesticide residue analysis.

The assemblies were placed in individual chambers maintained at 100% relative humidity, sealed, and allowed to remain overnight. The purpose of the humidity chambers was to simulate humid weather conditions under which pesticide application is frequently performed.

Pesticide was pipetted onto the fabric surface and allowed to remain in the sealed humidity chambers for 5 h at ambient laboratory temperatures. The time interval of 5 h was specified in order to evaluate the performance of the test fabrics for a time period that would be representative of the time a grower might typically wear his work clothing before changing. One replication for each pesticide was tested at a time, with the fabrics and volume being randomized.

The fabric assemblies were harvested and disassembled, and the cotton gauze specimens were placed in scintillation vials containing the scintillation solution. A cocktail mixture of 4 g 2,5-diphenyloxazole (PPO) and 50 mg

[1,4-bis-2-(5-phenyloxazolyl)-benzene] (POPOP) per litre of toluene was used for the scintillation solution. The specimens were counted for 2 min in a Packard Tri-Carb liquid scintillation counter. All the data were converted to disintegrations per minute (dpm) by the channels ratio method.

Results and Discussion

The results from the statistical procedure analysis of variance are given in Table 1. Statistically significant differences were found for the pesticide, volume, and fabric. In addition, the three two-way interactions were also determined to be significant. Therefore, before the main effects could be determined, the interactions were plotted and examined.

Two-Way Interactions

A clear-cut trend was observed for the pesticide-by-volume interaction plot, with disintegrations per minute increasing as volume increased. This was true for all the pesticides tested. However, the increase was not uniform for pesticide or for volume.

The fabric-by-volume interaction plot also showed a clear trend. In general, disintegrations per minute increased as volume increased, with two exceptions. The disintegrations per minute data for Laminate 1 were similar for Volumes 1, 2, and 3. Also, for Laminate 2, similar disintegrations per minute data were found for Volumes 2 and 3, which were slightly greater than the disintegrations per minute data for Volume 1. All the other fabrics were less effective barriers to Volume 2 than to Volume 1, and even less effective barriers for Volume 3.

The fabric-by-pesticide interaction plot did not show a clear trend as the two other plots had. There was a tendency, however, for the disintegrations per minute by fabric to be highest for parathion and lowest for paraquat. Five

TABLE 1—*Split-plot ANOVA table.*

Source	Degrees of Freedom	Mean Square	F-Value
Pesticide	3	$2\,526 \times 10^6$	18.12[a]
Within pesticide	8	139×10^6	0.89
Volume	2	$6\,237 \times 10^6$	39.65[a]
Fabric	6	$13\,302 \times 10^6$	84.59[a]
Volume × fabric	12	$1\,274 \times 10^6$	8.10[a]
Pesticide × volume	6	680×10^6	4.32[a]
Pesticide × fabric	18	974×10^6	6.20[a]
Pesticide × volume × fabric	36	280×10^6	1.78
Experimental error	160	157×10^6	...

[a]Indicates significance at the 0.001 level.

test fabrics showed least-barrier resistance to parathion. Denim and the coated poplin were the exceptions. All of the test fabrics but Laminate 3 exhibited the greatest resistance to paraquat.

Thus, two interaction plots clearly showed that volume was a critical variable in determining the effectiveness of the test fabrics. This has implications in terms of a methodology to assess performance of protective fabrics and a practical significance for the grower who experiences dermal exposure.

Pesticide Effect

Table 2 presents the results of Duncan's multiple range test for pesticides. Examination of Table 2 indicates that there were significant differences in the disintegrations per minute between paraquat and the other three pesticides and between dinoseb and the other three pesticides. There were no significant differences between Guthion and parathion. Fabrics, in general, were more effective barriers to paraquat, with an average of 3.210 dpm found on the gauze layers. Almost three times as much dinoseb permeated and penetrated the test fabrics as paraquat. The rates for Guthion and parathion were found to be almost five times that of paraquat.

Volume Effect

Table 3 displays the results of Duncan's multiple range test for volume. For Volume 1 (25 μL), an average of almost 4000 dpm was found. For Volume 2 (50 μL), 9000 dpm was found, and for Volume 3 (75 μL) almost 21 000 dpm. The data suggest that volume is a critical variable when determining the barrier effectiveness of protective fabrics. Also, the data show that disintegrations per minute for volume are not linear.

TABLE 2—*Duncan's multiple range test results for disintegrations per minute by pesticide.*

Variable	Mean dpm	Number	Duncan's Grouping[a]
Parathion	17 443	63	A
Guthion	14 942	63	A
Dinoseb	9 341	63	B
Paraquat	3 210	63	C

[a]Means with the same letter are not significantly different at the 0.05 level.

TABLE 3—*Duncan's multiple range test results for disintegrations per minute by volume.*

Variable	Mean dpm	Number	Duncan's Grouping[a]
Volume 1	3 954	84	A
Volume 2	9 002	84	B
Volume 3	20 748	84	C

[a]Means with the same letter are not significantly different at the 0.05 level.

Fabric Effect

Table 4, which presents Duncan's multiple range test for fabric, shows that three significantly different groupings were determined. The fabric offering the least protection was the 100% chambray shirting-weight fabric. A mean of almost 54 000 dpm was determined for chambray. The second group of fabrics that offered greater protection than chambray included the 100% cotton denim, the 65 polyester/35 cotton coated poplin, and the 100% spun-bonded olefin. Mean disintegrations per minute data ranged from 10 505 dpm for the poplin to almost 6000 dpm for the spun-bonded olefin. There were no significant differences between these three fabrics.

The fabrics offering the greatest protection constituted the third grouping and included the three laminate variations and the spun-bonded olefin. Although examination of the data would seem to suggest that differences were substantial between the three laminates and the olefin, the difference was not statistically significant because of large variability within the data.

TABLE 4—*Duncan's multiple range test results for disintegrations per minute by fabric.*

Variable	Mean dpm	Number	Duncan's Grouping[a]
Chambray	53 708	36	A
Coated poplin	10 505	36	B
Denim	8 407	36	B
Spun-bonded olefin	5 914	36	B, C
Laminate 3	52	36	C
Laminate 2	42	36	C
Laminate 1	11	36	C

[a]Means with the same letter are not significantly different at the 0.05 level.

Conclusions

The findings of this research support the assertion that a complex set of factors influences the barrier effectiveness of fabrics to pesticides. Statistically significant differences were found for fabric, pesticide, volume, and the two-way interactions. This strongly suggests the need for increased research to continue to identify the pertinent variables, to obtain a larger data base on these variables, and to investigate the possible interactions between these variables. The method proposed offers an efficient, cost-effective way of examining a large number of specimens.

From a practical standpoint, the findings also indicate that chambray shirts and denim jeans are not effective barriers against the pesticides tested. Thus, the need for protective clothing seems to be supported by these data. Further investigation of the laminates, particularly Laminate 1, appears warranted.

Acknowledgments

The authors wish to acknowledge that Michigan State University Experiment Station and the North Central Region Pesticide Impact Assessment Program provided financial support for this research. We also wish to thank W. Warde and M. H. Branson for their cooperation in providing statistical and computer analyses, and E. I. du Pont de Nemours and Co. and W. L. Gore and Associates for providing some of our test fabrics.

References

[*1*] Freed, V. H., Davies, J. E., Peters, L. J., and Parveen, F., *Residue Review*, Vol. 75, 1980, pp. 33-50.

[*2*] Orlando, J., Branson, D. H., Ayers, G. S., and Leavitt, R., *Journal of Environmental Science and Health*, Vol. 5, 1981, pp. 617-628.

[*3*] Davies, J. E., Freed, V. H., Enos, H. F., Barquet, A., Morgade, C., and Danauskas, J. X., *Residue Review*, Vol. 75, 1980, pp. 9-20.

Glyn A. Lloyd[1]

Efficiency of Protective Clothing for Pesticide Spraying

REFERENCE: Lloyd, G. A., **"Efficiency of Protective Clothing for Pesticide Spraying,"** *Performance of Protective Clothing, ASTM STP 900*, R. L. Barker and G. C. Coletta, Eds., American Society for Testing and Materials, Philadelphia, 1986, pp. 121-135.

ABSTRACT: Measurements in the United Kingdom of the exposure of operators to pesticide sprays were used to indicate the performance required of protective clothing. Laboratory tests, designed to simulate maximum levels of contamination, as far as practicable, were then applied to protective clothing materials and finished items to facilitate the development of national and international standards of performance. Studies of the relative efficiencies of protective garments in field spraying operations were also undertaken to assess the value of laboratory test data for the selection of effective spray suits.

High protection factors, based on the ratio of the amounts of spray chemical found on and inside the clothing, were usually recorded in the field studies. The correlation of performance data with laboratory test data was only fair. The laboratory tests of permeation and penetration posed high levels of contamination, however, and tended to err therefore on the side of safety in assessments of the potential efficiency of protective clothing.

KEY WORDS: evaluation, protective clothing, standards, pesticides, exposure, spraying

The major risk to operators spraying pesticides is usually associated with absorption of chemicals through the skin. Improvements to spraying equipment and handling procedures may reduce contamination, but for adequate protection of workers, there often remains a need for efficient protective clothing. The user faces several problems, however, in the selection of suitable items [*1*]. The clothing may, for example, be uncomfortable to wear, incompatible with other essential personal protective equipment, or not fit well. Moreover, the potential wearer may have little or no information with which to judge which of the available items on the market provides protection to a level considered to be safe.

The provision of national standard specifications for chemical protective clothing, preferably supported by a testing and certification scheme and au-

[1]Head of Operator Protection Research Group, Harpenden Laboratory of the Ministry of Agriculture, Fisheries and Food, Harpenden, AL5 2BD England.

thoritative advice notes on use and maintenance [2], should overcome many of the outlined problems. Laboratory test methods are therefore required to establish categories of protective efficiency against harmful chemicals for the guidance of both manufacturers and users of protective clothing. Features of design and physical properties of the materials used to make up clothing must also be considered in order to improve comfort factors and the level of acceptability to workers of made-up items.

One field study in the United States [3] suggested that the choice of protective clothing is often a matter of compromise between acceptability and safety, but lightweight garments are likely to be acceptable to all spray operators. More studies of this nature are required to ensure that any laboratory tests of efficiency and comfort of protective clothing relate as far as possible to operational conditions. As climatic factors, spraying practices, and the scale of operations with pesticides can vary widely from one country to another, the results of evaluations of exposure to the chemicals and the means of protection may not be universally applicable. No attempt has been made therefore in this paper, in the absence of suitable comparative standards or reference points, to review or equate data from other countries on the protective efficiency of clothing.

An ongoing program of work in the United Kingdom has therefore been established to provide a national data base on relative operator exposure levels to pesticides applied in different ways [4]. One application of the results of these studies is to indicate the form of protective clothing required for adequate protection of spray operators under conditions applicable to the United Kingdom and possibly other countries with similar climates, spraying practices, and controls over the use of pesticides.

This paper chiefly summarizes the results to date of pesticide exposure measurements on spray operators and associated studies of the efficiency of protective garments available in the United Kingdom or under development for their use. The relationships are examined between the results of laboratory and field assessments of selected items of clothing.

Experimental Procedures

Screening Tests for the Selection of Materials

Simple standardized screening tests (described in the two following subsections) were developed and applied to a wide range of protective clothing materials on the British market. This led to the selection of a small group of materials for further study in the form of garments for spray operators.

Measurement of Breakthrough Time ("Air-Impermeable" Clothing Materials)—The elapsed time between the application of a liquid chemical to the appropriate surface of a protective clothing material and its first appearance on the opposite surface at the molecular level is customarily referred to as the breakthrough time. In these studies only major differences in performance between specimens of the materials were sought, and as the breakthrough time of a liquid pesticide formulation was found to be dominated by that of the formulating solvent, qualitative screening tests with selected organic solvents [British Standards Institution (BSI) Method of Test for Resistance of Air-Impermeable Clothing Materials to Penetration by Harmful Liquids, British Standard (BS) 4724: 1971, under revision in 1984] provided a convenient means of ranking the performance of materials on the market.

Quantitative measurements of the amounts of chemicals diffusing through samples at various times were required occasionally to differentiate between the performance of materials showing similar breakthrough times. A standardized diffusion cell procedure, based on collaborative tests, was used [International Organization for Standardization (ISO) Draft Proposal 6529—Subcommittee ISO/TC94/SC13].

Measurement of the Ability of Woven and Other Porous Materials to Shed Spray Liquids—Protective clothing made from a porous fabric is likely to be more acceptable to spray operators, but clearly a material that is permeable to air may not provide adequate protection of the skin, when contaminated with a sprayed liquid, unless the surface of the fabric has been treated in some way to reduce penetration to an acceptable level. The abilities of potential trial clothing materials to repel a simulated dilute aqueous spray liquid were therefore compared conveniently by a simple screening test (International Organization for Standardization, Clothing for Limited Protection Against Dangerous Liquid Chemicals—Resistance to Penetration—Marking, ISO 6530-1980). This test involves release of the liquid onto a specimen of the material resting in an inclined gutter. Measurement of the volume of liquid that runs off its surface and examination for its penetration to the underside complete the test.

Screening Tests for the Selection of Protective Garments

Two forms of laboratory tests for checking the integrity of a finished garment are under consideration by the aforementioned BSI and ISO technical committees. These tests were developed by the Operator Protection Research Group chiefly to check that, where a permeation-resistant material has been used to make up a protective garment, the potential efficiency is not reduced by penetration of the chemical through other parts of the clothing. In this context, *penetration* is defined as the passage of a chemical through essential protected openings and overlaps, zippers, seams, or any pores or imperfections in a material.

Water adjusted to a low surface tension and containing a suitable tracer provides a convenient and safe means of assessing potential resistance to penetration by sprayed fluids under the conditions of the laboratory tests, which involve controlled applications of either a jet or spray to the whole surface of a test garment on a human subject or dummy figure. Penetration of water through defects in the garment is observed by the presence of a tracer (for example, visible or fluorescent dye) on the appropriate surfaces underneath.

Measurements of Operator Contamination and Spray Penetration of Clothing in Field Spraying Operations

Determination of the spray deposits on the inside and outside of an item of contaminated clothing was considered to provide a reasonable estimate of its protective efficiency and the potential level of exposure to the spray chemical. Absolute measurements were thought to be difficult to achieve, however, in that some of a chemical contaminant may evaporate or be absorbed and thus be difficult to recover quantitatively from exposed surfaces.

Absorbent patches, on and under the clothing [*5*] were used initially in exposure studies and assessments of protective clothing, but, as the patches tended to interfere with the distribution and penetration of spray and gave only approximate values in terms of whole body surface contamination, the practice was discontinued in favor of the use of whole items of clothing as the sampling media.

Rinsing or swabbing of skin with solvents [*6*] was considered as a means of measuring penetration of a spray through clothing, but this procedure was thought to be potentially harmful and impractical in these field studies. Comparable data on the efficiency of protective clothing were produced more conveniently, when operational conditions allowed a lightweight disposable suit to be worn underneath the clothing being tested for the measurement of penetration.

Analysis of Samples

Conventional methods were used to extract spray deposits from the various sampling materials and associated spiked samples for validation of the overall analytical procedure. Appropriate analytical methods were selected for each pesticide in use. Whenever practicable, a dye tracer was incorporated in the spray liquid to facilitate the analytical work involved with exposure measurements and assessments of the efficiencies of protective garments.

Selection of Protective Clothing Materials for Field Studies

Four materials (A through D in Table 1) were selected initially for further study when made up into protective garments. The selection process included consideration of the results of permeation tests [*7*], other screening test data (Tables 1 through 3), current or potential use for the manufacture of spray operators' clothing in the United Kingdom, and the potential for improved protection or comfort to the wearer.

TABLE 1—*Trial clothing and materials.*

Fabric Properties	Garment Reference and Fabric Type				
	A [Woven: Polyester/Cotton (65/35)]	B [Woven Nylon (Siliconized)]	C (Microporous PTFE Membrane Laminated Between Two Layers of Nylon Fabric)	D (Woven Nylon Coated with Neoprene)	E (Experimental Polyester Film Bonded to Nylon Fabric)
Weight, (g/m^2)	250	130	150	270	216
Tensile strength, warp, kg	93	110	90	125	240
Hydrostatic head, cm (water gage)	<2	89	>150	>150	>150
Jacket					
Weight, kg	0.70	0.30	0.38	0.68	0.59
Lining	woven nylon	none	bonded knitted nylon	none	none
Ventilation	none	none	none	none	vents under armpits, back vent with flap over
Trousers					
Weight, kg	0.55	0.28	0.30	0.45	0.40
Lining	woven nylon	none	bonded knitted nylon	none	none
Seams					
Type	overlap	overlap	overlap	overlap	overlap
Seal	none	none	none	"painted"	taped

TABLE 2—*Approximate breakthrough times of selected liquid chemicals through clothing materials from Suits A through E.*

Liquid Chemical	Breakthrough Time, at 30°C, for Suits A through E, h				
	A	B	C	D	E
Butanone	<0.02	<0.02	<0.02	0.1	>6.0
*Iso*octane	<0.02	<0.02	<0.02	0.6	>6.0
Xylene	<0.02	<0.02	<0.02	0.1	>6.0
Herbicide formulation (aqueous concentrate)	<0.02	<0.02	<0.02	>6.0	>6.0
Herbicide formulation (oil concentrate)	<0.02	<0.02	<0.02	>6.0	>6.0
	porous materials (A, B, C): test not usually applicable				

TABLE 3—*Initial penetration and repellency of the trial clothing Materials A through E.*

Liquid Chemical	Penetration and Repellency of Specimens[a]				
	A	B	C	D	E
Water + wetter	P	R	R	R	R
RI, %[b]	57	97	97	98	99
Aqueous concentrate	P	R	R	R	R
RI, %	87	99	99	99	99
Aqueous spray (dilute)	P	R	R	R	R
RI, %	74	97	97	98	99
Oil-based concentrate	P	R	R	R	R
RI, %	57	97	97	94	93
Oil spray (dilute emulsion)	P	R	R	R	R
RI, %	60	99	97	94	97

[a]P = penetrated; R = resistant to penetration.
[b]RI = repellency index.

Material A was a conventional woven fabric (polyester/cotton) used in work wear for many purposes but relying essentially on absorption for protection against liquids.

Material B was a conventional woven fabric (nylon) that had been "proofed" with a silicone, and thereby similar to many products used in rainwear with the potential for "shedding" spray liquids.

Material C comprised a three-layer laminate—a microporous polytetrafluoroethylene (PTFE) membrane sandwiched between two layers of nylon fabrics (woven and knitted). This product had the potential for resistance to penetration by spray liquids along with improved comfort to the wearer.

Material D comprised a woven nylon fabric coated with neoprene. This product is virtually impermeable to air and is used commonly in the construction of general-purpose chemical protective clothing.

An experimental material, Material E, also impermeable to air, was included chiefly for its exceptional resistance to permeation by organic solvents. The performance and durability of the material was of particular interest in that it comprised a very thin polyester film (0.01 mm) bonded to the inside surface of a woven nylon fabric.

Design of Field Studies for the Assessment of Protective Clothing

Exposure measurements (Table 4) had demonstrated that the operators of hand-held spraying equipment were likely to be exposed to relatively high levels of contamination, whereas operators of tractor-based sprayers, often equipped with air-conditioned cabs, would usually experience the least exposure. Back-mounted (knapsack) hydraulic sprayers and hand-held rotary-disk atomizers were therefore selected for use by spray operators in field trials for assessment of protective clothing.

To ensure uniformity of conditions for applications of the sprays, areas of similar scrubland in forest plantations were selected for the wearer trials. The conditions of spraying were typical of normal practice and not controlled as precisely as a laboratory test. Full records were made therefore of all the possible influencing operational and meteorological factors. Several groups of workers were used as spray operators—the work schedule being arranged so that each worker wore each of the five suits on at least two occasions with use of different spraying equipment.

The levels of contamination incurred by each operator could not be controlled precisely in the field, but, where it was practicable to include a visible dye tracer in the spray, overexposure to the spray liquid could be observed and obvious measures taken to safeguard wearers of the less efficient spray suits. As far as practicable, each suit was exposed to a wide range of contamination levels.

The operation of opening the pesticide containers and pouring the contents into spray tanks for mixing was undertaken by personnel other than the spray operators to exclude haphazard contamination of the trial suits by small splashes of concentrated chemicals. The trial clothing was therefore exposed only to spray contamination.

The selected clothing materials (Table 1) were made up into two-piece suits of similar style (Fig. 1). Covered vents were introduced, however, into the polyester laminate suit (see Table 1, Material E) in an attempt to improve the comfort to the wearer without serious loss of protective efficiency. As the PTFE laminate material (Material C) was porous, yet resistant to penetration (Tables 1 through 3), it appeared to have potential advantages over more conventional materials for the construction of spray suits. Garments made from

TABLE 4—*Observed contamination of operators' clothing by sprays emitted from handheld equipment and tractor-mounted hydraulic sprayers in field spraying.*[a,b]

Spray Equipment	Number of Studies	Spray Liquid: Concentration of Pesticide, %	Spray Liquid: Diluent	Exposure to Spray Liquid, mL/h: Mean	Exposure to Spray Liquid, mL/h: Range
Knapsack sprayer (clothing trials, hydraulic)	17	0.4 to 3.7	water	48	1.8 to 260
Knapsack sprayer (hydraulic, general)	(25)[b]	(0.5 to 1)		(60)	(3 to 260)
Concentrate sprayer (clothing trials, rotary disk)	14	40 to 50	water, oil, or emulsion	16	0.7 to 103
Concentrate sprayer (rotary disk, general)	(52)	(>10)		(43)	(9 to 260)
Tractor sprayer (general, without cab)	(204)	(0.1 to 1)	water	(18)	(2 to 520)
Tractor sprayer (general, with cab)	(46)	(0.1 to 1)	water	(6)	(<0.1 to 16)

[a]From "Operator Exposure to Pesticides in the United Kingdom," Internal Report No. OH/1/84, Ministry of Agriculture, Fisheries, and Food, Harpenden, Herts., England, 1984.

[b]The results in parentheses refer to studies of contamination of operators in trials unconnected with the trials involving assessment of Garments A through E.

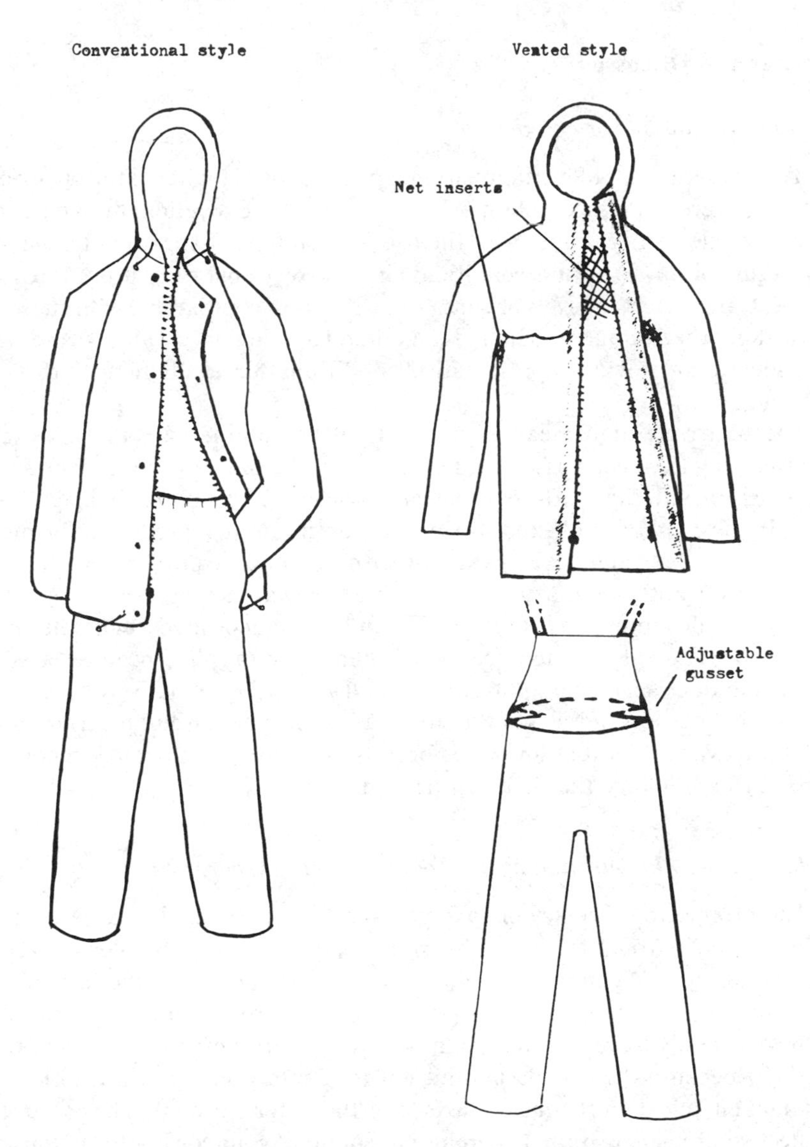

FIG. 1—*Styles of the trial spray suits.*

this material (Material C) were therefore used for the control or reference item in each trial to study the physiological responses of wearers and protective efficiency of the clothing.

Results and Discussion

Contamination of Spray Operators

To compare levels of contamination, spray deposits found on the operators' clothing were expressed volumetrically (Table 4). Exceptionally high levels were recorded on occasion, with the use of each type of spray equipment, as the result of adverse wind conditions or incorrect operating procedures. In general, the observed levels of contamination were comparable with those recorded in other studies to date in the United Kingdom (see Table 4) and were considered to provide representative conditions for evaluation of the test clothing.

The relatively high mean exposure level by volume with the knapsack sprayer may have been associated chiefly with the higher output of spray, in comparison with that of the concentrate sprayers. The picture changed, however, in the expression of exposure in terms of the active ingredient. The mean exposure levels sustained in operations with the concentrate sprayers then became significantly greater than those with the knapsack sprayer.

In general, more spray liquid was found to be deposited below the waist than above, except for a few occasions when the sprays had been released at about shoulder height for application to tall vegetation. Relatively heavy contamination of the lower legs of the suits was often experienced, partly because the spray was released at low levels but also because the operators tended to walk through foliage that had just been sprayed.

Relative Spray Penetration of the Test Clothing—Wearer Trials

The protective efficiency of each garment was calculated as a percentage value (Table 5) from the ratio of the spray deposit found on its outer surface to the overall level of contamination, which was assumed to be the total of the deposits found on the outer garment and the inner garment. The efficiency values may only be regarded as a guide to the relative efficiencies of the trial clothing because the calculation did not take into account any absorbed or residual chemical on the inside surface of the outer garment. The values for protective efficiency are not therefore absolute. As no constant relationship was apparent between penetration of clothing and the degree of contamination, only mean figures are reported (Table 5).

The results (Table 5) suggest that the design of the garment has more influence on its protective efficiency than the nature of the material from which it is made. Garments A through D are similar in design and appear to offer

TABLE 5—*Relative protective efficiencies of trial Garments A, B, C, D, and E described in Table 1.*

Spray Liquid	Mean Spray Deposit on Clothing, mL/h	Approximate Mean Protective Efficiency of Garment, %				
		A	B	C	D	E
Dilute (aqueous)	48	>99	98	93	>99	85
Concentrate	16	>99	99	>99	>99	86

almost the same initial degree of protection against spray contamination. Further contamination, before the garments are washed, is expected to reduce the effectiveness of absorbent clothing (for example, Garment A), especially as the point of saturation is approached, whereas the efficiencies of garments made from relatively impermeable materials would generally show little change.

Experimental Suit E appeared to be the least efficient garment, perhaps as a consequence of the vents provided to promote air circulation within the jacket. Inspection of the inside surfaces, after spraying fluids to which a dye tracer had been added, showed that leakages through poorly taped seams and small areas where the protective film had separated from the base fabric were chiefly responsible for the reduction in efficiency.

Laboratory Evaluation of Factors Associated with Protective Efficiency

The skin of spray operators may be exposed to pesticides as a result of molecular diffusion through "solid" protective clothing materials, "wicking" through porous fabrics, increased pressure of contact with spray liquids—for example, when walking through sprayed foliage—or penetration through stitch holes, closures (for example, zippers), and overlaps. The majority of these factors can be represented and examined individually in the laboratory.

Permeation of Protective Clothing Materials

The porous and microporous materials (from Garments A, B, and C) were, not unexpectedly, penetrated instantly by a selection of test liquids (Table 4) on application of a screening test for the measurement of relative breakthrough times (British Standard 4724: 1971, under revision). This form of test would not normally be applied to porous materials, but because one of them (PTFE Laminate C) had a high resistance to penetration by water (the hydrostatic head test, in Table 1), the author decided to check resistance to permeation. Outstanding resistance to permeation by the two principal herbicide concentrates used in the field trials was presented by the experimental polyester laminate (Material E) and the neoprene-coated fabric (Material D). Diffusion of the spray chemical through this laminated material (Material E) was not responsible, therefore, for the lowered efficiency of the made-up garment (Table 5).

Relative Resistance of Clothing Materials to Penetration by Liquid Chemicals

The test method (ISO 6530-1980) for assessment of the resistance to penetration of the porous types of protective clothing material differentiated between the protective efficiencies of the materials used in the test group of garments a little more clearly than did the measurement of breakthrough time.

The procedure simply involved measurement of the proportion by volume of a liquid that ran off the surface of a material on application in an inclined gutter, and, not unexpectedly, high indices of repellency (Table 3) were recorded in tests with the "solid" clothing materials. The relatively low indices found for the woven, porous fabric (Material A) were attributed to partial adsorption of the liquid, which appeared to vary in degree with the wetting power and viscosity of the liquid.

The "gutter" test required penetration of a material to be assessed visually by inspection of filter paper placed underneath the specimen. As an extension to the method, the volume of liquid adsorbed by the filter paper was measured by a weighing technique in tests with the absorbent material (Material A). This indicated that at least 10% of the applied volume was likely to be transferred instantly to an underlying surface on exposure of the material to small splashes and drips.

Penetration of Spray Liquids Through Made-Up Clothing

The application of a jet of liquid under pressure (150 KPa) for 5 min from all angles to the whole surface of a made-up protective garment or combination of garments has been proposed (ISO/DIS 6529) to check the integrity of closures, overlaps, and seams. None of the trial garments was able to withstand penetration by the jet under these conditions. Further experience in testing a wider range of designs of protective clothing demonstrated that only those garments constructed in one piece, with elasticated reductions at the wrists and ankles and a single well-protected closure of a zipper covered with a double storm flap, were capable of passing the test. The jet test therefore appeared to be too severe, at least for evaluating spray operators' protective clothing, where a compromise between efficiency and comfort factors had to be achieved in order to produce a wearable garment. The value of hydrostatic head measurements also appeared to be questionable as a means of indicating potential resistance to penetration by spray liquids in field spraying conditions.

Comfort of the Trial Clothing

The consensus of subjective reports by wearers of the different garments was that the most permeable and the least permeable items (Garments A, D, and E) were generally hot to wear. Criticisms of the most porous garment (Garment A) were surprising, but it appeared that operators in general found it too heavy to wear for prolonged periods (see Table 1). Garments made from the lightweight (microporous) materials were universally preferred (Garments B and C), in common with the experience of Staiff et al [*3*] in similar studies in the United States.

Supporting physiological measurements [*8*] on wearers of the lightweight microporous laminated suit and the lightweight impermeable suit indicated there was a small physiological advantage when the "breathable" garment (C) was worn during arduous spraying operations, but an unacceptable level of heat strain was likely to be experienced when either suit was worn under warm environmental conditions.

Correlation Between Field and Laboratory Test Results on the Protective Efficiency of the Trial Clothing

Field tests of the efficiency of the trial clothing, on exposure to contamination by spray particles, indicated that well-constructed garments, whatever material was used, were likely to offer initial protection factors of greater than 95% (see Table 5). Laboratory measurements of the resistance to permeation of the clothing materials (Table 2) suggested, however, that the porous and microporous specimens (from Garments A, B, and C) were unlikely to provide any significant degree of initial protection, at least against the more concentrated spray liquids. Measurements of resistance to penetration of the same materials, also in a laboratory (Table 3), indicated that only those materials of a nonabsorbent nature (for example, materials from Garments B, C, D, and E) were likely to offer adequate protection in practice.

The lack of precise correlation between the protective efficiencies of the trial clothing (arranged in descending order), based on field test data (Garments A/D, B, C, E), measurements of breakthrough times (Garments E, D, A/B/C), and penetration (Garments E, B/C/D, A), suggests a marked bias in the assessment of performance from one or possibly all of the methods of evaluation. For instance, in the laboratory tests the whole of a specimen of material for measurement of breakthrough time was maintained in continuous contact with the liquid chemical, and the instant of breakthrough was detected at the molecular level, whereas in the field study the clothing was usually exposed intermittently to discrete droplets rather than to a continuous film of liquid, and a far less sensitive analytical procedure was employed for measurement of penetration. The form of contact with the liquid chemical and the sensitivity of the method of evaluation were also influencing factors in the laboratory test of penetration (Table 3).

The laboratory tests of permeation and penetration of materials and whole garments tended to underestimate, to a marked degree, the efficiency of protective clothing on exposure to spray particles and therefore erred greatly on the side of safety. The use of a laboratory test alone for the selection of spray operators' clothing may lead, however, to rejection of items that would probably perform adequately in practice. More than one test should be applied, therefore, and account should be taken of the circumstances of use and the potential hazard in the selection of protective clothing.

Conclusion

Standard specifications for guidance in the selection of chemical protective clothing should recognize that supporting laboratory test methods are likely to underestimate the protective efficiency of a garment by a considerable margin. The selection procedures should therefore allow different grades of performance to be indicated and specified in the standard.

A fair method of assessment in the laboratory of the efficiency of spray operators' clothing against contamination by pesticide sprays in general would appear to be provided by the ISO procedure (the gutter test: ISO 6530—1980), with an additional modification to measure the volume of liquid penetrating the specimen of material. A controlled form of spray test, at present under development in the author's laboratory, in which the clothing is tested on a human subject and exposed to stepped increases of spray deposition with intermediate assessments of penetration, may prove to be more suitable as the conditions of the test are closer to those experienced in the field.

Measurement of the breakthrough time by a liquid chemical is considered to be more appropriate for the evaluation of protective gloves because they are the most important form of defense, for example, when opening pesticide containers and dispensing undiluted chemicals.

The increasing practice of spraying pesticide concentrates may justify application of the permeation test to materials used in the construction of an operator's protective garment, but, since the exposure would usually take the form of low levels of contamination by volume per unit area of clothing, but high levels in terms of the dissolved pesticide, again, the proposed spray test may be more appropriate. The permeation test would seem to be more convenient and appropriate for testing the efficiency of a seam.

References

[1] Lloyd, G. A., *Proceedings of the 1979 British Crop Protection Conference Pests and Diseases*, Vol. 3, 1979, pp. 821-831.

[2] Schwope, A. D., Cosras, P. P., Jackson, J. O., and Weitzman, J. D., *Guidelines for the Selection of Chemical Protective Clothing, Vol. 1, Field Guide*, Arthur D. Little, Cambridge, MA, 1983, pp. 1-79.

[3] Staiff, D. C. *Archives of Environmental Contamination and Toxicology*, Vol. 11, No. 4, 1982, pp. 391-398.

[4] *Research and Development Report, Pesticide Science 1983*, Reference Book of the Ministry of Agriculture, Fisheries and Food, No. 252/83, London, Her Majesty's Stationery Office, London, 1984.
[5] "Field Surveys of Exposure to Pesticides—Standard Protocol," Technical Monograph No. 7, CH-1211, World Health Organization, Geneva, 1982.
[6] Durham, W. F. and Wolfe, H. R., "Measurement of the Exposure of Workers to Pesticides," Bulletin 26, World Health Organization, Geneva, Switzerland, 1962, pp. 75-91.
[7] Lloyd, G. A., "The Resistance of Air-Impermeable Clothing Materials to Penetration by Liquid Chemicals," OPG Report PC/1, Harpenden Laboratory, Ministry of Agriculture, Fisheries and Food, Harpenden, U.K., 1976, pp. 1-45.
[8] Cuff, D. J., Taylor, L. G., and Thomas, N. T., "Pilot and Field Studies on Outdoor Protective Clothing," Health and Safety Executive, Bootle, Lancashire, U.K., 1983, Parts 1 and 2, pp. 1-50.

Joan M. Laughlin,[1] *Carol B. Easley,*[1] *Roger E. Gold,*[1] *and Robert M. Hill*[1]

Fabric Parameters and Pesticide Characteristics That Impact on Dermal Exposure of Applicators

---

REFERENCE: Laughlin, J. M., Easley, C. B., Gold, R. E., and Hill, R. M., **"Fabric Parameters and Pesticide Characteristics That Impact on Dermal Exposure of Applicators,"** *Performance of Protective Clothing, ASTM STP 900*, R. L. Barker and G. C. Coletta, Eds., American Society for Testing and Materials, Philadelphia, 1986, pp. 136–150.

ABSTRACT: Fabric functional finish and formulation of pesticides are factors that contribute to pesticide wicking, wetting, and penetration. Fluorocarbon soil-repellent finishes inhibit contamination of the fabric and of sentinel pads. An undergarment layer offers better protection than does a single layer. Spun-bonded olefin offers protection of the same magnitude as soil-repellent finishes. Methyl parathion residues after laundering were similar for the unfinished fabric, the durable-press finished fabric, and the soil-repellent finished fabric, but the initial contamination of the soil-repellent finished fabric was only 20% of that of the other two fabrics.

KEY WORDS: functional finish, soil-repellent finish, durable-press finish, pesticide soiling, wicking, wetting, pesticide penetration, pesticide residues, laundering, cotton/polyester blends, protective clothing

Chemical-resistant apparel is available to pesticide applicators; however, its use is often forfeited because of factors that include thermal properties, comfort, availability, cost, and lack of appreciation for the benefits. Common fabrics used for work clothing—cotton, polyester, and a blend of both—continue to be used for protective apparel. Pesticides spilled onto fabric may move (wick) through the fabric, wet underlying layers of fabric and skin, and be dermally absorbed. Investigations of fabric parameters (fiber content and functional finishes) and pesticide characteristics (concentration and formula-

[1]Professor and instructor, Department of Textiles, Clothing, and Design, professor, Department of Entomology, and associate professor, Department of Agricultural Biochemistry, respectively, University of Nebraska, Lincoln, NE 68583-0802.

tion) will aid in understanding how protective clothing can minimize dermal exposure.

Accidental spilling of concentrated or diluted pesticides onto washable protective garments by urban commercial applicators, home gardeners, and agriculture workers has emerged as a concern. Current research has focused on the problems of penetration of pesticides impelled by air-blast sprayers [*1,2*], particulate matter attraction [*2*], and the difficulty of pesticide residue (spills) removal by laundering [*3–12*].

Generally, moisture can pass through textile layers such as those found in clothing in three ways: (1) by water vapor diffusion through the large pores and channels of the fabric, (2) by swelling of the fiber on one side and movement of this swelling water through the fabric, followed by final desorption to the environment on the other surface, and (3) by liquid transport through the capillary interstices within the yarn or fabric or along the fiber surface [*13*]. Wetting means that a liquid-solid interface replaces an original solid-gaseous phase boundary [*14*]. The fiber interstices of the textile represent a capillary system that takes up the liquid in the same manner as a bundle of parallel capillaries. The yarns act as wicks to carry moisture through the fabric.

Orlando et al [*1*] state that the penetration of pesticide is influenced by capillary forces. Fourt and Harris [*15*] theorize that the rate at which water vapor passed through a complex system such as layered fabric was determined not only by the fabric but also by the air layers between fabrics which contributed a large fraction of the total resistance of the fabrics to penetration. Little work has been done with pesticide penetration, but Finley et al [*9*] studied a two-layer assembly worn by cotton scouts in methyl parathion-sprayed fields. They found more than 50% of the contamination of the first layer of fabric passed through to the second layer of fabric. Factors such as fiber content, yarn and fabric geometry, and functional textile finish determine the response of a textile to soiling such as from liquid pesticides. The metrology of wicking and wetting, penetration to and through layers that represent garments or body surfaces, would enhance understanding of work clothing as protection. The role of a protective system such as spun-bonded olefin fabric merits investigation as an inhibitor to soiling by pesticides.

Fluorocarbon polymers alter the surface properties of fabrics so that oil as well as moisture has less tendency to wet the fabric surfaces, and wicking is reduced. Liquid soil is partially inhibited from wetting, wicking, or penetrating the fabric. However, soil removal can be a problem unless hydrophilic groups are incorporated in the finish [*16*].

Freed et al [*17*] and Orlando et al [*1*] conclude that textiles treated with fluorocarbon finishes afford significantly better protection to pesticide sprays than non-fluorocarbon-finished fabrics. However, Berch and Peper [*18*] warn that hydrophobic fluorocarbon soil-repellent (SR) finishes promote redeposition of soil in laundering. Berch et al [*19*] conclude that fluorocarbon finishes have a strong retentiveness for soil corresponding to a large tendency to become soiled in an aqueous system and that, generally, fabric systems that are readily soiled from a water medium also have a strong tendency to retain soil during laundering. Although SR fluorocarbon finishes may be more resistant to pesticide soiling from an SR-finished fabric than an unfinished fabric, no work to date has assessed whether pesticide soil is more difficult to remove from SR than from an unfinished fabric.

Application of resin finishes affects the water vapor absorption properties of cotton by producing a more rigid internal fiber structure that becomes less accessible to liquid water [*20*]. Durable-press (DP) resin finishes on cotton/polyester blends cause a greater increase in soiling of cotton than of polyester, because they increase the hydrophobic nature of cotton while decreasing that of polyester [*21*]. In addition, most additives in a DP finish exert an adverse influence on soil release [*22*]. Resin finishes that form intrafiber crosslinks reduce water absorbency because of less availability of cellulose hydroxyl for interaction with water, which reduces hydrophilicity [*23*].

The objectives of this research include the following: (1) to determine the extent of pesticide transport through and between fabrics and to assess whether methyl parathion (MeP) movement in fabric is dependent on fiber content, functional textile finish, pesticide formulation, or concentration; (2) to determine the retention of pesticide (MeP) in fabrics in areas of secondary exposure (that is, pesticide contamination by wicking, wetting, and penetration to an underlayer) and the effectiveness of laundering in removal of pesticide residues using procedures previously developed by the investigators of this research project; and (3) to make recommendations as to methods that can be used to reduce human exposure to pesticides through proper selection, treatment, and laundering of clothing used during application of pesticides.

Procedures

Fabrics

Three finishes on 50% cotton/50% polyester bottom-weight poplin fabrics were obtained from Testfabrics. These are described in Table 1. The finishes included (*a*) no finish (UN), (*b*) DP finish, and (*c*) UN with fluorocarbon SR finish. The SR finish was a consumer application with Scotchguard fabric protector, a product manufactured by the 3M Co.

The fabrics were initially stripped of warp sizing and manufacturer-applied fabric softeners by washing according to the American Association of Textile Chemists and Colorists (AATCC) Test Method for Dimensional Changes in Automatic Home Laundering of Woven and Knit Fabrics (135-1978, Revised 1982). The outer 10% of the fabric was removed for preparation of test specimens as described in the ASTM Test for Breaking Load and Elongation of

TABLE 1—*Description of fabrics.*

Fabric	Designation[a]	Test Fabric Number	Fabric Count, yarns/10 cm	Weight, g/cm^2
50% Fortrel polyester/ 50% cotton poplin, bleached and mercerized	UN	7428	480 by 200	210
50% Fortrel polyester/ 50% cotton poplin, bleached with durable-press finish	DP	7428 WRL	480 by 200	210
50% Fortrel polyester/ 50% cotton poplin, bleached and mercerized with consumer-applied soil-repellent finish	SR	7428	480 by 200	210

[a]UN = unfinished; DP = durable press; SR = soil repellent.

Textile Fabrics [D 1682-64 (1975)] to ensure consistency of the warp yarns under evaluation.

The sentinel pads [*17*] were the Absorbent Cotton Company's Surpad (HRI-8035-90-8110) with dimensions of 28.4 by 25.3 cm (718.5 cm^2). The gauze pads were twelve ply folded to 7 by 7 cm (49 cm^2). These cotton pads were made from Type VII (7.9 yarns/cm by 4.7 yarns/cm) gauze. Spun-bonded olefin was obtained from the Textiles Fiber Department of E. I. du Pont de Nemours and Co. The Tyvek was Style No. 1422A.

Contamination of the Fabric

Three formulations of MeP were investigated: emulsifiable concentrate (EC), wettable powder (WP), and encapsulated materials (ENC). Pesticide dilutions were prepared at 1.25% active ingredient (AI), a usual field strength concentration. Solutions were held in suspension during the contamination process by placing them on a magnetic stirrer. Two tenths of a millilitre were pipetted onto the fabric surface using a MicroLab P programmable micropipette. The fabrics were placed on a raised needle bed surface to minimize contact points during contamination. The micropipette unit was held in a padded ring stand, allowing a constant distance of 5 cm between the pipette tip and fabric surface.

Moisture-Related Fabric Properties

The contributions of fiber content and functional textile finish were examined for their effect on moisture-related fabric properties. Investigations of

MeP wicking, wetting, penetration through consecutive layers of fabric, and laundering of contaminated fabric were executed to determine the rate at which the MeP formulation moved through fabric. Distilled water served as the control liquid for the wicking and wetting operation. Three MeP formulations applied to three functional textile finishes were the variables under study. All the work was replicated three times. All the fabrics were conditioned in accordance with the ASTM Practice for Conditioning Textiles for Testing (D 1776-79) at 21 ± 1°C (70 ± 2°F) and a relative humidity of 65 ± 2% for a minimum of 48 h prior to testing.

Pesticide Wetting—The AATCC Test Method for Absorbency of Bleached Woven Cloth (79-1979) was used to determine the drop absorbency. This liquid absorption test was used to estimate the capillary-type penetration properties of fabrics. Fabric specimens, cut into 8-cm squares, were placed horizontally on a needled surface. A 0.2-mL drop of liquid (distilled water or pesticide solution) was pipetted onto the fabric surface, and the time required for the mirrorlike reflection properties of the drop to disappear was measured.

Pesticide Wicking—The wicking test involved migration of liquid through interfiber and interyarn capillaries of the fabric and measured the ability of a fabric to transport liquid [*20*]. The time required for liquid (distilled water or pesticide solution) to wick a distance of 3 cm on the specimen was measured using a protocol similar to that of Weirick [*24*]. Two stopwatches were started as the micropipetting began. One stopwatch was halted when the 3-cm spread was achieved (wicking), and the other stopwatch was stopped when the mirrorlike reflection disappeared (wetting). Because the SR finish repelled liquid from being absorbed into the fabric, the liquid bead on the surface was allowed to remain for 10 min, after which time the specimen corner was held vertically for 10 s to allow the pesticide bead to roll into a beaker.

Pesticide Penetration Through Fabric Layers—The amount of pesticide in solution that moved between layered fabrics was used to quantify pesticide penetration. Hypothetically, a liquid pesticide should move more readily through fabrics of high synthetic fiber content than through similar fabrics of more absorbent natural fibers.

Laundering of Contaminated Fabrics—Specimens of finished fabrics, 8 by 8 cm, were contaminated with EC and ENC MeP of 1.25% AI. Contamination procedures were as described previously.

The laundry variables included two water temperatures [60°C (140°F) wash and 49°C (120°F) rinse or a 49°C (120°F) wash and rinse] and two detergent formulations [heavy-duty liquid nonionic detergent (HDL) or AATCC Standard Detergent 124]. Both detergents had been shown to be effective in MeP removal at both water temperatures [*5*].

Extraction

The specimens were individually extracted in 100 mL of reagent-grade acetone on a mechanical shaker for 1.5 h at 120 rpm. The extract was decanted and replaced by an additional 100 mL of acetone for a second shaking. At the end of the 3-h shaking time, the fabric specimen was removed, and the two extracts were combined.

Gas Chromatographic Analysis

The extracts were either concentrated with nitrogen (N_2) stream evaporation or diluted with acetone to facilitate gas chromatographic analysis. A 1.0-mL aliquot of the adjusted extract was mixed with 9 mL of toluene and a known quantity of ethyl parathion (internal standard). One-microlitre aliquots were analyzed with a Varian 3700 gas chromatograph with CDS 111C data system using an electron capture detector. Separation was achieved on a 2 m by 2-mm glass column packed with 1.5% OV-17 and 1.95% OV-210 on 80/100 mesh Chromasorb WHP with a nitrogen flow of 30 mL/min. The injection, detector, and oven temperatures, respectively, were 220°C (428°F), 270°C (518°F), and 190°C (374°F).

Statistical Analysis

Statistical differences between the control and the layered specimen or laundered specimens were calculated using an analysis of variance (ANOVA) with an indication of significance at the $P \leq 0.05$ level. Means were then separated with a Duncan's multiple-range test [25].

Results and Discussion

Pesticide Wetting

The SR finish proved to be superior to the UN and DP finishes in minimizing wetting of the pesticide solution into the fabric. This occurrence is attributable to the SR finish converting the fabric to a hydrophobic system. Even after an allotted 10-min period, the solution remained in a bead formation atop the SR specimen surface; hence, the mirrorlike reflection was never altered because of the low-absorbency characteristics of the fabric. Because of these results, the SR finish data were omitted from statistical analyses. The DP and UN specimens were statistically different across all formulations ($F = 7.78$; df = 1,14; $P = 0.0139$), with the DP being more resistant to wetting (Table 2). This supports the findings of Chandler and Zeronian [20]. The unfinished fabrics had no modifications to affect absorption.

Among the three pesticide formulations included in this study, statistical differences were found only between the EC and ENC ($P = 0.0227$). The EC formulation most readily wetted the specimen because of the presence of higher levels of a surfactant or carrier solvent ingredient. The surface-reactive agent aided in breaking down surface tension of both the fabric and the liquid pesticide solution, thereby increasing their penetrability. WP formulations also contain a surfactant, but because of the presence of inert ingredients (such as clay, talc), pesticide wetting does not occur as rapidly. In the ENC formulation the pesticide is encapsulated in microscopic polymer beads; this bead structure inhibits immediate release of the pesticide into the fabric.

Pesticide Wicking

The SR finish inhibited pesticide absorption, penetration, and spreading so that the SR data were obviously superior to the UN and DP; hence, the data were omitted from statistical analysis (Table 2). Although no statistical difference was found between the UN and DP finishes, there was a trend for faster wicking on the DP fabric. Since DP finishes increase hydrophobicity, the mechanism for moisture transport of MeP is likely to be along and between fibers and yarns rather then by absorption into the internal structure of individual fibers. The measured wicking time for the ENC formulation was significantly slower than that for the EC ($P = 0.0017$) and WP ($P = 0.0477$) on both the UN and DP fabrics. This could also be attributable to (1) the microencapsulated composition of the formulation or (2) the lower percentage of surfactant in the ENC formulation.

TABLE 2—*Summary of wetting and wicking experiment results.*

Formulation and Fabric[a]	Wetting Time, s	Wicking Time, s
EC		
UN	5.5 ± 0.7	2.5 ± 0.1
DP	8.1 ± 1.0	2.2 ± 0.5
SR	+600.0	+600.0
WP		
UN	7.5 ± 0.4	2.9 ± 0.4
DP	9.4 ± 1.4	2.5 ± 0.3
SR	+600.0	+600.0
ENC		
UN	8.7 ± 0.6	3.1 ± 0.0
DP	12.9 ± 2.7	3.4 ± 0.1
SR	+600.0	+600.0

[a]EC = emulsifiable concentration; WP = wettable powder; ENC = encapsulated; UN = unfinished; DP = durable press; SR = soil repellent.

Pesticide Penetration

The penetration experiments examined whether differences attributable to fabric finish or pesticide formulation were observable in the amount of MeP that moved between layered fabrics. These experiments involved four phases: (1) the amount of pesticide taken up by the outer garment fabric, (2) penetration of pesticide through the outer garment fabric onto a dermal surface represented by a cellulose sentinel pad (fabric/pad), (3) penetration of pesticide through the outer garment fabric to an undergarment fabric and onto a sentinel pad surface (fabric/fabric/pad), and (4) pesticide penetration through a spun-bonded olefin to an outer garment fabric and onto a sentinel pad surface (spun-bonded olefin/fabric/pad).

Garment Fabric—To determine whether fabrics differentially sorbed pesticide, it was necessary to examine the fabric type and pesticide formulation. The sorption of MeP by the outer garment fabric then became the baseline for comparisons of pesticide retention or movement to other surfaces in the subsequent phases of the experiment. Although pesticide sorption ranged from 9.3 to 51.9 g/cm^2 (Table 3), the SR was consistently lower than the UN and DP, regardless of MeP formulation. However, the EC formulation was significantly higher than the WP or ENC formulation ($F = 5.381$; df = 2,32; $P \leq 0.05$) probably because of the surfactants in the EC and the particulate nature of the other two formulations. There were significant differences due to pesticide formulation, with less MeP sorbed when in an ENC formulation ($F = 6.586$; df = 2,130; $P \leq 0.05$). This observation was consistent across all the finishes.

TABLE 3—*Outer garment fabric sorption of methyl parathion.* [a]

Fabric Finish[b]	Formulation and Sorption, ng/cm^2 [c]		
	EC	WP	ENC
UN	46.77$^{+/1}$	51.90$^{+/1}$	34.30$^{+/2}$
DP	43.88$^{+/1}$	48.42$^{+/1}$	34.14$^{+/2}$
SR	23.52$^{0/1}$	16.34$^{0/2}$	9.30$^{0/2}$

[a]Means followed by same symbol within columns are not significantly different at $P = 0.05$, and means with the same number within rows are not significantly different at $P = 0.05$ (Duncan's multiple range test).

[b]UN = unfinished; DP = durable press; SR = soil repellent.

[c]EC = emulsifiable concentrate; WP = wettable powder; ENC = encapsulated.

Garment Fabric/Sentinel Pad—The outer garment fabric was placed over a cotton gauze sentinal pad which modeled a dermal surface [*17*]. Because the sentinel pad was a hydrophilic substrate, it was necessary to determine if this surface acted in a spongelike manner, drawing more chemical into the fabric system than normally would be found in the outer garment fabric. When the MeP in the outer garment fabric and sentinel pad were totaled (Tables 3 and 4) and compared with the outer garment fabric alone, no significant differences were found across fabric finishes or pesticide formulations ($F = 0.289$; df $= 1,16$). Therefore, the presence of the sentinel pad did not significantly alter the amount of MeP in the total fabric system. The amount of MeP found in the sentinel pads was small (0 to 3.54% of the pesticide in the total system), and there were no significant differences across fabric, finish, or formulation. Under these test conditions, the MeP moved through the outer garment fabric and contaminated the sentinel pad, emphasizing the need for additional dermal protection when the pesticide is a highly toxic or concentrated mixture, or both. Pesticide retained by the outer garment fabric/sentinel pad was less than the pesticide retained by the garment fabric alone except for the ENC formulation (Tables 3 and 4). A significant difference between the outer garment fabric alone and the fabric of the fabric/pad system was found only for the unfinished fabric when EC MeP had been used to contaminate the system ($F = 5.29$; df $= 1,16$; $P = 0.0034$). This was

TABLE 4—*Methyl parathion penetration through the outer garment fabric/sentinel pad system.*

	Formulation and Penetration, ng/cm² [b]		
Treatment [a]	EC	WP	ENC
UN			
Fabric	39.44	45.78	34.52
Pad	0.17	0.02	0.07
Total	39.61	45.80	34.59
% of total in pad	0.43	0.04	0.20
DP			
Fabric	40.01	37.17	38.51
Pad	1.42	0.05	0.14
Total	41.43	37.22	38.65
% of total in pad	3.54	0.13	0.36
SR			
Fabric	20.45	14.50	10.15
Pad	0.01	0.00	0.03
Total	20.46	14.50	10.18
% of total in pad	0.04	0.00	0.32

[a] UN = unfinished; DP = durable press; SR = soil repellent.
[b] EC = emulsifiable concentrate; WP = wettable powder; ENC = encapsulated.

because of the difficulty of extracting MeP from the two-fabric system, and it confirms the findings of Easley et al [3]. As was true for the outer garment fabric alone, the pesticide found in the outer garment fabric/sentinel pad system for SR fabrics was about half that recovered for the other two fabrics (UN, DP) ($F = 45.22$; df $= 2{,}130$; $P = 0.0001$). Thus, the SR fabric provided protection by limiting sorption of MeP into the fabric system.

Outer Garment Fabric/Undergarment Fabric/Dermal Pad—Among the three finishes under study, the SR-finished fabric afforded the greatest level of protection. In the multilayer trials, the pesticide found in the outer garment fabric layer with the SR finish was approximately 50% of the amount retained by the other two finishes. These findings are congruent with the residues of MeP recovered from the outer layer fabrics discussed earlier.

An additional layer of fabric that absorbs a liquid spill may assist in limiting a pesticide solution from contaminating a dermal surface. That is, the more moisture a fabric will sorb, the less liquid is available for dermal contamination; therefore, additional layers of fabric that take up pesticide may limit the contamination of a dermal surface.

The amount of pesticide that contaminated the second layer was small; however, there was evidence of greater contamination with the DP as the outer garment fabric. The finish had imparted a hydrophobic nature to the DP garment fabric so that the moisture take-up was reduced. The pesticide moved quickly through the DP outer garment fabric, resulting in greater contamination of the undergarment fabric. This was supported by the wicking/wetting data presented previously.

Similar to the findings for the fabric/pad, extremely small percentages of pesticide were found in the sentinel pad of the fabric/fabric/pad system (Table 5). Larger percentages of pesticide were found in the pad when the undergarment layer was a fishnet fabrication and the outer garment layer was a DP. Given the open construction of the fishnet fabric and the limited sorbability of a DP finish, pesticide movement could have been from outer garment layer to pad without involvement of the undergarment layer. Another possibility is that the undergarment layer retained less liquid because of the limited number of interstitial spaces causing the pesticide to move through the undergarment layer to the pad. It would be important to consider this type of undergarment fabric, because this is the surface next to the skin and contamination in this layer could be available for dermal sorption. Noteworthy is the small percentage of contamination of the garment fabric found in the undergarment layer regardless of the type of undergarment layer. These findings indicate the importance of wearing a second fabric layer.

The composition of that second layer is important. Generally, the sweatshirt or fleeced fabric as an undergarment retained more pesticide than did other undergarment fabrics evaluated in this study (Table 5). For the sweatshirt, the pesticide found in the undergarment layer ranged from 0.00 to 2.93% of the MeP in the outer garment layer. It appears that the sweatshirt

TABLE 5—*Methyl parathion adsorption through outer garment fabric/undergarment fabric/dermal, in nanograms per centimetre.*[a]

Treatments	Tee-Shirt			Fishnet			Sweatshirt		
	EC	WP	ENC	EC	WP	ENC	EC	WP	ENC
Treatment I									
Outer garment fabric (UN)	43.54	30.32	31.91	36.59	44.09	34.14	39.04	41.58	38.27
Undergarment fabric	0.04	0.01	0.04	0.40	0.01	0.04	0.99	0.02	0.08
Pad	0.09	0.00	0.03	0.03	0.01	0.04	0.03	0.00	0.04
Total	43.67	39.33	31.26	37.02	44.11	34.21	40.06	41.60	38.39
Pad % of total	0.21	0.00	0.10	0.08	0.02	0.12	0.07	0.00	0.10
Treatment II									
Outer garment fabric (DP)	43.97	37.91	36.97	38.53	43.88	33.16	34.81	58.66	36.31
Undergarment fabric	0.03	0.03	0.03	0.08	0.01	0.03	0.34	0.02	0.17
Pad	0.04	0.00	0.04	0.07	0.00	0.08	0.01	0.00	0.04
Total	44.04	37.94	37.04	38.68	43.89	33.27	35.26	58.68	36.52
Pad % of total	0.09	0.00	0.11	0.18	0.00	0.25	0.03	0.00	0.11
Treatment III									
Outer garment fabric (SR)	19.79	10.86	5.71	22.86	24.18	10.53	19.56	17.94	13.65
Undergarment fabric	0.00	0.01	0.03	0.29	0.00	0.38	0.57	0.00	0.05
Pad	0.00	0.04	0.03	0.00	0.00	0.04	0.00	0.00	0.04
Total	19.79	10.91	5.77	23.15	24.18	10.95	20.13	17.94	13.74
Pad % of total	0.00	0.36	0.52	0.00	0.00	0.32	0.00	0.00	0.03

[a]EC = emulsifiable concentrate; WP = wettable powder; ENC = encapsulated.

fabric enhanced movement of pesticide from the outer garment layer to the undergarment fabric. Contributing to the take-up of pesticide solution into the sweatshirt fabric was the acrylic/cotton fiber content, as both fibers are known to have good wicking tendencies and sorbency. Contamination of the tee-shirt layer was 0.00 to 0.52% that of the outer garment layer. For the fishnet, the contamination of the undergarment layer was 0.00 to 3.61% that of the outer garment layer. Because of the open network construction of the fishnet, the level of contamination varied greatly, contingent on whether the liquid soil contacted fabric or a space in the fabric.

The MeP in the total multilayer system was not significantly different from the amount of pesticide on the outer garment fabric alone. The pesticide in the outer garment fabric of the multilayer system was significantly different from that in the outer garment fabric alone in every instance except when the formulation was EC for UN over fishnet ($F = 8.39$; df = 1,16; $P = 0.102$) or sweatshirt ($F = 6.26$; df = 1,16; $P = 0.0221$) and DP over sweatshirt ($F = 16.11$; df = 1,15; $P = 0.0014$). Based on these data, tee-shirt fabric is recommended for the second layer.

Spun-bonded Olefin/Outer Garment Fabric/Sentinel Pad—Another mechanism for protecting the applicator is the use of disposable coveralls to be worn over work clothing when one is working with full-strength or highly toxic pesticides. The amount of pesticide moving to the garment fabric was greatly reduced (Table 6) when spun-bonded olefin was used as a protective system over outer garment fabrics. No significant differences were found in pesticide in the outer garment fabric, or second layer, because of the highly protective functions of the spun-bonded olefin, with its marked limiting of moisture penetration.

It is noteworthy that the amount of pesticide in the spun-bonded olefin/outer garment fabric/sentinel pad system was significantly less ($F = 19.97$; df $= 1{,}53$; $P = 0.0001$) than the amount of pesticide in the outer garment fabric/sentinel pad system in all instances except for the SR fabric and the ENC formulation. The amount of protection provided by the SR finish for a fabric was similar to the protection provided by the spun-bonded olefin protective system ($F = 0.799$; df $= 1{,}14$). The ENC formulation provided for limited sorption of pesticides in spun-bonded olefin/outer garment fabric/sentinel pad systems and outer garment fabric/sentinel pad systems.

Effectiveness of Laundering

Although many studies have examined factors affecting the completeness of pesticide residue removal in laundering, no work to date has examined the impact of functional textile finishes on the effectiveness of launderings. Contamination before laundering was significantly lower for the SR-finished fab-

TABLE 6—*Spun-bonded olefin/outer garment fabric/sentinel pad sorption (nanograms per centimetre of 1.25% methyl parathion).*

Treatments[a]	EC	WP	ENC
Treatment I			
spun bonded	28.73	12.64	21.64
garment fabric (UN)	0.03	0.08	0.27
pad	0.03	0.01	0.07
total	28.79	12.73	21.98
% of total in pad	0.10	0.08	0.32
Treatment II			
spun bonded	18.78	3.74	21.56
garment fabric (DP)	0.03	0.00	0.05
pad	0.03	0.00	0.01
total	18.84	3.47	21.62
% of total in pad	0.16	0.00	0.05
Treatment III			
spun bonded	18.78	13.71	20.86
garment fabric (SR)	0.00	0.00	0.04
pad	0.01	0.03	0.03
total	18.79	13.74	20.92
% of total in pad	0.05	0.22	0.14

[a]UN = unfinished; DP = durable press; SR = soil repellent.

ric than for the UN fabric or DP-finished fabric (Table 7). Contamination of UN and DP specimens was five times that of the SR specimen. The SR finish effectively limited pesticide retention.

No significant differences were found between detergent type or washing temperatures, although the removal was slightly greater at the higher temperature. Residues after laundering were markedly lower when the ENC formulation had been used to contaminate the fabric than when EC was the formulation. These findings are congruent with those of previous studies [*4-6*].

There were no significant differences in MeP residues after laundering attributable to the fabric finish of the specimen. Residues after laundering were 5.350, 5.787, and 5.163 ng/cm^2 for UN, DP, and SR, respectively. It is important to note that initial contamination of the SR specimen was 20% that of the UN and DP specimens. This confirms the findings of Bevan [*16*], Berch and Peper [*18*], and Berch et al [*19*].

It is apparent that pesticide soil removal is a greater problem for the SR-finished fabrics, although the residue level is in the same range as in the UN and DP fabrics. A certain irreducible amount of residue may remain after laundering, in such locations as the lumen of cotton. Further work is needed to confirm whether SR fluorocarbon finishes have a retentiveness for pesticides and to ascertain ways to optimize residue removal in laundering.

TABLE 7—*Methyl parathion residues in specimens after laundering.*

Treatment[a]	MeP, ng/cm^2	df	*F*	*P*
Initial contamination				
UN	82.737	2,18	4.875	0.0199
DP	80.411			
SR	16.598			
UN by DP		1,12	0.140	0.9169
UN by SR		1,12	8.931	0.0110
DP by SR		1,12	10.300	0.0075
Laundering temperature				
60°C	5.123	1,83	0.091	0.7604
49°C	5.743	1,83		
Detergent				
HDL	5.942	1,83	0.248	0.6256
Phosphate	4.924			
Formulation				
EC	12.363	1,83	59.129	0.0001
ENC	0.246			
Finish				
UN	5.350	2,82	0.032	0.9687
DP	5.787			
SR	5.163			

[a]UN = unfinished; DP = durable press; SR = soil repellent.

Conclusion

Encapsulated and emulsifiable concentrate formulations of methyl parathion move across and through fabrics, contingent on the functional finish applied to the fabric. Fluorocarbon SR finishes inhibit the movement of pesticide liquids through fabric, and contamination is approximately 18 to 20% that of unfinished or DP-finished fabrics. Based on these findings, recommendation is made for a soil-repellent finish for fabrics worn during mixing, handling, or application of pesticides.

Pesticide solutions move through fabrics to contaminate sentinel pads. DP finish is less desirable since MeP moves more readily through the fabric, thus allowing more pesticide to penetrate to the underlayer and to the sentinel pad. SR finish is highly desirable; spills move through the fabric more slowly, and less penetrates to undergarments or sentinel pads.

An undergarment layer offers better protection than does a single layer of clothing. The presence of a second layer does not contribute to movement to the sentinel pad. The contamination of the second layer is generally less than 1% that of the contamination of the outer garment layer; thus, the pesticide is not available for dermal absorption. A tee-shirt undergarment is recommended over other fabrics studied.

Spun-bonded olefin offers protection in the same magnitude as the SR finish. Based on these findings, recommendations are made for use of the disposable spun-bonded olefin garments or for SR finish applied to non-durable-press work clothing. Theoretically, the greatest protection may be realized in use of the disposable olefin garments worn during mixing, handling, and application of pesticides, in addition to SR finish on the usual work clothing.

Laundering variables of temperature and detergent type were not important in differences in pesticide residue after laundering. Differences in MeP residues after laundering were due to formulation of pesticide, with encapsulated formulations being more completely removed in laundering. Residues after laundering were similar across the unfinished fabric, the durable-press-finished fabric, and the soil-repellent-finished fabric. It is important to note that the initial contamination of the SR fabric had been only 20% that of the other two fabrics. The fluorocarbon polymer had rendered the fabric more hydrophobic, thus more soil resistant, but the finish promoted soil redeposition in laundering, and residue removal was a smaller percentage of contamination. Additional work is needed to assess the difficulty in dislodging pesticide residues from SR-finished fabrics through exploration of laundering factors that optimize pesticide soil removal.

Acknowledgments

Appreciation is extended to the Pesticide Impact Assessment Program for funding of the grant "Fabric Parameters and Pesticide Characteristics That

Impact on Dermal Exposure of Applicators." This research was supported in part by the Nebraska Agricultural Research Division Project 94-012, and it contributes to North Central Regional Research Project NC-170, "Limiting Dermal Exposure to Pesticides Through Effective Cleaning Procedures."

This paper is registered as Paper No. 7532, Journal Series, Nebraska Agricultural Experiment Station.

References

[1] Orlando, J., Branson, D., Ayers, G., and Leavitt, R., *Journal of Environmental Sciences and Health*, Vol. B16, No. 5, 1981, pp. 617-628.

[2] Serat, W. F. and Van Loon, A. J., "Some Factors Influencing the Design of Fabrics for Protective Garments for Pesticide Field Workers," Unpublished Report: Community Studies on Pesticide, State Department of Health, Berkeley, CA, 1978.

[3] Easley, C. B., Laughlin, J. M., Gold, R. E., and Tupy, D., *Bulletin of Environmental Contamination and Toxicology*, Vol. 27, No. 2, July 1981, pp. 101-244.

[4] Easley, C. B., Laughlin, J. M., Gold, R. E., and Schmidt, K., *Bulletin of Environmental Contamination and Toxicology*, Vol. 28, No. 2, Feb. 1982, pp. 239-244.

[5] Easley, C. B., Laughlin, J. M., Gold, R. E., and Hill, R. M., *Bulletin of Environmental Contamination and Toxicology*, Vol. 29, No. 4, Oct. 1982, pp. 461-468.

[6] Easley, C. B., Laughlin, J. M., Gold, R. E., and Hill, R. M., *Archives of Environmental Toxicology and Contamination*, Vol. 12, No. 1, Jan. 1983, pp. 71-76.

[7] Easter, E. *Textile Chemist and Colorist*, Vol. 47, No. 3, March 1983, pp. 29-33.

[8] Finley, E. L., Metcalfe, G. I., McDermott, F. G., Graves, J. B., Schilling, P. E., and Baker, F. B., *Louisiana Agriculture*, Vol. 9, No. 1, 1974, pp. 4-5.

[9] Finley, E. L., Bellon, J. M., Graves, J. B., and Koonce, K. L., *Louisiana Agriculture*, Vol. 20, No. 1, Spring 1977, pp. 8-9.

[10] Finley, E. L., Graves, J. B., and Hewitt, F. W., *Bulletin of Environmental Contamination and Toxicology*, Vol. 22, No. 4/5, July 1979, pp. 598-602.

[11] Laughlin, J. M., Easley, C. B., Gold, R. E., and Tupy, D., *Bulletin of Environmental Contamination and Toxicology*, Vol. 27, No. 4, pp. 518-523.

[12] Metcalfe, G. I., "The Absorption and Retention of Selected Chlorinated Hydrocarbons and Organic Phosphate Residues in Cotton and Cotton-Polyester Fabrics," Master's thesis, Louisiana State University, 1972.

[13] Mecheels, J. J., Demeler, R. M., and Kachel, E., *Textile Research Journal*, Vol. 36, 1966, pp. 375-384.

[14] Langmann, W., *Leverkusen Farbenfabriken Bayer*, Vol. 19, Nos. 13-19, 1970, pp. 69-71.

[15] Fourt, L. and Harris, M., *Textile Research Journal*, Vol. 17, 1947, pp. 256-263.

[16] Bevan, G., *Textiles*, Vol. 8, No. 3, pp. 69-71.

[17] Freed, V. R., Davies, J. E., Peters, L. J., and Parreen, F., *Residue Review*, Vol. 75, 1980, pp. 159-160.

[18] Berch, J. and Peper, H., *Textile Research Journal*, Vol. 33, 1963, pp. 137-145.

[19] Berch, J., Peper, H., and Drake, G. L., *Textile Research Journal*, Vol. 34, 1964, pp. 29-34.

[20] Chandler, J. and Zeronian, S. H., *Textile Chemist and Colorist*, Vol. 11, 1979, pp. 20-25.

[21] Bowers, C. A., and Chantrey, G., *Textile Research Journal*, Vol. 39, 1969, pp. 1-11.

[22] Das, T. K. and Kulshreshta, A. K., *Journal of Scientific and Industrial Research*, Vol. 38, 1979, pp. 611-619.

[23] Hebeish, A., Wally, A., Abou-Zeid, N. Y., and El-Alfy, E., *American Dyestuff Reporter*, Vol. 72, No. 7, 1983, pp. 15-21.

[24] Weirick, J. H. S., *Textile Testing*, Chemical Publishing, Brooklyn, NY, 1949.

[25] Duncan, D. B., *Virginia Journal of Science*, Vol. 2, 1951, pp. 171-189.

Nancy E. Hobbs,[1] *Billie G. Oakland,*[1] *and*
Melvin D. Hurwitz[1]

Effects of Barrier Finishes on Aerosol Spray Penetration and Comfort of Woven and Disposable Nonwoven Fabrics for Protective Clothing

REFERENCE: Hobbs, N. E., Oakland, B. G., and Hurwitz, M. D., **"Effects of Barrier Finishes on Aerosol Spray Penetration and Comfort of Woven and Disposable Nonwoven Fabrics for Protective Clothing,"** *Performance of Protective Clothing, ASTM STP 900,* R. L. Barker and G. C. Coletta, Eds., American Society for Testing and Materials, Philadelphia, 1986, pp. 151-161.

ABSTRACT: Woven and nonwoven fabrics with and without barrier finishes were tested for resistance to aerosol spray penetration and physical factors which contribute to comfort desirable for protective clothing for pesticide applicators. An aerosol spray test procedure was developed to test the following spray carriers: (1) water, (2) water/surfactant (48 : 1), and (3) unrefined cottonseed oil/surfactant (4 : 1). All the carriers contained 0.1% methylene blue dye as an indicator. Physical properties (for example, density, weight, thickness, water vapor, and air permeability) were examined as indicators of fabric comfort.

The presence of fluorocarbon-based finishes increased the resistance to aerosol spray penetration. Based on the physical test results and the aerosol penetration test, the spunlace nonwoven fabrics ranked highest for disposable protective garments.

KEY WORDS: aerosol spray, comfort, pesticides, protective garments, fluorocarbon finishes, disposable garments, nonwoven fabrics, surfactant, air permeability, water vapor permeability, protective clothing

Protection for farm workers against pesticides during mixing, loading, and application is recognized as a high-priority need. In 1983, the state of Califor-

[1]Assistant professor, chairman, and professor, respectively, Department of Clothing and Textiles, School of Home Economics, University of North Carolina at Greensboro, Greensboro, NC 27412.

nia's Department of Food and Agriculture reported 128 cases of skin-related illnesses and 220 cases of systemic illnesses among pesticide applicators and mixer/loaders. Each case was judged to provide adequate information on which to base an exposure/illness relationship [*1*]. The North Carolina Department of Human Resources, Environmental Epidemiology Branch, reported 39 cases of agricultural exposure-related illnesses in 1982 [*2*].

The purpose of this study was to test woven and nonwoven fabrics as barriers to aerosol spray penetration by methylene blue dye solutions. This study focused on nonwoven substrates which offered comfort and protection at a low enough cost to be potentially disposable after one wearing. Comfort, a relative term in summer field conditions in the South, may better be defined as less discomfort. Of particular interest were the nonwoven substrates developed for medical use, which are designed to resist penetration of lipids and aqueous-based formulations but allow transmission of air and water vapor. These properties are important factors in the body's evaporative cooling process. Woven fabrics similar to those commonly found in apparel worn by farm workers were also examined as controls.

The U.S. Department of Agriculture (USDA), U.S. Environmental Protection Agency (EPA), and National Institute for Occupational Safety and Health (NIOSH) recommend the wearing of protective apparel and equipment for all people involved in the handling and application of most pesticides [*3-5*]. The garments recommended by the EPA for general use, such as long-sleeved shirts and long-legged trousers or coverall-type garments, do not provide adequate protection when liquid or aerosol application methods are used. They become wet and wick the pesticide formulation, resulting in dermal exposure [*6*]. Liquid-proof garments, recommended by the EPA [*3*] for handling toxic or concentrated pesticides, are generally impermeable to dust, oil, water, and water vapor. A major impediment in the use of this type of protective apparel is a lack of comfort in wearing due to its low air and water vapor permeability and, at times, excessive weight [*7-10*].

Although laundering procedures have been developed for contaminant removal, protective clothing decontamination and reuse are controversial and unresolved issues in terms of safety and relative cost [*11-14*]. Finley [*14*] reported that the laundering of contaminated garments through three complete cycles did not remove all residues. Southwick et al [*13*] reported that laundering in either ionic or cationic detergents removed less than 50% of the parathion present in the fabric. The addition of a hypochlorite bleach, they noted, increased the removal to 81%. The USDA recommends disposal of pesticide-contaminated garments rather than laundering in cases in which garments have been wetted with highly toxic or concentrated pesticide formulations. On the assumption that lightweight nonwoven fabrics would offer greater comfort than plastic films, heavy wovens, or rubberized fabrics as garment substrates, a range of lightweight woven and nonwoven fabrics were examined for resistance to aerosol spray penetration and wicking. The nonwoven fabrics were specifically chosen because they are disposable.

Materials

The following nonwoven fabrics were selected for the study:

(*a*) spun-lace 100% polyester (90 g/m^2), unfinished and laboratory finished with Scotchgard,[2]
(*b*) spun-lace rayon/polyester (70 g/m^2), commercially finished for surgical gown use,
(*c*) spun-bonded 100% olefin (40 g/m^2), unfinished and laboratory finished with Scotchgard,
(*d*) spun-bonded polypropylene with melt-blown (MB) fibers (50 g/m^2), unfinished and finished with Scotchgard,
(*e*) spun-bonded polypropylene with melt-blown fibers (50 g/m^2), commercially finished for industrial use, and
(*f*) spun-bonded polypropylene with melt-blown fibers (90 g/m^2), unfinished and laboratory finished with Scotchgard.

The woven fabrics selected for the study were the following:

(*a*) twill weave, denim, 100% cotton (283 g/m^2), unfinished;
(*b*) twill weave, drill, 100% cotton (221 g/m^2), unfinished;
(*c*) shirting-weight muslin, 100% cotton, commercially finished with Scotchgard (157 g/m^2);
(*d*) shirting-weight muslin, 100% cotton, with a Quarpel (fluorocarbon) finish (113 g/m^2); and
(*e*) shirting-weight muslin, 35% cotton/65% polyester, with a Quarpel (fluorocarbon) finish (117 g/m^2).

The twill weaves were comparable to bottom-weight fabrics. The shirting-weight muslins were top-weight muslins similar to work shirt fabrics. The Quarpel-finished shirting-weight muslins were manufactured and finished by the USDA Southern Regional Research Laboratory in New Orleans, Louisiana.

The nonwoven and woven fabrics were subjected to the following aerosol sprays: (1) water, (2) water/surfactant (48:1), and (3) unrefined cottonseed oil/surfactant (4:1). All the carriers contained 0.1% methylene blue dye as an indicator. The surfactant was obtained from a pesticide manufacturer and is used in pesticide formulations (the formulation is proprietary). Filter paper (18.5 cm) was used as the absorbent backing substrate.

[2]The use of trade names does not imply endorsement of the products named or criticism of similar ones not mentioned.

Procedure

Physical Data

The following physical data for each test fabric were collected: weight, thickness, density, air permeability, and water vapor permeability. The fabrics were conditioned in accordance with ASTM standards (21 ± 1°C and 65 ± 2% relative humidity) for 24 h prior to testing. The ASTM standard testing procedures were followed to determine the weight [D 1910-64 (1978)] and thickness (ASTM Method for Measuring Thickness of Textile Materials) [D 1777-64 (1975)] measurements. Density was computed by dividing the specimen mass by the specimen volume to obtain the weight in kilograms per cubic metre. The ASTM Test for Water Vapor Transmission of Materials (E 96-80), Procedure B, was used to test the transmission of the water vapor. The ASTM Test for Air Permeability of Textile Fabrics [D 737-75 (1980)] was used to measure air permeability. Three replications were completed for each fabric tested.

Aerosol Spray Penetration

The design of the aerosol spray test was based on the ASTM Method of Salt Spray (Fog) Testing [B 117-73 (1979)], developed primarily for testing metallic and metal-coated specimens. The spray test procedure was designed to assess the resistance of fabrics to penetration of aerosol sprays. The testing procedure consisted of mounting a fabric specimen (18.5 cm by 18.5 cm) in a frame over an absorbent backing (18.5 cm in diameter filter paper) and placing the framed specimen upright in a containment box 30 cm in front of an airless spraying device. An area 12.5 cm in diameter was exposed to the aerosol spray. Three replications were completed for each fabric tested with each aerosol spray formulation. The spray time for the water-base solutions was 60 s, the spray time for the oil-base solution was 30 s. The increased concentration of surfactant in the oil-base formulation resulted in a superior wetting system, which reduced the spray time for specimen differentiation to 30 s. The average amount of spray applied to the surface of the test fabric per square centimetre was 0.25 g/cm^2 for the water and the oil-base solutions. Scotchgard aerosol spray, manufactured by the 3M Corp., was applied to the surface of unfinished nonwoven fabrics. The Scotchgard spray was applied, following label directions, to wet the fabric surface thoroughly and was allowed to dry for 24 h prior to testing.

The purpose of the research was to identify woven and nonwoven fabrics that resisted penetration of aerosol sprays. The results of the spray tests were recorded as either pass or fail, with no relative ranking of the degree of dye penetration. The presence of any methylene blue dye on the backing fabric within the test area was categorized as "fail." This procedure was selected because the objective of the research was to identify fabrics that offer "complete" protection; therefore, the gradation of aerosol dye spray test results were considered of little value.

Results and Discussion

The physical data for the test fabrics, including weight, thickness, density, air permeability, and water vapor permeability are in Table 1. The test fabric weights ranged from 283 g/m^2 for denim to 40 g/m^2 for spun-bonded olefin. The woven test fabric weights ranged from 283 to 113 g/m^2. Nonwoven fabrics weighed significantly ($P = 0.05$) less than the woven fabrics. The spun-bonded olefin was the thinnest test fabric, measuring 0.01 cm. The denim, at 0.057 cm, was the thickest. Thus, the thicknesses of the spun-bonded olefin fabrics were significantly lower than those of the other fabrics tested.

Fabric densities ranged from 162 kg/m^3 (spun-bonded/MB, 50 g/m^2) to 541 kg/m^3 (Scotchgard-finished 100% cotton shirting). The spun-bonded/MB fabrics were the least dense (162 to 179 kg/m^3). The density of spun-bonded olefin and the spun-lace fabrics ranged from 229 kg/m^3 (commercially finished spun-lace polyester/rayon) to 384 g/m^3 (spun-bonded olefin). The range of fabric density readings was highest for the woven fabrics. The least dense woven fabric tested was the cotton drill (381 kg/m^3). The most dense woven fabric tested was the commercially finished, Scotchgard-treated cotton fabric (541 kg/m^3). The results of the analysis of variance (ANOVA) indicated a significant ($P = 0.05$) difference between the woven and nonwoven fabrics. The results of the Duncan's multiple range test for the density of the fabrics are summarized in Table 2.

The air permeability readings (see Table 1) indicated that those of the polyester spun-lace fabrics were closest to those of the shirting-weight woven fabrics used as controls and would probably have a comfort rating similar to that of finished woven fabrics of similar density. The spun-bonded and spun-bonded/melt-blown fabrics exhibited low or no air permeability, 0.0044 (m^3 of air/s)/m^2, with the exception of the spun-bonded/MB (90 g), which had a slightly higher air permeability of 0.0370 (m^3 of air/s)/m^2. This low air permeability indicates low comfort in a work situation. Air permeability readings of 0.0044 (m^3 of air/s)/m^2 are defined as being impermeable. The one-way ANOVA indicated that there was a significant between-group variation at the 0.01 level of significance. The results of the air permeability tests can be summarized using the Duncan's multiple range test, as shown in Table 3.

The water vapor permeability, as defined by the total amount of water evaporated (measured by the specimen weight loss), indicated that the nonwoven spun-lace fabrics and the spun-bonded/MB (90 g) were similar to the woven fabrics. The total specimen weight loss for the woven fabrics ranged from 10.8 g (denim) to 12.1 g (drill). The total specimen weight loss recorded was 12.0 g for the spun-lace fabrics and 11.74 g for the spun-bonded/MB (90 g). The water vapor permeability for the spun-bonded olefin fabric and the

TABLE 1—*A comparison of physical properties of woven and nonwoven fabrics for protective clothing for pesticide workers.*[a,b]

Fabrics	Average Weight, g/m^2	Average Thickness, cm	Density, kg/m^3	Air Permeability, $(m^3/s)/m^2$		Water Vapor Permeability, $g/h/m^2$[c]	
				$\bar{X}$	SD	$\bar{X}$	SD
Cotton drill	221	0.058	542.9	0.434	0.05	12.1	0.2
Cotton, commercially finished, Scotchgard	157	0.029	540.7	0.338	0.01	11.7	0.1
Denim	283	0.057	495.6	0.054	0.001	10.8	0.3
Cotton, Quarpel	113	0.025	445.0	0.576	0.05	11.5	0.2
35/65 Cotton/polyester, Quarpel	117	0.024	490.0	0.342	0.03	11.8	0.4
Spun-lace polyester, unfinished	90	0.025	354.0	0.323	0.01	11.7	0.1
Spun-lace polyester/rayon, commercially finished	70	0.020	233.0	0.417	0.01	11.6	0.1
Spun-bonded olefin	40	0.010	383.7	0.0044	0	7.7	0.2
Spun-bonded/MB polypropylene	90	0.051	175.3	0.037	0	11.7	0.3
Spun-bonded/MB polypropylene	50	0.031	161.7	0.0044	0	9.6	0.1
Spun-bonded/MB polypropylene, commercially finished	50	0.028	179.0	0.0044	0	9.7	0.1

[a]The data represent physical properties of test fabrics without finishes applied in the laboratory.
[b]Key to terms: $\bar{X}$ = average; SD = standard deviation.
[c]The rate of water vapor transmission (WCV) is in grams per hour per metre squared.

TABLE 2—*Density of fabric test results for wovens and nonwovens using Duncan's multiple range test.*[a]

	Spun-Bonded/ MB 50 g/m^2, Unfinished, kg/m^3	Spun-Bonded/ MB 90 g/m^2, Unfinished, kg/m^3	Spun-Bonded/[b] MB 50 g/m^2, kg/m^3	Spun-Lace Polyester/ Rayon, Finished, kg/m^3	Spun-Lace Polyester, Unfinished, kg/m^3	Spun-Bonded Olefin, Unfinished, kg/m^3	Cotton, Quarpel,[b] kg/m^3	Cotton/ Polyester, 35/65, Quarpel,[b] kg/m^3	Denim, Unwashed, kg/m^3	Cotton, Scotchgard,[b] kg/m^3	Drill, kg/m^3
$\bar{X}$	161.7	175.3	179.0	233.0	354.0	383.7	445.0	490.0	495.6	540.7	542.9

[a]Any two means underscored by the same line are not significantly different. Any two means not underscored by the same line are significantly different at the 0.05 level.
[b]Commercially finished.

TABLE 3—*Air permeability test results for wovens and nonwovens using Duncan's multiple range test.*[a]

	Spun-Bonded Olefin, Unfinished, (m^3/s)/m^2	Spun-Bonded/ MB 50 g/m^2, Unfinished, (m^3/s)/m^2	Spun-Bonded/[b] MB 50 g/m^2, Finished, (m^3/s)/m^2	Spun-Bonded/ MB 90 g/m^2, Unfinished, (m^3/s)/m^2	Denim, Unwashed, (m^3/s)/m^2	Spun-Lace Polyester/ Rayon, Unfinished, (m^3/s)/m^2	Cotton, Scotchgard, (m^3/s)/m^2	Cotton/ Polyester,[b] 35/65, Quarpel, (m^3/s)/m^2	Spun-Lace[b] Polyester/ Rayon, (m^3/s)/m^2	Drill, (m^3/s)/m^2	Cotton, Quarpel, (m^3/s)/m^2
$\bar{X}$	0.0044	0.0044	0.0044	0.0370	0.0540	0.3230	0.3380	0.3420	0.4170	0.4340	0.5760

[a]Any two means underscored by the same line are not significantly different. Any two means not underscored by the same line are significantly different at the 0.05 level.
[b]Commercially finished.

50-g/m^2 spun-bonded/MB fabrics ranged from 9.0 g (spun-bonded olefin) to 9.7 g (commercially finished spun-bonded/MB). A one-way ANOVA of the results of the water vapor test indicated that the between-group variation was significant at the 0.01 level. The results of the Duncan's multiple range test indicated that spun-bonded olefin and spun-bonded/MB (50 g/m^2) fabrics were significantly lower in water vapor permeability (0.05 level) than the other fabrics. The spun-lace fabrics had air permeability readings similar to those of the woven shirting fabrics. The results of the water vapor permeability tests are summarized in Table 4 using the Duncan's multiple range test.

The results of the aerosol spray test are shown in Table 5. The nonwoven fabric specimens, either commercially or laboratory-finished with fluorocarbon-based finishes, passed the aerosol spray tests. The commercially finished spun-lace fabrics passed all the spray tests. All of the Scotchgard laboratory-finished nonwoven fabrics passed. The unfinished cotton drill failed to pass any of the spray tests. The commercially (Scotchgard) finished cotton shirting-weight muslin passed the water-based spray tests but failed to pass the oil-based spray test. The unwashed denim tested with Solution I (water) failed two of the three replications; therefore, three additional replications using Solution I were made, and the results were combined. Three of the six replications passed the aerosol spray test using Solution I on unwashed denim. Because of the mixed test results of the unwashed specimen, the denim was laundered three times to ensure the removal of manufacturing sizes. After laundering one and three times, following the laundering procedure described in the American Association of Textile Chemists and Colorists (AATCC) Test Method 124-1978 (Appearance of Durable Press Fabrics after Repeated Home Launderings), the denim failed all the aerosol spray tests, indicating that the removal of sizings applied during manufacturing decreased its resistance to aerosol spray penetration.

The Quarpel-finished cotton and cotton/polyester blend fabrics passed the water-based spray tests but not the oil-based spray solution test.

Summary and Conclusions

The purpose of this research was to identify nonwoven fabrics that are resistant to oil-based and water-based aerosol spray penetration and to assess the value of these fabrics for potential use in disposable, protective garments. In addition, various physical properties, such as fabric density, weight, thickness, moisture vapor permeability, and air permeability, were examined as indicators of fabric comfort.

An evaluation of aerosol spray penetrability of finished and unfinished woven and nonwoven fabrics indicated that the presence of a fluorocarbon finish did increase the resistance to aerosol penetration. The woven fabrics tested failed to meet the criterion of being resistant to oil-based spray penetration. The finished nonwoven fabrics, with the exception of the commercially

TABLE 4—*Water vapor permeability test results for wovens and nonwovens using Duncan's multiple range test.*[a]

	Spun-Bonded Olefin, Unfinished, g/h/m^2	Spun-Bonded/MB 50 g/m^2, Unfinished, g/h/m^2	Spun-Bonded/[b] MB 50 g/m^2, Finished, g/h/m^2	Denim Unwashed, g/h/m^2	Cotton, Quarpel, g/h/m^2	Spun-Lace Polyester, g/h/m^2	Spun-Lace[b] Polyester-Rayon, Finished, g/h/m^2	Spun-Bonded/MB 90 g/m^2, Unfinished, g/h/m^2	Cotton,[b] Scotchgard, g/h/m^2	Cotton/Polyester,[b] 35/65, Quarpel, g/h/m^2	Drill, g/h/m^2
$\bar{X}$	7.7	9.6	9.7	10.8	11.5	11.6	11.7	11.7	11.7	11.8	12.1

[a]Any two means underscored by the same line are not significantly different. Any two means not underscored by the same line are significantly different at the 0.05 level.
[b]Commercially finished.

TABLE 5—*Aerosol dry spray penetration results categorized by pass/fail performance.*

	Aerosol Spray Test Performance		
Test Fabrics	Solution I, Water	Solution II, Water/ Surfactant (4:1)	Solution III, Cottonseed Oil/ Surfactant (4:1)
Category I	pass	pass	pass
Spun-lace polyester, Scotchgard	pass	pass	pass
Spun-lace polyester, commercially finished	pass	pass	pass
Spun-bonded olefin, Scotchgard	pass	pass	pass
Spun-bonded/MB (90 g/m^2), Scotchgard	pass	pass	pass
Spun-bonded/MB (50 g/m^2), Scotchgard	pass	pass	pass
Category II	pass	pass	fail
Commercially finished (Scotchgard) cotton[a]	pass	pass	fail
Cotton, Quarpel	pass	pass	fail
Cotton/polyester, 35/65, Quarpel finished	pass	pass	fail
Category III	pass	fail	fail
Spun-bonded olefin	pass	fail	fail
Spun-bonded/MB (90 g/m^2)	pass	fail	fail
Spun-bonded/MB (50 g/m^2)	pass	fail	fail
Spun-bonded/MB (50 g/m^2), commercially finished[a]	pass	fail	fail
Category IV	fail	fail	fail
Denim, unwashed[b]	fail	fail	fail
Denim, 1 wash	fail	fail	fail
Denim, 3 washes	fail	fail	fail
Cotton drill	fail	fail	fail
Spun-lace polyester, unfinished	fail	fail	fail

[a]Compositions of the commercial finishes are proprietary.
[b]Represents six replications for Solution I.

finished spun-bonded/MB, passed the spray tests using both the oil-based and water-based solutions (Table 5).

Comparison of comfort properties of air permeability, water vapor permeability, and density indicated that the spun-lace class of nonwoven fabrics was very similar to the woven fabrics: the spun-lace fabrics are denser or more "clothlike" than many of the spun-bonded fabrics. In an earlier study Oakland et al [*15*] indicated that disposable coveralls of the nonwoven fabrics used in field testing required a manufactured cost per garment of about $2.20. Protection of the nonwovens examined against the oil-based system was superior to that of all the wovens, and protection against aqueous systems was also superior to the best of the wovens, denim, after a single washing. Based on the test results, the fluorocarbon-barrier-finished spun-lace polyester fabrics and polyester/rayon fabrics are the best choice for use in disposable garments for protection against pesticides.

Acknowledgment

This paper is No. 9805 in the Journal Series of the North Carolina Agricultural Research Service, and was funded by the Southern Regional Project, S-163, and the North Carolina Agricultural Research Service, and Ciba-Geigy Corp., Agricultural Division.

References

[*1*] *Summary of Reports from Physicians of Illnesses that Were Possibly Related to Pesticide Exposure During the Period January 1-December 31, 1983 in California*, HS-1186, California Department of Food and Agriculture, Division of Pest Management Protection and Worker Safety, Worker Health and Safety Unit, Sacramento, CA, 1984.

[*2*] *1982 Summary of Pesticide Poisoning Cases*, North Carolina Department of Human Resources, Division of Health Services, Environmental Epidemiology Branch, Raleigh, NC, 1983.

[*3*] "Apply Pesticides Correctly, A Guide for Commercial Applicators," U.S. Department of Agriculture and Environmental Protection Agency, U.S. Government Printing Office, Washington, DC, 1976, pp. 623-656.

[*4*] *NIOSH Criteria for a Recommended Standard Occupational Exposure to Malathion*, HEW Publication No. (NIOSH) 76-205, U.S. Department of Health, Education, and Welfare, Public Health Service, Center for Disease Control, National Institute for Occupational Safety and Health, Washington, DC, June 1976.

[*5*] *NIOSH Criteria for a Recommended Standard Occupational Exposure During the Manufacture and Formulation of Pesticides*, DHEW (NIOSH) No. 78-174, U.S. Department of Health, Education, and Welfare, Public Health Service, Center for Disease Control, National Institute for Occupational Safety and Health, Washington, DC, July 1978.

[*6*] Wolfe, H. R., "Workers Should Be Protected from Pesticide Exposure," *Weeds, Trees, and Turf*, Vol. 12, April 1973, pp. 36-37, 52-53.

[*7*] Hollies, N. R. S. and Goldman, R., *Clothing Comfort, Interaction of Thermal, Ventilation, Construction and Assessment Factors*, Ann Arbor Science Pub., Ann Arbor, MI, 1974.

[*8*] Spencer-Smith, J. L., "The Physical Basis of Clothing Comfort Part 3: Water Vapour Transfer Through Dry Clothing Assemblies," *Clothing Research Journal*, Vol. 5, 1977, pp. 82-100.

[*9*] Spencer-Smith, J. L., "Some Aspects of Tropical Clothing" in *Proceedings of the Third Shirley International Seminar Textiles for Comfort*, Shirley Institute, Manchester, England, 1971, pp. 1-12.

[*10*] Slater, K., "Comfort Properties of Textiles," *Textile Progress*, Vol. 9, 1977, p. 4.

[*11*] Lillie, T. H., Livingston, T. M., and Hamilton, M. A., "Recommendations for Selecting and Decontaminating Pesticide Applicator Clothing," *Bulletin of Contamination and Toxicology*, Vol. 27, 1980, pp. 716-723.

[*12*] Serat, W. F. and Van Loon, A. J., "Some Factors Influencing the Design of Fabrics for Protective Garments for Pesticide Fieldworkers," Unpublished report to State of California Department of Health Service, Berkeley, CA, 1978.

[*13*] Southwick, T. W., Mecham, H. D., Cannon, P. M., and Gortatowski, M. T., "Pesticide Residues in Laundered Clothing," *Proceedings of Third Conference of Environmental Chemicals, Human and Animal Health*, Colorado State University, Fort Collins, CO, 1974, pp. 125-131.

[*14*] Finley, E. L., Metcalf, G. I., McDermott, F. G., Graves, J. B., Schilling, P. E., and Bonner, F. L., "Efficiency of Home Laundering in Removal of DDT, Methyl Parathion and Toxaphene Residues from Contaminated Fabrics," *Bulletin of Environmental Contamination and Toxicology*, Vol. 12, 1974, pp. 268-274.

[*15*] Oakland, B. G., Hurwitz, M. D., and Gahagan, C. O., "Cost-Benefit Analysis of Woven-Nonwoven Protective Clothing," Paper presented at the American Industrial Hygiene Conference, Detroit, 20-25 May 1984.

Julie L. Keaschall,[1] *Joan M. Laughlin,*[1] *and Roger E. Gold*[1]

Effect of Laundering Procedures and Functional Finishes on Removal of Insecticides Selected from Three Chemical Classes

REFERENCE: Keaschall, J. L., Laughlin, J. M., and Gold, R. E., **"Effect of Laundering Procedures and Functional Finishes on Removal of Insecticides Selected from Three Chemical Classes,"** *Performance of Protective Clothing, ASTM STP 900,* R. L. Barker and G. C. Coletta, Eds., American Society for Testing and Materials, Philadelphia, 1986, pp. 162-176.

ABSTRACT: Eleven pesticides from three chemical classes were chosen for comparison of residues after laundering. Pesticide was introduced to the fabric surface of one of three fabrics—unfinished (UN), renewable consumer applied fluorocarbon finished (RF), and commercially applied fluorocarbon finished (CM) fabric. The fabrics were laundered using one of three laundry treatments—a heavy-duty liquid detergent (HDL) alone, an HDL with a prewash spray, and an HDL with an agriculturally marketed pretreatment.

Gas chromatographic analysis showed that the fluorocarbon-finished fabrics absorbed only 10% of the pesticide absorbed by the UN fabrics. Residues after laundering were significantly different both among and within classes. Although both fluorocarbon finishes reduced absorption of pesticide, they did not facilitate removal of the contaminant through laundering. Laundry additives significantly aided residue reduction.

KEY WORDS: pesticide, pesticide residue, functional finish, soil-repellent finish, durable-press finish, surfactants, protective clothing, laundering

A major concern of agriculturalists today is contamination of clothing during mixing and application of pesticides. Pesticide contaminants can be difficult to remove from clothing. General recommendations have been made for the "best" methods of removal of pesticides from clothing in laundry [*1-4*].

The greatest imminent health risk from pesticides comes through primary

[1]Graduate research assistant and professor, Department of Textiles, Clothing, and Design, and head, Department of Entomology, University of Nebraska, Lincoln, NE 68583-0802.

exposure. Making persons exposed to pesticides more aware of dangers, of how to avoid contamination, and of how to deal with contaminated clothing may reduce the potential for adverse effects.

The effectiveness of laundering in removing pesticide is important in making contaminated clothing safe to wear. After the clothing has been laundered it is generally considered by most applicators to be clean; however, when the wearer is not aware of residues remaining in the clothing, contaminated garments provide a medium for exposing the skin to the pesticide [*5,6*], making dermal absorption possible [*7,8*].

A majority of studies that evaluated laundering variables included only one or two pesticides [*9–16*]. Based on these one or two-chemical studies, recommendations to maximize removal are being made for pesticides across the board [*1–3*]. In one of the few studies involving several chemicals, Lillie et al [*17*] found that differences in removal tend to occur among insecticides and that the best laundry procedures may not be best for all pesticides.

Highly concentrated active ingredient (AI) in the undiluted formulated material may be impossible to remove entirely from contaminated clothing. Easley et al [*6*] found that in fabrics contaminated with 54% AI emulsifiable concentrate methyl parathion (MeP), only 66.68% of the material was removed after ten machine washings.

Hot water [60°C (140°F)] contributed to increased removal of fonofos and alachlor on heavy fabric substrates [*15*]. Easley et al found decreased residues with elevated temperatures [49°C (120°F) and 60°C (140°F)] when working with 2,4-dichlorophenoxyacetic acid and amine (2,4-D) [*12*] and methyl parathion (MeP) [*11*]. Finley et al [*18*] also recommended hot water [60°C (140°F)] based on studies of MeP, toxaphene, and dichlorodiphenyltrichloroethane (DDT). Easter [*13*] found Guthion and captan removal to increase with temperature; however, Lillie et al [*19*] found removal of chlordane to be inversely related to water temperature.

Few attempts have been made in a single laboratory to compare pesticide residues after laundering within the same chemical class and between classes. In the work with diazinon, malathion, bromacil, propoxur, and chlordane, differences were found, but the solutions for each pesticide were at different concentrations of active ingredient, and an insufficient number of pesticides were tested to observe the effect of pesticide class as a causal factor in completeness of removal [*17*].

Livingston [*20*] also examined diazinon, chlordane, and carbaryl insecticides and the herbicide prometron for removal tendencies from all-cotton and all-polyester coveralls. In these studies of different concentrations of active ingredients, carbaryl and prometron were completely removed; diazinon and chlordane were more persistent. Chlordane was the most difficult to remove of the chemicals tested, with approximately 90% removal after a single laundering. Finley et al [*18*] have shown that pesticide soil removal is more difficult for fabrics contaminated with different pesticides and washed in the same wash load. The Livingston study [*20*] was done with all the contaminated fabrics laundered together.

The pesticide class was theorized to have a significant influence on removal in a laundry study involving fonofos and alachlor [*15*]. Easter [*13*] found differences in removal between captan and Guthion. Captan residue was more difficult to remove from denim, while Guthion was more difficult to remove from Gore-Tex.

Pesticide removal by laundering has been paralleled to soil removal. Specific formulations are more responsive to procedures directed toward those selected soil types. Treatments for oily soils have been suggested to be more efficient in the removal of specific formulations [*11,13*].

Use of fabrics with a water-repellent finish may provide protection from chemicals [*21*]. In studies by Gold et al [*22*] and Leavitt et al [*23*], it was observed that the skin was exposed to 5 to 6% of the carbaryl coming in contact with conventional clothing worn by workers. A fluorocarbon finish was found to make a significant difference in the permeability of clothing to pesticides [*24*]. Kawar et al [*7*] found that penetration by dust was reduced by 60% after the addition of a fluorocarbon finish on the fabric. Neither of these studies addressed issues related to laundering.

The primary objective of these laundry studies has been to determine the most effective laundry practices. If these procedures are to be generalized to be used with more pesticides than those directly under study, bases for such generalizations must be established. The first phase of this two-phase study examined the similarities and differences in pesticide removal both between and within insecticide classes. The second phase evaluated the effectiveness of fluorocarbon finishes in providing protection after repeated launderings. This was an interdisciplinary study involving the departments of Textiles, Clothing and Design, Environmental Programs, and Agronomy at the University of Nebraska.

Procedures

Fabrics

Fabrics for the study were obtained from the Southern Regional Research Center (SRRC) of the U.S. Department of Agriculture (USDA). The fibers were blended, the yarns spun, and the fabrics woven by the Fabric Engineering and Development Research Group in the Cotton Textile Processing Laboratory of the SRRC under the direction of George Drake. All were 122.5-g/m^2, 50% cotton/50% polyester fabrics; the thread counts were 28 threads per centimetre in the warp and 32 threads per centimetre in the filling. The unfinished fabric (UN) had been boiled off and bleached during wet finishing.

One portion of the UN fabric was sprayed with a renewable consumer-applied fluorocarbon finish (RF) to achieve a dry weight gain of 0.55 to 0.80%.

The third fabric was finished with a commercially applied soil-repellent fluorocarbon finish (CM) designed to repel moisture. The commercial fluorocarbon formulation (Quarpel) was composed of Zepel D fluoropolymer (7.0%), Norane F fluorochemical extender (10.0%), Nykon NRW3 wetting agent (0.4%), and water (82.6%). It was pad applied at a net pickup of approximately 70%.

Fabric stripping was eliminated to allow evaluation of functional finishes. The outer 10% of the fabric in the warp direction was discarded [ASTM Test for Breaking Load and Elongation of Textile Fabrics (D 1682-64 [1975]). Eight-centimetre square specimens were cut from each of the three fabrics. The specimens were conditioned [at 65 ± 2% relative humidity and 21 ± 1°C (70° ± 2°F)] [ASTM Practice for Conditioning Textiles for Testing (D 1776-79)] prior to contamination.

Insecticides

Insecticides are divided into chemical classes according to their components. The three classes of insecticides included in this study were organophosphates, carbamates, and organochlorines. These three classes are the most widely used and present the greatest hazard in terms of human toxicity [*25*]. Because of their wide use, emulsifiable concentrate pesticide formulations were used in this study, except for carbofuran, which was available only as a flowable formulation.

A 1.0% field strength solution was prepared according to label directions, for each of the following insecticides: *organophosphates*—chlorpyrifos, diazinon, dichlorvos (DDVP), dimethoate, malathion, and methyl parathion; *carbamates*—carbofuran and propoxur; *organochlorines*—aldrin, chlordane, and lindane.

Specimen Contamination—A hand-held micropipette with microprocessor was used to apply a 0.2-mL aliquot of pesticide solution to each specimen. The entire aliquot was absorbed by the unfinished fabric; however, both fluorocarbon finishes made the specimen hydrophobic. After 10 s the specimen was held in a vertical position and the unabsorbed solution was allowed to roll off the fabric surface and into a waste receptacle. Prior to analysis, all the specimens were allowed to air dry (18° to 22°C, ambient air, until weight loss reached equilibrium).

Phase I

Phase I was designed to compare similarities and differences of insecticide residues on fabric specimens after laundering. The effects of fabric finishes on removal tendencies were observed, as were the effects of laundry additives.

Six specimens from each fabric were contaminated with each of the eleven insecticides. Three of the specimens served as controls, while the other three were subjected to one of three laundry treatments. All work was replicated three times.

Laundry Treatments—The research design included three laundry treatments. Using procedures modified from American Association of Textile Chemists and Colorists (AATCC) standard 61-1980, Colorfastness to Washing, Domestic; and Laundering Commercial Accelerated, the accelerated method was adjusted to simulate a single laundry cycle. A 12-min wash cycle using a 0.2% detergent solution was followed by 5 and 3-min rinses. Agitation was provided by 25 steel balls; all the cycles used 49°C (120°F) distilled water. Teflon liners were placed between the rubber gaskets and the canisters to prevent absorption of pesticide by the gasket [*16*]. All three laundry treatments[2] included this standard laundry procedure. The two additional treatments included laundry additives of a prewash spray, or an agriculturally marketed degreaser pretreatment. These laundry pretreatments were applied with the micropipette programmed to deliver 0.225 mL of additive to each specimen.

Phase II

The longevity of the effectiveness of the soil-repellent finishes in limiting sorption of pesticide was analyzed in Phase II of this study. Clean specimens with no functional finish (UN), a renewable consumer-applied fluorocarbon finish (RF), and a commercially applied fluorocarbon finish (CM) were laundered zero through five times prior to contamination using the standard laundry procedure described earlier. All the work was replicated three times. Comparison of contamination levels for the UN, RF, and CM specimens after zero through five launderings determined the useful life of each fluorocarbon finish.

Extraction and Analysis of Pesticide

Each fabric specimen was extracted in 500-mL glass bottles with Teflon-lined lids in two 100-mL aliquots of glass-distilled acetone. Each specimen with solvent was shaken for 1½ h and decanted, the procedure was repeated, and the two aliquots were combined. The acetone extracts were analyzed using a Hewlett-Packard gas chromatograph Model 5850A with automatic injection, and a nitrogen-phosphorus-specific thermionic detector and dedicated microprocessor. All the separation columns used were glass of 2.0 mm inside diameter. Nitrogen was used as a carrier gas. The column tempera-

[2]Three laundry treatments are to be abbreviated Tx_1, Tx_2, and Tx_3. Tx_1 used Dynamo as a detergent. Tx_2 used Dynamo with Spray 'n Wash, a prewash spray. Tx_3 used Dynamo with Super D, an agriculturally marketed degreaser pretreatment. (Use of registered trade names does not imply endorsement of a specific product.)

tures differed with each chemical (Table 1). All OV columns were packed on Chromasorb W HP; Apiezon L columns were packed on Supelcoport.

Statistical Analysis

The total amount of contaminant in the specimen was expressed in micrograms per square centimetre; these were computed for the eleven pesticides for each of the three fabric finishes (UN, RF, and CM). The data were arc sine converted, and general linear model factorial experiment analysis of variance was performed with a decision level of $P \leq 0.05$. Least significant means tests were performed to determine whether or not differences between pairs of means were significant.

Results and Discussion

The UN specimen retained ten times the pesticide (26.3 $\mu g/cm^2$) of the RF (2.8 $\mu g/cm^2$) and CM (2.8 $\mu g/cm^2$) specimens at initial contamination. The fluorocarbon soil-repellent finishes effectively reduced the hydrophilicity of the specimens so that the liquid pesticide beaded on the surface of the finished fabric and reduced initial contamination (Table 2). There were no sig-

TABLE 1—*Column temperatures used in detection by gas chromatography.*

Insecticide	Column Temperature, °C	Column Packing
Organophosphates		
Chlorpyrifos	194	3% OV-25
Dichlorvos	152	3% OV-25
Diazinon	180	3% OV-25
Dimethoate	189	3% OV-25
Malathion	196	3% OV-25
Methyl parathion	210	3% OV-25
Carbamates		
Carbofuran	165	5% Apiezon L
Propoxur	170	3% OV-25
Organochlorines		
Aldrin	184 or 197[a]	3% OV-25
Chlordane	195	5% Apiezon L
Lindane	163 or 193[a]	1.5% OV-17 + 1.95%
		1.95% OV-210

[a]Two column temperatures were used with aldrin and lindane. Because of extreme low residue levels, solvent extracts were evaporated to minute amounts. Interference by coextractions necessitated a lower temperature to achieve peak separation on chromotograms.

TABLE 2—*Residue remaining after laundering, in micrograms per square centimetre.*

Pesticide	Treatment[a]	Residue Remaining on Finish, $\mu g/cm^2$		
		UN	RF	CM
Organophosphates				
Chlorpyrifos	Un_1	25.61	3.63	3.44
	Tx_1	6.92	0.94	1.23
	Tx_2	1.65	0.24	0.93
	Tx_3	1.07	0.13	0.72
DDVP	Un_1	18.40	1.46	1.68
	Tx_1	0.01	0.01	0.01
	Tx_2	0.01	0.00	0.00
	Tx_3	0.01	0.00	0.00
Diazinon	Un_1	24.79	3.66	2.79
	Tx_1	0.96	0.07	0.05
	Tx_2	0.25	0.03	0.03
	Tx_3	0.27	0.03	0.03
Dimethoate	Un_1	27.60	2.58	2.58
	Tx_1	0.00	0.00	0.00
	Tx_2	0.00	0.00	0.00
	Tx_3	0.00	0.00	0.00
Malathion	Un_1	26.63	3.10	3.08
	Tx_1	0.03	0.00	0.00
	Tx_2	0.02	0.00	0.00
	Tx_3	0.01	0.00	0.01
Methyl parathion	Un_1	31.65	4.99	4.45
	Tx_1	0.15	0.04	0.02
	Tx_2	0.06	0.01	0.01
	Tx_3	0.06	0.01	0.03
Carbamates				
Carbofuran	Un_1	35.11	2.57	2.48
	Tx_1	0.05	0.00	0.00
	Tx_2	0.04	0.00	0.00
	Tx_3	0.04	0.00	0.00
Propoxur	Un_1	37.93	5.46	5.33
	Tx_1	0.03	0.06	0.01
	Tx_2	0.04	0.00	0.00
	Tx_3	0.02	0.00	0.00
Organochlorines				
Aldrin	Un_1	27.24	2.28	3.12
	Tx_1	2.48	0.13	0.52
	Tx_2	1.41	0.08	0.60
	Tx_3	1.64	0.06	0.65
Chlordane	Un_1	11.44	0.78	0.79
	Tx_1	1.96	0.00	0.00
	Tx_2	0.65	0.00	0.00
	Tx_3	0.73	0.00	0.00
Lindane	Un_1	22.86	1.76	2.15
	Tx_1	0.12	0.05	0.21
	Tx_2	0.07	0.04	0.12
	Tx_3	0.13	0.07	0.15
Mean for finishes for the laundered specimens		.63	.06	.16

[a]Key to abbreviations:
Un_1 = unlaundered.
Tx_1 = detergent alone.
Tx_2 = detergent with prewash spray.
Tx_3 = detergent with agricultural degreaser.

nificant differences in initial contamination based on pesticide ($F = 1.12$; df $= 2,10$). (All references to 0% residue remaining may be interpreted as below the minimum amount detectable with the gas chromatographic analysis used.)

Phase I

Residues After Laundering—Contamination of specimens after laundering was determined for each pesticide and for each fabric finish. Residues after laundering were markedly lower in most instances for fabrics with fluorocarbon finishes (Table 2). Amounts of residue after laundering for dimethoate, dichlorvos, and malathion were at the boundary of detection. Residues for lindane on the CM were slightly higher than residues on the UN.

The means across replication were contrasted to determine the residue remaining after laundering as a percentage of the initial contamination. Pesticide residues after laundering ranged from 0.00 to 35.80% of the initial contamination (Table 3). The residues after laundering expressed as percentages of contamination were not always lower in the fluorocarbon-finished specimens (lindane), indicating that although the finish reduced the sorption of pesticide, these residues were not necessarily easier to remove.

Effect of Pesticide Class—When the pesticides were grouped by class, the interaction between the effect of pesticide class and the fabric finish was significant (Table 4). Because of the statistical significance of this interaction effect, main effects must be examined with caution. Class contributed to differences in residues remaining after laundering (Table 4). Organochlorine (OCl) insecticides were the most difficult to remove, followed by organophosphates (OP) and carbamates (CARB) (Table 5). The residues remaining after laundering for each class were 5.56%, 3.49%, and 0.10% for OCl, OP, and CARB, respectively.

Effect of Finish—The pesticide class-fabric finish interaction significantly affected the percentage of residue remaining after laundering (Table 4). When calculated by averaging across all pesticides (Table 3), the residues remaining for the RF specimen were the smallest (2.03%), followed by the UN (2.89%) and CM (5.27%) specimens. It is important to note that the impact of finish in repelling the pesticide solution, and thus the level of contamination (in micrograms per square centimetre) before laundering of the RF and CM finishes was about one tenth that of the UN (Table 2).

Effect of Laundry Treatment—Although there was no significant interaction between class and laundry treatment in pesticide residues after laundering, there was a trend for differences related to laundering procedures (Table 4). The laundry pretreatments designed to operate as degreasing agents, that is, prewash spray (Tx_2) and agricultural degreaser (Tx_3), worked in similar ways and reduced the percentage of residues remaining following laundering.

TABLE 3—*Percentage of residue remaining after laundering compared with the control fabrics of identical finishes.*

Pesticide	Treatment[b]	Percentage of Residue Remaining on Finish[a]		
		UN	RF	CM
Organophosphates				
Chlorpyrifos	Tx_1	26.78[a]	25.76[ac]	35.80[d]
	Tx_2	6.41[b]	6.73[b]	27.13[a]
	Tx_3	4.20[b]	3.27[b]	20.95[c]
Dichlorvos	Tx_1	0.06[e]	0.56[e]	0.73[e]
	Tx_2	0.05[e]	0.75[e]	0.13[e]
	Tx_3	0.03[e]	0.15[e]	0.05[e]
Diazinon	Tx_1	3.74[fh]	2.06[fgh]	1.84[fghi]
	Tx_2	1.13[ghi]	0.69[ghi]	0.98[ghi]
	Tx_3	1.07[ghi]	0.66[gi]	1.21[ghi]
Dimethoate	Tx_1	0.02[j]	0.00[j]	0.00[j]
	Tx_2	0.02[j]	0.00[j]	0.00[j]
	Tx_3	0.02[j]	0.00[j]	0.00[j]
Malathion	Tx_1	0.13[k]	0.06[k]	0.02[k]
	Tx_2	0.07[k]	0.05[k]	0.02[k]
	Tx_3	0.07[k]	0.02[k]	0.01[k]
Methyl parathion	Tx_1	0.51[m]	0.86[m]	0.46[m]
	Tx_2	0.20[m]	0.25[m]	0.30[m]
	Tx_3	0.20[m]	0.24[m]	0.56[m]
Carbamates				
Carbofuran	Tx_1	0.15[n]	0.02[n]	0.00[n]
	Tx_2	0.10[n]	0.04[n]	0.00[n]
	Tx_3	0.11[n]	0.02[n]	0.00[n]
Propoxur	Tx_1	0.08[q]	1.02[r]	0.09[pq]
	Tx_2	0.10[q]	0.03[q]	0.00[pq]
	Tx_3	0.06[q]	0.04[q]	0.00[pr]
Organochlorines				
Aldrin	Tx_1	6.27[st]	5.74[st]	16.72[v]
	Tx_2	5.16[st]	3.47[stu]	19.23[v]
	Tx_3	6.03[st]	2.70[tu]	20.89[v]
Chlordane	Tx_1	17.17[w]	0.00[y]	0.00[y]
	Tx_2	5.67[x]	0.84[y]	0.00[y]
	Tx_3	6.39[x]	0.00[y]	0.00[y]
Lindane	Tx_1	0.54[z]	2.84[z]	9.64[o]
	Tx_2	0.32[z]	2.44[z]	5.57[a]
	Tx_3	0.58[z]	4.24[a]	6.88[a]
Mean for finishes		2.89	2.03	5.27

[a]Numbers within chemicals followed with the same letter were not statistically different at $P \leq 0.05$. UN = unfinished; RF = renewable finish; CM = commercial finish.

[b]Key to abbreviations:

Tx_1 = detergent alone.
Tx_2 = detergent with prewash spray.
Tx_3 = detergent with agricultural degreaser.

TABLE 4—*ANOVA of the effects of pesticide class, fabric finish, and laundry treatment on the percentage of residue remaining after laundering.*

Effect	Degrees of Freedom	F Value	$P < 0.05$
Pesticide class	2	21.39	...
Fabric finish	2	3.09	...
Treatment	2	2.92	...
Pesticide class by fabric finish	4	2.77	...
Pesticide class by treatment	4	0.48	ns[a]
Finish by treatment	4	0.11	ns
Class by finish by treatment	8	0.10	ns

[a]ns = not significant.

TABLE 5—*Mean residues remaining based on the percentage of contaminant absorbed by fabric of like finish for each pesticide.*

Pesticide	Pesticide Class	$\bar{X}$ % of Residue Remaining	LS of Mean Group[a]
Chlorpyrifos	OP	17.50	A
Aldrin	OCl	9.62	B
Lindane	OCl	3.67	C
Chlordane	OCl	3.34	D
Diazinon	OP	1.48	D
Methyl parathion	OP	0.40	E F
Dichlorvos	OP	0.28	E F G
Propoxur	CARB	0.16	F G H
Malathion	OP	0.05	F G H
Carbofuran	CARB	0.05	G H
Dimethoate	OP	0.01	G H

[a]Letters indicate statistical similarity.

Across all the pesticides and finishes, the prewash spray was generally more effective than the agricultural degreaser.

Effect of Pesticide (Within Class)—Given the difference between classes in residue removal in laundering and the observation that great variability was observed in the after-laundering residues of the organophosphates, it then became important to ascertain whether differences in removal were attributable to differences within chemical classes. Three-way and two-way ANOVA analyses showed interaction effects for pesticide/fabric finish/laundry treatment ($F = 6.61$; df $= 40$; $P \leq 0.05$) and two-way interaction effects for pesticide/fabric finish ($F = 8.19$; df $= 20$; $P \leq 0.05$), with main effects for pesticide ($F = 245.69$; df $= 10$; $P \leq 0.05$), finish ($F = 27.18$; df $= 2$; $P \leq$

0.05), and laundry treatment ($F = 29.65$; df $= 2$; $P \leq 0.05$). Least-square (LS) means were used to explore these interactions, as they contribute to the main effects. In some cases the residue differences between finishes or laundry treatments were small but substantial enough to make an interaction significant. Such differences are not crucial and may be considered minute for practical purposes. For example, dimethoate residues were near or below a detectable level regardless of the treatment or finish (Table 2). As the residues approached zero, there was little effect attributable to finish or treatment. Interactions that cause a reduction or increase in removal may be indicators of phenomena that should be pursued further.

Scrutiny of the LS means showed that specimens with large percentages of residue remaining after laundering were contaminated with different pesticides; they also were from different treatment groups and frequently of dissimilar finishes. The interaction of the combination of effects may act uniquely for an individual specimen, resulting in differences in residues attributable to no single factor. LS means were used to separate pesticides (Table 5).

Fabric finish played a twofold role in the study. First, it limited initial contamination, and second, it was a facilitator or inhibitor of residue removal during laundering. Pesticide and laundry treatment in combination produced residues after laundering that were specific for the pesticide (Tables 2 and 3).

To make recommendations for protection of agricultural workers, the total contamination, in micrograms per square centimetre, for each pesticide and the toxicity of that pesticide must be examined. By combining the amount absorbed by the fabric, and its tendency for retaining residues, a model can be built from among the variables studied for optimizing protective conditions.

The authors recommend that protocols be established and exceptions noted prior to disseminating recommended procedures for limiting absorption and reducing residues for specific pesticides. The objective of this phase of the study was to determine the effects of several variables on residues remaining after contamination and after laundering. The interactions discussed illustrate inaccuracies that can occur if one laundry treatment is recommended as advantageous for all pesticides or pesticide classes.

Phase II

Longevity of Soil-Repellent Finishes—The pesticide sorption after repeated laundering of the fabrics increased from 25.37 to 31.05 $\mu g/cm^2$ for UN, from 5.07 to 17.52, $\mu g/cm^2$ for RF, and from 4.57 to 7.80 $\mu g/cm^2$ for CM (Fig. 1). The LS means showed no significant difference in pesticide residues for the CM specimens across all the launderings. The CM finish was not significantly affected by five launderings.

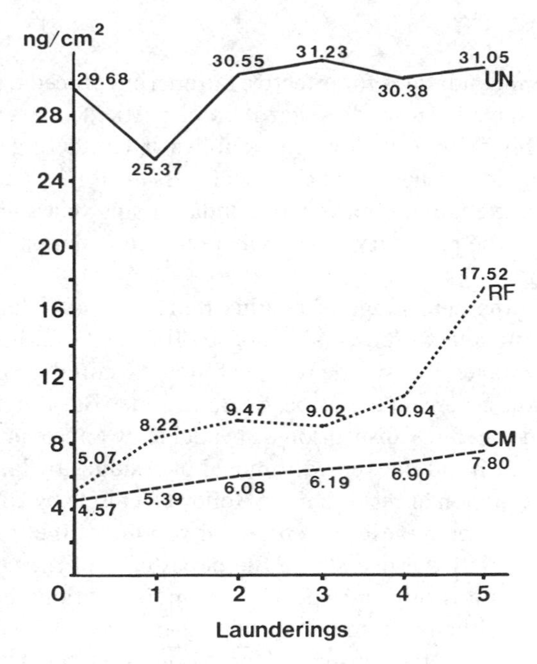

FIG. 1—*Amount of pesticide residue on fabrics.*

The RF finish was effective in reducing absorption through four launderings. After four launderings there were no significant differences between the unlaundered and the laundered specimens. The fifth laundering resulted in a significant difference in pesticide sorption.

Fluorocarbon finishes work significantly better ($F = 363.95$; $P \leq 0.05$) than UN fabrics in limiting sorption of pesticide. Knowledge of longevity of the soil-repellent finishes is necessary for practical understanding of the use of finishes in protective apparel systems. These data showed the amount of pesticide sorbed by the fluorocarbon-finished specimens to be related to the number of launderings to which the fabric has been subjected.

It is not anticipated that a consumer would expect to reapply the CM finish. The garment may no longer be protective after the initial finish is rendered ineffective by repeated launderings or by abrasion during wear. Further work is needed to determine if and when the pesticide repellency of the CM finish decreases significantly.

The RF finish functions effectively through four launderings. At that point reapplication of the finish is recommended to achieve continued protection. By thoroughly reapplying the finish after four launderings, the consumer can be confident that the protective clothing will remain protective for another series of launderings.

Conclusion

To date, recommendations for effective laundering procedures for contaminated clothing have been made regardless of pesticide class and chemical composition. This study was done to establish if laundering recommendations could be made irrespective of chemical class or, if that was not possible, whether appropriate laundering recommendations per class could be established, or if it will be necessary to establish recommendations for individual pesticides.

Differences in the percentage of residue remaining after laundering were found both among and within classes and by finishes. OCl insecticides had the largest percentages of pesticide residue following laundering. Carbamates showed the smallest percentage of pesticide residue. Both fluorocarbon finishes were found to reduce absorption of pesticide by approximately 90%. In most cases the finish did not retard removal of residues by laundering. The RF facilitated reduction of residue most, followed closely by UN fabrics. The CM fabric retained the largest percentage of residue of the three fabric finishes; this was attributed especially to the persistence of the OCl.

As a result of this study, no specific recommendations based on class should be made. Although removal is related to pesticide class, which has practical implications to the consumer, there is no predominant trend for differences in removal based on class. Water solubility of the active ingredient may be a more reliable indicator of the relative ease of removal than chemical class.

The residues, expressed in micrograms per square centimetre, were less for fabrics with fluorocarbon finishes. The micrograms per square centimetre were slightly lower for the RF than for the CM in some cases. The laundry pretreatments also aided in residue reduction. The prewash spray was slightly more effective than the agriculturally marketed product.

Phase II showed that the amount of pesticide absorbed by the fluorocarbon-finished fabrics was inversely related to the number of launderings to which the specimen has been subjected prior to contamination with pesticide. The CM was not affected by five machine launderings. The RF functioned effectively through four launderings. The authors recommend that the finish be reapplied after four wash cycles to achieve continued protection. Although the CM finish remained effective longer than the RF, recommending it over the RF is questionable. This study did not determine how long the CM would remain effective. In recommending the nonrenewable CM, one cannot predict performance beyond five launderings; therefore, until the outer limits for effective functioning of the CM can be established over a greater number of launderings, the RF is recommended. The recommendations of the RF for protection are accompanied by suggestions to reapply the finish at a minimum rate of 0.80% dry weight gain. This study did not incorporate the effects of abrasion, sunlight, heat, and other variables on the finishes; therefore, renewal of the finish should occur at least after every fourth laundering. This study did not examine whether or not the soil-repellent treatment does make subsequent pesticide soil more difficult to remove by laundering. Such a study is suggested.

The implications from these findings must be added to previous recommendations. Previous studies have suggested multiple launderings of contaminated garments in large water volumes at hot temperatures of 60°C (140°F). Few items per load, as recommended by Easley et al [*3*] in "Laundering Pesticide Contaminated Clothing," will also reduce contamination levels. Large water volumes limit pesticide transfer to other items and to the laundering apparatus. Heavy-duty liquid detergent has been recommended as the detergent of choice in previous studies [*11,12*]. Additional consumer recommendations based on results of this study include the use of a renewable fluorocarbon-finished fabric and the use of a laundry pretreatment before laundering.

Acknowledgments

Appreciation is extended to the Pesticide Impact Assessment Program for funding of the grant "Organophosphate, Carbamate, and Organochlorine Insecticide Launderability from Contaminated Clothing" for 1983-84. The research was supported in part by the Nebraska Agricultural Research Division Project 94-011, and it contributes to Southern Regional Research Project S-163, "Effects of Functional Textile Finishes on Comfort and Protection of Consumers." Acknowledgment is also due to the industries that cooperated through the contribution of necessary information and products—3M and Mary M. Hurlocker, senior chemist, for contributing Scotchgard and FMC Chemical Co. for chemical acquisition—and to the U.S. Department of Agriculture Southern Regional Research Center for the fabrics and commercial soil-repellent finish.

This paper is registered as Paper No. 7686, Journal Series, Nebraska Agricultural Experiment Station.

References

[*1*] Baker, D. E., *Homeowner Chemical Safety: Science and Technology Guide*, University of Missouri-Columbia, Extension Division, Columbia, MO, May 1979.

[*2*] Branson D. and Henry M., "Take Cover!" *Michigan Cooperative Extension Bulletin*, July 1981.

[*3*] Easley, C. B., Laughlin, J. M., and Gold, R. E., "Laundering Pesticide Contaminated Clothing," Home Economics NebGuide HEG81-152, Cooperative Extension Service, Institute of Agriculture and Natural Resources, University of Nebraska, Lincoln, NE, November 1981.

[*4*] "Laundering Pesticides-How Ag Applicators Can Avoid Wash Day Woes," *FAA General Aviation News*, July/Aug. 1982, p. 12.

[*5*] Southwick J. W., Mecham, P. M., and Gortatowski, M. J., "Pesticide Residues in Laundered Clothing," *Environmental Chemicals: Human and Animal Health Proceedings*, 1974, pp. 125-131.

[*6*] Easley, C. B., Laughlin, J. M., Gold R. E., and Hill, R. M., "Laundry Factors Influencing

Methyl Parathion Removal from Contaminated Denim Fabric," *Bulletin of Environmental Contamination and Toxicology*, Vol. 29, 1982, pp. 461-468.

[*7*] Kawar, N. D., Gunther, G. A., Serat, W. F., and Iwata, Y., "Penetration of Soil Dust Through Woven and Nonwoven Fabrics," *Journal of Environmental Science and Health*, Vol. 13, No. 4, 1978, p. 401.

[*8*] Gehlback, S. H., Williams, W. A., and Freeman, J. I., *Archives of Environmental Health*, Vol. 34, March/April, 1979, p. 111.

[*9*] Finley, M. L. and Rogillio, J. R. B., "DDT and Methyl Parathion Residues Found in Cotton and Cotton/Polyester Fabrics Worn in Cotton Fields," *Bulletin of Environmental Contamination and Toxicology*, Vol. 4, No. 6, 1969, pp. 343-351.

[*10*] Easley, C. B., Laughlin, J. M., Gold, R. E., and Tupy, D., "Methyl Parathion Removal from Denim Fabrics by Selected Laundry Procedures," *Bulletin of Environmental Contamination and Toxicology*, Vol. 27, No. 2, July 1981, pp. 101-108.

[*11*] Easley, C. B., Laughlin, J. M., Gold, R. E., and Schmidt, K., "Detergents and Water Temperature as Factors in Methyl Parathion Removal from Denim Fabrics," *Bulletin of Environmental Contamination and Toxicology*, Vol. 28, No. 2, 1982, pp. 239-244.

[*12*] Easley, C. B., Laughlin, J. M., Gold, R. E., and Tupy, D. R., "Laundering Procedures for Removal of 2,4-Dichlorophenoxyacetic Acid Ester and Amine Herbicide from Contaminated Fabrics," *Archives of Environmental Contamination and Toxicology*, Vol. 12, 1983, pp. 71-76.

[*13*] Easter, E., "Removal of Pesticide Residues from Fabrics by Laundering," *Textile Chemist and Colorist*, Vol. 47, No. 3, March 1983, pp. 29-33.

[*14*] Finley, E. L., Graves, J. B., Summers, T. A., Schilling, P. E., and Morris, H. F., "Some Facts about Methyl Parathion Contamination of Clothing in Cotton Fields and Its Removal by Home Laundering," Lousiana State, Baton Rouge, LA, 1977.

[*15*] Kim, C. J., Stone, F. F., and Sizer, C. E., "Removal of Pesticides Residues as Affected by Laundering Variables," *Bulletin of Environmental Contamination and Toxicology*, Vol. 29, No. 1, July 1982, pp. 95-100.

[*16*] Laughlin, J. M., Easley, C. B., Gold. R. E., and Tupy, D., "Methyl Parathion Transfer from Contaminated Factors to Subsequent Laundry Equipment," *Bulletin of Environmental Contamination and Toxicology*, Vol. 27, No. 4, Oct. 1981, pp. 518-523.

[*17*] Lillie, T. H., Hamilton, M. A., Livingston, J. M., and Porter, K., "The Effects of Water Temperature on Decontamination of Pesticide Applicator Clothing," USAF Occupational and Environmental Health Laboratory Report, Sept. 1980, Paper 80:35, U.S. Air Force, Wright-Patterson Air Force Base, OH, 1980.

[*18*] Finley, E. L., Metcalfe, G. I., and McDermott, F. G., "Efficacy of Home Laundering in Removal of DDT, Methyl Parathion, and Toxaphene Residues from Contaminated Fabrics," *Bulletin of Environmental Contamination and Toxicology*, Vol. 12, No. 3, 1974, pp. 268-274.

[*19*] Lillie, T. H., Livingston, G. M., and Hamilton, M. A., "Recommendations for Selecting and Decontaminating Pesticide Applicator Clothing," *Bulletin of Environmental Contamination and Toxicology*, Vol. 27, 1981, pp. 716-723.

[*20*] Livingston, J. M., "Evaluation of Cotton and Polyester Coveralls for Protection for Pesticide," TR 78-75, USAF Occupational and Environmental Health Laboratory Technical Report, U.S. Air Force, Wright-Patterson Air Force Base, OH, Aug. 1978, p. 75.

[*21*] Davies, J. E., Enos, H. F., Barquet, A., Margade, C., Peters, L. J., and Danauskas, J. X., "Protective Clothing Studies in the Field—An Alternative to Re-entry," *Abstracts of the American Chemical Society*, Vol. 180, No. 2, Aug. 1980, p. 65.

[*22*] Gold, R. E., Leavitt, J. R. C., Holcslaw, T., and Tupy, D., "Exposure of Urban Applicators to Carbaryl," *Bulletin of Environmental Contamination and Toxicology*, Vol. 11, 1982, pp. 63-67.

[*23*] Leavitt J. R. C., Gold, R. E., Holcslaw, T., and Tupy, D., "Exposure of Professional Pesticide Applicators to Carbaryl," *Archives of Environmental Contamination and Toxicology*, Vol. 11, 1982, pp. 57-62.

[*24*] Freed, V. H., Davies, J. E., Peters, L. J., and Parveen F., "Minimizing Occupational Exposure to Pesticides: Repellency and Penetrability of Treated Textiles to Pesticide Sprays," *Residue Reviews*, Vol. 75, 1980, pp. 159-167.

[*25*] Ware, G. W., *The Pesticide Book*, San Francisco, Freeman, 1978.

Karen K. Leonas[1] and Jacquelyn O. DeJonge[1]

Effect of Functional Finish Barriers on Pesticide Penetration

REFERENCE: Leonas, K. K., and DeJonge, J. O., **"Effect of Functional Finish Barriers on Pesticide Penetration,"** *Performance of Protective Clothing, ASTM STP 900,* R. L. Barker and G. C. Coletta, Eds., American Society for Testing and Materials, Philadelphia, 1986, pp. 177-186.

ABSTRACT: The purpose of this study was to evaluate how functional finishes and different levels of laundering affect a fabric's ability to prevent or inhibit pesticide penetration. In this study, three fabrics (cotton, cotton/polyester, and polyester) treated with selected functional finishes (durable press, soil release, and water repellent) and laundered at predetermined levels (0, 10, 30, and 50 times) were exposed to methyl parathion spray. The amount of pesticide that moved through the fabric was then measured. The amount measured was compared for different finishes and between launderings to determine the effect of these variables. The application of the pesticide was completed using an enclosed spray chamber that simulated actual field conditions encountered during air blast spraying. Water-repellent and soil-release finishes were found to increase the amount of protection provided by fabrics. Fabrics treated with the durable-press finish permitted more pesticide penetration than untreated fabrics, resulting in decreased protection. Methyl parathion penetration increased with the number of launderings of the water-repellent-treated polyester fabrics; however, laundering did not significantly affect the cotton and cotton/polyester fabrics' protective characteristics.

KEY WORDS: pesticide, penetration, permeation, fabrics, durable press, water repellent, soil release, methyl parathion, finishes, protective clothing

Interest in the development of garments that are effective in preventing pesticide penetration and in barrier fabrics has been increasing, since the 1960s, when research verified a connection between exposure to pesticides and health hazard. Wolfe [*1*] found that dermal absorption rather than inhalation or ingestion was the primary route by which pesticides enter the body. Of the pesticide that enters the body, 97% is absorbed through the skin [*2*]. Different parts of the body absorb pesticide at different rates. By eliminating

[1]Graduate research assistant and department head, respectively, Department of Textiles, Merchandising and Design, College of Home Economics, The University of Tennessee, Knoxville, TN 37996-1990.

contact between the chemical and the body, the opportunity for dermal absorption to occur is eliminated. Protective clothing, as a barrier, can be the single most important factor in minimizing pesticide exposure of those agricultural workers who handle pesticides in mixing, application, or cleanup situations.

The development of accepted garments that limit dermal exposure of agricultural workers involves many areas of research. Some of these are collection of information from agricultural workers on garment characteristics, patterns of use, fabric preferences [*3*], cost [*4*], deposition of pesticides during field work [*5*], deposition of pesticides during air blast spraying [*6*], thermal acceptance levels of garment design [*7*], and removal of pesticides from fabrics when laundered [*2*].

In addition to design, acceptance, and care characteristics, the fabrics' barrier properties have also been evaluated in field [*4*] and laboratory [*8*] studies. From the studies on penetration, innovative barrier fabrics such as Gore-Tex and Tyvek have been found to have good barrier properties in preventing the penetration of the pesticide Guthion. However, little research evaluating traditional fabrics treated with durable, functional finishes has been completed.

The barrier property of a fabric refers to the fabric's ability to prevent (or inhibit) the movement of a chemical through the fabric. Two basic concepts are used to describe this movement—permeation and penetration. These terms have been defined by ASTM Subcommittee F23.30 on Chemical Resistance, a subcommittee of ASTM Committee F-23 on Protective Clothing [*9*]. Permeation is the "process by which a chemical moves through clothing on a molecular level." Penetration is the "flow of chemicals through closures, porous materials, seams, and pinholes, or other imperfections in clothing material." The first step in both these processes is sorption of the chemical at the substrate surface. Altering the surface characteristics alters the degree of sorption. Chemical means of fabric surface alteration include the application of finishes. The purpose of water-repellent finishes is to alter the surface tension of the fabric. Other chemical finishes, such as soil release and durable press, also alter the fabric's surface characteristics.

In addition to sorption, permeation and penetration are influenced by chemical movement through the material. This movement is controlled by the structure of the fabric. Laundering affects the dimensional stability of the fabric and yarn structure. As a consequence of these changes, the barrier properties of laundered and unlaundered fabrics may differ.

Experimental Work

Fabrics and Finishes

This research examined the influence of selected functional finishes and levels of laundering on the penetration of methyl parathion (MeP) through fabrics. The exposure took place under laboratory conditions that simulated field conditions of air blast spraying. Three fabrics, three finishes, and four levels of laundering were considered. The fabrics included 100% cotton, 50% cotton/50% polyester blend, and 100% polyester fiber contents. The fabrics treated were 130 g/m^2 (3.5-oz/yd^2) print cloth. The thickness and yarn counts, shown in Table 1, were similar. All the yarns were produced from staple fibers. The functional finishes selected for this project were a Quarpel water-repellent, acrylic acid soil-release finish, and a durable-press finish. All the fabrics were prepared at the University of Tennessee Agricultural Experiment Station, Southern Region, in New Orleans, Louisiana, for Project S-163.

The formulation for the water-repellent-finished fabric was 7.0% Zepel D, 10% Norane F, 0.4% Mykon NRW3, and 83.6% water. The fabrics were processed using two dips and two nips padding to obtain a wet pickup of ap-

TABLE 1—*Fabric yarn count and thickness.*

Fabric	Treatment	Launderings, No.	Yarn Count, ends/cm[a] Warp	Fill	Thickness, mil[b]
Cotton	none	0	33	30	12
		50	32	28	12
	water repellent	0	33	28	10
		10	32	28	11
		30	32	28	12
		50	32	28	12
	soil release	0	34	30	10
	durable press	0	32	29	10
Cotton/polyester	none	0	34	29	11
		50	33	28	12
	water repellent	0	33	29	10
		10	32	28	11
		30	31	29	11
		50	31	28	12
	soil release	0	29	26	11
	durable press	0	34	29	10
Polyester	none	0	33	29	10
	water repellent	0	33	30	10
		10	33	29	10
		30	32	28	10
	soil release	0	32	27	9

[a]Fabric yarn count was determined using ASTM Test for Fabric Count of Woven Fabric (D 3775-79). The results are based on three replications. The mean is reported.

[b]Fabric thickness was determined using the ASTM Method for Measuring Thickness of Textile Materials [D 1777-64 (1975)]. Results are based on three replications. The mean is reported.

proximately 70%. Treated fabrics were dried at 160°C for 3 min using a tenter frame. The acrylic acid soil-release formulation was 9.4% Emory Soft 7777, 4.4% Industrial ML-14, 15.6% Acrysol ASE-60, 15.6% Glorez CP-7, and 55.0% water. The fabrics were padded two dips, two nips to obtain a net pickup of 75%. They were then dried at 110°C for 2 min using a tenter frame. The formulation for the durable-press finish was 20.0% Protocol C, 5.0% Curite HC, 2.8% Protolube HD, 2.8% Protolube L-20, 0.25% Protowet 100, and 69.15% water. The processing procedure was two dip, two nip padding to achieve a net pickup of 75%. The treated fabrics were then dried at 70°C for 1.5 min and cured at 165°C for 1.5 min on a tenter frame. All the fabrics were prewashed before testing to remove any loose auxiliary chemicals remaining on the fabric surface.

To determine the effect of laundering, the treated and untreated specimens were laundered 0, 10, 30, and 50 times, in accordance with the American Association of Textile Chemists and Colorists (AATCC) Test for Colorfastness to Washing, Domestic; and Laundering, Commercial: Accelerated (61-1980, IIA). Finishes are considered durable if they are functional throughout the life of the garment [*10*]. The level of laundering (50 times) was chosen as a maximum point in this study to simulate laundering over the life of the garment. The intervals selected (between 0 and 50) were based on the level of acceptance used in evaluating flame-retardant-finished fabrics. To determine the initial repellency of all fabrics, the AATCC Test for Water Repellency: Spray Test (22-1980) was used. From these tests, the soil-released (SR) and durable-press-(DP) finished fabrics showed no change in repellency at 0, 10, 30, or 50 launderings; therefore, only the unlaundered specimens of the SR and DP-treated fabrics were used through the remainder of the study. Three replications of each fabric/finish/laundering combination chosen were exposed to pesticide and analyzed in this study. These combinations are shown in Fig. 1.

Exposure

Methyl parathion, an organophosphate, was the chemical used in this study. An emulsifiable concentrate formulation was prepared in water to make a 0.12% solution. An enclosed spray chamber designed to simulate field conditions of air blast spraying was used. This method allows the specimen to be exposed to a controlled amount of pesticide. The pesticide is applied using a tee-jet fan spray nozzle No. 730023 obtained from SprayCo Products which transverses above the fabric. The nozzle height remained constant at 30.48 cm above the fabric specimens surface. One pass of a 6.2-s duration from edge to edge of each specimen delivers 0.59 mg of pesticide.

The exposed specimen was a three-layer composite, 15.24 by 15.24 cm square, consisting of the test fabric, a collector layer, and foil. In this study, the test fabric is being evaluated for its ability to prevent pesticide penetra-

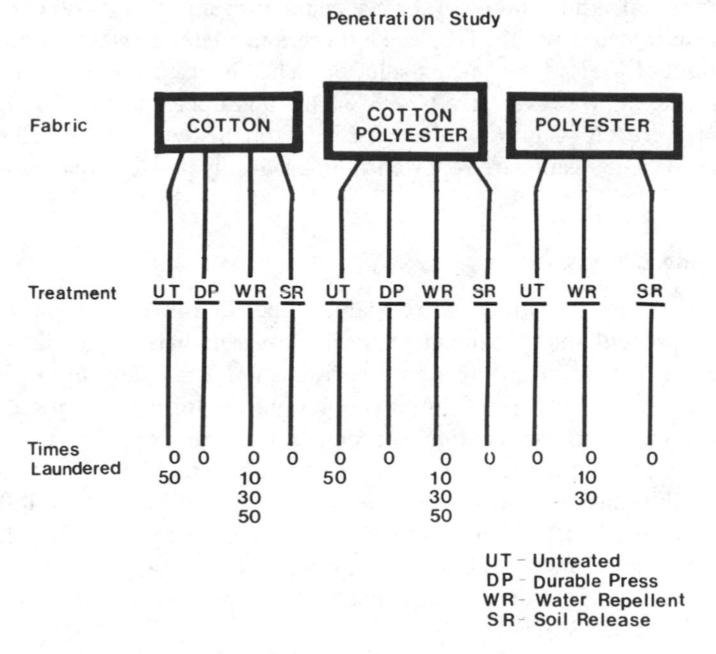

FIG. 1—*Fabrics used in the penetration study.*

tion. The collector layer was a 50% cotton/50% polyester blend tee-shirt knit fabric supplied by Standard Knitting Mills of Knoxville, Tennessee. The three layers were combined, face fabrics upward, using double stick tape at each edge and between each layer and masking tape around the perimeter. This results in a stable composite specimen and prevents any transfer of pesticide around the edge of the fabric. A random order of exposure was determined statistically, using an incomplete block design to minimize any systematic errors in the spray system.

After exposure, the specimens were removed from the chamber and allowed to dry for 1 h. Although temperature and humidity were not monitored, conditions for exposure and drying (that is, location and date) were the same for all specimens. The layers were then separated and a 7.62 by 2.54-cm swatch was cut from the center of each layer for analysis. Using the center eliminates any edge effects occurring.

Analysis

Each swatch was analyzed using gas chromatography (GC) techniques to determine the amount of pesticide residue on each layer of the specimen. The pesticide was extracted from the fabric specimen using an adaptation of the extraction method established by Easter [11]. Pesticide was extracted from each of the foil, knit, and test fabric specimen layers. A 2-μL sample of the extract was injected into the GC, and a thermionic detector was used to detect the amount of MeP in the extract solution. The GC used was a Varian 3700. Column packing was 1.5% 2250/1.95% 2401 on Supelcoport 100/120. Temperatures were 220°C for the injector, 270°C for the detector, and 200°C for the column. The results are reported in micrograms per square centimetre.

Results and Discussion

The purpose of this study was to evaluate specific fabrics in terms of their ability to prevent the movement of pesticide through them. In this study, whether permeation or penetration occurred is not important, but rather the protectiveness of the fabric. From this point on, the term penetration will be used to discuss both permeation and penetration concepts. Each layer was analyzed to determine the amount of pesticide. The amount of pesticide detected on the foil layers was negligible, ranging from 0.000001 to 0.000006 $\mu g/cm^2$. Since interest lies in the amount of pesticide that penetrated through the fabric, it is the amount of pesticide extracted from the collector layer that is discussed here. The amounts of pesticide extracted from each collector layer are listed in Table 2.

Fabrics and Finishes

Each of the finishes used in this study had an effect on the barrier properties of the fabrics (Fig. 2). The water-repellent finish was effective in reducing the amount of pesticide penetration on all three fabrics. However, this finish did not prevent all penetration, and pesticide was extracted from the knit layer of those multilayer specimens. The amount of penetration was reduced by approximately 50% in each case.

The soil-release finish also improved the barrier property characteristics of the fabrics, resulting in reduced pesticide residue extracted from the knit layer of the multilayer specimen. The hydrophilic/hydrophobic nature of the finish accounts for this improvement in barrier properties. This type of finish (acrylic acid) is also known for forming a film on the fabric, which would alter the surface characteristics, resulting in a change in sorption.

The durable-press finish resulted in increased pesticide penetration. The cotton fabric treated with dimethyl dihydroxyethylene urea (DMDHEU) showed only a slight increase in the penetration. However, the cotton/polyester blend fabric treated with DMDHEU showed a much more sizable increase in the amount of pesticide penetrating the fabric. This result—decreased protection of DMDHEU-treated fabrics—is of concern since the majority of work clothes used by agricultural workers are durable-press-finished fabrics.

TABLE 2—*Pesticide extracted from collector layer.*

Fabric	Treatment	Launderings, No.	Mean Amount Penetrated, ($\mu g/cm^2$)	SD Between Replicates[a]
Cotton	none	0	0.0011	0.0004
		50	0.0013	0.0004
	water repellent	0	0.0003	0.0004
		10	0.0003	0.0003
		30	0.0016	0.0008
		50	0.0012	0.0002
	soil release	0	0.0006	0.0002
	durable press	0	0.0012	0.0005
Cotton/polyester	none	0	0.0015	0.0010
		50	0.0015	0.0005
	water repellent	0	0.0008	0.0004
		10	0.0010	0.0006
		30	0.0008	0.0007
		50	0.0004	0.0000
	soil release	0	0.0007	0.0002
	durable press	0	0.0023	0.0009
Polyester	none	0	0.0042	0.0006
	water repellent	0	0.0014	0.0005
		10	0.0022	0.0007
		30	0.0047	0.0101
	soil release	0	0.0021	0.0009

[a]SD = standard deviation.

Effects of Laundering

The cotton fabric showed a slight decrease in protection when laundered 50 times. The water-repellent cotton fabrics showed no change in protection between the unlaundered treated specimen and the treated specimen laundered 10 times. There was a large decrease in protection after 30 launderings. The water-repellent-treated cotton fabric laundered 50 times provided protection similar to that of the untreated fabric.

In the cotton/polyester untreated fabric specimens there was no change in protection between the unlaundered specimen and those specimens laundered 50 times. The water-repellent-treated fabric laundered 50 times provided the greatest amount of protection compared with all other cotton/polyester fabrics evaluated. This increase in protection could be explained by the physical degradation of the staple fiber yarn that occurs after repeated launderings. This results in increased fiber ends projecting from the yarn, which act as a filtering mechanism reducing the flow of the pesticide.

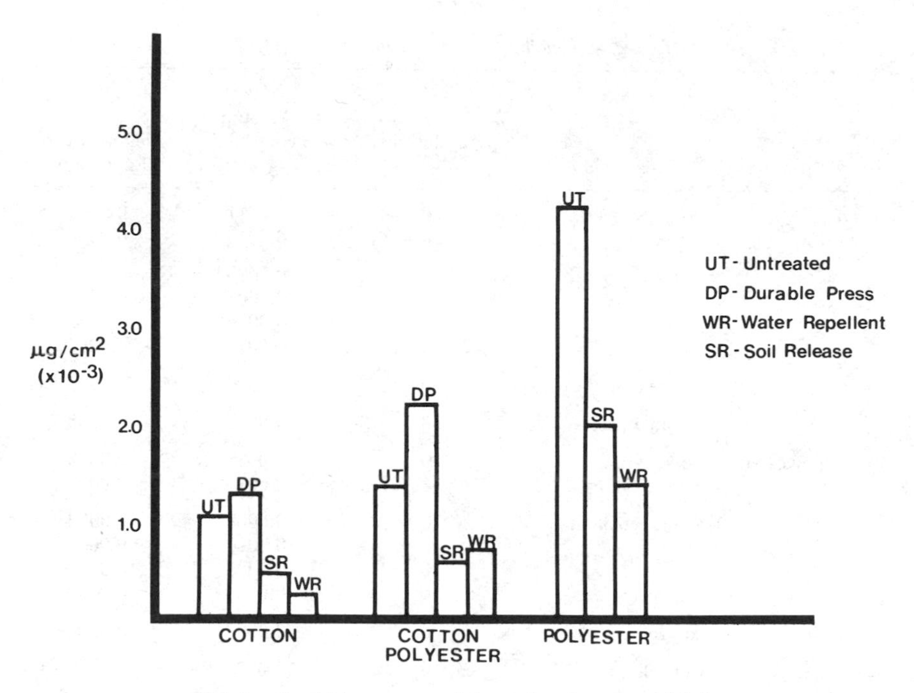

FIG. 2—*Pesticide penetrated through unlaundered fabrics.*

The protection provided by the polyester fabric treated with the water-repellent finish was very sensitive to laundering (Fig. 3). The treated, unlaundered specimen provided almost three times as much protection as the same specimen laundered 30 times. Noticeably, the polyester fabric performs differently from the cotton/polyester fabric after laundering. The polyester fiber, being uncrimped and having a smooth surface with a circular cross section, reduces the amount of friction occurring between fibers in a yarn. As the polyester is laundered, the yarns abrade and loss of fiber occurs. In the cotton/polyester fabric, the cotton fibers have natural crimp, nonsmooth surfaces, and crescent-shaped cross sections, which results in increased contact between fibers and increased friction, reducing fiber loss.

Statistical Analysis

Statistical analysis was completed to evaluate the significance between various finishes and laundering levels and pesticide penetration. Using the *t*-test and analysis of variance, no significant differences were found in finishes or numbers of launderings. It was determined that this lack of significance could be due to several factors, including the number of replications, concentration of the pesticide solution, and the amount of pesticide applied. How-

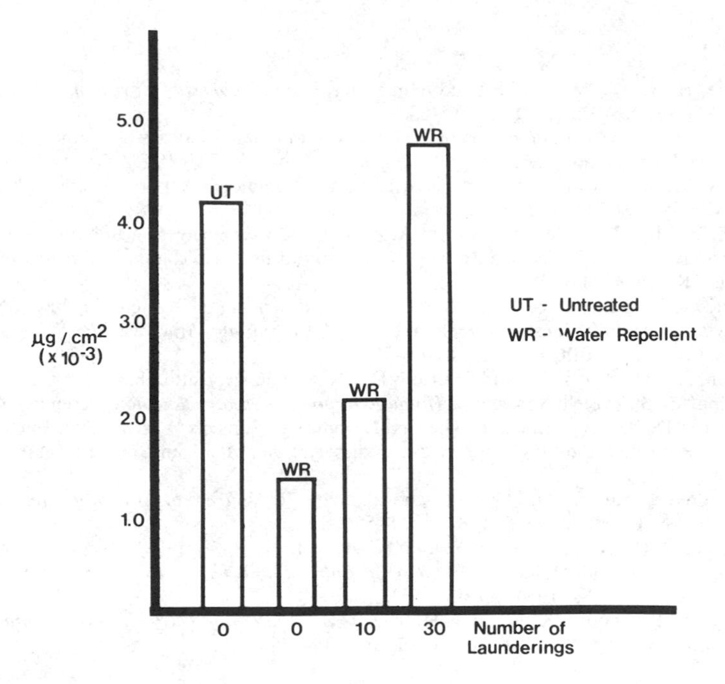

FIG. 3—*Pesticide penetrated through laundered, water-repellent-treated and unlaundered, untreated polyester fabric.*

ever, trends were detected showing a decrease in pesticide penetration when fabrics were treated with soil-release or water-repellent finishes.

Conclusion

Treatment with selected durable, functional finishes of Quarpel water-repellent and acrylic acid soil release increased the capability of cotton, cotton/polyester, and polyester fabrics to provide protection against the penetration of the pesticide methyl parathion. DMDHEU-treated cotton and cotton/polyester fabrics resulted in decreased barrier properties.

Laundering did influence the barrier properties of all the fabrics tested, but the extent of this effect was dependent on the fiber. The effect of laundering was most pronounced on the polyester fabrics, with a noticeable decrease in protection after laundering.

Acknowledgment

This research was funded in part by the Agricultural Experiment Station, University of Tennessee, Knoxville, as part of the Southern Region Project S-163.

References

[1] Wolfe, H. R., Durham, W. F., and Armstrong, J. F., *Archives of Environmental Health*, Vol. 14, No. 4, April 1967, pp. 622–633.
[2] Easter, E. P., "Decontamination of Pesticide Contaminated Fabrics by Laundering," unpublished dissertation, University of Tennessee, Knoxville, TN, 1982.
[3] Henry, M. S., "Users' Perceptions of Attributes of Functional Apparel," unpublished thesis, Michigan State University, East Lansing, MI, 1980.
[4] King, R. M., "The Effect of Garment Attributes and Risk Factors on Adoption/Purchase Decisions for Pesticide Protective Apparel," unpublished dissertation, University of Tennessee, Knoxville, TN, 1984.
[5] Serat, W. F. and Van Loon, A., "Some Factors Influencing the Design of Fabrics for Protective Garments for Pesticide Field Workers," *Community Studies of Pesticides*, State Department of Health, Berkeley, CA, 1979.
[6] DeJonge, J. O., Ayres, G., and Branson, D., "Pesticide Deposition Patterns on Garments During Air Blast Field Spraying," *Home Economics Research Journal*, December 1985.
[7] Branson, D. H., "Assessment of Thermal Response of Subjects Wearing Functionally Designed Protective Clothing," unpublished dissertation, Michigan State University, East Lansing, MI, 1982.
[8] Orlando, J., Branson, D., Ayres, G., and Leavitt, R., *Journal of Environmental Science and Health*, Vol. B16, No. 5, 1981, pp. 615–628.
[9] Showpe, A. D., *ASTM Standardization News*, Vol. 11, No. 7, July 1983, pp. 19–22.
[10] Wingate, I. B. and Mohler, J. F., *Textile Fabrics and Their Selection*, Prentice Hall, Englewood Cliffs, NJ, 1984, pp. 152–153.
[11] Easter, E. P., Leonas, K. K., and DeJonge, J. O., *Bulletin of Environmental Contamination and Toxicology*, Vol. 31, Dec. 1983, pp. 738–744.

S. Kay Obendorf[1] *and Camille M. Solbrig*[1]

Distribution of Malathion and Methyl Parathion on Cotton/Polyester Unfinished and Durable-Press Fabrics Before and After Laundering as Determined by Electron Microscopy

REFERENCE: Obendorf, S. K. and Solbrig, C. M., **"Distribution of Malathion and Methyl Parathion on Cotton/Polyester Unfinished and Durable-Press Fabrics Before and After Laundering as Determined by Electron Microscopy,"** *Performance of Protective Clothing, ASTM STP 900*, R. L. Barker and G. C. Coletta, Eds., American Society for Testing and Materials, Philadelphia, 1986, pp. 187-204.

ABSTRACT: The distributions of the organophosphorus pesticides malathion and methyl parathion on 50% cotton/50% polyester fabric with and without a durable-press finish were studied by electron microscopy. Osmium tetroxide was used to tag the pesticides for the study of their distribution on the yarns and fibers by backscattered electron imaging and energy-dispersive X-ray microanalysis. Distributions of malathion and methyl parathion on the unlaundered specimens were similar, with high concentrations present on the surfaces of both polyester and cotton fibers and in the cotton lumen. No pesticide was observed inside the polyester fibers. Laundering removed 60 to 70% of each pesticide by cleaning the surfaces of the fibers but had little effect on the concentration of pesticide in the cotton lumen. Laundering removed more malathion than methyl parathion from the surfaces of the polyester fibers. The durable-press finish had no effect on the total amount retained or on the distribution of malathion after laundering, while slightly more methyl parathion was observed on the unfinished fabric.

KEY WORDS: malathion, methyl parathion, pesticides, cotton fibers, polyester fibers, washing, electron microscopy, protective clothing, decontamination, chemical analysis

The retention of pesticides on clothing is an occupational concern for pesticide applicators and field workers, since many pesticides are readily absorbed into the body through the skin. The protective clothing often recommended

[1]Professor and research support specialist, respectively, Department of Textiles and Apparel, New York State College of Human Ecology, Cornell University, Ithaca, NY 14853-4401.

consists of a "clean" long-sleeved shirt and long-legged trousers of closely woven fabric [1]. Although laundering procedures are recommended for decontamination, the effectiveness of various laundering conditions is not clearly defined. The penetration and retention of a variety of pesticides on different fabrics have been studied under a range of end-use conditions [2-9]. The purpose of this study was to investigate the microscopic distribution of the organophosphorus pesticides malathion and methyl parathion on unfinished and durable-press-finished 50% cotton/50% polyester fabric before and after laundering.

Procedure

Pesticides

The organophosphorus pesticides used to contaminate the fabrics were malathion and methyl parathion (Fig. 1). The physiochemical parameters of these pesticides, which can be absorbed dermally by the body, thus causing systemic poisonings and illnesses [10], are listed in Table 1 [11,12].

Fabrics

Plain woven 50% cotton/50% polyester fabric unfinished and treated with a durable-press finish was supplied by the U.S. Department of Agriculture

MALATHION

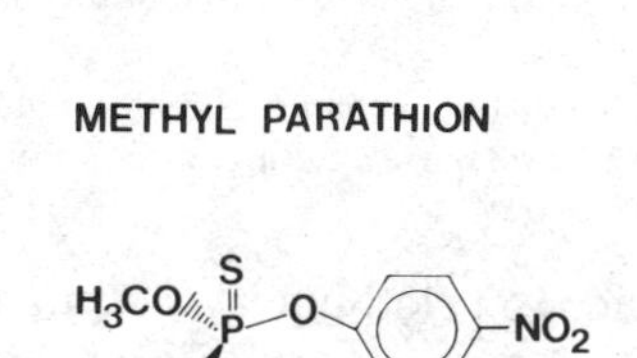

METHYL PARATHION

FIG. 1—*Chemical structures of the organophosphorus pesticides used to contaminate the fabrics.*

TABLE 1—*Physiochemical parameters of malathion and methyl parathion* [11,12].

Parameter	Malathion	Methyl Parathion
IUPAC name	S-1,2-*bis*[ethoxycarbonyl]ethyl *0,0*-dimethylphosphorodithioate	*0,0*-dimethyl *0*-4-nitrophenyl phosphorothioate
Melting point	276 K	308 K
Vapor pressure	1.33 MPa at 293 K	1.29 MPa at 293 K
Solubility		
water	145 mg/cm^3 at 293 K	55 to 60 mg/cm^3 at 298 K
organic solvents	highly soluble, except in alkanes	soluble in most organic solvents
Decomposition	below pH 5 and above pH 7	by alkali (above pH 7)

(USDA) Southern Regional Research Center. Cotton fibers 2.86 cm in length were blended with 3.81-cm staple polyester fibers using weigh-pan hoppers. The blended fabric was desized, scoured, bleached, and soured prior to finishing, as well as being heat set for 15 s at 461 K. The portion of the fabric with a durable-press finish was treated with 20% Protocol C [dimethyloldihydroxyethyleneurea (DMDHEU)], 5% Curite HC [a mixed catalyst of magnesium chloride ($MgCl_2$) and aluminum chloride ($AlCl_3$)], 2.8% polyethylene softener, 2.8% fatty acid softener, 0.25% nonionic wetting agent, and 69.15% water. The average wet pickup was 72%. The fabric was dried for 90 s at 338 K and cured for 90 s at 442 K.

The fabric was characterized in accordance with the ASTM Test for Fabric Count of Woven Fabric (D 3775-84) and the ASTM Test for Weight (Mass) per Unit Area of Woven Fabric (D 3776-84). Values for yarn number and yarn twist are based on cotton count data and the twist multipliers supplied with the fabric. The parameters are as follows: weight = 122 g/m^2; fabric count = 75 by 81; yarn number = warp, 20 mg/m; weft, 16 mg/m; twist = warp and weft, 800 turns/m.

All the fabric specimens were given an afterwash treatment consisting of a presoak in 0.5% (weight/volume) sodium carbonate for 21.6 ks, Soxhlet extraction with deionized, distilled water for 21.6 ks, air drying, and Soxhlet extraction in toluene for 43.6 ks followed by air drying.

Contamination Procedures

Before contamination, fabric specimens (8 by 8 cm) with machine zigzag stitched edges were conditioned for 86.4 ks, in accordance with the ASTM Practice for Conditioning Textiles for Testing (D 1776-79). A fixed-volume Eppendorf pipette was used to deliver either 10 mm^3 of malathion (Cythion) (91% active ingredient) or 100 mm^3 of a solution of 111 g/cm^3 methyl parathion (99.8% active ingredient) in anhydrous methanol onto the fabric specimen, which was held taut in a Plexiglas frame under a fume hood. The contaminated specimens were air dried for 3.6 ks under the fume hood before they were stored in a foil-lined glass desiccator containing a saturated solution of magnesium acetate, which produced an atmosphere of 65% relative humidity in the enclosed container. The fabric specimens in the sealed desiccator were placed in the conditioning room (294 K) for 259 ks before laundering or extraction of the pesticide residues.

Determination of the Quantity of Pesticide Residue

The pesticides were solvent extracted from the unlaundered and laundered fabric specimens using ethanol for malathion and methanol for methyl parathion. Each fabric specimen was placed in an Erlenmeyer flask and shaken on an automatic laboratory shaker with 40 cm^3 of alcohol for 3.6 ks. After a second extraction, the two aliquots were combined in a 100-cm^3 volumetric flask and brought to the total volume with additional solvent.

The quantity of malathion was determined by a colorimetric method using four replications for each test condition [*13–15*]. A measured aliquot of the pesticide extract was transferred to a separatory funnel and diluted to a 25-cm^3 volume with anhydrous ethanol. A 2-cm^3 volume of 0.5 *N* sodium hydroxide was swirled into the solution and set aside for 120 s to allow the malathion to hydrolyze, after which 75 cm^3 of an acidified 0.02% (weight/volume) ferric chloride solution [8 cm^3 concentrated hydrochloric acid (HCl) to 1000 cm^3 of solution] was added. The solution was allowed to stand for 300 s before 50 cm^3 of carbon tetrachloride and 2 cm^3 of a 1% copper sulfate solution were added to the solution in the separatory funnel. After the mixture was shaken vigorously for 60 s, the two phases were allowed to separate, and a portion of the yellow carbon tetrachloride layer was decanted into a cuvette for colorimetric analysis. A blank consisting of 25 cm^3 anhydrous ethanol, 2 cm^3 of 0.5 *N* sodium hydroxide, 75 cm^3 ferric reagent, 50 cm^3 carbon tetrachloride, and 2 cm^3 of 1% copper sulfate solution was used as a reference. A standard curve of absorbence at 420 nm versus malathion concentration was prepared using solutions of known malathion concentrations. The absorptions at 420 nm measured using a Coleman Junior spectrophotometer were utilized in determining the amount of malathion residue on each fabric specimen.

The quantity of methyl parathion on a single specimen for each test condition was determined by electron capture gas chromatography. The extract was diluted in hexane at a ratio of 1000 to 1. A 1-mm^3 aliquot of the hexane solution was injected into a Sigma 3B Perkin-Elmer gas chromatograph equipped with a nickel-63 electron capture detector. The gas chromatograph with a 1.8-m column packed with 3% OV-17 on Gas Chrom Q was operated at 473 K using nitrogen as the carrier gas.

Laundry Procedure

After a 259-ks storage period, each specimen for the pesticide retention analysis was laundered individually in the Atlas Launder-Ometer with 200 cm^3 of a 0.175% (weight/volume) solution of a powdered detergent, containing anionic surfactant with carbonate and zeolite builders (Tide), and ten steel balls for 720 s at 311 K. After two 300-s rinses at 294 K using 200 cm^3 of water, the specimens were air dried overnight in the conditioning room.

Electron Microscopy

The distribution of the pesticides on laundered and unlaundered fabrics was determined using backscattered electron imaging and X-ray microanalysis [*16,17*]. To tag the pesticide with osmium, warp yarns extracted from the center of the contaminated unlaundered or laundered specimens were placed in 2 cm^3 of 2% (weight/volume) osmium tetroxide (OsO_4) in water for 10.8 ks. The specimens were then rinsed in water for three 1200-s periods and air dried.

Yarns were embedded in Spurr low-viscosity resin and cured for 28.8 ks at 343 K. The thick, cross-sectional specimens (20 to 30 μm) were cut using glass knives on a Du Pont Sorvall MT-1 ultramicrotome. Several embedded cross sections and one longitudinal yarn specimen for each treatment were placed on a carbon stub with double-stick tape. The edges of the tape and the ends of the yarn were sealed with carbon paint. Microscopy specimens were coated with carbon in a Veeco VE400 high vacuum evaporator.

Backscattered electron images and energy-dispersive X-ray data were obtained using a Japan Electron Optics Laboratory (JEOL) JSM-35CF scanning electron microscope equipped with a Tracor Northern energy dispersive X-ray analyzer. The accelerating voltage was 2.4 fJ (15 keV), and the specimen current was 2.0 nA. Counts of X-ray emission from osmium were recorded at 0.304 fJ (1.90 keV) for seven morphological locations on the cotton fibers [*16*] and for three locations with three observations each on the polyester fibers. Six randomly selected cross-sectional specimens of each treatment were taken from three separate warp yarns of two fabric specimens; two cotton and two polyester fibers were analyzed for each cross section selected. One of these fibers was located in the center of the yarn cross section, whereas the other fiber of the same type was located near the yarn surface.

Statistical Analyses

Analyses of variance were calculated for the relative concentrations of pesticide residue using the X-ray data. Tukey's studentized range test at the 0.1 alpha level was used to compare multiple means.

Results and Discussion

When exposed to splashes or spills during pesticide usage, a field worker's clothing can become contaminated. The amount of pesticide absorbed by clothing has been observed to be higher for cotton fabrics than for polyester fabrics [*5*]. On contaminated and unlaundered fabrics, we observed that distributions of the organophosphorus pesticides malathion and methyl parathion were very similar within the yarn and fiber structures (Figs. 2 and 3). Malathion was uniformly distributed on the fiber surfaces across the yarn bundle of the contaminated, unlaundered fabric (Figs. 4 and 5). Although the concentrations of methyl parathion on the polyester fiber surfaces were similar over the yarn cross section ($P = 0.74$) (Fig. 4 and Table 2), higher concentrations were observed on the cotton fibers located near the yarn surface ($P = 0.0003$) (Fig. 5 and Table 2). This observation could be related to one or more factors, including the use of a methyl parathion solution for contamination, the absorbent nature of cotton fibers, and the tendency of less mature cotton fibers to migrate toward the surfaces of blended yarns.

High concentrations of both malathion and methyl parathion were detected on the surfaces of cotton and polyester fibers and in the lumen of the cotton fibers from the contaminated, unlaundered fabric (Figs. 2, 3, 6, and 7). Little or no pesticide was observed inside the polyester fibers (Figs. 2, 3, and 6). This observation is in accordance with our previous studies of the distribution of body oils (oleic acid, monoolein, and triolein) on polyester/cotton blended fabrics [*16,18,19*].

To assess the incompatibility of the pesticides with polyester in terms of cohesive energy and the concept of solubility, the solubility parameters of malathion and methyl parathion were calculated based on their chemical structures [*20*] for comparison with the solubility parameters of polyester [*21*] (Table 3). Both malathion and methyl parathion are observed to fall just outside the region of the Hansen solubility parameter plot, in which Knox et al [*21*] observed solvent-induced swelling of drawn polyester fibers that resulted in 3% or more shrinkage after 90 days. The solubility parameter analysis is in agreement with our experimental observation that the malathion and methyl parathion are not deposited in the polyester fibers.

Laundering once with a built, powdered detergent (pH 11) removed 60 to 70% of the malathion and methyl parathion from the fabrics (Table 4). Microscopic analyses showed that the pesticides were removed from the surfaces of the fibers (Figs. 2, 3, and 6 through 9). However, little change in the concentration of methyl parathion on the surfaces of polyester fibers was observed (Figs. 4 and 6). The observed removal of malathion and methyl parathion by laundering may be enhanced by hydrolysis of these pesticides in alkaline medium (Table 1).

With laundering, malathion was removed preferentially from the surfaces of cotton fibers that were located near the surfaces of the yarns (Fig. 5). The

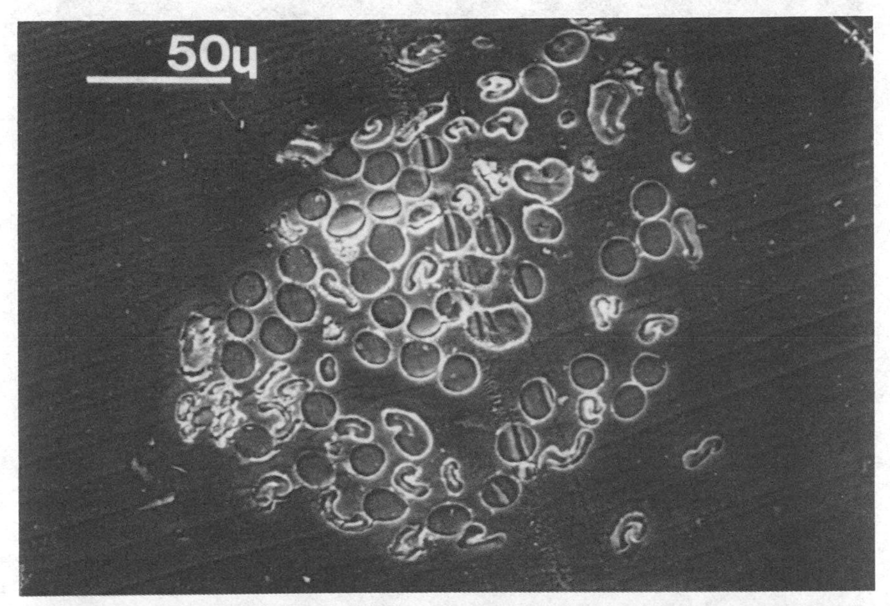

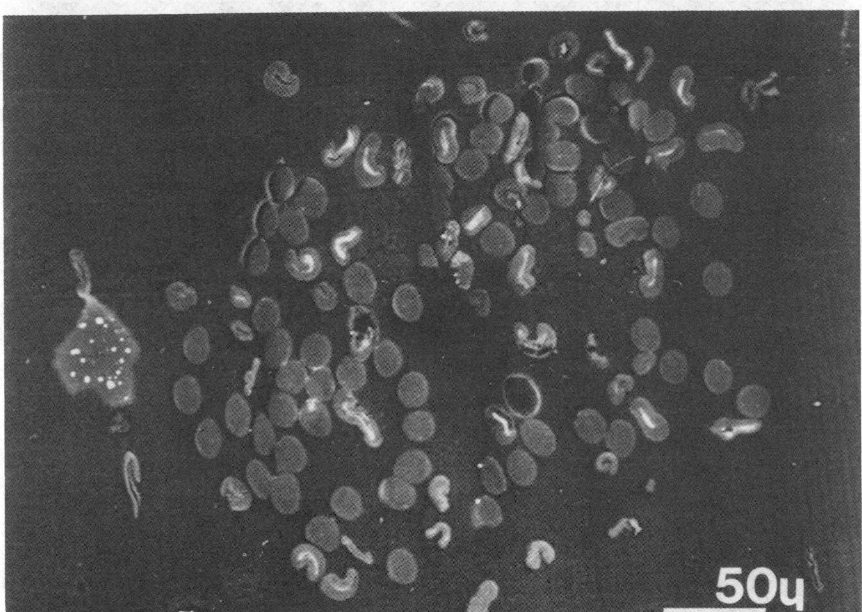

FIG. 2—*Backscattered electron images of the malathion distribution on warp yarns from durable-press (DMDHEU) 50% cotton/50% polyester fabrics before* (top) *and after* (bottom) *laundering with a built, powdered detergent at 311 K.*

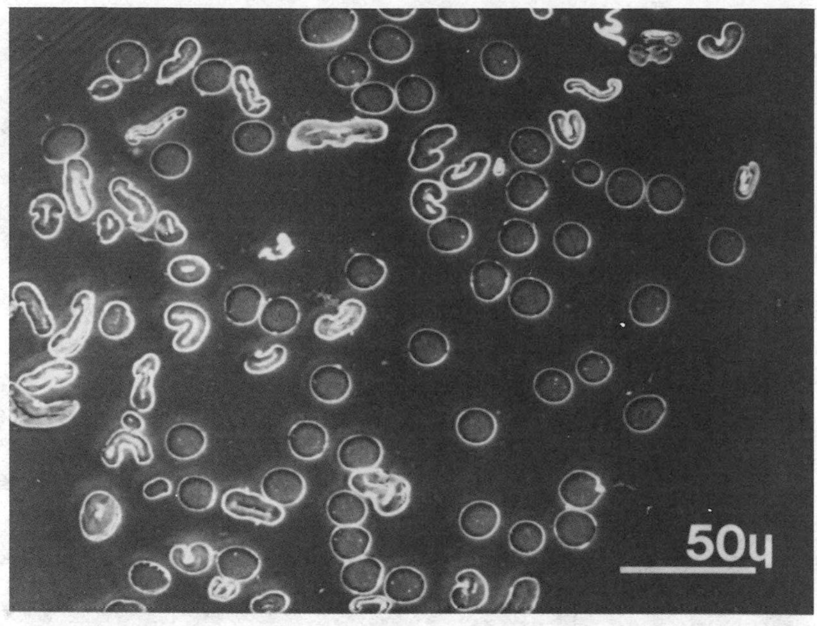

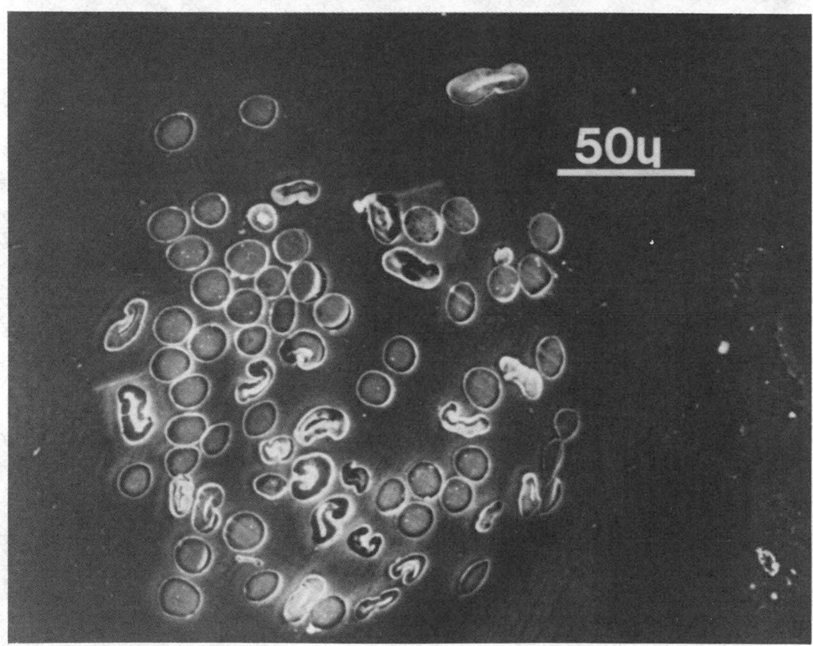

FIG. 3a—*Backscattered electron images of the methyl parathion distribution on unfinished 50% cotton/50% polyester fabric before* (top) *and after* (bottom) *laundering with a built, powdered detergent at 311 K.*

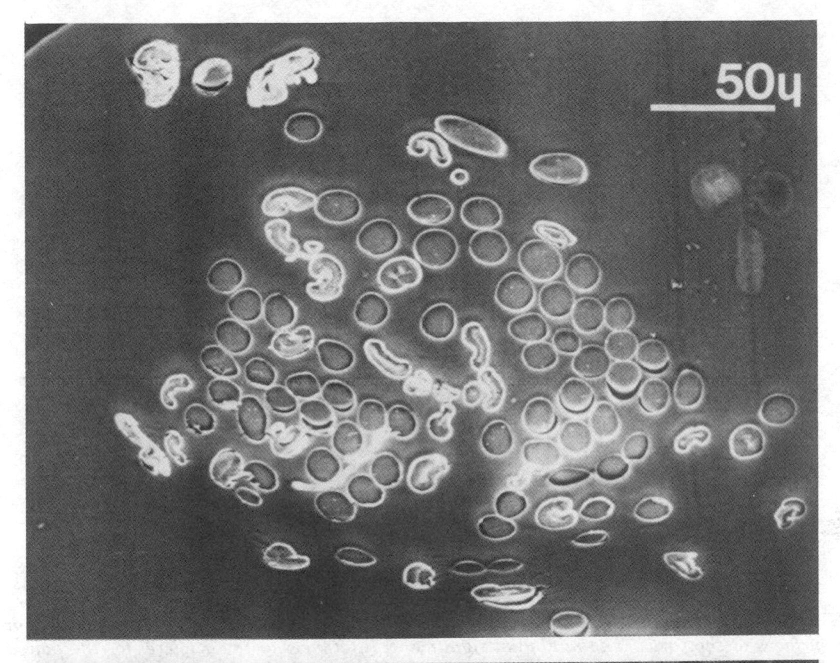

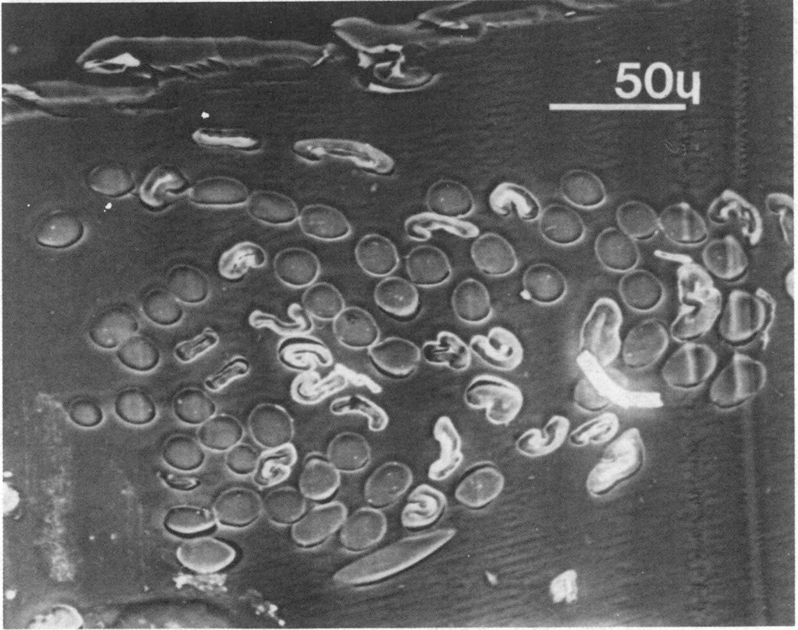

FIG. 3*b*—*Backscattered electron images of the methyl parathion distribution on durable-press 50% cotton/50% polyester fabric before* (top) *and after* (bottom) *laundering with a built, powdered detergent at 311 K.*

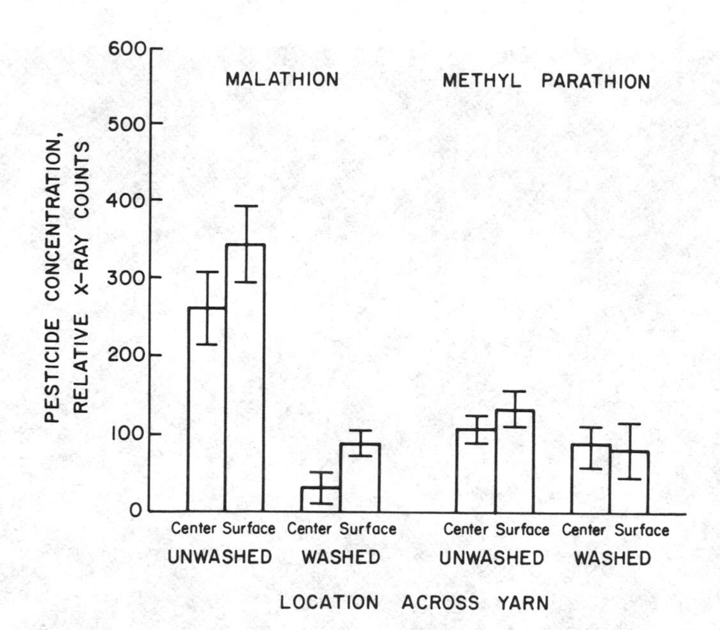

FIG. 4—*Retention of pesticides on the surfaces of polyester fibers located in the center or near the surfaces of the yarns. The minimum significant differences at the 0.1 level for the paired means, from left to right in the figure, are 109, 46, 38, and 77.*

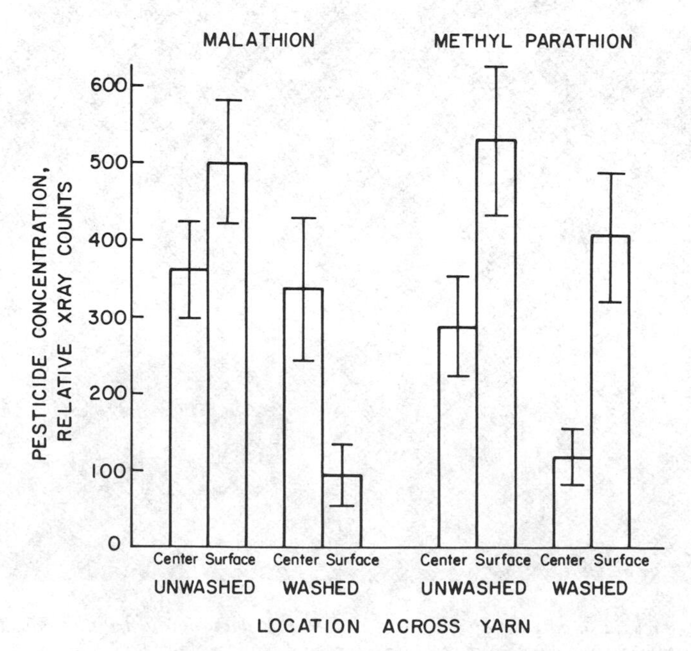

FIG. 5—*Retention of pesticides on the surfaces of cotton fibers located in the center or near the surfaces of the yarns. The minimum significant differences at the 0.1 level for the paired means, from left to right in the figure, are 168, 164, 196, and 153.*

TABLE 2—*Summary of analyses of variance arranged by fiber and pesticide.*

Source	Cotton				Polyester			
	Malathion		Methyl Parathion		Malathion		Methyl Parathion	
	F^a	$P > F^b$	F	$P > F$	F	$P > F$	F	$P > F$
Selected fiber areas	1.43	0.24	4.85	0.009	31.78	0.0001	8.25	0.0004
Laundry	9.42	0.003	4.21	0.04	76.25	0.0001	2.06	0.15
Finish	1.29	0.26	3.30	0.07	0.18	0.67	9.00	0.003
Location in yarn	0.56	0.46	13.68	0.0003	6.17	0.014	0.11	0.74
Fiber area by laundry	4.35	0.015	0.99	0.37	14.84	0.0001	0.49	0.61
Fiber area by finish	0.98	0.38	1.59	0.21	0.19	0.82	0.59	0.56
Fiber area by location in yarn	0.96	0.39	0.48	0.62	0.23	0.79	0.30	0.74

[a]Calculated F ratio.
[b]$P > F$ = probability that the calculated F ratio is greater than the standard F distribution.

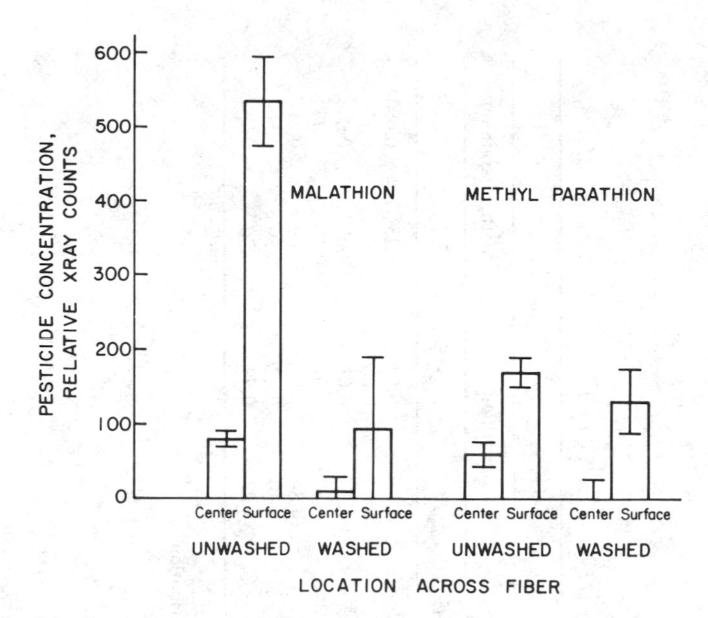

FIG. 6—*Distribution of the pesticides on the surface and within the polyester fibers. The minimum significant differences at the 0.1 level for the paired means, from left to right in the figure, are 125, 69, 53, and 114.*

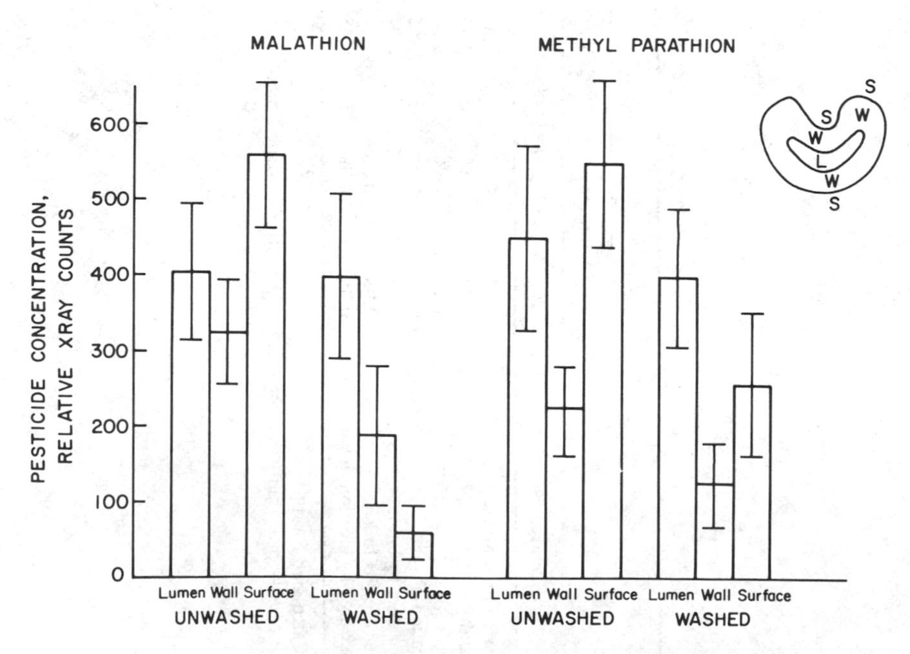

FIG. 7—*Distribution of the pesticides on the surfaces and within the cotton fibers. The minimum significant differences at the 0.1 level for the grouped means, from left to right in the figure, are 257, 251, 300, and 244.*

TABLE 3—*Solubility parameters, δ(J/cm³)^(1/2)*.[a]

	δ Total	δ Dispersive	δ Polar	δ H-Bonding
Malathion	23.73	15.20	15.71	9.25
Methyl parathion	28.88	23.89	11.95	11.00
Cellulose	27.53	18.70	19.27	6.12
Polyethylene terephthalate				
aromatic segment	20.05[b]	19.55	0.21	4.38
aliphatic segment	24.75[b]	21.93	5.44	10.10

[a]Values were calculated by the method of van Krevelen and Hoftyzer [*20*].
[b]Values are quoted from Knox et al [*21*].

TABLE 4—*Amounts of pesticides retained by fabrics after laundering.*

Finish	Pesticide	Unwashed,[a] mg	Washed, mg	Method of Analysis
Unfinished	malathion	11.2	3.6 (0.3)[b]	colorimetric
Durable press, DMDHEU	malathion	11.2	3.7 (0.4)[b]	colorimetric
Unfinished	methyl parathion	11.1	4.4[c]	gas chromatography
Durable press, DMDHEU	methyl parathion	11.1	3.8[c]	gas chromatography

[a]The amount of pesticide on unwashed fabric was calculated from the known density and the volume of pesticide applied to the fabric.
[b]Standard errors of the mean are in parentheses.
[c]The amount of methyl parathion was determined on a single specimen in a contracted testing laboratory.

cleaning of malathion from the surfaces of polyester was more uniform across the yarn (Fig. 4), as was the removal of methyl parathion from the cotton fiber surfaces (Fig. 5).

Although one laundering greatly reduced the amount of malathion on the surfaces of both cotton and polyester fibers and the amount of methyl parathion on the surfaces of the cotton fibers, it did not reduce the concentration of either malathion or methyl parathion in the cotton lumen (Fig. 7). However, there did appear to be some removal of both pesticides from the secondary wall of the cotton fibers (Fig. 7). Repeated launderings have been reported to remove successive amounts of methyl parathion, although when contaminated with undiluted methyl parathion, residual pesticide was observed even after ten laundry cycles [*6*]. Removal of residual pesticide from the lumen of the cotton fibers may be a slow and possibly incomplete process with an aqueous laundry system.

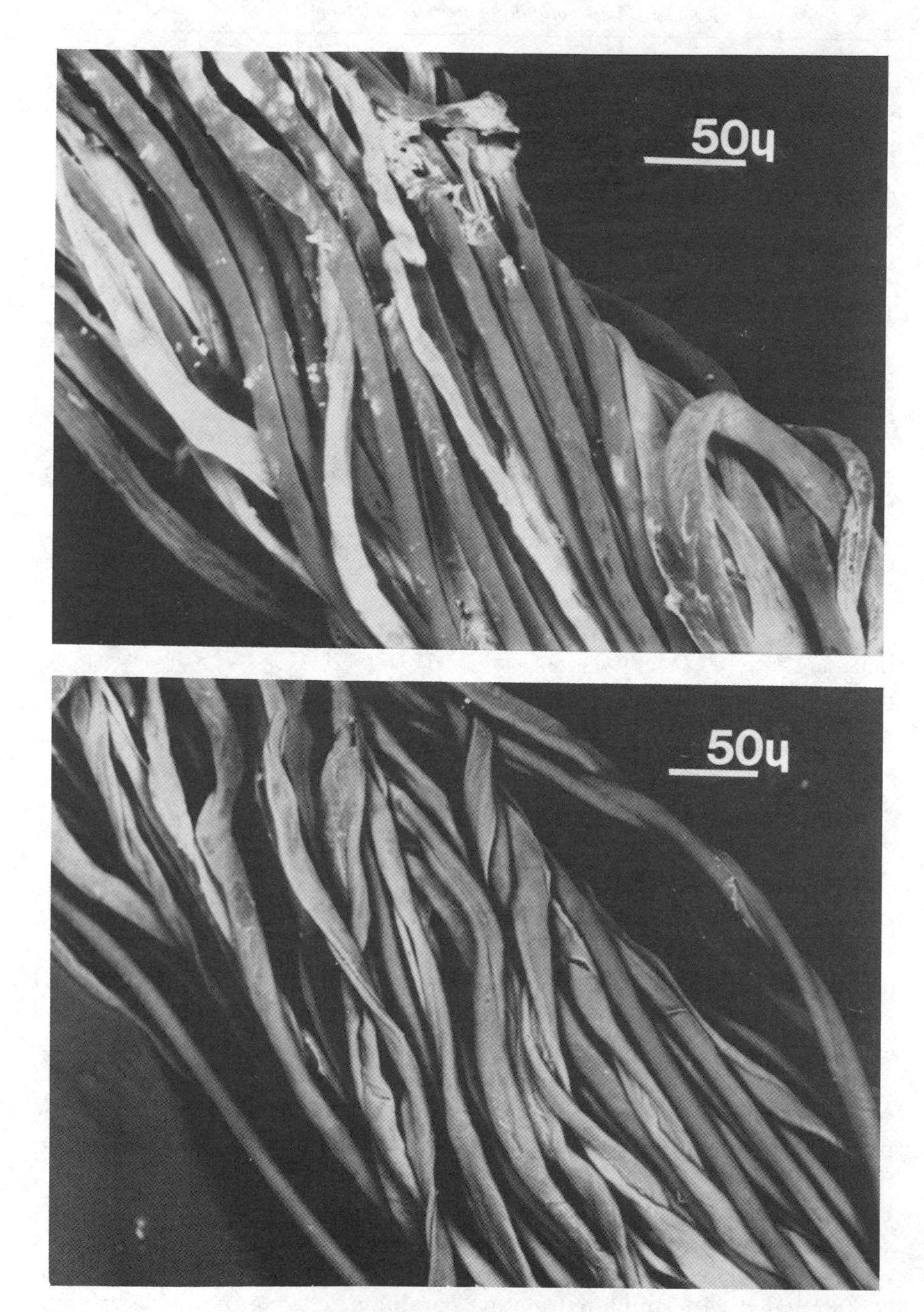

FIG. 8—*Backscattered electron images of the methyl parathion distribution on the surfaces of warp yarns from the durable-press 50% cotton/50% polyester fabric before* (top) *and after* (bottom) *laundering with a built, powdered detergent at 311 K.*

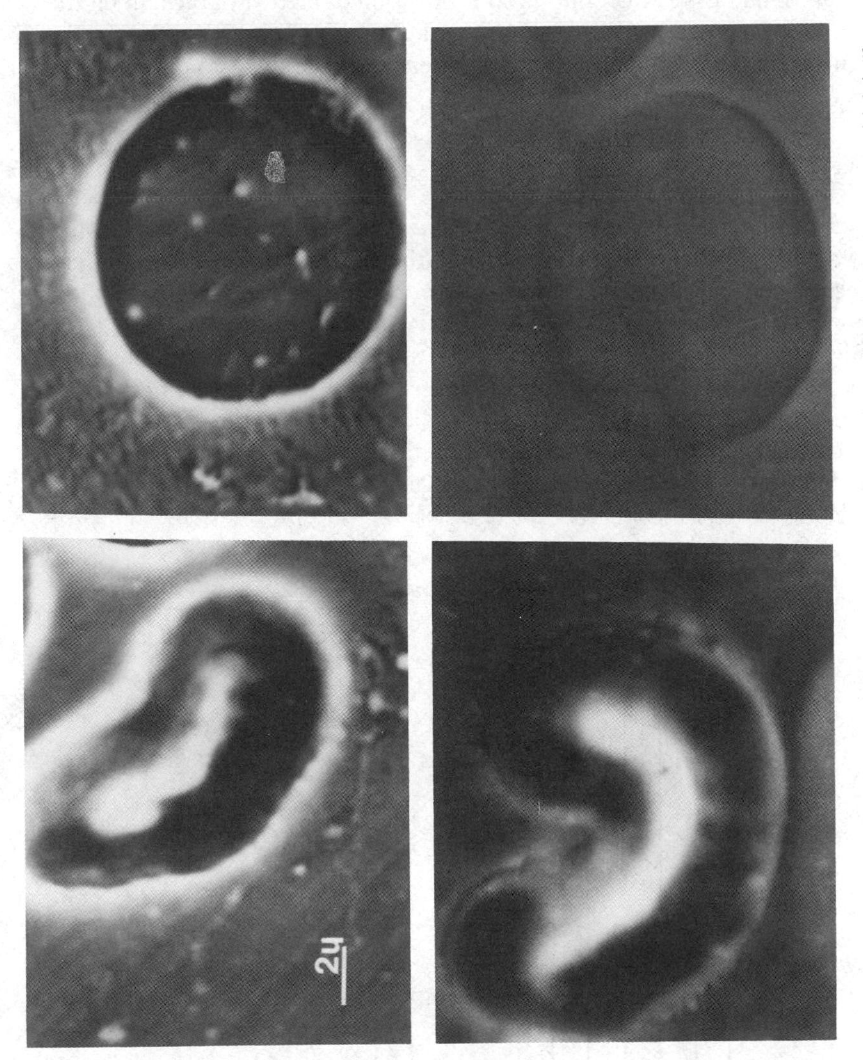

FIG. 9—*Backscattered electron images of the retention of malathion on cotton* (left) *and polyester* (right) *fibers from the durable-press 50% cotton/50% polyester fabric before* (top) *and after* (bottom) *laundering with a built, powdered detergent at 311 K.*

The durable-press (DMDHEU) finish did not affect the amount or the distribution of malathion on the fabric (Table 4; Figs. 10 and 11). On the unlaundered specimens, higher concentrations of methyl parathion were observed on the surfaces of the polyester fibers from the unfinished fabric than from the durable-press-finished fabric (Fig. 10). Laundering of the unfinished fabric had little effect on the concentration of methyl parathion on the surfaces of the polyester fibers (Fig. 10); however, a large reduction in the concentration of methyl parathion was observed on the surfaces of the cotton fibers (Fig. 11).

Conclusions

Electron microscopic techniques using backscattered electron imaging and X-ray microanalysis are useful in studying the effectiveness of decontamination procedures for protective clothing worn by workers handling pesticides. Detailed information about the penetration and retention of these toxic chemicals on fabrics can be obtained. The retention of pesticides is observed on the surfaces of the cotton and polyester fibers and in the lumen of the cotton fibers. One laundry cycle removed a large portion of the pesticide from the fiber surfaces with the exception of methyl parathion on the surfaces of the polyester fibers. No reduction of the concentration of pesticides in the cotton lumen was detected after one laundering. This information can be uti-

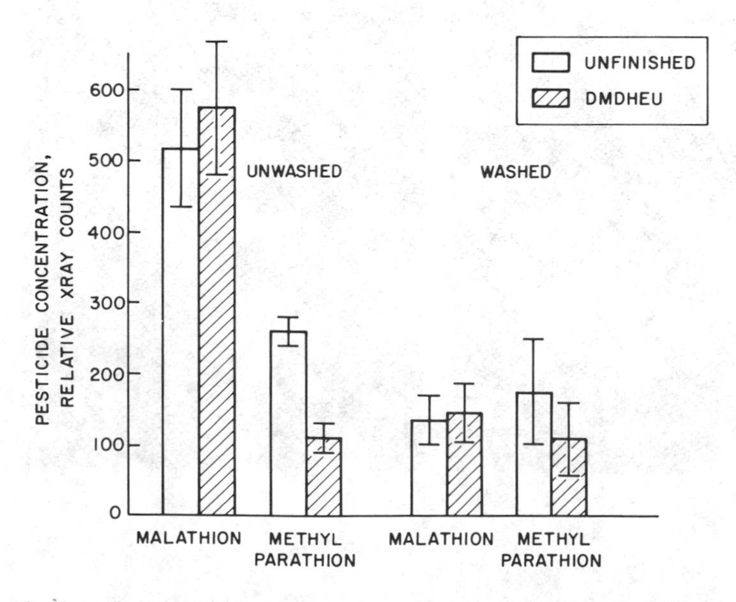

FIG. 10—*Retention of pesticides on the surfaces of polyester fibers from the 50% cotton/50% polyester fabric with and without durable-press finish. The minimum significant differences at the 0.1 level for the paired means, from left to right in the figure, are 216, 49, 94, and 162.*

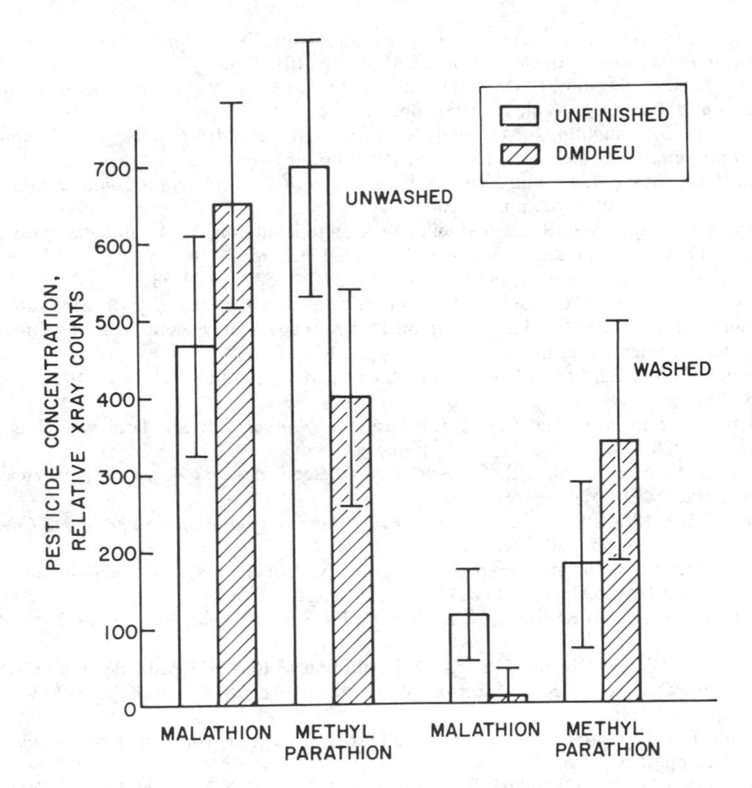

FIG. 11—*Retention of pesticides on the surfaces of cotton fibers from the 50% cotton/50% polyester fabric with and without durable-press finish. The minimum significant difference at the 0.1 level for the paired means, from left to right in the figure, are 339, 381, 119, and 326.*

lized in designing higher-performance disposable and reusable protective clothing.

Acknowledgments

We thank Ellen Uyenco and Vasudha Ravichandran for technical assistance, Yarrow Namasté for the statistical analyses, and Donald Lisk for the chemical analyses of methyl parathion. This research is part of the U.S. Department of Agriculture Cooperative Regional Research S-163.

References

[*1*] *Federal Register*, Vol. 39, 1974, pp. 16888–16891.
[*2*] Finley, E. L. and Rogillio, J. R. B., *Bulletin of Environmental Contamination and Toxicology*, Vol. 4, 1969, pp. 343–351.
[*3*] Finley, E. L., Metcalfe, G. I., McDermott, F. G., Graves, J. B., Schilling, P. E., and Bonner, F. L., *Bulletin of Environmental Contamination and Toxicology*, Vol. 12, 1974, pp. 268–274.

[4] Easley, C. B., Laughlin, J. M., Gold, R. E., and Tupy, D. R., *Bulletin of Environmental Contamination and Toxicology*, Vol. 27, 1981, pp. 101-108.
[5] Lillie, T. H., Livingston, J. M., and Hamilton, M. A., *Bulletin of Environmental Contamination and Toxicology*, Vol. 27, 1981, pp. 716-723.
[6] Easley, C. B., Laughlin, J. M., Gold, R. E., and Hill, R. M., *Bulletin of Environmental Contamination and Toxicology*, Vol. 29, 1982, pp. 461-468.
[7] Kim, C. J., Stone, J. F., and Sizer, C. E., *Bulletin of Environmental Contamination and Toxicology*, Vol. 29, 1982, pp. 95-100.
[8] Lillie, T. H., Hampson, R. E., Nishioka, Y. A., and Hamilton, M. A., *Bulletin of Environmental Contamination and Toxicology*, Vol. 29, 1982, pp. 89-94.
[9] Easter, E., *Textile Chemists and Colorists*, Vol. 15, 1983, pp. 29-33.
[10] Task Group on Occupational Exposure to Pesticides, *Occupational Exposure to Pesticides*, Report to the Federal Working Group on Pest Management, Federal Working Group on Pest Management, Washington, DC, 1974, pp. 120-127.
[11] Windholtz, M., Ed., *The Merck Index*, 9th ed., Merck and Co., Rahway, NJ, 1976, pp. 740, 796.
[12] Martin, H. and Worthing, C. R., *The Pesticide Manual*, 5th ed., Droitwich, England, 1977, pp. 326, 400.
[13] Gunther, F. A. and Blinn, R. C., *Analysis of Insecticides and Acaricides*, Interscience, New York, 1955, pp. 475-478.
[14] Norris, M. V., Vail, W. A., and Averell, P. R., *Journal of Agricultural and Food Chemistry*, Vol. 2, 1954, pp. 570-573.
[15] World Health Organization Expert Committee on Insecticides, *Specifications for Pesticides*, World Health Organization, Geneva, 1961, pp. 65-69.
[16] Obendorf, S. K. and Klemash, N. A., *Textile Research Journal*, Vol. 52, 1982, pp. 434-442.
[17] Solbrig, C. M. and Obendorf, S. K., "Distribution of Residual Pesticide Within Textile Structures as Determined by Electron Microscopy," *Textile Research Journal*, Vol. 55, 1985, pp. 540-546.
[18] Obendorf, S. K., Namasté, Y. M. N., and Durnam, D. J., *Textile Research Journal*, Vol. 53, 1983, pp. 375-383.
[19] Weglinski, S. A. and Obendorf, S. K., *Book of Papers, 1983 National AATCC Technical Conference*, American Association of Textile Chemists and Colorists, Research Triangle Park, NC, 1983, pp. 385-390.
[20] Van Krevelen, D. W. and Hoftyzer, P. J., *Properties of Polymers, Their Estimation and Correlation with Chemical Structure*, Elsevier, New York, 1976, pp. 129-155.
[21] Knox, B. H., Weigmann, H. -D., and Scott, M. G., *Textile Research Journal*, Vol. 45, 1975, pp. 203-217.

Risk Assessment of Chemical Exposure Hazards in Selecting Protective Clothing

S. Zack Mansdorf[1]

Risk Assessment of Chemical Exposure Hazards in the Use of Chemical Protective Clothing—An Overview

REFERENCE: Mansdorf, S. Z., **"Risk Assessment of Chemical Exposure Hazards in the Use of Chemical Protective Clothing—An Overview,"** *Performance of Protective Clothing, ASTM STP 900*, R. L. Barker and G. C. Coletta, Eds., American Society for Testing and Materials, Philadelphia, 1986, pp. 207-213.

ABSTRACT: Occupational exposure of the skin to toxic chemicals is a recognized health problem. The magnitude of the problem is evidenced by the fact that dermatitis is the major occupational disease in the United States. Perhaps more important, the occurrence of a dermatitis of unknown etiology may be indicative of a more serious and insidious systemic exposure. Very little research has been published on the significance of dermal exposures as they relate to the incidence of occupational disease from toxic chemicals. This is evidenced by the lack of any published standards or recommendations for "safe" levels of skin exposure to most toxic chemicals.

The majority of workers having a potential for skin contact with hazardous materials are protected by utilizing some form of chemically resistant clothing or gloves. These applications range from simple latex gloves to totally encapsulating suits. However, the selection, use, and maintenance of these protective garments have largely been determined empirically. More important, the commonly accepted belief that most rubber garments are "impermeable" has been recently dispelled through the development and use of standard resistance tests by ASTM and others. It is now known that most chemical protective clothing has some level of permeation to toxic chemicals. Additionally, a steady-state rate of release, with the garment base material acting as a toxic reservoir, is possible. The author has termed this phenomenon "matrix release." The question therefore arises, "What is an acceptable level of dermal exposure?" Obviously, many variables will distinguish the level of risk to the worker. These include the toxicity of the chemical or chemicals, permeability through the skin, location of the exposure site, rate and extent of contact, synergistic or additive effects of other routes of exposure, and other factors. Confounding the evaluation of an acceptable level of skin exposure to toxic chemicals are the pragmatic questions of the effects of reuse, situations requiring entry into atmospheres in which no dermal contact is acceptable, and the potential effects of garment failure.

Pragmatically, the previously discussed assessment criteria must be scrutinized within

[1]Principle health scientist, S. Z. Mansdorf & Associates, Cuyahoga Falls, OH 44223.

cost/benefit constraints. These constraints, plainly expressed, could be the choice between using a $2 glove that leaks a little or a $12 glove that leaks a little less.

This paper suggests a need for 1) a series of recommended dermal exposure levels within the framework of atmospheric exposure limits and 2) a field-validated method to measure exposure quantitatively inside the protective garment or glove.

KEY WORDS: chemical protective clothing, exposure assessment, decontamination, skin exposure, protective clothing, protective gloves, chemical protective clothing testing, chemical skin exposures, matrix release

Occupational exposure of the skin to toxic chemicals is a national health problem. The magnitude of this problem is evidenced by the fact that dermatitis is the major occupational disease in this country [*1–4*]. The Bureau of Labor Statistics (BLS) reported that, for 1980, skin diseases accounted for 43% of all occupational illnesses reported by the private sector [*5*]. The BLS report probably underestimates the true impact of this problem since the reporting rules require only the listing of diseases that result in medical treatment beyond first aid or that result in lost work time. In addition, many workers may well be hesitant to report minor rashes or skin irritations.

The most significant concern in a dermatitis of unknown etiology is the potential for system damage from adsorption through the skin of a toxic chemical. Such deleterious effects can range from systemic intoxication with vital organ damage to promotion of cancer. This can be more insidious than the outward appearance of a dermatitis might otherwise indicate. Very little research has been reported on this aspect of occupational health; hence, there are no published standards for skin exposures to most toxic chemicals.

The use of chemical protective clothing has increased tremendously in the last decade. This is due in part to 1) regulations and standards resulting in a greater reliance on the use of personal protective equipment 2) the apparent cost-effectiveness of using personal protective equipment over that of engineering controls, and 3) the need for "totally protective" suits for use in asbestos removal, hazardous waste disposal sites, and certain governmental and industrial operations in which exposure is considered especially hazardous.

Historically, most protective clothing was considered to be "impermeable" and an absolute safeguard for the worker. Application restrictions were generally based on the salesperson's recommendations or simple immersion tests by the manufacturers. These historical test methods have been shown to be wholly inadequate for evaluating the level of protection provided. More important, these tests may have encouraged the use of inadequate protection. Hence, both employer and employees may have been given a dangerously false sense of security.

Both routine and emergency chemical handling may result in direct exposure to toxic chemicals. Examples of such situations include the following:

(*a*) handling of liquid chemicals during manufacture,
(*b*) maintenance and quality control activities for chemical processes,
(*c*) acid baths and other treatments in electronics manufacture,
(*d*) application of pesticide and agricultural chemicals,
(*e*) chemical waste handling,
(*f*) emergency chemical response, and
(*g*) equipment leaks or failures.

The lack of suitable test methods for determining the resistance of chemical protective clothing led to the formation of ASTM Committee F-23 on Protective Clothing in 1977. This committee has promulgated the ASTM Test for Resistance of Protective Clothing Materials to Permeation by Hazardous Liquid Chemicals (F 739-81). As a result of this work and the efforts of several public and private laboratories, quantitative, reliable test data are being generated [*6–15*]. The ASTM permeation test method has been validated through extensive round-robin testing and is currently an accepted standard for evaluating permeation through chemical clothing.

ASTM Committee F-23 has developed a companion method for testing the resistance of protective clothing to penetration by liquid chemicals, which is currently undergoing ballot by ASTM [*16*]. In addition, another companion method for determining the resistance of protective clothing to degradation is presently in committee. Both of these methods are being validated in round-robin laboratory trials, under the sponsorship of the National Institute for Occupational Safety and Health (NIOSH).

ASTM Test F 739-81 provides data on breakthrough time and rate, as a steady-state rate, while the penetration and proposed degradation tests are principally of a pass–fail nature. In either case, these data must be applied by a qualified individual to a specific end use with an associated risk assessment of skin contact by a hazardous chemical.

The Occupational Safety and Health Administration (OSHA) requires that "all personal protective equipment shall be of a safe design and construction for the work to be performed" [*17*]. However, the publication of data demonstrating permeation or penetration of protective clothing and gloves that have been generally believed to be impermeable has led to a professional problem—namely, determining what is an acceptable or safe skin exposure.

This paper provides an overview of the boundaries of the acceptable exposure question and additional issues of importance that govern the proper selection of chemical protective clothing.

Protective Clothing Use

The majority of workers with a potential for skin contact by hazardous materials are protected by utilizing some form of chemically resistant clothing or gloves. These applications range in use from single latex gloves to totally encapsulating suits. The recommendations for selection of an appropriate level of protective clothing to be used in any particular situation should be based on a number of objective and subjective factors, including the following:

(*a*) the potential effects of skin contact with the chemical (for example, corrosiveness, toxicity, physical damage, allergic reaction),
(*b*) the exposure period (that is, time of contact),
(*c*) the body zone of potential contact (for example, hands, feet, arms, legs, face, chest, back),
(*d*) the permeability or penetration potential of the protective garment or glove (that is, breakthrough time and steady-state rate),
(*e*) the characteristics of potential contact (for example, splash, immersion),
(*f*) the additive or synergistic effects of other routes of exposure (that is, inhalation and ingestion),
(*g*) the physical properties required of the protective garment or glove (for example, flexibility, puncture and abrasion resistance, thermal protection), and
(*h*) cost (that is, based on single or multiple use and acceptable exposure).

Presently, only limited data are available for objective evaluation of the toxic potential of skin contact. This is especially critical since toxic potential should be the major criterion for selection of protective garments or gloves. Dermal toxicity assessments have been constrained by limited data on the permeability of most chemicals through intact skin *in vivo*, the systemic toxicity of chemicals permeating the skin, and the difficulty in applying laboratory data generated through the ASTM permeation method to the human model. A few skin exposure studies have been conducted for some organic solvents such as benzene and selected pesticides [*18*]. However, animal studies on hairless mice and pig skin, as well as rate-controlling studies of dermal drug delivery, may improve our level of present knowledge in the next few years.

Exposure Assessment

The exposure period calculation for most work situations should be fairly straightforward. Coupled with a breakthrough time and steady-state permeation rate, an approximation of dermal dose could be calculated by

$$\text{dose} = (T_c - T_b)(J A_e) \quad (1)$$

where

dose = the skin dose, mg,
T_c = the chemical contact period, s
T_b = the time to breakthrough, s,

J = the steady-state flux of the permeating chemical, $mg/m^2 \cdot s$, and
A_e = the area exposed, m^2.

However, the accuracy and value of this dose estimation would be limited by a number of factors including the following:

(*a*) the influence of chemical mixtures,
(*b*) the quality of manufacture of the protective garment,
(*c*) the probability of physical failure of the barrier,
(*d*) individual work practices of the employee that could lead to contamination or changes in the exposure period,
(*e*) improper or inadequate decontamination techniques, and
(*f*) the difficulty of applying a dermal dose estimation to the risk of systemic toxicity.

The obvious solution is to recommend a totally indestructible and impermeable garment for any chemical with toxic properties and some skin permeation potential. Even if such a garment or glove were commercially available, however, cost/benefit considerations would rule out this choice for most situations. Additionally, any garment or glove that is totally impermeable is likely to create a significant use problem by preventing normal skin respiration and body heat displacement.

Decontamination and Reuse

A key element in the cost/benefit equation is the potential for decontamination and reuse of garments or gloves.

Effective decontamination of protective clothing can be critical, as shown in a recent study by the Vermont Health Department and NIOSH, which indicated elevated mercury levels in the children of workers employed in a mercury thermometer manufacturing plant. These workers were wearing their work clothing home for laundering [*19*].

Single-use items are cost-effective for some low-risk or nonroutine jobs, but for routine exposures or highly toxic exposures requiring expensive, fully encapsulating suits, decontamination and reuse may be critical. No standardized and validated tests for the effectiveness of various decontamination procedures for chemical protective clothing are currently published. Furthermore, establishment of a toxic reservoir in elastomeric suits and gloves is a possibility. Research performed in the area of controlled release technology has demonstrated that organic compound reservoirs can be created in elastomers which will slowly release over exceptionally long periods of time [*20*]. In addition, this "matrix release" could be enhanced by liquid contact from body perspiration. This assumption of potential matrix release after decontamination could be tested by evaluating protective clothing materials after several uses and decontamination in a standard ASTM permeation test cell with deionized water as the challenge liquid.

An additional factor in evaluating the effectiveness for chemical protective clothing is the possibility of entrapment of a substance against the skin. Specifically, pinholes, poorly sealed seams, or other material imperfections could allow penetration of a hazardous liquid chemical into the clothed skin area. The protective garment would then act as a barrier to increase the gradient for movement of the liquid into the skin, thereby increasing rather than decreasing the risk. This situation could also occur through poor work practices in which the garment or glove is not properly sealed against splash.

The foregoing discussion suggests a need for 1) a series of recommended dermal exposure levels similar to existing atmospheric exposure limits and 2) a field-validated method to measure exposures quantitatively inside the protective garment or glove. Without a safe exposure limit and method to measure it, only garments or gloves with no breakthrough could reasonably be used for most exposure situations. A no-exposure situation is totally impractical and not cost-effective in most industrial operations. Additionally, even the latest published literature on selection of personal protective clothing does not address mixtures and is limited to only 300 chemicals [*21*].

There are a number of possible approaches to the simpler of the two needs, field measurement of exposure. For example, an activated charcoal strip, similar to those found in a passive dosimeter, could be worn on the skin and removed for analysis. Hopefully, research will be conducted in this area. The more difficult of the two needs, establishment of "safe" exposure limits for dermal exposures, must still be addressed. It is this author's opinion that this toxicological research is fundamental to the establishment of adequate worker protection within reasonable cost/benefit constraints.

References

[*1*] *The Prevention of Occupational Skin Diseases*, 3rd Printing, Soap and Detergent Association, New York, 1981.

[*2*] *A Summary of the NIOSH Open Meeting on Chemical Protective Clothing*, held in Rockville, MD, 3 June, 1981, National Institute for Occupational Safety and Health, Cincinnati, OH, 1981.

[*3*] Mansdorf, S. Z. and Miles, B., "A Protective Dermal Film System," Paper presented at the American Industrial Hygiene Association Conference, Los Angeles, CA, May 1978.

[*4*] "Report of the Advisory Committee on Cutaneous Hazards," Report to the Assistant Secretary of Labor, Occupational Safety and Health Administration, Washington, DC, 19 Dec., 1978.

[*5*] "Occupational Injuries and Illness in the United States by Industry, 1980," Bulletin 2130, Bureau of Labor Statistics, Washington, DC, 1982.

[*6*] Coletta, G. C., Schwopen, A. D., Arons, I. J., King, J. W., and Sivak, A., "Development of Performance Criteria for Protective Clothing Used Against Carcinogenic Liquids," DHEW (NIOSH) Publication No. 79-106, National Institute of Occupational Safety and Health, Arthur D. Little, Cambridge, MA, Oct. 1978.

[*7*] Sansome, E. B., and Tewori, Y. B., "The Permeability of Laboratory Gloves to Selected Solvents," *American Industrial Hygiene Association Journal*, Vol. 39, 1978, pp. 169-174.

[*8*] Williams, J. R., "Permeation of Glove Materials by Physiologically Harmful Chemicals," *American Industrial Hygiene Association Journal*, Vol. 40, 1979, pp. 877-882.

[*9*] Williams, J. R., "Chemical Permeation of Protective Clothing," *American Industrial Hygiene Association Journal*, Vol. 41, 1980, pp. 884-887.

[10] Henry, N. W., and Schlatter, C. N., "The Development of a Standard Method for Evaluating Chemical Protective Clothing to Permeation by Hazardous Liquids," *American Industrial Hygiene Association Journal*, Vol. 42, 1981, pp. 202-207.
[11] Nelson, G. O., Lum, B. Y., Carlson, G. J., Wong, C. M., and Johnson, J. S., "Glove Permeation by Organic Solvents," *American Industrial Hygiene Association Journal*, Vol. 42, 1981, pp. 217-225.
[12] Williams, J. R., "Evaluation of Intact Gloves and Boots for Chemical Permeation," *American Industrial Hygiene Association Journal*, Vol. 41, 1980, pp. 884-887.
[13] Weeks, R. W. and McLeod, M. J., "Permeation of Protective Garment(s) by Liquid Benzene and Tritiated Water," *American Industrial Hygiene Association Journal*, Vol. 43, 1982, pp. 201-211.
[14] Abernathy, R. N., Cohen, R. B., and Shirtz, J. J., "Measurements of Hypergolic Fuels' and Oxidants' Permeation Through Commercial Protective Materials—Part I: Inhibited Red Fuming Nitric Acid and Unsymmetrical Dimethylhydrazine," *American Industrial Hygiene Association Journal*, Vol. 44, 1983, pp. 505-513.
[15] Berardinelli, S. P., Mickelsen, R. L., and Roder, M. M., "Chemical Protective Clothing: A Comparison of Chemical Permeation Test Cells and Direct-Reading Instruments," *American Industrial Hygiene Association Journal*, Vol. 44, 1983, pp. 886-889.
[16] Mansdorf, S. Z., Swick, R., and Brinton, S. J., "Development of a Standard Test Method for Resistance of Protective Clothing Materials to Penetration by Liquids," Paper presented at the American Industrial Hygiene Association Conference, Philadelphia, May 1983.
[17] OSHA, *General Industry Standards*, 29 CFR 1910.132(c), Occupational Safety and Health Act of 1970 (84 Stat. 1593), Revised 1978.
[18] Libich, S., To, J. D., Frank, R., and Sirons, G. J., "Occupational Exposure of Herbicide Applicators to Herbicides Used Along Electric Power Transmission Line Right-of-Way," *American Industrial Hygiene Association Journal*, Vol. 45, 1984, pp. 56-62.
[19] *Occupational Safety and Health Reporter*, Vol. 14, Bureau of National Affairs, Washington, DC, July 1984, p. 145.
[20] Mansdorf, S. Z., "Slow Release: Development of Aquatic Herbicides and Carrier Systems," *Proceedings*, Controlled Release Pesticides Symposium, Controlled Release Society, University of Akron, OH, 1974-75, pp. 121-127.
[21] Schwope, A. D., Costas, P. P., Jackson, J. O., and Weitzman, D. J., *Guidelines for the Selection of Chemical Protective Clothing*, Vols. I and II, Arthur D. Little, Cambridge, MA, 1983.

Anders Boman[1] and Jan E. Wahlberg[1]

Bioassay for Testing Protective Glove Performance Against Skin Absorption of Organic Solvents

REFERENCE: Boman, A. and Wahlberg, J. E., "**Bioassay for Testing Protective Glove Performance Against Skin Absorption of Organic Solvents,**" *Performance of Protective Clothing, ASTM STP 900*, R. L. Barker and G. C. Coletta, Eds., American Society for Testing and Materials, Philadelphia, 1986, pp. 214-220.

ABSTRACT: Solvents were administered into closed skin depots on guinea pigs, and the resulting blood concentrations were measured in samples repeatedly withdrawn through a catheter in the carotid artery. The exposure areas were protected with glove membranes of different kinds. Results from experiments with three solvents and various gloves are presented. The authors conclude that this animal model seems suitable for screening purposes and is capable of discerning between various commercial products.

KEY WORDS: protective gloves, percutaneous absorption, toluene, 1,1,1-trichloroethane, *n*-butanol, guinea pigs, blood concentrations, gas chromatography, protective clothing

The major risk to the skin in the industrial environment is the risk of developing chemically induced contact dermatitis of an irritant or allergic type.

Irritant contact dermatitis may be caused by a number of chemicals. These range from strongly lipophilic chemicals, such as organic solvents, to mild tensides and water. Allergic contact dermatitis is mainly caused by about 30 well-known chemicals. They are of low molecular weight and may be inorganic, such as metals (chromium, cobalt, nickel), or organic, such as preservatives, perfumes, plastics, or rubber chemicals. In most cases the hands and lower parts of the arms are affected [*1*].[2]

Absorption of industrial chemicals through the skin is affected by a num-

[1]Research engineer, and professor, respectively, National Board of Occupational Safety and Health, Research Department, Department of Occupational Dermatology, S-171 84 Solna, Sweden.

[2]J. E. Wahlberg, Karolinska Hospital, Department of Occupational Dermatology, Stockholm Sweden, unpublished results, 1984.

ber of factors, of which the exposed site on the body surface, status of the exposed skin, and size of the exposed area are the most important.

Some chemicals may be hazardous on skin contact. Chemicals with high percutaneous toxicity include pesticides and certain organic solvents.

The rate of skin absorption and percutaneous toxicity of organic solvents varies over a wide range. Percutaneous toxicity of ten solvents studied in the guinea pig showed considerable variability, ranging from no effect, to minor inhibitory effects on weight gain, to total lethality within 24 h in the exposed group [*2*].

In industry, solvents with varied percutaneous toxicity are used, but acute systemic toxicity after percutaneous absorption in an industrial environment is not a major occupational hazard because the main exposure to solvents is through the respiratory tract. However, as solvents have a potent local effect of defatting the skin and inducing edema and erythema [*3,4*], and because they are used in many various applications in industry and are often carelessly handled, resulting in accidental or voluntary skin contact [*5*], they represent a considerable occupational health hazard and warrant specific studies. The skin absorption that occurs, although small, may add to the total body burden of solvents that are inhaled during work, and the local irritation may be a starting point for a chronic dermatitis and a facilitated route of entry for sensitizing chemicals.

To reduce skin exposure to locally and systemically harmful chemicals, synthetic barriers are being developed and evaluated for efficacy. One method to evaluate glove materials is technical material testing according to the ASTM Test for Resistance of Protective Clothing Materials to Permeation by Hazardous Liquid Chemicals (F 739-81) or an equivalent method. But as the alternatives to gloves—barrier creams—are also claimed to be effective, and as it is essential to evaluate and compare the effects of protective materials *in vivo*, a biological method for monitoring the protective effect is of great importance. The literature contains a number of reports on *in vivo* tests of protective effects of gloves to different chemicals. These tests are generally performed on humans, either as laboratory investigations [*6*], as field studies in the real work environment [*7,8*] or as clinical investigations [*9–13*].

Although the best and most accurate values are generated in human studies, there is a certain risk for the subjects [*7,8*].

In order to reduce the need for human experiments, an animal model has been developed to investigate the absorption of organic solvents through the skin. The method is also applicable for testing the protective effect of glove materials or other protective devices [*14*].

Materials and Methods

Skin absorption of three organic solvents (toluene, 1,1,1-trichloroethane, and *n*-butanol) was studied in guinea pigs with and without protective barriers. Female albino or pigmented guinea pigs weighing 500 to 700 g were used in the experiment. The animals were anesthetized intraperitoneally with pentobarbital sodium (Mebumal, ACO, Solna), 39 mg/kg body weight. A polyethylene catheter was inserted into the carotid artery. The hair on the back was cut with electric clippers, and care was taken not to injure the skin. Glove membranes [polyvinyl chloride (PVC), 0.45 mm; butyl rubber, 0.60 mm; and latex rubber, 0.17 mm] were glued to the bottom of a glass ring, inner diameter of 20 mm, and then tested for leakage. The glass depot with glove membrane, inside facing down, was attached to the skin with a cyanoacrylate glue (Cyanolit 201). The glass deport was covered with a pierced cover glass. Solvent was administered into the glass depot, and it was then sealed to exclude inhalation (Fig. 1). Exposure was continuous for 6 h. Animals with unprotected skin were exposed in the same manner for comparison.

During the whole experimental period, blood samples (0.5 mL) were taken repeatedly from the catheter and put into glass vials, which were sealed. The blood loss was compensated for with Macrodex (Pharmacia, Uppsala).

The blood concentration of the studied solvents was analyzed with a gas chromatograph according to a head space technique [*15*] (Varian 3700), using a flame ionization detector (FID) and nitrogen (N_2) as the carrier gas and methyl silicone SE-30 20% on Chromosorb as the stationary phase.

Results

The results from the studies with three solvents are summarized in Figs. 2 through 4.

Figure 2 shows the skin absorption of *toluene* across normal and protected skin (latex rubber, PVC, and butyl rubber). The time of exposure is shown along the abscissa, and the blood concentration of solvent is shown along the ordinate.

A notable difference between protected and unprotected skin in the blood concentrations can be seen during the first hour of exposure. For PVC and butyl rubber, very low blood concentrations were recorded during the first 20 min. At the later part of the exposure, the latex rubber and the PVC membranes were less efficient than the butyl rubber.

Figure 3 shows similar experiments with *1,1,1-trichloroethane*. A notable

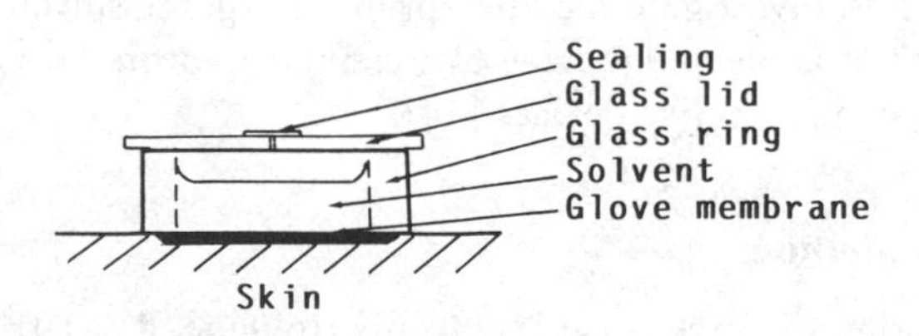

FIG. 1—*Schematic drawing of exposure chamber.*

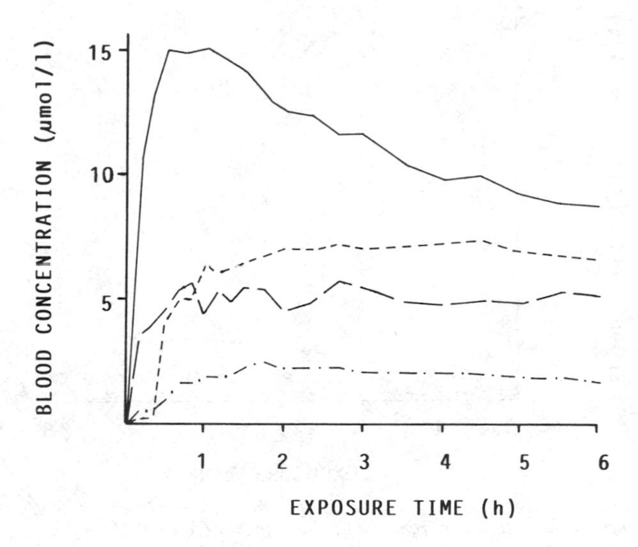

FIG. 2—*Blood concentration of toluene on exposure of normal skin (———) and normal skin with 0.45-mm PVC (-----), 0.6-mm butyl rubber (—·—·—·), and 0.17-mm latex rubber (———) membranes.*

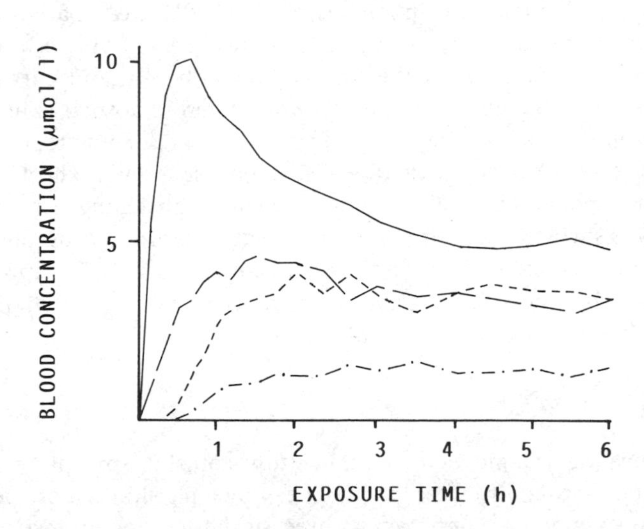

FIG. 3—*Blood concentration of 1,1,1-trichloroenthane during exposure of normal skin (———) and normal skin with 0.45-mm PVC (-----), 0.6-mm butyl rubber (—·—·—·), and 0.17-mm latex rubber (———) membranes.*

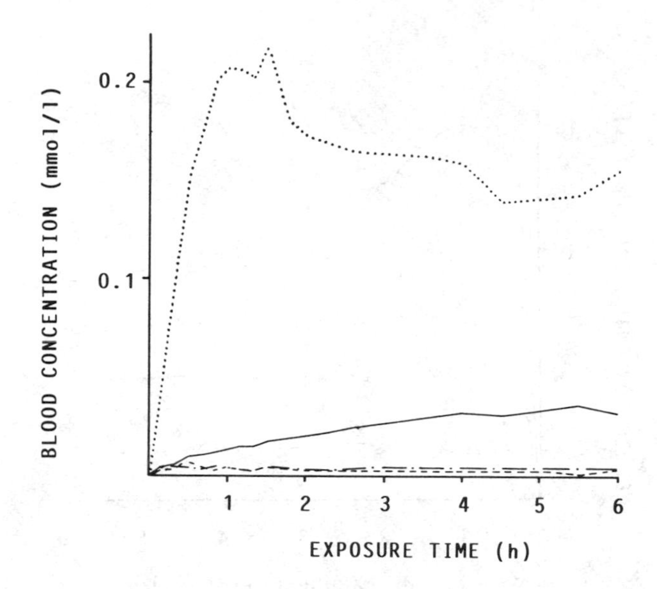

FIG. 4—*Blood concentration of* n*-butanol during exposure of normal skin (———), stripped skin (----------), stripped skin with 0.45-mm PVC (-----), and stripped skin with 0.6-mm butyl rubber (—-—-—-) membranes.*

protective effect during the first hour of exposure for all three materials and less effective protection for thin latex rubber and PVC, compared with butyl rubber, during the later phase of exposure were observed. As for toluene, low blood values were seen during the first 20 min with PVC and butyl rubber.

Figure 4 shows the absorption of *n-butanol* across normal skin and skin that has been deprived of the horny layer. This was done by repeated treatments of the skin surface with tape—so-called stripping. The horny layer sticks to the tape, and after about ten treatments a glistening wet surface can be seen. This surface is a poor barrier to the absorption of *n*-butanol, as can be seen from the great increase in the blood concentration. If stripped skin is exposed to *n*-butanol with a glove membrane of PVC or butyl rubber between, the barrier function is reestablished.

Discussion

The guinea pig as a model for man in studies on skin absorption of chemicals has been thoroughly investigated. The guinea pig skin is more permeable than human skin by a factor of two to three for the compounds tested [*16,17*]. The higher absorption rate is of minor importance, as the proposed animal model compares skin absorption through normal skin with skin protected with a glove membrane. The correlation of absorption rates in guinea pig skin to the human skin is thus less important.

The use of an animal model for evaluation of the protective effect of glove materials against skin absorption of chemicals complements a purely technical material test according to ASTM Test F 739-81 or an equivalent method. The biological model takes into account factors that are not considered in the technical test, among which are local interaction on and in the skin of the permeating substance and the occlusive effect of the glove on the skin, and enables the results to be compared for absorption kinetics. For the physician, the skin is a part of a human being and, in deciding on its protection, he cannot regard it as an isolated phenomenon.

The results from the described bioassay show that a considerable reduction in the blood concentration of solvents occurs when the skin is protected with glove membranes. This decrease is most marked in the early phase of the exposure. The method can discern the biological significance of the protective effect of various materials, as can be seen when the results of protection with PVC, latex rubber, and butyl rubber are compared. The results indicate that a PVC or butyl rubber glove gives sufficiently good protection against skin absorption of toluene and 1,1,1-trichloroethane during the first 20 min of exposure. The PVC glove is less efficient during longer exposure times, and the latex rubber glove is hardly effective; the solvents were already detectable after 10 min. However, even if no detectable amounts of solvent are found in blood during the first period of exposure, there is nevertheless a risk that the solvent will defat and irritate the skin [*14*].

From the experiments with *n*-butanol, it can be seen that an impaired barrier function can easily be restored with the use of a suitable glove.

No permeation data, according to ASTM Test F 739-81, on the materials and solvents used in this investigation were found in the literature. However, data on the same type of polymers and solvents show that PVC and natural rubber latex are less effective than butyl rubber [*18,19*].

Acknowledgment

The skillful technical assistance of G. Hagelthorn and B. Berkevall is greatly acknowledged. The study was supported by the Swedish Work Health Fund, 80/266.

References

[*1*] Emmett, E. A., "Occupational Skin Disease," *Journal of Allergy and Clinical Immunology*, Vol. 72, 1983, pp. 649–656.
[*2*] Wahlberg, J. E. and Boman, A., "Comparative Percutaneous Toxicity of Ten Industrial Solvents in the Guinea Pig," *Scandinavian Journal of Work Environment and Health*, Vol. 5, 1978, pp. 345–351.
[*3*] Wahlberg, J. E. "Edema-Inducing Effects of Solvents Following Topical Administration," *Dermatosen in Beruf und Umwelt*, Vol. 32, 1984, pp. 91–94.
[*4*] Wahlberg, J. E., "Erythema-Inducing Effects of Solvents Following Epicutaneous Admin-

istration to Man—Studied by Laser Doppler Flowmetry," *Scandinavian Journal of Work Environment and Health*, Vol. 10, 1984, pp. 159-162.

[5] Anderson, I.-M., Janbell, H., and Rosén, G., "Degreasing with Organic Solvents," Investigation Report 1981:20 ("Kallavfettning med organiska lösningsmedele: En undersökning av arbetsmetoder och lösningsmedele exponering," Undersökningsrapport 1981:20), National Board of Occupational Safety and Health, Research Department, S-171 84 Solna, Sweden (in Swedish), 1981.

[6] Hogstedt, C. and Ståhl, R., "Skin Absorption and Protective Gloves in Dynamite Work," American Industrial Hygiene Association Journal, Vol. 41, 1980, pp. 367-372.

[7] Brooks, S. M., Anderson, L., Emmett, E., Carson, A., Tsay, J. Y., Elia, V., Buncher, R., and Karbowsky, R., "The Effects of Protective Equipment on Styrene Exposure in Workers in the Reinforced Plastics Industry," *Archives of Environmental Health*, Vol. 35, 1980, pp. 287-294.

[8] Lauwerys, R. R., Kivits, A., Lhoir, M., Rigolet, P., Houbeau, D., Buchet, J. P., and Roels, H. A., "Biological Surveillance of Workers Exposed to Dimethylformamide and Influence of Skin Protection on Its Percutaneous Absorption," *International Archives of Occupational and Environmental Health*, Vol. 45, 1980, pp. 189-203.

[9] Lidén, C., "Occupational Dermatoses in a Film Laboratory," *Contact Dermatitis*, Vol. 10, 1984, pp. 77-87.

[10] Moursiden, H. T. and Faber, O., "Penetration of Protective Gloves by Allergens and Irritants," *Transactions of the St. John's Hospital Dermatological Society*, Vol. 59, 1973, pp. 230-234.

[11] Pegum, J. S. and Medhurst, F. A., "Contact Dermatitis from Penetration of Rubber Gloves by Acrylic Monomer," *British Medical Journal*, Vol. 2, 1971, pp. 141-143.

[12] Pegum, J. S., "Penetration of Protective Gloves by Epoxy Resin," *Contact Dermatitis*, Vol. 5, 1979, pp. 281-283.

[13] Wall, L. M., "Nickel Penetration Through Rubber Gloves," *Contact Dermatitis*, Vol. 6, 1980, pp. 461-463.

[14] Boman, A., Wahlberg, J. E., and Johansson, G., "A Method for the Study of the Effect of Barrier Creams and Protective Gloves on the Percutaneous Absorption of Solvents," *Dermatologica*, Vol. 164, 1982, pp. 157-160.

[15] Jakobson, I., Wahlberg, J. E., Holmberg, B., and Johansson, G., "Uptake Via the Blood and Elimination of 10 Organic Solvents Following Epicutaneous Exposure of Anaesthetized Guinea Pigs," *Toxicology and Applied Pharmacology*, Vol. 63, 1982, pp. 181-187.

[16] Andersen, K. E., Maibach, H. I., and Anjo, M. D., "The Guinea Pig—An Animal Model for Human Skin Absorption of Hydrocortisone, Testosterone and Benzoic Acid?" *British Journal of Dermatology*, Vol. 102, 1980, pp. 447-453.

[17] Wester, R. C. and Maibach, H. I., "In Vivo Percutaneous Absorption" in *Dermatotoxicology*, 2nd ed. F. M. Marzulli and H. I. Maibach, Eds., Hemisphere, Washington, DC, 1983.

[18] Nelson, G. O., Lum, B. U., Carlson, G. J., Wong, C. M., and Johnson, J. S., "Glove Permeation by Organic Solvents." *American Industrial Hygiene Association Journal*, Vol. 42, 1981, pp. 217-225.

[19] Nolen, R. J. "Evaluation of Protective Clothing Materials: Permeation, Degradation, Penetration and Particulate Analysis," Radian Corp., Austin, TX, 1984.

Arthur D. Schwope[1]

Permeation of Chemicals Through the Skin

REFERENCE: Schwope, A. D., **"Permeation of Chemicals Through the Skin,"** *Performance of Protective Clothing, ASTM STP 900,* R. L. Barker and G. C. Coletta, Eds., American Society for Testing and Materials, Philadelphia, 1986, pp. 221-234.

ABSTRACT: An overall risk/benefit analysis relative to the use of chemical protective clothing for protection from potentially harmful chemicals must ultimately consider the barrier effectiveness of the skin to the chemicals in question. For example, there may be cases in which some permeation of chemical through the clothing is tolerable if the skin itself is unaffected by and acts as a barrier to the chemical. An overview of the structure of skin, methods for assessing skin permeation, and examples of published data are presented. The article is directed toward those with little or no knowledge of chemical permeation through skin and provides references for further research.

KEY WORDS: skin, permeation, protective clothing, chemicals

One mechanism for the potential exposure of a worker wearing chemical protective clothing to chemicals involves the following sequence of steps:

(*a*) contact of the outer surface of the clothing with the chemical;
(*b*) absorption of the chemical by the clothing;
(*c*) diffusion of the chemical through the clothing material;
(*d*) desorption of the chemical from the clothing material;
(*e*) transfer across the air interface separating the protective clothing from the skin. In some cases, this could involve transfer through undergarments and street clothing; and
(*f*) absorption and transfer through the skin.

This model assumes that no penetration of the chemical occurs through seams, closures, pinholes, or other opening.

Many recent publications by those involved in clothing testing have investigated Steps *a* through *d* [*1-5*]. In addition, many papers reporting perme-

[1]Unit leader, Arthur D. Little, Inc., Cambridge, MA 02140.

ation rates and breakthrough times for a variety of chemicals and clothing materials have been included as part of this volume. Such work is critically necessary and is directly applicable to assessing the potential for some forms of contact dermatitis. However, such testing and data are only part of the answer to the question of determining the amount and rate of chemical that actually comes in contact with and enters the worker's body. These latter factors are integral to any overall risk/benefit analysis relative to worker health and clothing utilization.

The purpose of this paper is to provide a broad overview of the function and behavior of skin as a barrier material.

The paper begins with a brief discussion of risk assessment from the perspective of skin permeation. Next, the composition of and mechanisms for transport through skin are described. Skin permeation data from several sources are then presented. The paper concludes with a series of questions intended to stimulate thinking and laboratory work toward developing an understanding of the critical factors in the mechanism just described. It is the author's hope that this information will be a starting point for those involved with clothing testing and selection, but with little background on the subject of skin, to begin to consider the role and importance of the skin in protecting our bodies.

Risk

In performing a risk/benefit analysis, one attempts to quantify the risks and benefits. The focus of the following discussion is the risk component of such an analysis.

A risk assessment requires input for the (1) probability of exposure, (2) the amount of exposure, and (3) the toxicity of the chemical. These are combined to yield the risk estimate

$$\text{probability of exposure} \times \text{amount of exposure} \times \text{toxicity} \rightarrow \text{risk}$$

Each factor in this equation can be quantified with some degree of uncertainty, leading to an overall quantifiable risk value with its own degree of uncertainty. If the level of uncertainty is too high, then attempts to perform the risk/benefit analysis will be futile. To reduce uncertainty, one applies judgment and data.

Protective clothing is used with the presumption that it reduces risk, and it probably does. But by how much? And for the same dollar spent, could risk be reduced by a greater amount by another approach?

To answer these questions, one estimates risk for scenarios with and without clothing. Each component is analyzed separately. The toxicity is the same with or without clothing; clothing, however, can affect the probability of exposure and the amount of exposure.

The remainder of this paper addresses the amount of exposure, that is, the amount of chemical that permeates the skin.

Skin

An adult male of average size is covered by approximately 1.7 m^2 (19 ft^2) of skin. In fact, skin is the largest organ of our bodies. Skin is critical to life because it maintains body fluids and body temperature and acts as a barrier to invasion by elements of the outside world. No other organ is so greatly exposed to the external environment.

Because skin is a critical barrier between ourselves and the outside world and because skin is the part of us that others see, feel, and touch, it has been the subject of considerable research. Cosmetics producers for millenia have searched for and sold us products which enhance our appearance or make our skin softer to the touch. Both health care houses and folk medicine offer a virtually unlimited array of ointments and salves for what ails us. Nitroglycerin is now administered to angina patients by means of transdermal drug delivery patches. Other researchers are attempting to develop synthetic skin.

Skin is generally described as consisting of three functional layers: the dermis, the epidermis, and the stratum corneum (Fig. 1) [*6*]. The dermis is the innermost layer, and it is interlaced with blood vessels and viable cellular tissue. Hair follicles, sebaceous glands, and sweat ducts originate in the dermis.

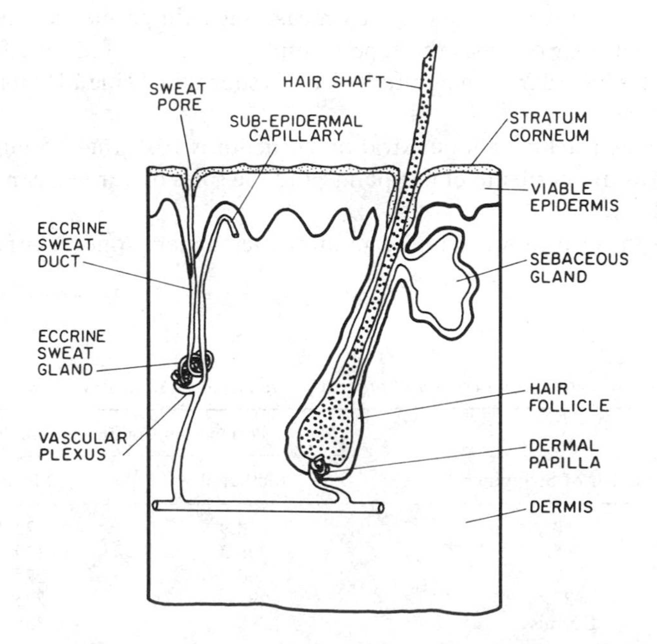

FIG. 1—*Schematic cross section of human skin* [6].

These are discussed later, since they are potential routes for chemical penetration.

The epidermis is the middle layer. It contains no blood vessels, but it is biologically active and is maintained so by the exchange of body fluids with the dermis. Cells of the epidermis eventually move outward to form the stratum corneum.

The stratum corneum is actually the topmost layer of the epidermis and consists of dehydrated, flattened epidermal cells. The stratum corneum is metabolically inactive. The surface of the stratum corneum is covered with a thin layer of oils and waxes secreted by the sebaceous glands.

When one analyzes the key functions of the skin, its role as a protector becomes obvious. Each function is aimed outward:

1. Oils from the sebaceous glands are secreted outward.
2. Perspiration flow is outward.
3. The cellular layer of the epidermis is continually moving to the stratum corneum, where it is ultimately sloughed off. On the average, mature adults regenerate their entire stratum corneum every two weeks.

After considerable debate that persisted into the 1960s, it was finally proven and agreed that the barrier properties of skin are largely determined by the stratum corneum [*7*].

This can be readily demonstrated by the so-called tape stripping experiment. Adhesive tape applied to and removed from the skin has been found to take with it stratum corneum cells. By measuring skin permeation as a function of the number of times the tape is applied and removed, one finds the rate increases by orders of magnitude. This is shown in Table 1 [*8*] for methanol and phenol.

The average thickness of our stratum corneum is 10.4 μm (0.5 mil). However, the stratum corneum of our palms and the soles of our feet can be 40 to 60 times thicker.

Much of the effort toward understanding the barrier properties of skin has

TABLE 1—*Permeation as a function of tape stripping of mouse skin,* in vitro, *at 37°C* [8].

Number of Strippings	Permeation Coefficient, cm/h	
	Methanol	Phenol
0	2.9	22.4
5	48.3	120
10	260	277
25	291	275
Dermis	395	301

focused on measuring the moisture vapor transmission rate (MVT) through skin. Such studies have produced the following findings:

(*a*) the MVT rate varies from location to location on our bodies (Table 2);
(*b*) the MVT rate varies with age;
(*c*) the MVT rate varies with degree of hydration of stratum corneum; and
(*d*) the MVT rate is a strong function of skin—abrasion, eczema, psoriasis.

Chemical Penetration

Chemical penetration of skin proceeds by three routes:

(*a*) the stratum corneum,
(*b*) the sweat ducts, and
(*c*) follicular and sebaceous openings.

Penetration through sweat ducts, hair follicles, and sebaceous pores is usually not significant for most nonpolar liquids since these routes represent only 1/1000 of the surface area, and the diffusion of nonpolar liquids in the stratum corneum is relatively rapid. For highly polar compounds, however, the pore routes can dominate; penetration by these routes is shunted directly to the dermal layer [*9*].

Chemical permeation through the stratum corneum occurs by mechanisms analogous to that through polymeric sheets and films and involves the following processes:

(*a*) sorption at the surface,
(*b*) diffusion through the membrane, and
(*c*) desorption into a collecting medium—in this case, body fluids of the epidermis and ultimately the dermis.

TABLE 2—*Moisture vapor transmission in human skin at 30°C* [7].

Skin Region	Flux, mg/cm²/h
Back	0.29
Forearm	0.31
Abdomen	0.34
Back of hand	0.56
Forehead	0.85
Palm	1.14
Sole	3.9

Diffusion within the stratum corneum is generally thought to occur along two paths:

(*a*) directly through the desiccated, keratinized cells, and
(*b*) through intercellular lipid material.

As with diffusion through polymeric films, diffusion through the stratum corneum is a function of the following factors:

(*a*) the concentration gradient, which is dependent on the partition coefficients on both sides of the skin,
(*b*) the diffusion coefficient,
(*c*) the thickness, and
(*d*) the temperature.

It should be noted that the partition coefficient, thickness, and diffusion coefficient can be strongly influenced by the degree of hydration of skin. Researchers have found that the permeation rate of the chemical warfare agent sarin increased by factors of up to ten as the skin became fully hydrated [*10*].

One difference between skin and simple polymeric materials is that skin can metabolize penetrants. This can affect the concentration gradient. Furthermore, in determining the toxicity portion of the risk assessment equation, the metabolic products must be considered.

In summary, the stratum corneum appears to behave, in general, as a polymeric film. It often exhibits Fickian behavior and has breakthrough times and permeation rates dependent on the chemical, the temperature, and its thickness.

Test Methods

Over the years, researchers have investigated scores of procedures for measuring skin permeability to chemicals. These can be divided into two approaches: *in vivo* and *in vitro*.

In vivo approaches involve living organisms, typically rodents, rabbits, pigs, and monkeys. Human testing would also fall into this category. Testing with humans provides the only truly definitive results. Obviously, however, safety and practical considerations such as the small number of test subjects, the poor control over the test subjects, and the variability of the test subjects preclude human testing in most cases. Therefore, researchers have sought to develop animal models that closely approximate human experience. This is a difficult task since the skin of each species has certain unique characteristics that can influence barrier properties. Furthermore, some of the species, for example, monkeys, are difficult to control and are expensive.

At present, guinea and miniature swine [for example, Hormel strain, Food and Drug Administration (FDA) breeding colony], pigs, and hairless mice [for example, C3H strain, National Institute of Health (NIH), Bethesda, MD] are considered to be the best animal models.

Common techniques for assessing permeation *in vivo* are listed in Table 3.

In vitro testing involves the use of skin excised from the animal species just described and humans. The excised skin is put in a permeation cell, the challenge chemical is added to the chamber facing the stratum corneum, and a collection medium is added to the dermal side chamber. Saline or distilled water is used as the collecting medium in order to prevent the skin from drying out.

The basic rationale for such testing is that the stratum corneum is the layer that controls permeation, and since this layer is composed of essentially dead tissue, the results will closely approximate *in vivo* findings. This has been well substantiated for water-soluble compounds, but there can be problems with chemicals with very low solubility in aqueous medium, as shown in Table 4 [*11*].

Skin availability and uniformity are important considerations in *in vitro* tests. Although human skin is available from cadavers and amputations, it is usually scarce, and it is never of uniform quality. Thus, it is difficult to replicate experiments. For this reason, considerable effort has gone into breeding

TABLE 3—In vivo *procedures.*

Monitoring urine, feces, expired air
Monitoring blood
Tissue analysis
Monitoring symptons:
(*a*) Topical
(*b*) Systemic
Cold hexane rinse
Ultraviolet fluorescence

TABLE 4—In vitro *versus* in vivo *permeation rates in rat skin at 25°C* [11].

Substance	Time Lag, h	Rate, $\mu g/cm^2/h$	
		In Vitro	*In Vivo*
Dichloromethane	0.5	5810	6580
1,2-Dichloroethane	0.5	1710	2880
1,1,2-Trichloroethane	0.6	335	1040
Trichloroethylene	1.1	67.6	...
Tetrachloroethane	1.4	23.2	544
Tetrachloroethylene	2.5	5.0	240

animals especially for their skin. Again, the most successful have been the hairless mouse and a particular type of pig.

In general, the ranking of excised skins of various sources in relation to their closeness to living human skin is as follows:

(*a*) human cadaver,
(*b*) hairless mouse/pig,
(*c*) monkey,
(*d*) rat, and
(*e*) rabbit.

Permeation Data

The data pertaining to the permeation of several chemicals through animal and human skin, both *in vivo* and *in vitro*, are summarized in Tables 5 [*12,13*], 6 [*14*], 7 [*15*] and in Figs. 2, 3, and 4. The data are presented in terms of permeation rate (milligrams per square centimetre per hour), lag time (minutes, where lag time is the time value obtained by extrapolation of the permeation rate curve back to zero permeation rate), and permeability coefficient (centimetre per hour or centimetre per minute). The permeability coefficient is determined from experimental data obtained with a simple permeation cell and is defined as

$$P = \text{permeability coefficient} = \frac{VdC/dt}{A\Delta C}$$

where

V = volume of the collecting medium,
A = surface area of the skin,
dC/dt = the rate of change of the permeant concentration in the collecting medium, and
ΔC = the concentration gradient across the skin membrane.

TABLE 5—*Permeation through human skin* in vivo [12,13].

Substance	Rate, mg/cm^2/h
Aniline	0.2 to 0.7
Nitrobenzene	0.2 to 0.3
Benzene	0.4
Xylene	4.5 to 9.6
Styrene	9 to 15
Carbon disulfide	9.7
Toluene	14 to 23

TABLE 6—*Comparison of mouse skin with human skin* in vitro *at 25°C* [14].

Substance	$P \times 10^6$, cm/min	
	Hairless Mouse	Human
Methanol	15.5	8.3
Ethanol	20.5	13.3
n-Butanol	73.3	20.0
n-Hexanol	143.3	216.7
n-Octanol	525.0	866.7

TABLE 7—*Permeation of non-electrolytes from aqueous solutions in rabbit skin at 33°C* [15].

Substance	Lag Time, min	$P \times 10^6$, cm/min
Ethyl iodide	6.25	91
Methanol	7.75	42
Ethanol	13.75	44
Thiourea	31.0	2.9
Glycerol	45.0	3.9
Urea	64.3	2.4
Glucose	109.3	0.9

Multiplication of the permeability coefficient by the concentration gradient yields the permeation rate. For example, with $P = 10^{-6}$ cm/min and $C = 0.1$ g/cm^3, the permeation rate equals 6 μg/cm^2/h.

The permeation rates for several common aromatic liquids and carbon disulfide through human skin (*in vivo*) are listed in Table 5. In comparison, the permeation rates at 20 to 25°C for benzene through Viton (0.24 mm thickness), butyl rubber (0.56 mm), and neoprene (0.46 mm) have been reported as 0.006, 5.4, and 48 mg/cm^2/h, respectively.

Table 6 summarizes the permeability coefficients for a series of alcohols and for both hairless mouse and human skin *in vitro*. The hairless mouse data are replotted in Fig. 2 as a function of alkyl chain length, along with corresponding data for 20 and 37°C. The coefficient of permeability increased by two orders of magnitude over the series of alcohols from methanol to octanol. Other researchers have shown that the coefficients peak and then drop as the alkyl chain length increases beyond that of octanol. For this reason high octanol/water partition coefficients are considered strong indicators of potentially high skin permeation rates. The octanol/water partition coefficient is the ratio of the solubility of the chemical in octanol to its solubility in water. Time lag data for the same group of alcohols are shown in Fig. 3. Time lag is seen to decrease with temperature and increase with the length of the chain. The latter point is consistent with breakthrough time trends for synthetic polymers. Increased chain length hinders the ability of the permeant molecule to diffuse

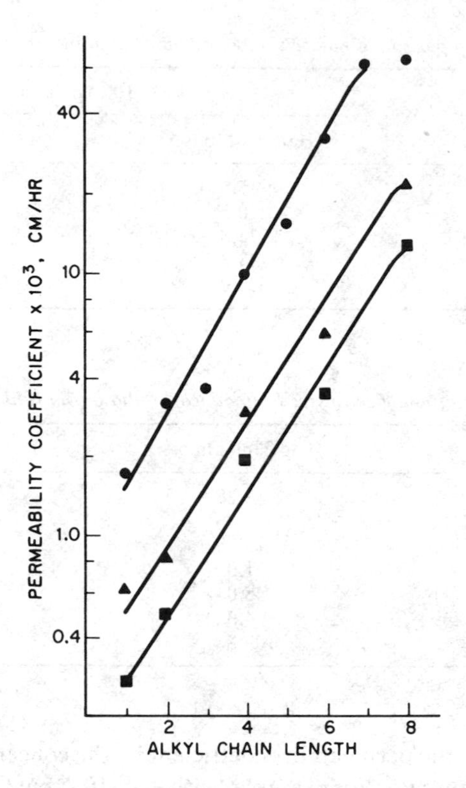

FIG. 2—*Alcohol permeation in the hairless mouse* in vitro *at 20 (■), 25 (▲), and 37°C (●)* [14].

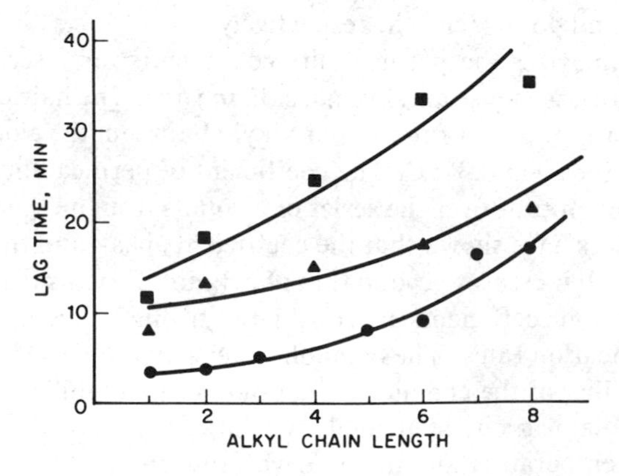

FIG. 3—*Alcohol permeation in the hairless mouse* in vitro *at 20 (■), 25 (▲), and 37°C (●)* [14].

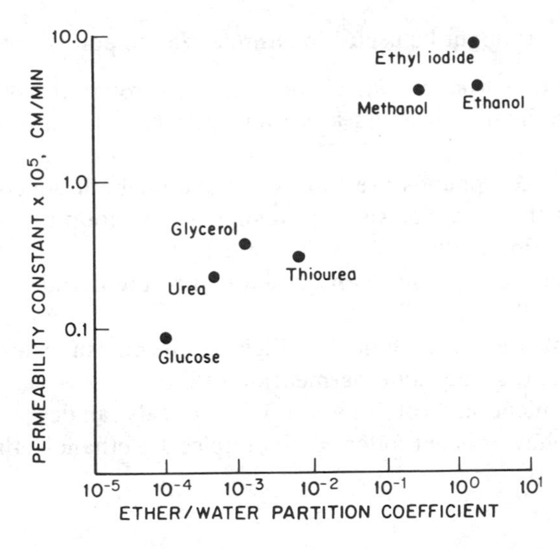

FIG. 4—In vitro *permeation at 33°C in rabbit skin* [15].

through the macromolecular network of the polymer, and the diffusion coefficient decreases. The overall permeation rate increases, however, because the solubility of the longer-chain alcohols increases to a greater degree than the diffusion coefficient decreases.

Table 7 and Fig. 4 contain additional permeation data, in this case for rabbit skin exposed to aqueous solutions of the listed chemicals. Similar to the alcohol data, higher-permeability coefficients were found for those compounds which exhibit both water and organic solvent (ether) solubility.

Another topic worth mentioning, although no data are presented, is that of permeation enhancers. These are chemicals that alter (in a reversible way) the barrier properties of the stratum corneum. By so doing, they significantly increase the rate of permeation of other chemicals. Dimethyl sulfoxide (DMSO) is probably the best known enhancer and has been demonstrated in some cases to increase rates by 25 times [*16,17*].

Several other potent enhancers are known and are of keen interest to companies involved in the development of transdermal drug delivery systems. They also represent a particular challenge to the industrial hygienist responsible for those who work with such chemicals.

Finally, for completeness, it must be mentioned that soaps, surfactants, and detergents in dilute aqueous solutions can significantly reduce the barrier properties of skin. On the other hand, barrier creams and emollients have been developed to provide additional protection to the skin.

At this point, it might be useful to summarize the points covered so far:

1. The stratum corneum controls permeation through the skin.
2. Excised animal skin offers a convenient means for investigating skin permeation.
3. Permeable compounds are those that exhibit both aqueous and lipid solubility, with the lipid side slightly favored. This group includes ketones, intermediate-chain-length alcohols, and aldehydes.
4. Highly polar compounds can be shunted to the dermis through sweat ducts and hair follicles.
5. Skin permeation often follows a Fickian mechanism exhibiting breakthrough times and steady-state permeation rates.
6. Some chemicals and combinations of chemicals can degrade the skin so that Fickian behavior is not followed. Examples are ethanol/ether and chloroform/methanol.

Conclusions

In light of the preceding discussion, how can breakthrough time and permeation rate data for protective clothing materials be used to estimate and reduce exposure and risk?

One way to reduce risk is to use clothing with breakthrough times significantly greater than any possible exposure time. Theoretically, this will reduce the exposure to a very low level, although not necessarily to zero. This is the approach being used by those who select clothing solely on the basis of breakthrough time data.

The breakthrough time approach may not be reasonable in cases in which the toxicity of the chemical or the probability of exposure is low. In these cases, some exposure may be permissible, and there is interest in estimating the amount of exposure. This can be done directly by any of the *in vivo* techniques described earlier or by attempting to combine skin and clothing permeation testing. Techniques for this direct approach have been described by Boman [*18*].

The indirect approach of estimating exposure from skin and clothing permeation data is not sufficiently advanced to permit one to confidently make such predictions with a reasonable level of uncertainty. Several questions must be answered:

1. Given a breakthrough time and permeation rate, what is the concentration inside an item of clothing? This will be dependent on unquantified and highly variable volumes. Furthermore, air exchange may occur through seams, closures, and openings. This may be due to natural convection or pumping resulting from flexing and movement or forced air systems.
2. What is the surface area challenged, and by what mode of challenge? Adults hands have several hundred square centimetres of surface area. A

fraction of this skin will be touching the glove, while other parts will be separated from the glove by an air gap. The transfer mechanisms will be different in both situations. Furthermore, the air/skin partition coefficient will differ from that of the clothing/skin.

3. Are data obtained from liquid challenge of skin applicable to challenges that are likely to be gaseous in nature?

4. What is the extent of hydration of skin under protective clothing? What is the hydration/permeation correlation for the particular chemical of concern?

5. What is the temperature of the skin? It will be a function of location on the body, outside temperature, type of clothing, and work load.

Until we can answer these and related questions from the point of view of protective clothing utilization, those responsible for worker health should take the conservative, prudent approach of "overprotecting" the worker in order to ensure reduced risk.

References

[1] Cheron, J., Guenier, J. P., Moncelon, B., and Lima, "Resistance des Gants de Protection Aux Solvents Industriels—Tableaux Recapitulatifs," *Travail et Securite*, No. 573, L'Institut National de Recherche et de Securité, Paris, France, Dec. 1976.

[2] Coletta, G. C., Schwope, A. D., Arons, I., King, J., and Sivak, A., *Development of Performance Criteria for Protective Clothing Used Against Carcinogenic Liquids*, DHEW (NIOSH) Pub. No. 79-106, Department of Health, Education and Welfare, Washington, DC, Oct. 1978.

[3] Henry, N. W. and Schlatter, C. H., "The Development of a Standard Method for Evaluating Chemical Protective Clothing to Permeation by Liquids," *American Industrial Hygiene Association Journal*, Vol. 42, 1981, p. 202.

[4] Nelson, G. O., Lum, B., Carlson, G., Wong, C., and Johnson, J., "Glove Permeation by Organic Solvents," *American Industrial Hygiene Association Journal*, Vol. 42, 1981, p. 217.

[5] Schwope, A. D., Costas, P. P., Jackson, J. O., and Weitzman, D. J., *Guidelines for the Selection of Chemical Protective Clothing*, American Conference of Governmental Industrial Hygienists, Cincinnati, OH, 1983.

[6] Barry, B. W., *Dermatological Formulations—Percutaneous Absorption*, Marcel Dekker, New York, 1983.

[7] Scheuplein, R. J. and Bronaugh, R. L., "Percutaneous Absorption" in *Biochemistry and Physiology of the Skin*, L. A. Goldsmith, Ed., Oxford University Press, New York, 1983, pp. 1255–1295.

[8] Behl, C. R., Linn, E. E., Flynn, G. L., Pierson, C. L., Higuchi, W. I., and Ho, N. F. H., "Permeation of Skin and Eschar by Antiseptics I: Baseline Studies with Phenol," *Journal of Pharmaceutical Sciences*, Vol. 72, April 1983, p. 391.

[9] Webster, R. C. and Maibach, H. I., "Cutaneous Pharmokinetics: Steps to Percutaneous Absorption" in *Drug Metabolism Reviews*, Vol. 14, 1983, p. 169.

[10] Wurster, D. E., Ostrenga, J. A., and Matheson, L. E., "Sarin Transport Across Excised Human Skin I: Permeability and Adsorption Characteristics," *Journal of Pharmaceutical Science*, Vol. 68, 1979, p. 1406.

[11] Tsuruta, H., "Percutaneous Absorption of Trichloroethylene in Mice," *Industrial Health*, Vol. 16, 1978, p. 145.

[12] Dutkiewicz, T. and Tyras, H., *Skin Absorption of Toluene, Styrene, and Xylene by Man*, Department of Toxicological Chemistry and Industrial Toxicology, Medical Academy, Lodz, Poland, 1968.

[13] Dutkiewicz, T. and Tyras, H., "Studies on the Skin Absorption Properties of Styrene in Human Beings," *Gigiena Truda; Professional nye Zabolevaniia*, Vol. 12, No. 4, 1968, p. 35.
[14] Durrheim, H., Flynn, G. L., Higuchi, W. I., and Behl, C. R., "Permeation of Hairless Mouse Skin I: Experimental Methods and Comparison with Human Epidermal Permeation by Alkanols," *Journal of Pharmaceutical Science*, Vol. 69, 1980, p. 781.
[15] Treherne, J. E., "The Permeability of Skin to Some Non-Electrolytes," *Journal of Physiology*, No. 133, 1956, p. 171.
[16] Stoughton, R. B. and Fritsch, W., "Influence of Dimethylsulfoxide (DMSO) on Human Percutaneous Absorption," *Archives of Dermatology*, Vol. 90, 1964, p. 512.
[17] Chandrasekaran, S. K., Campbell, P. S., and Michaels, A. S., "Effect of Dimethyl Sulfoxide on Drug Permeation through Human Skin," *A.I.Ch.E. Journal*, Vol. 23, 1977, p. 810.
[18] Boman, A., Wahlberg, J. E., and Johnansson, G., "A Method for Studying the Effect of Barrier Creams and Protective Gloves on the Percutaneous Absorption of Solvents," *Dermatologica*, Vol. 164, 1982, pp. 157-160.

Gerard C. Coletta[1] *and Mark W. Spence*[2]

Managing the Selection and Use of Chemical Protective Clothing

REFERENCE: Coletta, G. C. and Spence, M. W., **"Managing the Selection and Use of Chemical Protective Clothing,"** *Performance of Protective Clothing, ASTM STP 900,* R. L. Barker and G. C. Coletta, Eds., American Society for Testing and Materials, Philadelphia, 1986, pp. 235–242.

ABSTRACT: Five key elements important to the success of a chemical protective clothing management program are discussed and illustrated with a case study. These elements are (1) determining the likelihood of skin exposure, (2) identifying the consequences of direct skin contact, (3) establishing the levels of protection provided by available protective clothing, (4) making an appropriate selection and documenting the basis for the selection, and (5) training employees in the proper use of the selected items. These elements form the basis of an ongoing risk/benefit analysis intended to ensure maximum employee protection within constraints imposed by operating and business considerations.

KEY WORDS: protective clothing, chemical protective clothing, permeation testing, risk/benefit analysis, risk assessment, protective clothing limitations, protective clothing selection, protective clothing use

Interest in chemical protective clothing has grown tremendously during the past few years. In part, this is due to a greater emphasis on the use of personal protective equipment rather than on the use of engineering controls for worker protection. New base materials for protective clothing have been developed and, with attendant claims for greater levels of protection, have been marketed aggressively. Further, this trend is being supported by an increasing demand for "totally protective" suits for use in hazardous waste disposal sites, in responding to chemical spills or fires, and in high-hazard industrial operations.

Historically, the selection of protective clothing has been determined empirically and with little consistency. Many manufacturers and users have depended on simple immersion tests to establish the degradation resistance of

[1]President, Risk Control Services, Inc., San Francisco, CA 94920.
[2]Industrial hygienist, Dow Chemical U.S.A., Midland, MI 48674.

clothing materials to a large number of chemicals. Unfortunately, such test procedures have not been designed to simulate occupational exposures closely and, therefore, may not provide accurate information on levels of protection. Further, these procedures often vary from company to company, so that direct comparisons of results are difficult at best.

Selection and Use

With recent efforts by ASTM and others in developing standardized, representative test methods, employers have the opportunity to ensure high levels of protection for their employees. To take full advantage of this opportunity, a well-thought-out and actively supported protective clothing management program is essential. Such a program should be based on balancing the risks associated with skin contact by individual chemicals against the benefits of protection levels provided by individual items of protective clothing.

A successful protective clothing management program is the result of planning, conscientious execution, and continuous follow-up. Five key elements are important for a program's long-term success. These are (1) determining the likelihood of skin exposure, (2) identifying the consequences of direct skin contact, (3) establishing the levels of protection provided by available protective clothing, (4) making an appropriate selection and documenting the basis for the selection, and (5) training employees in the proper use of the selected item or items. These elements should be considered the cornerstones required to support any effective program.

This paper discusses these five key elements. If pursued with diligence, these elements will maximize the benefits to be derived from the use of chemical protective clothing.

Likelihood of Skin Exposure

Determining the likelihood of skin exposure to individual chemicals in a work environment is not an easy job. It involves assessing normal operations and maintenance, as well as estimating the potential for spill or leak resulting from equipment failure or human error. This task is particularly important when using highly toxic substances.

Assessing normal operations begins with an inventory of chemicals used and produced, focuses on defining tasks involving those chemicals, and concludes with an evaluation of individual chemical handling procedures to determine a duration for skin exposure. Physical characteristics important in protective clothing used for these operations are also identified (for example, cut resistance, wet grip, or manual dexterity). Normal operations might include reactor sampling, filter changes, drumming operations, and maintenance.

Maintenance is an important consideration that is often neglected when workplace hazards are reviewed. Maintenance tasks tend to be sporadic, less predictable, and frequently, more hazardous than routine operating procedures.

Estimating the potential for a chemical spill or leak requires a second, even more pragmatic review of equipment, conditions, and procedures associated with handling and containing the chemicals of interest. For example, the integrity of engineering controls and the thoroughness of preventive maintenance are both important in minimizing accidental exposures. Also important are individual work practices and emergency response procedures.

In-house accident statistics, workers' compensation experience reports, and maintenance and repair records provide excellent sources of information that will either confirm or refute the assessment of operations and the estimate of spillage or leakage. By building such experience factors into this process and modifying the first results accordingly, one can generate a realistic determination of the potential for skin exposure. Combined with an evaluation of the consequences of direct skin contact by individual chemicals, this risk factor forms a primary "tool" to be used in the final selection of proper chemical protective clothing.

Consequences of Direct Skin Contact

The consequences of direct skin contact by individual chemicals can be ascertained from both published literature and in-house experience. The primary question to be answered is whether or not skin contact could result in significant illness or injury. Are surface effects, such as irritation or minor dermatitis, the only consequences of skin contact, or can sensitization develop? Can the chemical be absorbed through the skin, and, if it is absorbed, what systemic effects will result?

Available Levels of Protection

As users of protective clothing continue to become aware of the value of testing clothing materials and designs under simulated occupational conditions, the old approach of basing selection decisions on the results of simple degradation (immersion) tests will no longer be accepted. Materials and designs not supported by quantitative test data will succumb to those in the marketplace that are. Occupational exposures and individual protective clothing materials and designs should be viewed as matched pairs. The notion among users that there are all-purpose materials is a myth.

ASTM's Committee F-23 on Protective Clothing has made significant progress in developing such laboratory test methods for chemical protective clothing. These methods encourage the establishment of quantified levels of performance by measuring specific parameters such as materials' resistance to permeation, to degradation, and to penetration. The permeation test method, for example, has already resulted in the generation of a large body of data describing protective clothing resistance to very low level permeation. These data have been of considerable value in showing that some commonly used items do not provide expected protection under simulated exposures. Manufacturers, users, and contract laboratories are all sources of testing expertise and data.

Laboratory test methods are becoming the foundation for predicting the performance of protective clothing materials and designs in the field. They provide the best technical and legal base for selection decisions, especially in this era of implied liability for employee health and safety. A well-founded protective clothing management program will incorporate, as much as possible, input from laboratory tests in its selection protocol.

But because laboratory tests are expensive and time-consuming, not all combinations of chemicals and clothing material can, or should, be tested. The next generation of test methods should see the development of supplementary field tests. By definition, these would be simpler and faster, and perhaps would focus on chemical analogy to expand the use of existing data. The extensive number of variables associated with the use of protective clothing, including process conditions and physical hazards, should also encourage the use of field tests to build on base-level data generated by carefully selected laboratory tests. At the present time, field tests are conducted on a "home-grown," laissez-faire basis. Most have been developed to accommodate individual, highly specific needs with little effort toward standardization.

But whether by laboratory test or by a combination of field and laboratory tests, levels of protection provided by chemical protective clothing can be established only through the generation of quantitative performance data representing as closely as possible the actual occupational environment.

Most Appropriate Selection

The final effort in selecting appropriate chemical protective clothing is actually a case-by-case best judgment based on the factors outlined earlier. Since protective clothing cannot provide complete protection indefinitely, a balance must be established among the likelihood of the exposure, the consequences of actual skin contact, available levels of protection, and budgetary or other business constraints. Above all, the bases for selection, including process descriptions, failure analyses, protective clothing test data, and recommended use life should be documented for easy retrieval and updating.

Combining the potential for skin exposure with toxicological information on the consequences of actual skin contact permits the development of in-house criteria for protective clothing performance. Such criteria should be as specific as possible in defining the levels of protection required under both "normal" and emergency conditions that would lead to chemical spills or leaks. These criteria might include such parameters as breakthrough time and steady-state permeation rate for permeation resistance, or retention of flexibility and puncture resistance for degradation resistance.

These performance criteria are compared with the levels of protection provided by available protective clothing. Through this process, potential candidates are identified for further scrutiny in terms of initial cost, ability to be cleaned or decontaminated for reuse, comfort, and durability. With all of this information, a sound business decision can be made that will result in optimum levels of protection for employees.

It is possible that protective clothing currently available in the marketplace will not meet the initial in-house performance requirements. In such a situation, either process conditions or, more important, employee job requirements and procedures should be reviewed and modified to permit a successful match.

Employee Training

An important part of any chemical protective clothing management program is formal training of all affected employees. Training should be a focused, ongoing process that addresses the whys and hows of using the selected protective clothing and should include some hands-on instruction. Normal operations as well as emergency conditions should be covered.

Protective clothing training is best designed to incorporate information and data on the following:

(*a*) hazards associated with the expected chemical exposures,
(*b*) protection provided by the selected chemical protective clothing,
(*c*) proper care of the protective clothing, and
(*d*) limitations in using the protective clothing.

The segments on care and limitations should include procedures for regular inspection of individual items of protective clothing as well as for cleaning, decontamination, and reuse. After the initial training has been completed, a brief refresher should be planned on an annual basis. Of course, as new or additional items of protective clothing are introduced into the work environment, additional comprehensive training should be conducted.

A Case Study

In the following paragraphs, a hypothetical clothing management program is illustrated for a plant that produces styrene polymer. The management program is structured to highlight each of the five key program elements as used under normal operating conditions.

Likelihood of Skin Exposure

Three chemicals found in the plant have high potential for skin contact. These three styrene, ethyl benzene, and 1,1,1-trichloroethane—are encountered routinely in a number of operations, as shown in Table 1.

Each operation has been characterized for extent of skin exposure as follows:

Process sampling is a daily operation lasting approximately 5 min. Exposure is moderate as 100-mL samples are drawn for laboratory analyses.

Filter changes require 30 min of relatively high-level exposure on a monthly basis. Mechanical strength of gloves is important in draining and opening filter housing prior to removing or inserting elements.

Maintenance is the most unpredictable set of operations. A given operation may last from 5 min to 8 h. Operations such as pump repair, line repair, and reactor seal replacement often require cleaning parts in chlorinated solvents such as 1,1,1-trichloroethane. Again, mechanical strength in gloves is important during times of high-level exposures.

Laboratory analyses involve frequent 1 to 2-h tasks such as withdrawing small-volume aliquots for analysis from the 100-mL process samples. Usually this operation represents a low potential for exposure. Manual dexterity is critical.

Consequences of Direct Skin Contact

The skin can be a significant route of exposure for styrene and ethyl benzene but is of less significance for 1,1,1-trichloroethane. With all three chemicals, irritation and dermatitis result from the defatting action of prolonged skin contact [*1*].

Available Levels of Protection

Some permeation resistance data for a number of glove materials are available from manufacturers' literature; these are shown in Table 2. Since very

TABLE 1—*Chemicals found in styrene polymer production.*

	Chemical		
Operation	Styrene	Ethyl Benzene	1,1,1-Trichloroethane
Process sampling	X	X	
Filter changes	X	X	
Maintenance	X	X	X (parts cleaning)
Laboratory analyses	X	X	

TABLE 2—*Permeation resistance of selected glove materials.*[a]

Gloves		Breakthrough time, min[b]				
			Styrene/Ethyl Benzene Analogs			
Material	Cost Per Pair, $	Styrene[c]	Divinyl Benzene	Toluene	Xylene	1,1,1-Trichloroethane
PVA	20	>480	...	15	>480	60
NBR[d]	3	...	...	10	75	120
Viton	20	...	>1020	>960	...	>900
Butyl	8	...	132	...	...	...
PVC	...	degrades	...	...	...	...
Neoprene	...	degrades	...	...	...	...

[a]Product data are supplied by the Edmont Division of the Becton-Dickinson Co., Coshocton, OH, and by the Norton Co., Safety Products Division, Watertown, MA.

[b]Breakthrough times were determined using ASTM Test for Resistance of Protective Clothing Materials to Permeation by Hazardous Liquid Chemicals (F 739-81).

[c]Based on 1984 manufactures list prices.

[d]Superior mechanical strength.

few data are available for styrene permeation rates and even fewer for ethyl benzene rates, the data are listed for three compounds which are similar in structure.

Most Appropriate Selections

Based on the potential for skin exposure, the consequences of direct skin contact, and the permeation data shown previously, the following recommendations could be made:

1. Because of their low cost, superior mechanical strength, and performance with xylene (taken to be a styrene/ethyl benzene analog), nitrile-butadiene rubber (NBR) gloves are recommended for the following operations and replacement intervals:

   (*a*) process sampling—replace gloves weekly,
   (*b*) filter changes—replace gloves after each change, and
   (*c*) maintenance of process equipment—replace gloves every 4 h.

2. For the parts cleaning maintenance operation involving 1,1,1-trichloroethane, Viton gloves are recommended because immersion of the hands is involved. These gloves should be replaced once a month and kept exclusively in the 1,1,1-trichloroethane bath area.

3. Manual dexterity is the overriding factor in the laboratory analysis oper-

ations. For this reason, as well as the very low potential for any significant contact, thin gloves of a stretchy material like latex rubber or polyvinyl chloride (PVC) should be used. These gloves should be replaced every time they are removed and as soon as possible after any known contact.

(All the replacement times are maximum. Gloves should be replaced sooner if degradation is apparent.)

Employee Training

These recommendations, as well as the background information and reasoning leading to them, should be documented as a section in the plant operating procedures.

Employees should be trained by the plant industrial hygienist during plant safety meetings. Subjects include which gloves to use for each operation, when and how to dispose of them, and the proper way to remove gloves without touching the outside surface. The importance of inspecting gloves before each use should be emphasized. One monthly safety meeting each year should be devoted to protective clothing.

Summary

The five key elements described in this paper are important components of a comprehensive protective equipment and clothing program. As illustrated in the example cited, these elements can be used to arrive at well-thought-out, documented protective clothing selection and use procedures. With periodic management review, employee protection can be maximized in environments with continuous potential for chemical exposure.

Reference

[1] Proctor, N. H. and Hughes, V. P., *Chemical Hazards of the Workplace*, Lippincott, Philadelphia, 1978.

Krister Forsberg[1]

Selection of Chemical Protective Clothing Using Permeation and Toxicity Data

REFERENCE: Forsberg, K., "**Selection of Chemical Protective Clothing Using Permeation and Toxicity Data,**" *Performance of Protective Clothing, ASTM STP 900*, R. L. Barker and G. C. Coletta, Eds., American Society for Testing and Materials, Philadelphia, 1986, pp. 243–249.

ABSTRACT: Recommendation of chemical protective clothing should be based on chemical resistance test data, assessment of the chemical in question, and the time workers are at risk of chemical exposure.

The availability of the ASTM standard test method has increased the amount of permeation data in the literature. By measuring breakthrough time, the method can be used to estimate the maximum protection time from chemical exposure under conditions of continuous contact. The second variable in permeation data is permeation rate, that is, the method can be used to identify protective clothing materials that limit chemical exposure to acceptable skin contact levels.

In this paper materials have been grouped in different classes based on permeation data. Three intervals of breakthrough time and five levels of steady-state permeation rate are proposed. The authors also recommend applying a labeling and classification system to provide information for risk assessment of chemical exposure hazards in protective clothing.

Risk assessment and recommendation of chemical protective clothing materials have to be carried out by a competent person.

KEY WORDS: risk assessment, chemical exposure, chemical toxicity, permeation data, permeation class, risk information, chemical protective clothing, protective clothing

There is a wide range in the hazardous properties of chemical substances that can be found in the workplace. The potential for exposure to these chemicals can vary from minutes to hours during a normal day, depending on the characteristics of the process or operation involved.

There are many variables to consider when selecting chemical protective

[1]Researcher, Department of Work Science, The Royal Institute of Technology, S-100 44 Stockholm, Sweden.

clothing to protect against chemical contact. Each chemical interacts differently with chemical protective clothing materials.

A recent publication, *Guidelines for the Selection of Chemical Protective Clothing*, addresses this issue [*1*]. The guidelines presented specifically address the chemical resistance of protective clothing materials. The overall goal is to provide users with the necessary information to make informed judgments relative to selecting clothing and to provide some consensus on the most appropriate clothing for chemicals of interest.

Recommendation of chemical protective clothing is a serious matter. An industrial hygienist should be the most competent person to assess the risks and recommend chemical protective clothing materials.

The purpose of this paper is to describe a labeling and classification system for risk assessment and to propose grouping permeation data into classes.

Risk Information

In the selection of protective clothing materials there is a risk that the chemical may have hazardous effects, which can range from dry skin and irritation to sensitization and corrosion. Effects on internal organs from skin penetration cannot be excluded.

A labeling and classification system is included in the Swedish guidelines [*2*]. The hazardous properties of substances are presented as risk information (Table 1). The risk code indicates the general nature and severity of the observed effects. The risk phrase describes the risk in greater detail. The risk code and risk phrase follow the Product Control Board's recommendations concerning application of the labeling ordinance [*3*]. The purpose of the labeling ordinance is to ensure that anyone handling a product hazardous to health receives adequate information on what risk such handling may involve, the size of the risk, and what safety precautions may be taken. Descriptions of the risk code and risk phrase are presented in Table 2.

TABLE 1—*Risk information for ten chemicals (examples).*

Chemical Name	Risk Code[a]	Risk Phrase[a]
Acetaldehyde	X	38
Acetic acid	Cx	35
Acetone	V	313
Acrylonitrile	T	24 and 45
Allyl alcohol	Tx	24
Allyl chloride	T	24 and 34
Ammonium hydroxide	C	34
Amyl acetate	V	313
Amyl alcohol	X	313
Aniline	T	24

[a] As illustrated in Table 2.

TABLE 2*a*—*Risk information: description of risk codes.*

Risk Code	Code Designation
Tx	very toxic chemicals
T	toxic chemicals
X	harmful and irritant chemicals
Cx	highly corrosive chemicals
C	corrosive chemicals
V	other chemicals for which labeling is obligatory
—[a]	not harmful chemicals in normal use

[a]This code is not included in the labeling and classification system developed by the Products Control Board.

TABLE 2*b*—*Risk information: description of risk phrases.*

Risk Phrase	Risk in Contact with Skin (relevant examples for skin contact)
R 27	very toxic in contact with skin
R 24	toxic in contact with the skin
R 35	highly corrosive
R 34	corrosive
R 21	harmful in contact with the skin
R 45	may cause cancer after often-repeated exposure
R 340	some risk for cancer cannot be excluded after often-repeated exposure
R 43	may cause sensitization in contact with the skin
R 38	irritating to the skin
R 321	may be harmful in contact with the skin
R 334	corrosive at repeated contact with the skin
R 313	dries the skin

Following are the definitions of the Swedish ratings when acute toxicity is the primary concern:

Substances with *very high acute toxicity*, i.e., substances that even in small doses can cause death or very serious injury, are assigned to the category -Tx-.

Substances with *high acute toxicity*, i.e., substances that in small doses can cause serious injury, are assigned to the category -T-.

Substances with *medium-to-moderate toxicity*, i.e., substances that in relatively small doses can cause injury, are assigned to the category -X-.

Substances with *low acute toxicity*, i.e., substances that only in relatively large doses can cause injury, are assigned to the category -V-.

As a rule of thumb, substances are graded according to their acute toxicity to animals, as follows:

Level	Median Lethal Dose (LD_{50}), Dermal Rat or Rabbit, mg/kg
Very high acute toxicity	<50
High acute toxicity	50 to 400
Medium-to-moderate acute toxicity	400 to 2000
Low acute toxicity	. . .

The chronic toxicity scale uses many of the same categories with somewhat different definitions:

Substances with *high chronic toxicity*, i.e., substances that cause serious injuries to, for example, the kidneys, liver, nerves or brain upon prolonged or repeated exposure due to contact with skin in small doses, are assigned to the category -T-.

Substances with *moderate chronic toxicity*, i.e., substances that cause injuries to, for example, the kidneys, liver, nerves or brain upon prolonged or repeated exposure due to contact with skin in relatively small doses, are assigned to the category -X-.

Substances that can cause injuries after repeated or prolonged exposure in larger doses are assigned to the category -V-. These include, for example, certain organic solvents which can—in connection with repeated handling—cause such injuries as pronounced defatting of the skin.

The following is the rating scale for corrosive substances:

Highly corrosive substances are those that damage tissue and cause severe burns (even if the symptoms are delayed) upon brief contact (up to three minutes) with the skin. These substances are assigned to the category -Cx-.

Corrosive substances are those that, without being strongly corrosive, damage tissue and cause burns after more prolonged exposure (up to 4 hours). These substances are assigned to the category -C-.

Irritant substances are those that, without being corrosive, cause local inflammation upon direct, prolonged or repeated contact with living tissue (skin). These substances are assigned to the category -X-.

Certain substances can cause skin irritation such as itching and burning without producing local inflammation. Such substances should be assigned to the category -V-.

The final two definitions cover allergenic and carcinogenic substances:

Allergenic substances are those that can cause sensitization or photosensitization in non-insignificant frequency among those exposed to the substance. These substances are assigned to the category -X-.

> Substances that entail such a risk of cancer in connection with their handling that special labeling is warranted to protect the individual should, as a rule, be assigned to the category -T-. If the probability of the substance causing cancer is very low, however, it may be assigned to the category -X-.

As explained earlier, risk phrases describe the risk in greater detail. For example, Risk Phrases 34, 38, and 313 represent local effects. The risk information is limited to the most hazardous properties of the substance, that is, Phrase 38 covers Phrase 313. If there is a risk of systemic effects, for example, cancer and injuries to nerves, Phrases 45 and 21 are also used.

This model of using risk code and risk phrases is a simple and practical way to provide information for risk assessment of chemical exposure hazards in protective clothing.

Permeation Classes

Permeation testing has been standardized under ASTM Test for Resistance of Protective Clothing Materials to Permeation by Hazardous Liquid Chemicals (F 739-81). The method can be used to estimate the duration of maximum protection from chemical exposure under conditions of continuous contact, that is, measuring breakthrough time. A second variable—permeation rate at steady-state—can be calculated and used to identify protective clothing materials that limit chemical exposure to acceptable skin levels.

The amount of permeation data is increasing, and there is a need to classify and organize all the data. Therefore, a proposal to classify steady-state permeation rates and breakthrough times has been developed (Table 3). The

TABLE 3—*Permeation class: classification of permeation rate and breakthrough time.*[a]

Class	Level of Permeation Rate	Permeation Rate, $mg/m^2/min$	Breakthrough Time, h	Comment
0			>4	best selection
1	very low	<10	>1	
2	low	10 to 100	<1	second choice
3	moderate	100 to 1 000	<1	poor selection
4	high	1 000 to 10 000	<1	due to large permeation
5	very high	>10 000	<1	rate or use only versus possible splashes

[a]The division into different classes is arbitrary. The classification is valid if the test is to be carried out in >4 h, in case breakthrough does not occur. For some chemicals it is difficult to select a protective clothing material because the breakthrough time is <1 h. In those cases it is important to change clothing more often. The permeation rate is a question of exposure to acceptable skin levels. The permeation test is performed during continuous contact of the challenge chemical with the garment material. Under actual usage the permeation rate may be lower than the result of the material testing; that is a safety factor. No acceptable skin levels are published, such as the threshold limit value for skin alone.

classification of permeation rates is similar to that of the Edmont Chemical Resistance Guide (Table 4) [*4*]. The permeation class can be used to guide the industrial hygienist to make better decisions in the selection of chemical protective clothing. In this proposal, Class 0 is equal to the most resistant material.

Permeation Class 2 can be accepted, if the chemical of interest has hazardous properties of low toxicity, that is, Risk Phrases 38, 321, 334, and 313. Nevertheless, how long workers are at risk of chemical exposure should be a determining factor in risk assessment and recommendation.

In general, Permeation Classes 3, 4, and 5 are not recommended for protective clothing materials. Materials with breakthrough times of less than 1 can also be included in this group.

Work with extremely toxic chemicals requires chemical protective clothing materials with breakthrough times far longer than the risk times at chemical exposure. The toxicity of the chemical involved is also an important factor in the reuse of chemical protective clothing.

All numbers in the different permeation classes are arbitrary. A more detailed analysis of permeation data with reference to permeation class may improve the accuracy in the proposed numbers.

Recommendation of protective clothing materials against multicomponent liquids makes risk assessment more difficult. One component with low toxicity may act as a pilot substance and allow more harmful substances to reach the skin. This phenomenon must be pointed out to the industrial hygienist. In those cases, multicomponent permeation test data should be analyzed against the chemical toxicity of all substances. Such chemicals can be more harmful in combination than any single component.

The permeation classes are valid for initial tests and also for reuse tests. This means that a combination of chemical and material may be in a different class after the first use. At reuse you must take into consideration the possible risk of chemical exposure resulting from retained chemical from the first use.

TABLE 4—*Scale for permeation rates developed by Schlatter at Edmont* [*4*].[a]

Key to Permeation Rate	Permeation Rate, $mg/m^2/min$
ND—none detected during a 6-h test	
E—excellent	<9
VG—very good	10 to 90
G—good	91 to 900
F—fair	901 to 9 000
P—poor	9 001 to 90 000
NR—not recommended	$>90\ 000$

[a]Converted to the units of measurement used in this report.

Discussion

Criteria for selection of chemical protective clothing are based on permeation data and risk assessment.

Risk information in the form of risk code and risk phrases can easily be included in a handy reference for industrial hygienists. This would be a valuable tool for risk assessment.

The risk phrases cannot be compared with the *NIOSH/OSHA Pocket Guide to Chemical Hazards* [5]. However, risk assessment can be improved by using the *NIOSH/OSHA Pocket Guide* and the toxicity data presented in this paper.

Permeation classes guide the industrial hygienist in the selection of protective clothing materials. The classification into five different levels of steady-state permeation rate gives a better picture of the calculated values. Three intervals of breakthrough times simplify the selection criteria to apply to three areas, namely, best selection, second choice, and poor selection.

Degradation ratings are not included in this proposal. However, Permeation Classes 4 and 5 are indications of degradation. High interaction is needed for high permeation rates. The need to recommend protective clothing materials against mixtures of chemicals makes risk assessment more difficult.

The described risk information and the proposal for permeation classes should be discussed at the development of selection criteria. Those criteria must be relevant and easy to understand when assessing the risk of chemical exposure hazards in protective clothing.

References

[1] Schwope, A. D., Costas, P. P., Jackson, J. O., and Weithzman, D. J., *Guidelines for the Selection of Chemical Protective Clothing*, Arthur D. Little, Cambridge, MA, March 1983; 2nd ed., March 1985 (available through the American Conference of Governmental Industrial Hygienists).

[2] Forsberg, K. and Olsson, K. G., *Guidelines for the Selection of Chemical Protective Gloves*, The Royal Institute of Technology, Stockholm, May 1985 (in Swedish) (available through Föreningen för Arbetarshydd, Kungsholm Hamnplan 3, 11220 Stockholm, Sweden).

[3] The Products Control Board, *Application of the Labelling Ordinance: The National Swedish Environment Protection Board, Advice and Guidelines*, March 1983 (available through Liber Distribution, Förlagsorder, 16289 Stockholm, Sweden).

[4] *Edmont Chemical Resistance Guide: Edmont Gloves and Protective Equipment*, Edmont Division Becton, Dickinson & Co., Coshocton, OH, 1983.

[5] *NIOSH/OSHA Pocket Guide to Chemical Hazards*, DHEW (NIOSH) Publication No. 78-210, National Institute of Occupational Safety and Health/Occupational Safety and Health Administration, Cincinnati, OH.

Stephen P. Berardinelli[1] *and Michael Roder*[1]

Chemical Protective Clothing Field Evaluation Methods

REFERENCE: Berardinelli, S. P. and Roder, M., **"Chemical Protective Clothing Field Evaluation Methods,"** *Performance of Protective Clothing, ASTM STP 900,* R. L. Barker and G. C. Coletta, Eds., American Society for Testing and Materials, Philadelphia, 1986, pp. 250–260.

ABSTRACT: This is a concept paper. The reasons for conducting field evaluations as well as chemical resistance field test methods are discussed. The field tests methods include chemical resistance of protective clothing to degradation, penetration, or permeation by a liquid chemical.

KEY WORDS: chemical resistance test methods, chemical protective clothing, chemical permeation, penetration, degradation, protective clothing

This is a concept paper to acquaint the reader with chemical protective clothing (CPC) field evaluation methods.

Chemical protective clothing is defined as garments, gloves, boots, coveralls, aprons, or fully encapsulating suits that protect workers from dermal exposures to contacted chemicals. Exposure occurs through three modes: (1) material failure resulting from chemical or physical degradation or a combination of both, (2) bulk penetration of liquid chemical through pinholes, seams, closures, imperfections, and so on, and (3) permeation (molecular flow) of liquid chemical through the garment material. ASTM Committee F-23 on Protective Clothing, which develops consensus standard methods, has promulgated a standard test method for chemical permeation, ASTM Test for Resistance of Protective Clothing Materials to Permeation by Hazardous Liquid Chemicals (F 739-81), and is promulgating standard test meth-

Mention of a company name or product does not constitute endorsement by the National Institute for Occupational Safety and Health (NIOSH).

[1]Industrial hygiene chemist and industrial hygiene engineer, respectively, National Institute for Occupational Safety and Health, Centers for Disease Control, U.S. Department of Health and Human Services, Morgantown, WV 26505-2888.

ods for the other two transport modes. These standard methods are designed for the laboratory rather than for field use.

Candidate garment materials can be selected by using degradation, penetration, and permeation data, as well as physical properties. These materials must be evaluated under usage conditions for the following reasons:

1. Product formulations and processing conditions can vary from manufacturer to manufacturer or even from lot to lot. Neoprene, a common generic and widely used material, is available in many formulations using various fillers and active ingredients to provide slightly different properties.
2. All of the preceding evaluation techniques use cut specimens and not the entire garment. The quality of manufacture could produce significant differences, such as pinholes, thickness variations, and formulation variations.
3. Evaluations are generally first-time exposures made under laboratory conditions utilizing pure chemicals in continuous contact. In actual practice, chemicals are mixtures, exposures are intermittent, conditions such as temperature vary, and products are often reused with or without decontamination. For most garments, decontamination is the exception rather than the rule.

Field evaluations could include the following:

1. Degradation, penetration or permeation testing, or a combination of those, utilizing the actual chemicals to be encountered, specimens from the products being considered, and the conditions expected (contact sequence, temperature, and reuse).
2. Simple field evaluation observations could include many factors. Does the CPC introduce other hazards such as catching on moving equipment or loss of dexterity? Do workers accept and use the CPC? Does additional exposure occur from donning, recycling, or storing the CPC? Do exposure indications (skin disorders, symptoms, positive blood tests) increase or decrease? What are the economics and potential exposure risks with reuse of the garments (decontamination and reuse versus single use and disposal)?
3. Workers provide feedback to their industrial hygienist so that the selection process can be fine-tuned.

Experimental Procedure

Several field test methods are discussed in some detail.

Degradation Testing

An ASTM draft chemical degradation standard test method is amenable to field use. The cell is shown in Fig. 1. This cell is an ASTM Test F 739-81 permeation cell, but the collection side of the cell has been modified. This cell

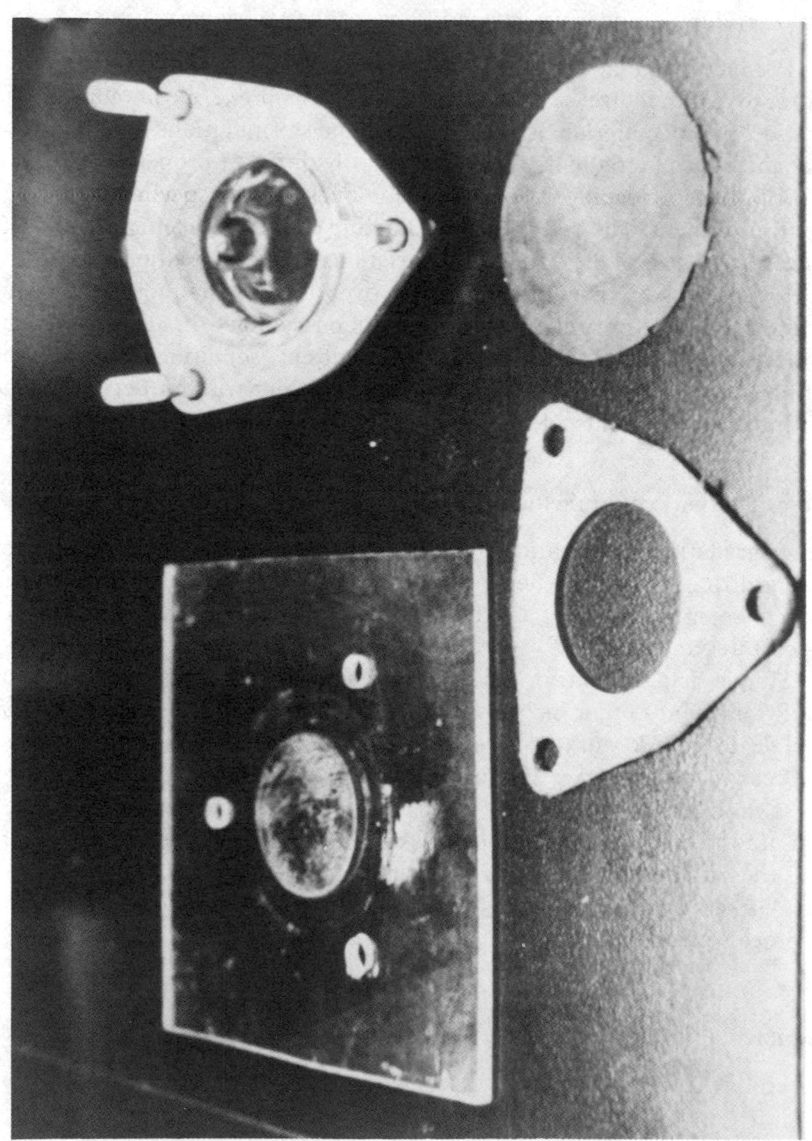

FIG. 1—*Chemical degradation test apparatus.*

is commercially available for about $350.00. Its important features are (1) the challenge liquid side of the cell, (2) the collection side of the cell, and (3) the garment fabric.

The procedure used is as follows:

1. Cut a circular specimen approximately 75 mm (3 in.) in diameter.
2. Weigh the specimen and determine its average thickness using a micrometer.
3. Mount the specimen outside of the glove to contact liquid in the test cell, and add the indicator to the undersurface [acid/base—pH paper, organics—oil red EGN indicator, which is available from Matheson, Coleman, and Bell (MCB), Cincinnati, Ohio].
4. The undersurface is monitored for permeation/penetration of the liquid chemical; the upper surface is monitored for chemical degradation.
5. Fill the cell and start the clock.
6. Using a magnifying glass, observe the upper section; using a lighted mirror, observe the lower section (undersurface).
7. If the upper section changes appearance (for example, becomes discolored, wrinkles, undergoes a chemical reaction), the undersurface should be monitored closely for appearance of the liquid chemical through the garment.
8. The test is usually run until either penetration or degradation is observed or, if there are no visual effects, until 4 h have elapsed. The elapsed time is recorded.
9. The cell is disassembled and the specimen is patted dry of liquid chemical and reweighed. The change in weight, if positive, is indicative of the amount of chemical in the glove that could cause further degradation, penetration, or permeation. If the weight change is negative, some chemical components (for example, plasticizers) of the glove have been removed or solubilized by the liquid chemical. This usually affects some physical property such as flexing. The time at which the chemical is first observed on the undersurface provides an estimation of penetration time, or the time it takes the chemical to penetrate from the outside to the inside under continuous exposure to the chemical. Degradation/penetration testing will not detect the molecular flow of a challenge liquid through the garment, that is, permeation; however, the test is suitable for bulk flow of a challenge liquid and chemical degradation of the garment material.

Penetration Testing

There is also a chemical penetration test method, the ASTM Test for Resistance of Protective Clothing Materials to Penetration by Liquids (F 903-84), which is also amenable to field use. Chemical penetration testing can determine bulk flow of liquid through porous materials, closures, pinholes, seams, or imperfections in protective clothing material. The test cell in this method is a mechanical device which costs less than $1000. The specimen is mounted in the test cell, then charged with a challenge liquid. The specimen is observed after 5 min, at atmospheric pressure, then after 10 min at 13.8 kPa (2 psig). The appearance of a drop of liquid indicates failure.

The test system is shown in Fig. 2. The following are its salient features:

(*a*) air or nitrogen tank and pressure line,
(*b*) pressure regulator,
(*c*) pressure relief valve—pressure gage (0 to 34 kPa, 0 to 5 psig),
(*d*) air line connector,
(*e*) test cell with cover and drain valve, and
(*f*) safety enclosure with transparent shield.

The test procedure is as follows:

1. Cut circular specimens, approximately 75 mm (3 in.) in diameter.
2. Condition for 24 h at 21 ± 3°C, relative humidity 30 to 80%.
3. Measure the thickness of each specimen, making ten measurements, which are averaged. In the study, the actual closure or seam was measured and not the supporting fabric.
4. Test for visibility of the liquid chemical—a 10-μL droplet is placed on the surface of a scrap piece of the garment material; the droplet should be visible to the naked eye. Indicators can be used to enhance the visibility.
5. Mount the conditioned specimen in the cell, torque the four bolts to 3.4 to 5.1 N·m (12 lb/in.). An indicator may be applied with a small brush.
6. Charge the cell through the air line connector with the liquid chemical. A 50-mL syringe with needle is used. Approximately 60 mL is needed. The specimen, liquid chemical, and test cell should all be at the temperature selected for testing, usually room temperature.
7. Connect the air line, set the air regulator to 0 kPa, and close the cell vent valve.
8. Start the clock, and watch for a droplet. If none appears after 5 min, apply 13.8 kPa (2 psig) pressure for 10 min more and observe.
9. If no droplet is detected after 10 min, the specimen passes. If a droplet is seen during the 15-min test [5 min at atmospheric pressure, 10 min at 13.8 kPa (2 psig)], the specimen fails.

This test is made portable by using a smaller air cylinder. The equipment assembled could be contained in a 1-m^3 case. ASTM Test F 903-84 has been used successfully in our laboratory to investigate penetration of chemical liquid through bulky seams and closures.

Permeation Testing

Chemical permeation field methods are discussed next. Typically, breakthrough time and steady-state permeation rates are two measurements of

FIG. 2—*Chemical penetration test apparatus.*

chemical resistance that are reported for protective clothing materials. Breakthrough time is defined as the time between the initial contact of the liquid chemical with the outside surface of the protective clothing and the detection of the chemical on the inside surface, as observed by some analytic method. Therefore, a highly sensitive analytic method will yield an earlier breakthrough time than one which is less sensitive. The steady-state permeation rate is defined as the net flow of liquid chemical through the protective clothing on a molecular level after breakthrough and under steady-state conditions. Bulk flow through pinholes, imperfections, and so on would be penetration rather than permeation. Theoretically, doffing of chemical protective clothing before breakthrough results in no exposure. After chemical breakthrough and under steady-state equilibrium conditions, the chemical exposure is known. The situation, however, becomes complex since multidisciplinary knowledge is needed to relate the chemical exposure to harmful human effects. As there are no threshold limit values (TLVs) or exposure guidelines for skin alone, only the current TLVs, which assume all routes of entry, are available. However, for a TLV with a skin notation, a significant route of entry could be percutaneous.

Therefore, breakthrough time and steady-state permeation rates for chemicals with skin permeation/penetration potentials are especially important to the industrial hygienist in selecting the most appropriate chemical protective clothing.

A standard test method for resistance of protective materials to permeation is used to estimate dermal exposure to potentially hazardous liquids. However, chemical resistance standard methods evaluate only a part of the whole protective clothing. Usually, a specimen is cut from the palm, back, or gauntlet. In the past, whole glove chemical resistance testing was performed by measuring a physical parameter such as its weight, filled with the challenge liquid; after a specified time, the specimen is drained of the liquid and patted dry, and the physical parameters are remeasured. Recently, the chemical permeation of whole gloves and that of the patch or pieces were determined by Williams to be equivalent [*1*].

A whole glove chemical permeation test method is discussed next. The breakthrough time and the challenge liquid concentration at steady state of specific glove sites can be evaluated. In a National Institute of Occupational Safety and Health (NIOSH) study, acetone was selected as the challenge liquid, with latex neoprene gloves as the protective clothing. In summation, the thinnest part of the gloves, which is between the fingers, back, and palm, had the shortest breakthrough time and the largest steady-state concentration. The thickest part of the gloves, the fingertips, had the longest breakthrough time and the lowest steady-state concentration.

A photoionization detector (H-Nu PI-101, available from H-Nu Systems Inc., Newton, Massachusetts) equipped with a recorder is used to measure the permeant concentration (Fig. 3). The detector response time was deter-

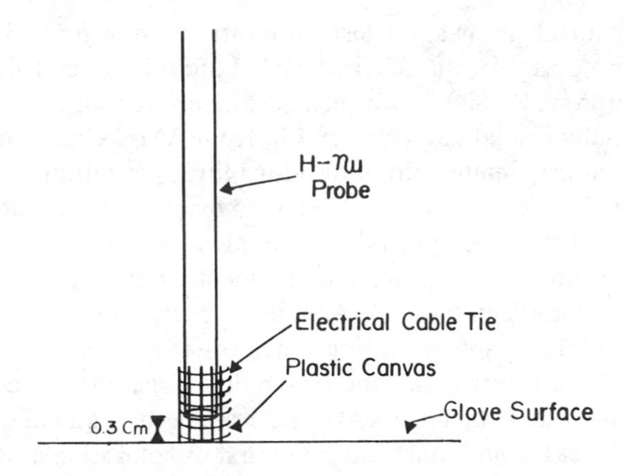

FIG. 3—*Probe tip of photoionization detector.*

mined to be under 10 s. The photoionization detector and recorder are calibrated as a unit.

The thickness of each glove site was measured to ±0.01 mm using a dial gage. The mean of ten readings for each site is recorded as the mean site thickness. The glove is turned inside out and approximately 100 mL of challenge liquid is added, and the neck of the glove is tied off with an electrical cable tie. The tied glove neck is evaluated for a tight seal with the photoionization detector probe; then the probe is moved to monitor the glove sites. The sites are noted on the recorder chart.

Breakthrough time is calculated from the recording's trace, as the recorder chart speed is a constant. The mean challenge concentration under steady-state equilibrium at all specific sites is calculated by averaging four concentrations after steady-state equilibrium is achieved.

This method can be used in the field, provided the chemical has sufficient vapor pressure and gives a signal on the photoionization detector. This is easily determined by passing the probe tip over the liquid. Should it not respond, another quick-response analytical instrument could be substituted. A calibration curve for quantitative evaluation can be determined.

Field test methods are intended to complement ASTM Test F 739-81 (chemical permeation test method). A field test method provides rapid, semiquantitative information upon which initial chemical protective clothing selections are based. ASTM Test F 739-81, on the other hand, being more precise and versatile, permits thorough, rigorous testing.

One limitation of these field permeation test methods is the vapor pressure (volatility) of the challenge liquid. The challenge liquid must volatilize permeation through the clothing specimen into the air collection stream. Polychlorinated biphenyls have a very low vapor pressure ($<10^{-3}$ torr). Hence, this permeation method should not be used for these challenge chemicals.

Another chemical permeation test apparatus is shown in Fig. 4. All the components are portable, quickly assembled and disassembled, and easy to operate. A Du Pont P-4000 pump pushed air through an AMK permeation cell (commercially available for under $40 from AMK Glass Co., Vineland, New Jersey) at approximately three standard litres per minute. The chemical resistance of neoprene to acetone was selected for study, since the ASTM Test F 739-81 protective clothing standard chemical resistance test method has recently been validated using acetone permeation through neoprene. This was used as a comparison method. Acetone concentrations were determined with an H-Nu PI-101 photoionization detector (PID) connected to a Varian A-2 recorder. Also, Draeger acetone 100/b direct-reading gas detector tubes (range of measurement 100 to 12 000 ppm) were used as an alternative means of measuring breakthrough time using the first discoloration as the indicator. The specimens' thicknesses to ±0.01 mm were measured using a micrometer, and the mean thickness was computed. Next, the specimen, in the cell, was exposed to American Chemical Society (ACS) reagent-grade acetone. The breakthrough time and steady-state equilibrium concentration were determined.

The mean breakthrough time for the PID was 738.0 s; for the detector tubes it was 840.0 s. The precision of breakthrough times for the PID are in good agreement, as shown by the 0.4 standard deviation and 3.3% coefficient of variation. The precision of breakthrough times for the detector tubes is not nearly as good; the standard deviation is 1.3 and the coefficient of variation is 9.2%. The detector tube breakthrough times, compared with those obtained with the PID, are statistically different when analyzed by an unpaired replication *t*-test. The mean breakthrough time can be compared with 738.0 ± 12.0 s, which was obtained using acetone/neoprene by ASTM Test F 739-81 and gas chromatography/flame ionization detector (GC/FID) analysis. There is excellent agreement between the breakthrough times determined by the field method and those determined by ASTM Test F 739-81. A similar comparison for the detector tube results yields poor agreement. This can be explained by the relative sensitivity of the PID compared with the detector tube. The PID is much more sensitive and, hence, can detect breakthrough sooner than the detector tube. However, detector tubes are inexpensive, easy to use, and readily available. Although the precision is not as good as that of the PID, order-of-magnitude breakthrough times can be estimated—for example, rapid breakthrough versus long breakthrough times.

Steady-state permeation rates were determined using the PID only. The mean PID steady-state permeation rate is 55.10 ± 2.47 $mg/s/m^2$. This means the steady-state permeation rate is 60% greater than the 33.15 ± 6.32 $mg/s/m^2$ rate of ASTM Test F 739-81. A possible explanation for this disparity is the variation of the steady-state permeation rate with the collection gas flow rate. This phenomenon has been observed previously by other researchers. No proposed explanation for this phenomenon is convincing.

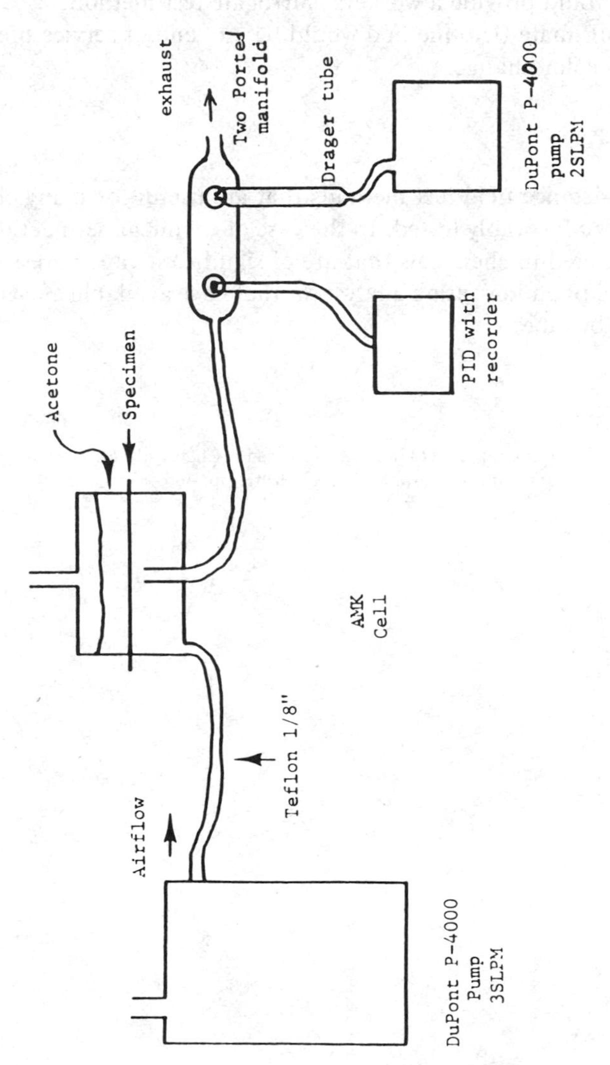

FIG. 4—*Permeation field test method.*

What about the future? Field tests evaluating actual industrial usage are forecast. These field tests could use passive dosimeters or charcoal pads inside the workers' garments, which would be analyzed at the end of the work shift. Alternatively, field test methods could model industrial usage of chemical protective clothing, for example, intermittent exposure, as in the cleaning of parts. This would provide a workers' surrogate test method.

The logical ultimate field method would be an "end of service life indicator," such as a color change.

Conclusion

Chemical resistance field test methods that are useful for many challenge liquids have been favorably tested. In the case of chemical permeation, these methods are limited to chemicals that are of significant vapor pressure, that are amenable to photoionization analysis or that have available an acceptably sensitive detector tube.

Reference

[*1*] Williams, J., "Evaluation of Intact Gloves and Boots for Chemical Permeation," *American Industrial Hygiene Association Journal*, Vol. 42, 1981, pp. 468-471.

Testing the Chemical Resistance of Seams, Closures, and Fully Encapsulated Suits

Stephen P. Berardinelli[1] *and Len Cottingham*[1]

Evaluation of Chemical Protective Garment Seams and Closures for Resistance to Liquid Penetration

REFERENCE: Berardinelli, S. P. and Cottingham, L., **"Evaluation of Chemical Protective Garment Seams and Closures for Resistance to Liquid Penetration,"** *Performance of Protective Clothing, ASTM STP 900*, R. L. Barker and G. C. Coletta, Eds., American Society for Testing and Materials, Philadelphia, 1986, pp. 263–275.

ABSTRACT: Chemical penetration is defined as bulk flow of liquid through porous materials, closures, pinholes, seams, or imperfections in protective clothing material. The objective of this research was to evaluate bulky seams and closures for resistance to liquid penetration, using the ASTM Test for Resistance of Protective Clothing Materials to Penetration by Liquids (F 903-84). The test cell in this method is a mechanical device which costs less than $1000. A modification of the test cell was used throughout the course of this study. The test cell and cell modification are explained in detail. Penetration of the challenge chemical was determined by visual inspection of the protective clothing material in the cell. The specimen was mounted in the test cell, then charged with a challenge liquid. The specimen was observed after 5 min at atmospheric pressure, and then after 10 min at 13.8 kPa (2 psig). The appearance of a drop of liquid indicated failure. Eight different bulky seams and closures were evaluated against several liquids by this test method. Chemical penetration data for control specimens (specimens without seams or closures), seams, and closures demonstrate the utility of this test method. Problems associated with the sealing of these bulky items into the test cell, as well as with the solutions were critically evaluated.

KEY WORDS: chemical penetration, chemical protective garments, seams, closures, protective clothing

ASTM Committee F-23 on Protective Clothing promulgates protective clothing consensus standard test methods. This ASTM committee has developed the ASTM Test for Resistance of Protective Clothing Materials to Penetration by Liquids (F 903-84), which has been recently evaluated by means of a round robin test. Penetration is defined as bulk flow of liquid through

[1]Industrial hygiene chemist and engineering technician, respectively, National Institute for Occupational Safety and Health, Centers for Disease Control, U.S. Department of Health and Human Services, Morgantown, WV 26505-2888.

porous materials, closures, pinholes, seams, or imperfections. The resistance of protective clothing specimens to penetration by a liquid is evaluated by the visual appearance of the liquid through the material. A special cell for this test method has been designed by W. L. Gore and Associates, Elkton, Maryland.

Our study applied ASTM Method F 903-84 to bulky closures and seams using challenge liquids suggested by ASTM.

Experimental Procedure

The test specimen is shown in Fig. 1. The salient features are the following:

(*a*) air or nitrogen tank and pressure line,
(*b*) pressure regulator,
(*c*) pressure relief valve,
(*d*) pressure gage 0 to 34.5 kPa (0 to 5 psig),
(*e*) air line connector,
(*f*) test cell with cover and drain valve, and
(*g*) safety enclosure with transparent shield.

The test cell was Teflon[2] with an aluminum cover mounted on a steel plate. The test cell is shown in more detail in Fig. 2. The specimen was mounted in the Teflon cell by means of an aluminum cover. Four bolts secured the cover to the cell. In the original design, the four bolts clamping the cover and cell together were cemented into the Teflon. We modified the design so that the four bolts pass through the Teflon and are secured to the steel back plate. This design change was warranted because the cemented bolts separated from the Teflon block after several uses. Further, the revised design allows more torque on the four bolts. This is very important for a good seal of the closures or seam in the test apparatus.

The test procedure is as follows:

1. Cut circular specimens, approximately 76 mm (3 in.) in diameter.
2. Condition for 24 h, at 21 ± 3°C, with a relative humidity of 30 to 80%.
3. Measure the thickness of each specimen, making ten measurements per specimen, from which the arithmetic mean is calculated. In the study, the actual closure or seam was measured and not the supporting fabric.
4. Test for visibility of the liquid chemical—a 10-μL droplet was placed on the surface of a scrap piece of the garment material; the droplet should be visible to the naked eye. Indicators can be used to enhance visibility. The droplet is viewed under the same lighting as the test.
5. Mount the conditioned specimen in the cell, torque the four bolts to 3.4 to 5.1 N · m (30 to 45 in. · lb). Mounting of the specimen is discussed in detail in Step 10. An indicator may be applied with a small brush.
6. Charge the cell through the air line connector with the liquid chemical. A 50-mL syringe with needle was used. Approximately 50 mL is needed. The

[2]Mention of a company name or product does not constitute endorsement by the National Institute for Occupational Safety and Health (NIOSH).

FIG. 1—*Penetration test apparatus.*

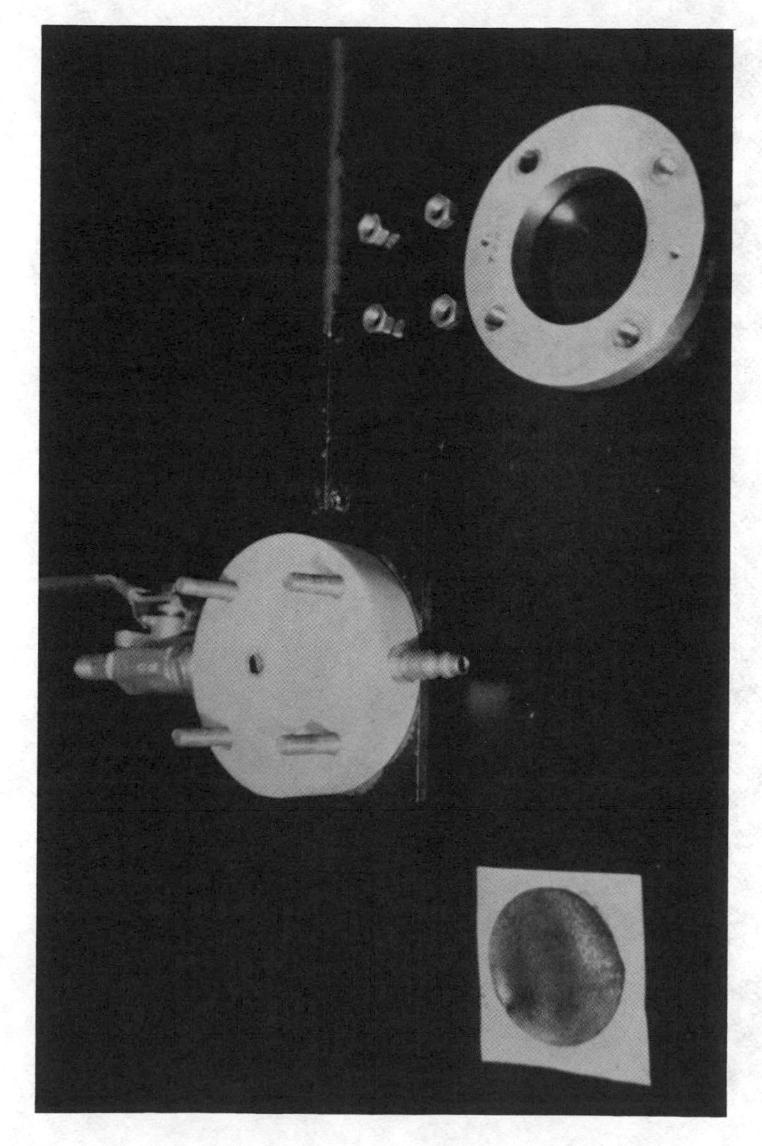

FIG. 2—*Penetration test cell.*

specimen, liquid chemical, and test cell should all be at the temperature selected for testing, usually room temperature (20 to 25°C).

7. Connect the air line, set the air regulator to 0 kPa (0 psig), and close the cell vent valve.

8. Start the clock and watch for a droplet. If none appears after 5 min, apply a pressure of 13.8 kPa (2 psig) for 10 min more and observe the specimen.

9. Look for a droplet. If there is none, the specimen passes; if a droplet is seen during the 15-min test [5 min at atmospheric pressure, 10 min at 13.8 kPa (2 psig)], the specimen fails.

10. Shut off the gas tank regulation and open the cell vent valve to release the pressure. The drain valve is opened next and the liquid discharged into a beaker.

Proper mounting of the specimen is essential for accurate results. Gore-Tex-expanded polytetrafluoroethylene (PTFE) is an excellent gasket material. Several layers around the specimen circumference are needed as well as short transverse pieces, as indicated in Fig. 3. Obtaining a good seal is a question of technique. The best seal is obtained using expanded PTFE; however, silicone rubber and neoprene caulking compounds were also evaluated. Trichloroethylene softened both neoprene and silicone rubber gaskets, so that they leaked within the 15-min test time. These gasket materials are not as chemically resistant as PTFE. The liquid chemicals and indicators are the following:

1. Oil red EGN (Matheson, Coleman, and Bell), was used as an indicator of penetration.

2. The following challenge liquids were used:

(*a*) distilled, deionized water (H_2O),

(*b*) methyl ethyl ketone (MEK), American Chemical Society (ACS) reagent grade,

(*c*) *iso*octane (IO), ACS reagent grade,

(*d*) trichloroethylene (TCE), ACS reagent grade, and

(*e*) for routine testing, an aromatic compound such as toluene or xylene.

The garments' constructions and chemical compositions are as described in the following paragraphs.

Specimen A is a disposable garment. A copolymer of vinylidene chloride and vinyl chloride is laminated to a base material of spun-bonded polyethylene. Five different seam constructions and two different zipper constructions were tested.

The seam constructions were as follows:

(*a*) stitched inside (SI),

(*b*) stitched outside, sealed inside (SO/SdI),

(*c*) stitched inside and with additional overlaying material (SIO),

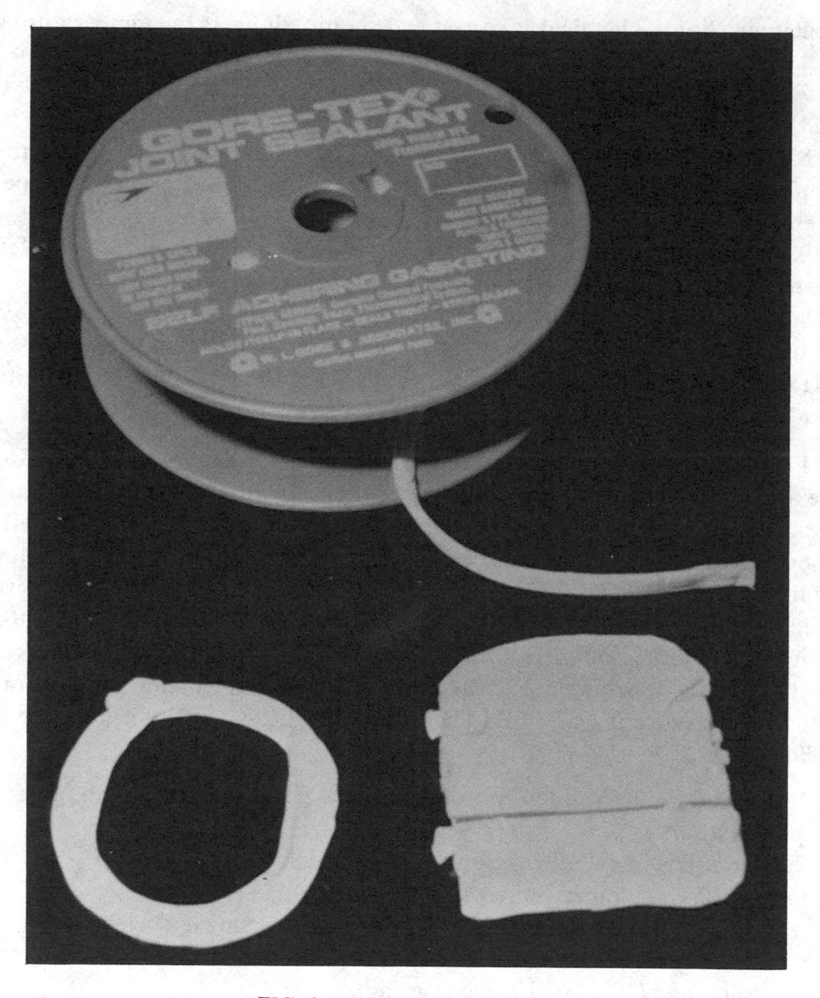

FIG. 3—*Cell sealant material.*

(*d*) sealed inside and then stitched with additional overlaying material (SdI + SIO), and

(*e*) stitched inside, with nylon reinforced strapping tape on both sides (SN).

The zipper constructions were as follows:

(*a*) plastic zipper on a woven fabric sewn to the garment (ZIP) and
(*b*) the same construction but with a taped, protective flap (Z/F).

Specimen B is a disposable garment—a laminate of melted polypropylene between spun-bonded polypropylene. The seams are sewn; the plastic zipper, on a woven fabric, is sewn to the garment.

Specimen C is expanded, microporous polytetrafluoroethylene laminated to a knit or woven fabric. The seams are double-sewn and armored on the inside of the fabric. The zippers were not tested. An armored seam uses a reinforcing strip of garment material overlayed on the seam.

Specimen D is plasticized polyvinyl chloride coated on both sides of a cotton-base fabric. The seams are double-sewn and armored on both sides.

Specimen E is plasticized polyvinyl chloride on a cotton base with a double-sewn and armored seam.

Specimen F is butyl rubber with a sewn and cemented seam.

Specimen G is Viton elastomer with a sewn and armored seam.

Specimen H is chlorinated polyethylene, with a double-track closure.

Results and Discussion

The results of the penetration test for Specimen A appear in Table 1. Three specimens without seam or zipper were tested as controls. Specimens with a zipper or a seam were tested in triplicate, where practical. In many instances, the challenge liquid immediately penetrated the seamed or zippered specimens before the clock could be started. If this happened, only one test was conducted. Specimen A controls passed both atmospheric and pressure penetration tests with all chemicals. In general, the construction of Specimen A seams passed only the 5-min leak test with water. With the organic liquid (IO, MEK, TCE), the seam construction failed both the atmospheric and pressure penetration tests. The best seam construction proved to be a homemade design in which the inside and outside seams were covered by nylon strapping tape. The zipper flap passed the 5-min atmospheric water leak test only. All other combinations failed both parts of the test.

TABLE 1—*Penetration test results for Specimen A.*

Run	Specimen	Thickness, mm	Test Liquid	5-min Leak Test[a]	10-min Pressure Test	Comments
1	A	0.145	water	+	+	
2	A	0.149	water	+	+	
3	A	0.162	water	+	+	
1	A seam SI	0.284	water	−	−	
2	A seam SI	0.275	water	−	−	
3	A seam SI	0.280	water	−	−	
1	A zip	0.762	water	−	−	
2	A zip	0.635	water	−	−	
1	A	0.150	IO	+	+	
2	A	0.207	IO	+	+	
3	A	0.161	IO	+	+	
1	A seam SI	0.309	IO	−	−	
1	A zip	0.774	IO	−	−	

TABLE 1—*Continued.*

Run	Specimen	Thickness, mm	Test Liquid	5-min Leak Test[a]	10-min Pressure Test	Comments
1	A	0.158	MEK	+	+	
2	A	0.169	MEK	+	+	
3	A	0.165	MEK	+	+	
1	A seam SI	0.350	MEK	−	−	
1	A zip	0.784	MEK	−	−	
1	A zip	0.814	MEK	−	−	
1	A	0.154	TCE	+	+	
2	A	0.169	TCE	+	+	
3	A	0.147	TCE	+	+	
1	A seam SI	0.327	TCE	−	−	
1	A-SIO	1.297	water	+	−	
2	A-SIO	1.055	water	+	−	
3	A-SIO	1.063	water	+	−	
1	A-SIO	1.395	IO	−	−	
1	A-SIO	1.267	MEK	−	−	
1	A-SIO	1.302	TCE	−	−	
1	A-SdI+SIO	1.196	water	+	−	
2	A-SdI+SIO	1.277	water	+	−	
3	A-SdI+SIO	1.331	water	+	−	
1	A-SdI+SIO	1.172	IO	−	−	
1	A-SdI+SIO	1.362	MEK	−	−	
1	A-SdI+SIO	1.342	TCE	−	−	
1	A-SO/SdI	1.670	water	+	−	
2	A-SO/SdI	1.523	water	+	−	
3	A-SO/SdI	1.651	water	+	−	
1	A-SO/SdI	1.702	IO	−	−	
1	A-SO/SdI	1.728	MEK	−	−	
1	A-SO/SdI	1.646	TCE	−	−	
1	A-SN	1.566	water	+	+	
2	A-SN	1.576	water	+	+	
3	A-SN	2.046	water	+	+	
1	A-SN	1.791	IO	+	−	
2	A-SN	1.862	IO	+	−	
3	A-SN	1.737	IO	+	−	
1	A-SN	1.758	MEK	−	−	
2	A-SN	1.696	MEK	−	−	
3	A-SN	1.720	MEK	+	−	
1	A-SN	1.788	TCE	+	+	
2	A-SN	1.799	TCE	+	+	
3	A-SN	1.608	TCE	+	+	
1	A-Z/F	1.187	water	+	−	
2	A-Z/F	1.210	water	+	−	
3	A-Z/F	1.306	water	+	−	
1	A-Z/F	1.103	IO	−	−	adhesive dissolved
1	A-Z/F	1.149	MEK	−	−	adhesive dissolved
1	A-Z/F	1.123	TCE	−	−	adhesive dissolved

[a] + = passing; − = failure.

Specimen B results are shown in Table 2. Specimen B controls passed the water leak test only. One specimen, however, appeared to have pinholes. The construction of the seam and zippers is similar to that of Specimen A; however, Specimen B does not have chemical resistance against organic solvents.

For Specimen C, only the seams were tested (Table 3). This seamed material passed both the 5-min leak and the 10-min pressure tests except for a challenge using *iso*octane, in which case the material failed the 10-min pressure test. A degree of difficulty was encountered during attempts to seal these specimens into the test cell. Using two layers of gasket material on each side of the fabric proved to be a satisfactory way to ensure a proper seal.

Specimen D is a nondisposable garment. The inside and outside of the garment could not be determined easily; therefore, both permutations were tested. During water penetration testing, pinholes were suspected but not observed. The results shown in Table 4 indicate that the seams in one orientation passed both the leak and pressure tests and, in the other, failed the 10-min pressurized water penetration test. Whether this is due to the orientation of the side to the challenge or to imperfections in the Specimen D seams cannot be ascertained because of the limited amount of data. Regardless of the reason, this was not the intention of the present study; the intention was to

TABLE 2—*Penetration test results for Specimen B.*

Run	Specimen	Thickness, mm	Test Liquid	5-min Leak Test[a]	10-min Pressure Test	Comments
1	B	0.252	water	+	−	
2	B	0.286	water	−	−	pinholes
3	B	0.312	water	+	−	
1	B seam	0.374	water	−	−	
2	B seam	0.321	water	−	−	
1	B zipper	0.895	water	−	−	
2	B zipper	0.908	water	−	−	
3	B zipper	0.951	water	−	−	
1	B	0.264	IO	−	−	
2	B	0.279	IO	−	−	
1	B seam	0.625	IO	−	−	
1	B zipper	0.873	IO	−	−	
1	B	0.267	MEK	−	−	
1	B seam	0.386	MEK	−	−	
2	B seam	0.404	MEK	−	−	
1	B zipper	0.881	MEK	−	−	
2	B zipper	0.919	MEK	−	−	
1	B	0.276	TCE	−	−	
2	B seam	0.418	TCE	−	−	

[a]+ = pass; − = failure.

TABLE 3—*Penetration test results for Specimen C seam only.*

Run	Thickness, mm	Test Liquid	5-min Leak Test[a]	10-min Pressure Test
1	0.708	water	+	+
2	0.738	water	+	+
3	0.700	water	+	+
1	0.687	IO	+	−
2	0.723	IO	+	−
3	0.704	IO	+	−
4	0.718	IO	+	−
5	0.720	IO	+	−
1	0.721	MEK	+	+
2	0.737	MEK	+	+
3	0.735	MEK	+	+
1	0.745	TCE	+	+
2	0.732	TCE	+	+
3	0.734	TCE	+	+

[a] + = pass; − = failure.

evaluate the suitability of the test method for seams and closures. Specimen D material controls and seams passed the *iso*octane 5-min leak test but failed the 10-min pressure test. Two of six specimens failed the 5-min leak test.

Again, this may be due to specimen orientation in the cell or to a quality control problem. For MEK, the Specimen D control and seams passed all the tests. However, for TCE, the controls all passed but the seams all failed. Possibly the armor is cemented and TCE is a solvent for the cement.

Specimens E, F, and G were seamed specimens with enough material for only a very few tests. Only water was used; all the specimens passed (Table 5).

Specimen H, a double-tracked closure, was evaluated in detail (Table 6). Once the problem of a poor gasket seal was resolved, Specimen H passed all the tests except for MEK. This closure is not recommended against ketones. The ability to form a good seal is further demonstrated in reuse. The cell was drained of liquid and then refilled, and the test was rerun. In these cases, Specimen H passed.

Assuming no chemical effect, a physical property such as viscosity or surface tension might correlate with penetration. Viscosity is a measure of the resistance a fluid offers to an applied shear force. Surface tension is the force per centimetre of the surface which opposes expansion of the surface area. A liquid with a high viscosity or high surface tension should not penetrate as readily as a liquid with a low viscosity or low surface tension. The surface tension and viscosity values range from high to low in the following order: H_2O, TCE, MEK, and IO (Table 7).

TABLE 4—*Penetration test results for Specimen D.*

Run	Specimen	Thickness, mm	Test Liquid	5-min Leak Test[a]	10-min Pressure Test	Comments
1	D	0.512	water	+	−	pinholes suspected
2	D	0.514	water	+	+	
3	D	0.504	water	+	+	
4	D	0.498	water	+	+	
5	D	0.507	water	+	+	material reversed
6	D	0.528	water	+	+	material reversed
7	D	0.526	water	+	+	material reversed
1	D seam	0.769	water	+	−	1 drop penetration
2	D seam	0.719	water	+	−	2 drops penetration
3	D seam	0.751	water	+	−	1 drop in pressure
4	D seam	0.802	water	+	+	material reversed
5	D seam	0.784	water	+	+	material reversed
6	D seam	0.781	water	+	+	material reversed
1	D	0.507	IO	+	−	
2	D	0.519	IO	+	−	
3	D	0.514	IO	+	−	
4	D	0.531	IO	+	−	material reversed
5	D	0.508	IO	+	−	material reversed
6	D	0.518	IO	+	−	material reversed
1	D seam	0.754	IO	−	−	
2	D seam	0.736	IO	+	−	
3	D seam	0.776	IO	−	−	
4	D seam	0.743	IO	+	−	material reversed
5	D seam	0.766	IO	+	−	material reversed
6	D seam	0.801	IO	+	−	material reversed
1	D	0.513	MEK	+	+	
2	D	0.538	MEK	+	+	
3	D	0.495	MEK	+	+	
1	D seam	0.679	MEK	+	+	
2	D seam	0.664	MEK	+	+	
3	D seam	0.801	MEK	+	+	
1	D	0.539	TCE	+	+	
2	D	0.530	TCE	+	+	
3	D	0.535	TCE	+	+	
1	D seam	0.771	TCE	−	−	
2	D seam	0.788	TCE	−	−	
3	D seam	0.803	TCE	−	−	

[a] + = pass; − = failure.

For three of the seam constructions in Specimen A, water, which has the highest viscosity and surface tension, did not penetrate during the 5-min atmospheric pressure, whereas the other liquids did. Specimen C seams all passed the leak tests, except that for IO, which failed the 10-min pressure test. *Iso*octane has the lowest surface tension and a low viscosity. However, in Specimen D seams, TCE penetrated during the 5-min atmospheric leak test; MEK did not penetrate at all; and IO penetrated during the 10-min leak test.

TABLE 5—*Penetration test results for Specimens E, F, and G tested in water.*

Run	Specimen	Thickness, mm	5-min Leak Test[a]	10-min Pressure Test
1	E seam	1.056	+	+
2	E seam	1.061	+	+
3	E seam	1.135	+	+
1	F seam	0.461	+	+
2	F seam	0.460	+	+
1	G seam	0.362	+	+
2	G seam	0.361	+	+

[a] + = pass; − = failure.

TABLE 6—*Penetration test results for Specimen H closure only.*

Run	Thickness, mm	Test Liquid	5-min Leak Test[a]	10-min Pressure Test	Comments
1	8.555	water	+	+	
2	8.256	water	+	+	
3	8.217	water	+	+	
1	8.435	IO	+	−	expanded PTFE gasket leaked
2	8.578	IO	+	−	expanded PTFE gasket leaked
3	8.578	IO	+	−	reuse of No. 2—using silicone
4	8.397	IO	+	+	expanded PTFE gasket okay
5	8.397	IO	+	+	reuse of No. 4—cell drained, refilled
6	8.595	IO	+	+	expanded PTFE gasket okay
7	8.595	IO	+	+	reuse of No. 6
8	8.595	IO	+	+	second reuse of No. 6
9	8.793	MEK	+	−	not recommended against MEK
10	8.267	TCE	−	−	expanded PTFE gasket leaked
11	8.465	TCE	+	+	expanded PTFE gasket okay
12	8.210	TCE	+	+	expanded PTFE gasket okay
13	8.210	TCE	+	+	reuse of No. 12

[a] + = pass; − = failure.

TABLE 7—*Physical properties of liquid challenge chemicals.*

Liquid	Surface Tension at 20°A, dyne/cm[a]	Viscosity at 20°C, cP[a]
Water (H_2O)	73	1.0
Trichloroethylene (TCE)	29	0.58
Methyl ethyl ketone (MEK)	24	0.41
*Iso*octane (IO)	19	0.50

[a]One dyne/cm = 10^{-3} N/m; 1 cp = 10^{-3} Pa · s.

Thus, surface tension and viscosity are probably not reliable predictors of penetration.

Conclusions

The penetration test does accommodate bulky zippers and seams; further, it furnishes valuable information that has significance for industrial hygiene. In addition, the test indicates that, as suspected, seams and zippers may be the limiting factors when assigning protection values to chemical protective clothing.

Charles E. Garland,[1] *Lynn E. Goldstein*,[2] *and Campbell Cary*[1]

Testing Fully Encapsulated Chemical Suits in a Simulated Work Environment

REFERENCE: Garland, C. E., Goldstein, L. E., and Cary, C., **"Testing Fully Encapsulated Chemical Suits in a Simulated Work Environment,"** *Performance of Protective Clothing, ASTM STP 900*, R. L. Barker and G. C. Coletta, Eds., American Society for Testing and Materials, Philadelphia, 1986, pp. 276-285.

ABSTRACT: The ultimate performance objective of chemical protective clothing is protection of the wearer in his workplace environment. The chemical barrier properties of the materials used in the clothing are of prime importance; however, in fully encapsulated suits, it is mandatory that the total ensemble afford effective protection to the wearer as he performs his normal work routine. A quantitative fit-test booth for respirators was used to determine the protection factors (PF) of total ensembles. The PFs are measured for three exercises simulating typical work activities. A number of suit designs and concepts including limited-use disposable clothing have been tested, and modifications in design have been incorporated to improve performance. Based on the test results, we have been able to proceed to actual on-the-job testing to confirm acceptability with minimum risk to the wearers.

KEY WORDS: suits, encapsulated chemical suits, butyl, Saranex laminated Tyvek, simulated work environment, protection factor, total ensemble, ventilating vest, disposable, protective clothing

At the Chambers Works of E. I. du Pont de Nemours and Co., a large number of air-supplied one- and two-piece butyl suits are used. While butyl suits offer excellent protection, especially against cyanosis-causing chemicals, they have the following inherent problems:

(*a*) heavy weight,
(*b*) production of claustrophobia,

[1]Laboratory supervisor and retired manufacturing representative, respectively, Chambers Works, E. I. du Pont de Nemours and Co., Deepwater, NJ 08023.
[2]President, Durafab, Cleburne, TX 76031.

(*c*) noisiness,
(*d*) difficulty in decontamination,
(*e*) repair and testing problems, and
(*f*) high cost per wearing.

These problems have been recognized for some time, yet there was no known elastomer that would alleviate them and be effective as a barrier to a majority of our chemicals.

Permeation tests during the past two years have focused the authors' attention on Saranex[3] laminated Tyvek[4] as an inexpensive fabric that is an effective barrier to many of the chemicals used at the Chambers Works. Table 1 shows breakthrough times for a variety of challenge chemicals. The times vary from many hours for inorganic acids, bases, and polar organics to only a few minutes for some nonpolar organic solvents.

Saranex laminated Tyvek is a rather complex fabric. The outer layer is Saranex 23, a coextruded multilayered film 0.05 mm thick. It has an outside layer of low-density polyethylene, an inner layer of Saran,[3] a copolymer of vinylidene chloride and vinyl chloride, and the other outside layer of ethyl vinyl acetate/low-density polyethylene copolymer, which is used for bonding to the Tyvek. Tyvek is a spun-bonded sheet structure of high-density polyethylene fibers, 0.13 mm thick. Figure 1 illustrates the multilayered fabric as it is used in suits, from the outside, at the top of the diagram, to the inside, toward the wearer. The lamination combines the salient features of its component parts: a reasonably tough, inexpensive substrate combined with a film resistant to chemical permeation.

Since Saranex laminated Tyvek represented a potential solution to most of the butyl suit problems, it was decided to make a fully encapsulating suit from it. While some suits were available on the market, the authors did not consider them suitable for use at Chambers Works. At this stage, we designed and constructed a workable fully encapsulated protective suit. Figure 2 is a front view of the suit.

Among the key features in the suit design is sufficient space at the top for wearing a hard hat (Fig. 3). The face piece of Du Pont 0.10-mm Mylar[4] is of sufficient size and clarity to provide excellent visibility. It has three disadvantages: it is relatively stiff for a thin sheet; it clouds when exposed to strong mineral acids; and seams with the suit fabric are not easily sealed. Although these problems do not seriously impair the usability of the suit, we are actively investigating alternative face piece materials.[5]

Tie downs below the face piece permit vertical adjustment when the suit is

[3]Trademark of the Dow Chemical Co.

[4]Registered Du Pont Co. trademark.

[5]Since the original presentation of this paper, we have found that Teflon FEP fluorocarbon film is much more inert to chemical exposure than Mylar. It was also found that polyester tape can be used to cover needle holes where the face piece is sewn to the suit fabric.

TABLE 1—*Permeation data for Saranex laminated Tyvek versus a variety of challenge chemicals.*

Challenge Chemical	Breakthrough Time, min
Acetic acid	>4000
Acetone	33
Aniline	>300
Chlorine, 20 ppm	>480
Chloroform	<1
Chlorosulfonic acid	350
Cresols	>120
Dimethylacetamide	64
2-Ethoxyethanol	>480
Ethylene diamine	>480
Hydrochloric acid, 37%	>2800
Nitric acid, 70%	>2800
Nitric acid, 90%	107
Nitrobenzene	180
Oleum, 65%	37
Polychlorinated biphenyl (PCB)	120 to 180
Sodium hydroxide, 40%	>480
Sulfuric acid, 98%	>480
Tetraalkyl lead	>60
Titanium tetrachloride	>1000
Toluene	<5
o-Toluidine	>120

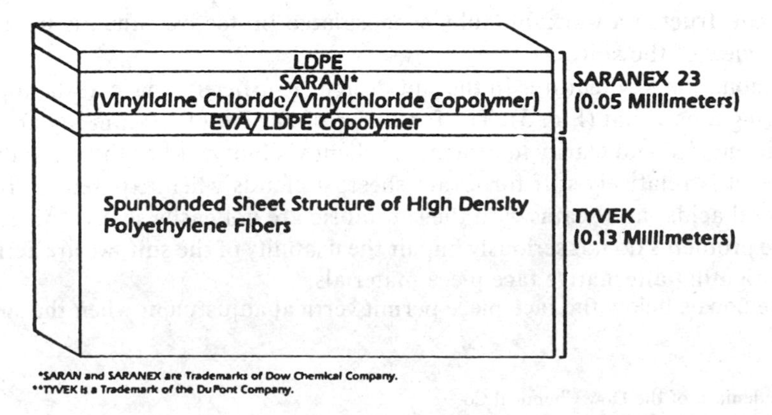

FIG. 1—*Layered structure of Saranex laminated Tyvek.*

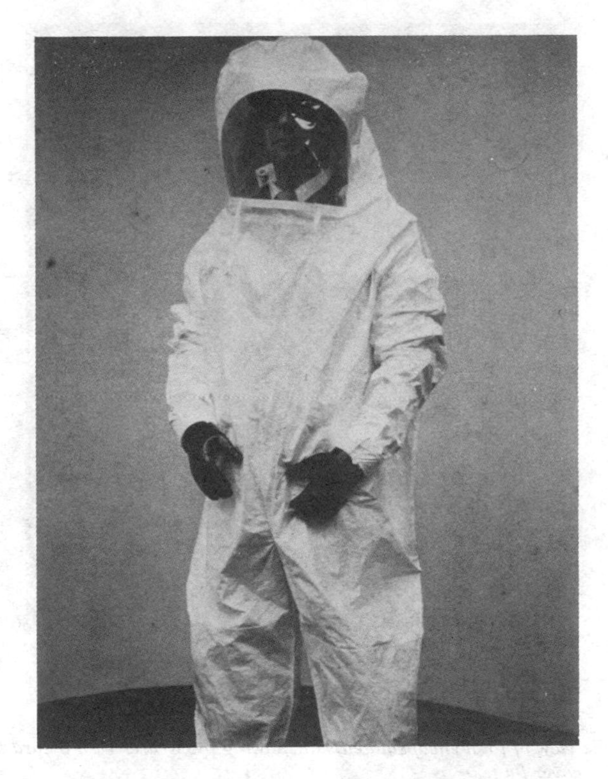

FIG. 2—*Total encapsulating suit of Saranex laminated Tyvek.*

inflated, so that the suit does not ride up and block the wearer's field of vision. Proper selection of one of the four available sizes and correct use of the tie downs will ensure overall fit with good visibility.

Suit entry is accomplished through the back (Fig. 3). The closure is a zipper, which can be sealed by a flap containing pressure-sensitive adhesive. Because the closure is located in the back, away from most splashes, a backup man must assist in getting in and out of the suit.

Openings at the wrists are closed with the addition of gloves (Fig. 4). A prerequisite to the selection of gloves is that they must be comparable in barrier properties to the Saranex laminated Tyvek and adaptable to the tapered plastic cuffs. The gloves are mounted on the plastic cuffs and then put on the hands. As the suit is donned, the elastic at the wrists rides up on the tapered plastic cuffs, forming a seal. Some tapering is also designed into the suit sleeves to improve this seal.

A second layer of fabric is attached over the knees as reinforcement and as a boot cover (Fig. 5). Booties are an integral part of the suit; however, butyl or neoprene boots are worn over them for added durability. The knee flap has an

FIG. 3—*Back view of total encapsulated suit showing the wearer with a hard hat, the zipper closure, and the cover flap.*

elastic cuff that fits snugly over the boot tops to prevent chemical contamination inside the boots.

The flap over the zipper, the exhaust port covers, and face piece are single-needle stitched into the garment with a 301-type stitch. The remaining seams are made according to a Durafab patent (USP 4,272,851) in which the seams are Saranex film ultrasonically sealed to Saranex film in an inside-out mode, then turned right-side out and bound with a 401-type stitch. This process greatly increases seam strength while reducing the potential for chemical penetration. The more common type of sewn seam leaves needle holes and gaps where the thread stretches.

An air hose is inserted through an elastic opening under the right arm and connected to an air distribution system inside the suit (Fig. 6). The best system evaluated to date is a Fyrepel ventilating vest with neck ring. Air enters the back and is distributed through holes inside the vest as well as in the neck ring. There are Velcro[4] closures in front. Air from the neck ring flows over the face piece keeping it free of fog. This positive air flow over the wearer's face also lessens his feeling of claustrophobia.

A problem occurred with the prototype suit when air was introduced. The

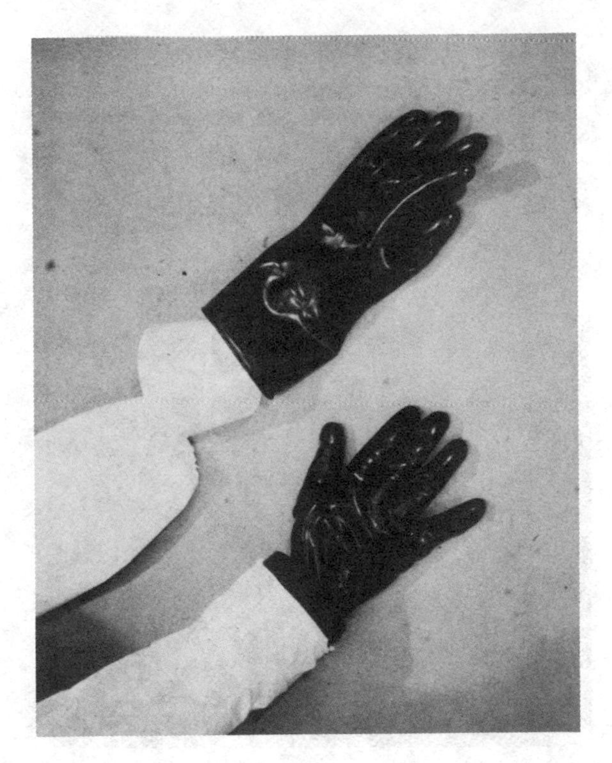

FIG. 4—*Attachment of gloves by means of tapered plastic cuffs.*

wearer found himself standing in the middle of a balloon, and it was nearly impossible to bend or stoop. While this was a good demonstration of the suit's integrity, it represented a condition that would greatly hamper the wearer's maneuverability. We tried to improvise an exhaust system by installing several respirator exhalation valves into the suit; however, they popped out when only moderate pressure was applied. Durafab personnel solved the problem by simply making holes in the suit at several locations and covering them with inverted pockets (Fig. 7). The pockets act as air valves while keeping chemicals out.

Early in the testing program, we invited some chemical operators to try on the suit and give us their comments. Reinforcement at the knees resulted from an operator suggestion. The most prevalent operator comments concerned lightness and quietness in comparison with the butyl suits they were accustomed to wearing. Table 2 compares the weights and noise levels of the three suits. Note that more than half of the new suit weight is due to the air distribution system. Noise increases with air flow. The new suit is the quietest, and the two-piece suit is the noisiest, requiring hearing protection.

Once we had a workable design, we wanted to measure the level of protec-

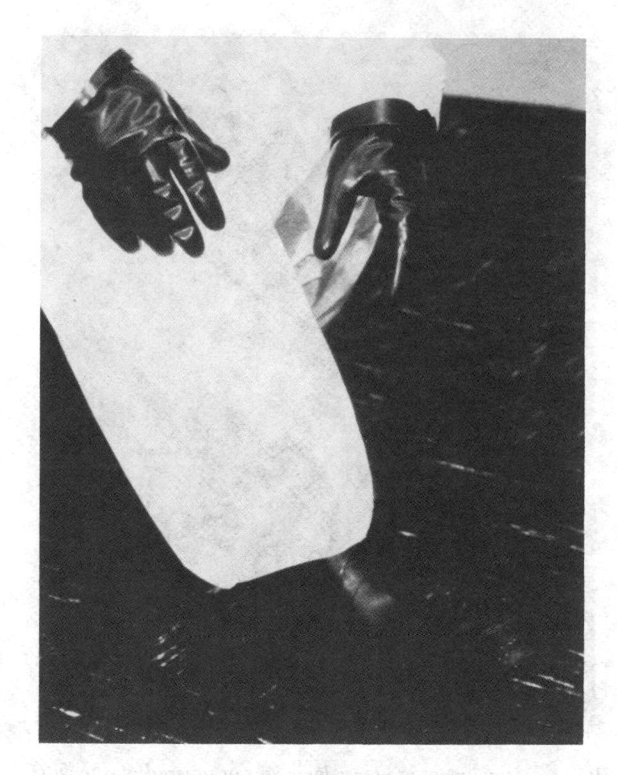

FIG. 5—*Extension of knee reinforcement covering boot top and held in place with an elastic cuff.*

tion offered by the total ensemble and compare it with the one and two-piece butyl suits. The basic principle of ensemble testing is the same as that for the quantitative fit testing of respirators. A probe is inserted into the face piece and connected to a Dynatech Frontier Model 260 aerosol test system that is outside of the fit test booth. This permits analysis of the air inside the suit. A controlled concentration of corn oil aerosol is maintained inside the booth. The protection factor (PF) is the ratio of booth concentration to suit concentration; that is, if the concentration of corn oil is 100 times greater in the booth than it is in the suit, the (PF) is 100.

A simple regimen of exercises was established to simulate working conditions. After entering the booth and connecting sampling probe and air line, the subject stands still until the system comes to equilibrium. Next, he puts his hands overhead and stretches. This exercise is continued for several minutes until the suit concentration becomes steady. The subject then does deep knee bends until a steady state is reached. Last, the subject stands again until the suit recovers to the original measurement obtained while standing.

Protection factors can be calculated for each step in the regimen or calculated as an average for the four steps. Table 3 shows a typical PF distribution

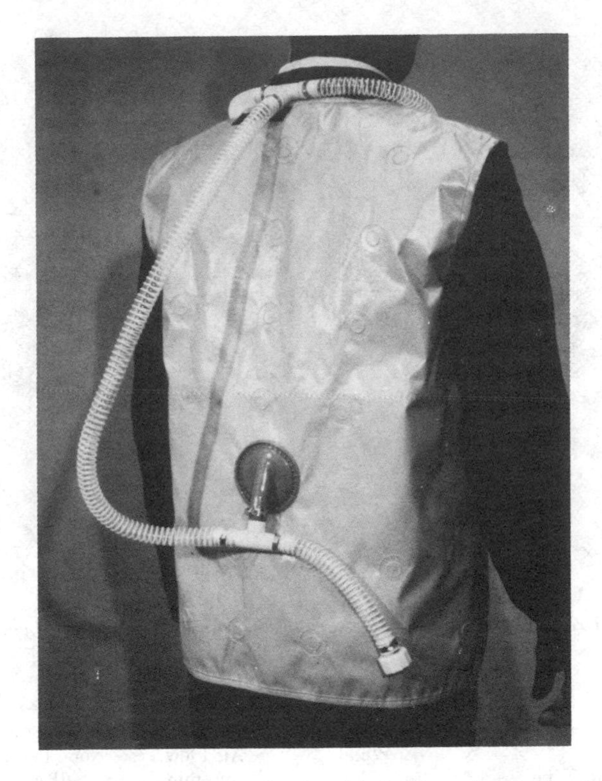

FIG. 6—*Back view of ventilating vest showing the air inlet and neck ring that provides air to the facial area.*

for the exercise regimen. The deep knee bends are the most difficult because they can cause short-duration negative suit pressure. Even though the PF is much lower during knee bends than during other steps in the regimen, the suit still provides considerable protection. The deep knee bends exaggerate a condition that would not often occur during normal working conditions.

Recovery time is determined during the last step in the regimen. It is the time required for corn oil introduced during knee bends to be flushed from the suit.

The PF average for the four steps is a convenient single-number index for comparing suits and is used in Table 4.

Table 4 illustrates the variability in protection factors and recovery time with rate of air flow to the suit. An air flow of about 0.40 m^3/min is recommended. Increased air flow markedly increases the PF and lowers the recovery time; however, there is more suit ballooning, which hampers operator maneuverability. The new suit provides almost the same protection as the one-piece butyl suit. The two-piece butyl suit sucks air in under the hood during exercising and offers very little protection.

The Du Pont company elected not to buy a large number of suits for indis-

FIG. 7—*Exhaust valve showing holes in the suit covered by an inverted pocket.*

TABLE 2—*Suit weight and noise level comparison.*

Suit Type	Weight, kg	Air Flow, m^3/min	Noise Level, dBA
Two-piece butyl[a]	6.4	0.42	94
One-piece butyl	4.3	0.42	78
		0.37	72
Saranex laminated Tyvek	2.0[b]	0.45	77
		0.37	69
		0.31	63

[a]Modified with Velcro seals.
[b]Includes 1.1-kg ventilating vest.

TABLE 3—*Protection factor variability during exercise regimen.*

Exercise	PF
Standing still	16 000
Hands overhead—stretch	14 000
Deep knee bends	2 000
Standing still[a]	16 000
Average	12 000

[a]Recovery time is determined at this step as the time required for corn oil, introduced during knee bends, to be flushed from the suit.

TABLE 4—*PF and recovery time variability with air flow.*

Suit Type	Air Flow, m^3/min	PF[a]	Recovery Time, s
Saranex laminated Tyvek	0.22	2 615	114
	0.31	4 274	54
	0.40	11 517	15
	0.46	25 175	9
One-piece butyl	0.42	15 833	22
Two-piece butyl[b]	0.44	105	49

[a]Average of four-step exercise regimen.
[b]Modified with Velcro seals.

criminate use. They are being introduced under controlled conditions into those areas where they will be most effective, and the field test results are being documented. During the test period, air will be monitored in the work area and inside the suits, prejob and postjob biological monitoring will be done where methods are available, suits will be inspected at the end of each shift, and the workers will be interviewed concerning fit, comfort, and wearability.

While Tyvek garments are often called disposables, the total encapsulating suit of Saranex laminated Tyvek is considered a limited-use garment. If the suit is contaminated with a toxic, tarry substance, it is safer and more economical to dispose of it than clean it. On the other hand, a suit that is only slightly contaminated can be cleaned and reused. Even small tears can be sealed with vinyl tape without adversely affecting the suit's protection factor, provided the tape is resistant to the chemicals being handled.

To summarize, we have designed a total encapsulating suit from Saranex laminated Tyvek that provides about the same protection as the one-piece butyl suit and much greater protection than the two-piece butyl suit. The new suit is less than half the weight of a one-piece suit and less than a third the weight of a two-piece suit. A lighter, less complex suit that provides a flow of air over the facial area is found to reduce claustrophobia problems. Noise attributable to air distribution is slightly less than that of the one-piece suit and appreciably less than that of the two-piece suit. Hearing protection is not necessary. Difficult decontamination and repair problems are avoided by disposing of the suit. Considering the limited life of a butyl suit and the relatively high maintenance costs, it is estimated that a butyl suit costs between $150 and $200 per wearing. The Saranex laminated Tyvek suit costs around $40. The indicated savings are more than $100 per wearing, and more savings are possible with multiple wearings. It is expected that the ventilating vest and plastic cuffs can be reused so many times that the cost per wearing is insignificant. Boots do not enter into the savings calculations because they are necessary with both kinds of suits.

Jeffrey A. Moore[1]

Survey of Use and Maintenance Procedures for Chemical Protective Total Encapsulation Garments

REFERENCE: Moore, J. A., **"Survey of Use and Maintenance Procedures for Chemical Protective Total Encapsulation Garments,"** *Performance of Protective Clothing, ASTM STP 900*, R. L. Barker and G. C. Coletta, Eds., American Society for Testing and Materials, Philadelphia, 1986, pp. 286–297.

ABSTRACT: This investigation essentially determined that all total encapsulation garments require basically the same use and maintenance procedures, regardless of base material or manufacturer. A few important differences exist between suits, however, and these differences tend to be concentrated in the closure systems, dump valves, glove to sleeve joints, and pressure sealing zippers. Other areas in which there is a slight disagreement or missing information include "buddy" systems, decontamination, laundering, and drying.

It also needs to be emphasized that input from users fell far below the hoped-for amounts; thus, this paper's input is heavily slanted in favor of the manufacturers.

KEY WORDS: total encapsulation garments, dump valves, pressure sealing zippers, "buddy" system, decontamination, protective clothing

The basic premise of this paper is that, even though there are six or seven different manufacturers of total encapsulation suits and three or four base materials commonly used, there should be *one* set of use and maintenance guidelines to aid users of these garments. Information was sought from manufacturers since they should know the most about their suits. Input was also sought from users, based on the premise that time and experience might have taught them a few "tricks" the manufacturers did not know. It was hoped that combining these sources would result in increased knowledge and insight on both sides. Unfortunately, the input from the users did not materialize, at least not in the hoped-for amounts. The data received from users is essentially input from a major metropolitan hazardous materials squad and one industrial user working with hydrofluoric acid in an alkylation process.

[1]Director, Research and Development, Standard Safety Equipment Co., Palatine, IL 60078.

Thus, readers of this paper should understand that input from users is lacking. However, the author still hopes that looking into generic use and maintenance procedures will enable the user to extend the useful life of his garment and reduce the overall cost of worker safety.

Use Procedures

Buddy System

If no other procedures are used, the buddy system is a must. Under no circumstances should a worker go into a hazardous environment alone. But the buddy system means a bit more than that.

Multiple-Person Donning—A man getting into a suit should be assisted by someone on the outside. This is to ensure that respiratory equipment and suit are put on properly, that the respiratory system works, and that all closure systems (that is, zippers) are fully closed. Zippers in particular would be difficult for the wearer to close properly.

Ready Personnel—The minimum number of people suited up and ready to go will vary with the conditions. It is always at least *two*. In industrial settings a distinction must be made between emergency situations and routine but potentially hazardous situations. Most industrial uses are routine but potentially hazardous. The hazard is known and quantifiable, and the job is done regularly. One man *could* do the required work, and the second is purely backup. If distances are not too great, the second man may only have to suit up to the waist. However, even in industry there can be real "emergency" situations in which the procedures used should be similar to the procedures for hazardous materials, with a third or fourth man ready.

In hazardous materials applications, the minimum number is four, because of the unpredictability of the hazard and the unknown length of time required to do the task. Two men are sent in initially, with two men suited up to the waist so they can be ready in 1 min.[2]

Planning the Mission

What is to be done should be decided ahead of time and what is known about the hazard summarized. Again, for industrial users, the task may be a routine practiced often and performed on a regular basis. For hazardous materials squads, the first trip in may simply be for hazard appraisal. Hazardous materials squads should also have at their disposal a "GATX tanker book," which contains descriptions and valve schematics of almost all tanker cars in use in the United States today. The hazardous materials squad leaders the

[2]John Eversole, Chicago Fire Department, personal communication, January 1984.

author talked to consider this book worth its weight in gold.[2] The planning should take into account the air supplies available [be it an air line or self-contained breathing apparatus (SCBA)] and the heat generated by the individuals in these suits. Most missions will be limited by the air supply and heat buildup.

Occassionally, someone is tempted to use pure oxygen instead of air to prolong work times: *Do not do this!* The enriched oxygen environment that would result in the suit would be extremely reactive in the presence of a spark and a hydrocarbon (possibly even in polymeric form). Flames and sparks are present in many situations. There is no guarantee that a dump valve venting an enriched oxygen and hydrocarbon mixture will not turn the worker into a walking bomb. A possible exception would be rebreather systems that limit the oxygen concentration in the suit to just above normal.

Communications

Since noise and respiratory equipment often prevent speech communication between workers in total encapsulation garments, a mode of communications needs to be agreed on. If men are used to working together, either in or out of suits, or if the task is known, this will lessen the need for formal training. However, even industrial users may find themselves working with hazardous materials teams in the event of a major problem. Thus, basic hand signal communications should be covered in *any* program. A good starting point is the National Fire Academy's eight signals (Fig. 1).[2] Some radio communications are available for total encapsulation garments, but cost and the lack of standardization limit their use. This equipment should be considered in addition to, *not* instead of, hand signals.

Decontamination

Ideally, before entry is allowed to a hazardous area, a decontamination site should be specified. Personnel performing these functions should also have adequate protective clothing. The workers exiting a contaminated site should go directly to the agreed-upon decontamination area and be extensively doused with water (there is no such thing as too much). If chemical deposits appear to remain, a soapy water solution is brushed all over the suit with a soft window-type brush. If chemical deposits still remain, the chemical is probably not water soluble and water will not do the job (for example, cresotic acid). At this point, you are left with a choice: use solvents to clean the material or leave it for a later laundering cycle. Solvents will attack most polymeric materials. There *seems* to be no consensus on solvent cleaning. A mild chlorinated solvent (1,1,1-trichloroethane) is used in the author's plant as a cleaning agent, applied with a rag. The solvent cleans the surface but evaporates quickly. So while this company, like most manufacturers, does not recom-

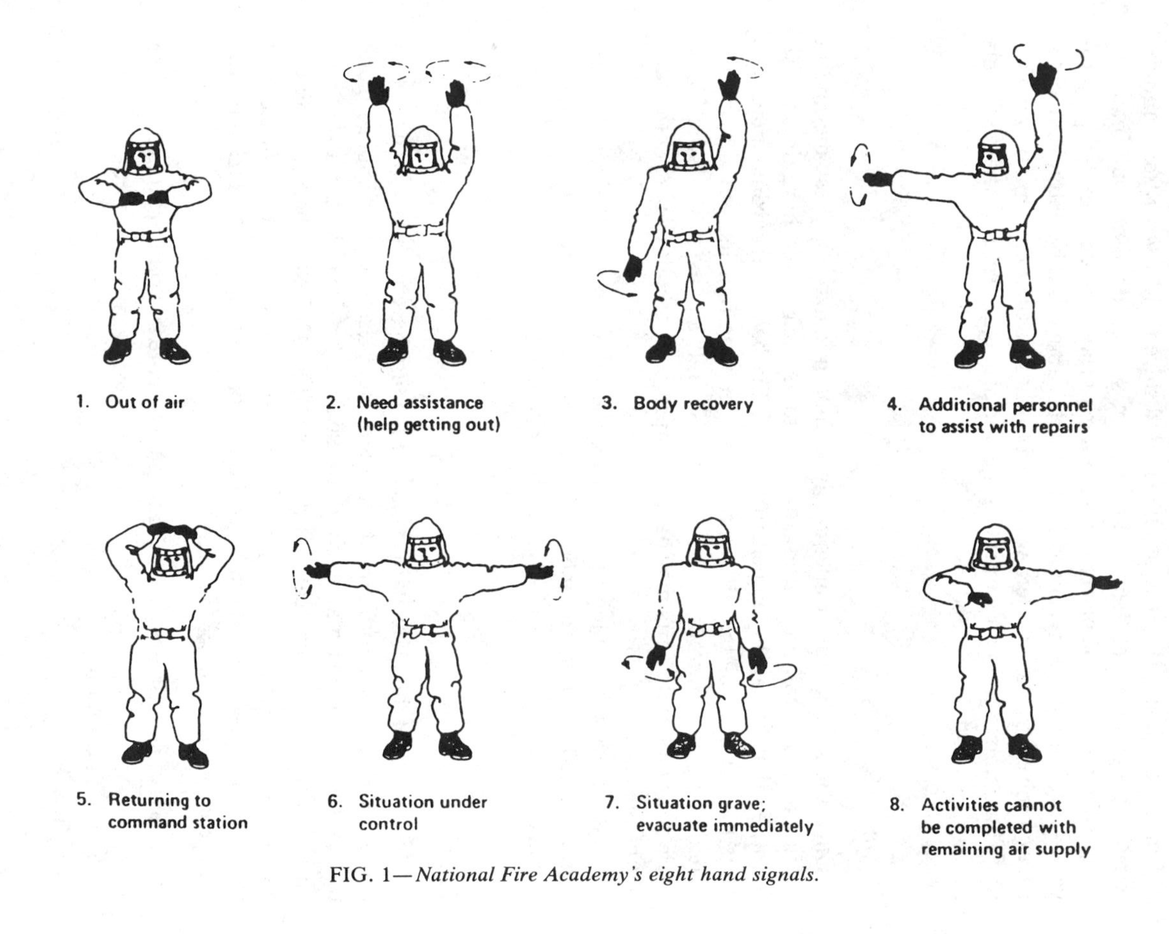

FIG. 1—*National Fire Academy's eight hand signals.*

mend the use of solvents or dry cleaning, I feel that leaving a deposit on the material will probably cause more damage than the effect of a *wiped-on solvent* (*not* full immersion). Note that this does not apply to clear windows of any *type*. Most solvents will cause clear plastics to haze.

I must emphasize that use of chlorinated solvents is a *last resort*, and extensive or regular use of them will reduce the service life of *any* suit material. Some solvents would be better for certain materials. For chlorinated polyethylene (CPE) and polyvinyl chloride (PVC), hexane would be better than chlorinated solvents. The problem with most of these is their extreme flammability. Thus, unless you can control usage and storage carefully, these solvents probably should be avoided.

Doffing

Once the suit has been adequately decontaminated, the best method of doffing it is to unroll the suit off the man to hip height and have the man sit down, so that he can be helped to take off the boots. Hazardous materials units have two problems: they have a harder time figuring out if a man is "adequately decontaminated," and there is rarely a place to sit down. Thus, hazardous materials units may have to improvise methods of taking the man out of the suit.

It also bears mentioning that whether the subject seems "adequately decontaminated" or not, the men helping a man out of a total encapsulation garment should wear suitable protective clothing.

Training

Men using total encapsulation garments must have a chance to wear and use these garments before an actual emergency. Users who have been using them for years are fortunate enough to have old "retired" suits for this purpose. In this case, it is critical that the word "TRAINER" be boldly displayed on the suit in order to prevent usage in an actual emergency. In any case, this allows men to practice tasks and communication skills, as well as train for the three contingencies most feared:

(*a*) loss of suit integrity because of puncture, tear, seam breakage, and so on,
(*b*) air cutoff, and
(*c*) fire.

Loss of Suit Integrity—As much as we would like to eliminate this, many sharp things are encountered that no polymeric material will stand up to. In case of puncture, the training should be to fold good suit material over the puncture, hold it with one hand, and leave the area immediately. Since the buddy system stays in effect, the *other* man must leave also. It also must be emphasized that careful decontamination must be practiced to avoid washing the chemical inside the suit. This is another reason why *two*-people buddy systems are sometimes insufficient. If you have only two suits, as is common in industry, when one tears you may be left with few good choices. I would suggest calling a local hazardous materials squad for backup.

An option that should *not* be considered is taping the cut. What looks like a great tape job may fall apart on exposure to chemicals.

Air Cutoff—If the suit is equipped with SCBA, low air warnings should alert the wearer to leave the area. If air is still cut off, the first action to take is to use the regulator bypass valve. This will dump the contents of the bottle (if any) into the mask. Should this not produce air, the mask should be lifted off the face, to break the seal. Even though dangerous levels of carbon dioxide will build up quickly, there is usually some breathable air inside the suit to allow the wearer to leave the hazardous area.

If the suit is equipped with an air line that is intended for use in areas immediately dangerous to life and health (IDLH), it should be equipped with an escape bottle. If not, the reaction to an air cutoff should be to breath the suit air while leaving the area.

On some suits, the wearer is capable of unzipping the pressure sealing zipper, or a second zipper is installed for emergency air. Since this breaks the seal to the environment, training should emphasis that this is a last resort.

Fire—No manufacturer advises the use of chemical protective total encapsulation garments in situations involving fire entry. (The exceptions are a few suits that are specifically designed for that purpose; these are usually multilayer suits, with one layer aluminized or heat resistant, or both, and an elastomer layer for chemical resistance.) Unfortunately, fire, or the threat of it, is present in many situations. Fire can be caused by cutting and welding, grinding, and even chemical reactions (fuming sulfuric acid and oil or pure oxygen and a hydrocarbon). Suit materials fall into two categories: self-extinguishing and non-self-extinguishing. *All* chemical protective suits will burn or melt when exposed to a hot enough flame for a sufficient length of time. Self-extinguishing suits will stop burning when removed from the source of the fire. Suits without this feature will continue to burn after removal from the source. The important factor in training is that the supervisor and the worker *know* which suits have this feature and which suits do not. This is especially true if the user has two or three types of suits. In one case, the worker may be able to put out the suit fire by simply moving from the source. In the other case, the worker may need help from his buddy, or others, in putting out the fire. In either case, the worker should slowly turn around 360° to let his buddy verify that the fire is out.

It bears mentioning that the self-extinguishing property works only if the suit material is the only fuel source for the fire. If a suit becomes covered in hydrocarbons, the fire will burn the hydrocarbon and not the suit. Thus, the self-extinguishing properties do not have a chance to operate until the fire hits the base material. In the meantime, heat may melt the suit. Thus, training should make clear that if a worker becomes covered with oil or grease, he should either leave the area with his buddy or steer clear of flame sources.

A word should also be said about chemical reactions during training. Suit reactions covering fire have already been discussed. Many chemical reactions are exothermic, generating heat. An example of this is mixing concentrated acids with water. Most chemical protective materials become poorer barriers as the chemical temperature rises. Thus, in the presence of oxidizing chemicals, care should be taken not to allow a wet suit into service or not to allow a worker to stand in pools of mixing oxidizers and water.

In training workers and supervisors should also be instructed that air hoses in air-fed suits should not be counted on to remove injured workers. If this option is desired, the suit must be reinforced, a waist belt added inside, and a metal lifeline loop added.

Maintenance

Laundering

Laundering is necessary to remove any residual chemicals and leave a clean, *dry* suit for inspection. A mild soap solution (≈1%) is the general rule. There is no consensus on whether to leave in the gloves and boots, and close the zipper to clean only the outside, or to take out the gloves and boots, open the zipper, and allow the inside to be cleaned. In the former case, water may seep through the dump valves. Also, if excessive sweating occurred in the last usage and is not cleaned off, odor and mold could occur. However, if the latter procedure is used, the suit will take longer to dry. Storing a wet or damp suit invites mold and mildew. Thus, the option may depend on how soon you need to resuit.

Some industrial users are able to combine the decontamination and laundering steps. An example of this is the hydrofluoric acid alkylation process. After a thorough water rinse, clothing is immersed in an alkaline solution (pH 10), which neutralizes and cleans the material. At this point in time, it has not been determined if this strongly basic solution hurts the material life. Although it does not appear to do so, the author would like more experience with this method before fully accepting it.

Drying

As stated earlier, all suits must be totally dry before they are stored. All manufacturers apparently specify hanging the garment and air drying it with the zipper open to ensure that the inside is dry. Some users hang suits in dryers to speed the process. The temperatures used range from 48 to 65°C (120° to 150°F). The major concerns with dryers are hot spots and the method of hanging. Elastomers elongate as the temperature rises, and improper hanging may cause a localized overstretched condition. A hanger with as wide a support surface as possible should be used. While this appears to work, care must be taken to maintain even temperatures and not let the suit touch any "hot spots." The author's company would like more experience with this technique before giving it an unconditional recommendation.

Something that is definitely *not* recommended is a commercial tumble dryer. Regardless of where the heat is set, these dryers have hot spots that are *much* hotter than the set temperature of the dryer. These hot spots can burn or melt holes in suits or create thin spots in the coatings.

Reassembly

Respiratory Equipment—A full paper could be written on respiratory equipment, but this must be said: care should be taken because this equipment is critical. Care should be taken not to "mix" different suppliers' equipment.

Dump Valves—Quite an assortment of dump valves is on the market, so I can only generalize. They must be checked for proper operation and to make sure the laundering process has not left any residues that might hinder performance (for example, undissolved soap inside the "dog house"). All dump valves should allow air to pass out of the suit but not allow venting into the suit. Some of the simpler "flapper valve" types are field replaceable and can be taken out for inspection. Dump valves that are not field replaceable are much more difficult to deal with in that the protective dog house—the protective cover over the dump valves—hinders inspection. If any doubt exists, the suit should be returned to the manufacturer, who is able to remove the dog house and inspect the valve.

Visual Inspection

The entire suit should be laid out on a table with excellent lighting and inspected. Extra effort should be expended on seams and closure systems, but *every* inch should be looked at. A variation of this is to put a light inside the suit, darken the room, and look for bright spots. Another variation is the inflation test; the suit can be inflated to a set pressure and then checked later for pressure loss.

The visual inspection provides an opportunity to look for areas of chemical attack. I do not profess to know all the ways chemical attack will show up, but at least two can be determined by "feel." These are when the material has been stiffened by the chemical and when the material is softened (plasticized) by the chemical. Thus, in the visual inspection, inspectors should be on the lookout for areas of material that are either much *stiffer* or much *softer* than the rest of the material. Also, "tacky" areas may signal chemical attack.

Soapy Water Inspection

This method is intended to look for pinholes or small leaks. First, the dump valves are taped over; then the user should blow up the suit to 12 to 25 mm (0.5 to 1 in.) of water pressure. The experience of one user is that *not much* pressure is necessary. Adequate pressure should just fill out the suit form. A *rich soapy* water solution is brushed over the seams and closure systems as well as over the whole suit. Any leaks should show up readily. The four major areas of leaks are (1) respiratory equipment pass-throughs, (2) glove to sleeve joints, (3) zippers, and (4) boot to leg joints.

Respiratory Equipment Pass-Throughs—Specifically, pass-throughs in the face area are susceptible to leaks. Upon reassembly, this gear may not be installed properly or not torqued down, and simply tightening the nut could stop the leak.

Glove to Sleeve Joints—The only commercial joint I am familiar with is the friction-fit sleeve. Thus, it is the only one I can discuss here. A cone is inserted inside the glove, and the assembly is pulled into an elastomeric cuff on the end of the sleeve. Several things can work to defeat this system:

1. The outside elastomeric cuff can lose its stretch. This can happen because of age, overstretching, or chemical exposure. Each manufacturer has a slightly different yardstick. The author's company recommends that the inside insert extend at least 10 mm (3/8 in.) beyond the outside cuff, but not more than 26 mm (1 in.). At that point, the cuff needs to be replaced.
2. Soap residues can be left on the inside of the elastomeric cuff. These can easily be wiped off with water.
3. Lubrication of any type will defeat the friction-fit type of cuff. Since friction-fit gloves are hard to install, some people apply light coats of a lubricant to ease assembly. This should not be done under any circumstances, because this can cause the glove to come out of the sleeve entirely. Any lubricant on the mating surfaces should be cleaned off completely, with a chlorinated solvent wipe if necessary. Some manufacturers recommend taping the joint between glove and sleeve; to my knowledge, this is to prevent seepage and is not necessary for air-tightness. Seepage is defined here as a wicking action between two surfaces. Thus, testing should occur without tape.

Zippers—Zippers of the lap type cannot be tested in this manner, because they were designed to keep vapor out, not air in. They should be closely inspected for any bulges or gaps. Compression-type zippers and extruded types should work under this test. Some poor seals could be the result of foreign matter on the seal surface or in the sealing channels. Water should be used to clean them, with a soft brush. Care should be taken not to leave brush bristles on the sealing surface.

Boot to Leg Joints—These joints can be of the friction-fit type, in which case the procedures described under Glove to Sleeve Joints should apply. Some are permanently sealed to the boots or to a "sock." Since these areas tend to take the most punishment, extra inspection time should be spent here.

When inspection is complete, wipe off the excess soapy water, and allow the suit to dry thoroughly. A variation of this technique is to completely submerge the suit in water. If you have the facilities to do this, and can seal off the dump valves well enough, it is an excellent choice.

Storage

The first item to consider in storage is the zipper.

Zippers—I consider this to be the single most important item in suit maintenance. Unfortunately, this is also the area of greatest divergence between manufacturers. I would like to focus on three zipper types: (1) the lap type (neoprene), (2) the compression type (neoprene), and (3) the extruded "zip-lock" type (CPE).

Lap type—This zipper used to be the standard in the industry, and may still be in production, so there are quite a few in the field. It is essentially a standard zipper with neoprene on the outside and two neoprene lips covering the teeth. One design had the slider underneath the lips, and another version had it on top. In either case, the teeth should get a light coat of zipper lubricant (silicone) before storage. This type of zipper should be stored in the completely *closed* position. If stored open, the lips may take a set and may not mate when zipped up. The suit should be stored by hanging it over a rod, folded at the waist. If folding is needed, one user reports that the zippers will stay good if the suit is rolled up loosely with the zipper on the outside of the roll.

Compression type—This type of zipper requires more effort to open and close, and care should be taken to pull always in the direction of the zipper teeth. A light film of beeswax/wool grease lubricant is recommended by the manufacturer for lubrication.

There is currently a debate as to whether to store this zipper open or closed. One school of thought says that storing the zipper closed will cause the neoprene to take a "compression set," thereby reducing the quality of the seal and the useful life of the zipper. The argument against leaving it open is that suits are harder to fold and pack, and using brute force to pull the suit out of a storage bag may exert lateral stress on the zipper, causing a distortion of the sealing surface. The supplier for the author's company recommends storing the suit with the zipper closed, and in the company's three-year experience with this zipper, there have been *no* reports of zipper failure due to compression set. However, this point will be investigated further.

Extruded type—This type of zipper, which has a double row of sealing lips, should be lubricated with silicone grease or petroleum jelly and be fully *closed* during storage.

Storage Conditions—Conditions of 20% humidity and 23°C (75°F) would be nice, but are impractical. Here are some general rules:

1. Make sure the suit is completely dry.
2. Hang the suit, or drape it over a rod, if possible, using hang-up loops if the suit has them.
3. If it must be stored folded, fold it loosely, watching the zipper, and insert it carefully into the storage bag.
4. Keep the suit out of direct sunlight, and do not expose it to wide swings in temperature or humidity.
5. Each suit has a low-temperature flex point. Storage should not be anywhere near this temperature for two reasons: (*a*) the suit will be stiff and difficult to put on when needed; (*b*) if the suit is pushed or stepped on while near the low-temperature point, cracks or stress points may form at the folds. If a suit must be stored in cold areas, you should probably get a low-temperature suit which will remain usably flexible below the storage temperature.

PVC and Color—PVC behaves as though it *loves* color. If materials of different color are put together, color changes will occur if the surfaces are smooth or if light pressure is applied. If your suit contains materials of different colors, the dissimilar sections should be wrapped with Saran or polyethylene to prevent color changes. While the color change is cosmetic, and does not affect the chemical resistance, it nonetheless may be undesirable. It should also be noted that newspaper is undesirable for the wrapping material. Newspapers use carbon black, and wrapping PVC with this may give you a legible article on your suit. To my knowledge, CPE will also do this under some circumstances.

Conclusions

In conclusion, I believe the data in general support the premise that use and maintenance procedures are similar for all current types of materials. While the base materials looked at showed similarities (PVC, CPE, butyl rubber), *important* differences show up in the closure systems, and more data need to be gathered in areas such as laundering, drying, and decontamination.

It also must be noted that some items from manufacturers' data were not covered. The user is advised to read his supplier's literature to cover the points not covered here.

Acknowledgment

I would like to thank all the people who contributed to this paper, especially John Eversole, of the Chicago Fire Department, Glen Davis, Mine Safety Appliances Co., James Gauerke, Fyrepel Products, Inc., Evan Hensley, ILC Dover, Don Schmidt, Wheeler Protective Apparel Co.—and others not listed.

Karl C. Ashley[1]

Polychlorinated Biphenyl Decontamination of Fire Fighter Turnout Gear

REFERENCE: Ashley, K. C., "**Polychlorinated Biphenyl Decontamination of Fire Fighter Turnout Gear,**" *Performance of Protective Clothing, ASTM STP 900*, R. L. Barker and G. C. Coletta, Eds., American Society for Testing and Materials, Philadelphia, 1986, pp. 298–307.

ABSTRACT: This study addresses two key issues: (1) a decontamination methodology to remove polychlorinated biphenyl (PCB) contaminants from fire fighter protective clothing and (2) application techniques for various types of protective clothing, so as to minimize damage to the garments subjected to the decontamination process. The occupational exposure situations evaluated include fire fighting, organic chemical manufacturing, and fueling of liquid fuel rockets. Removal efficiencies of up to 99+% for PCBs have been achieved using a Freon solvent technique.

With respect to physical damage to the garment, a tumble wash action proved satisfactory for rugged garments (that is, fire fighter turnout gear).

A surface spray-flush action in an enclosed chamber was required for butyl-covered fabric-type construction. The latter type of garment had internal construction and hardware items that tended to cause wear points and fabric gouging when subjected to tumbling action. Some protective clothing fabrics proved unsuitable for the Freon solvent because of plasticizer removal.

These decontamination techniques were originally developed to remove radioactive contaminants from protective clothing worn in the nuclear utility and radioactive materials industries. The Freon-based decontamination system is a closed system that flows solvent across the fabric surfaces; the solvent is "cleaned" using filtration, adsorption, and partial distillation. Removal of solvent and vapors from the garments and cleaning chamber is achieved by mechanical refrigeration.

KEY WORDS: decontamination, protective clothing, polychlorinated biphenyls (PCBs), Freon TF solvent, fire fighters, fabric sampling

Limited surveys of the literature and users of protective clothing have revealed varying approaches to decontamination. Technical emphasis has been placed on suit design, construction, and selection to meet challenges of

[1]Program manager, Quadrex HPS, Inc., Gainesville, FL 32606.

specific occupational exposures. After a given use or exposure, the owner is confronted with the problem of effectively removing the suit's exterior contaminants as well as body soils on the interior surface. In addition, the decontamination process should not impair the suit's performance, nor cause physical damage. For this discussion, two clarifications are in order. The first is that the term *decontamination* means the physical removal of contaminants from the suit's surfaces rather than neutralization or deactivation of these substances. Second, *protective clothing* refers to clothing designed for multiple use and market priced at a few hundred dollars or more per set. It is believed that protection levels for specific occupational exposures will escalate, resulting in the design and manufacture of protective clothing with a high initial first cost. A satisfactory method of decontamination will offset the higher initial cost of such a garment by permitting repeated use and thus reducing the cost per wearing.

A review of the requirements of Subpart Z, Toxic and Hazardous Substances, Occupational Safety and Health Administration (OSHA) Safety and Health Standards (29 CFR 1910), indicates that protective clothing must be properly disposed of, cleaned, laundered, or decontaminated after wear, as required for the specific occupational exposure. However, no recommended cleaning and decontamination techniques are listed, nor any evaluation of efficacy. A federal agency decontamination technique for fire fighter turnout gear exposed to a polychlorinated biphenyl (PCB) fire was found, after analysis, to apparently increase PCB levels and cause interior surface contamination.[2]

This paper presents limited data on the decontamination of fire fighter turnout gear (coat and trousers) worn in a recent PCB fire incident. In addition, a spray-flush technique is described for protective clothing whose design and hardware items are incompatible with tumble wash action.

PCB Decontamination of Fire Fighter Turnout Gear

In early 1984, a fire occurred at a facility that was used to temporarily store waste oils, solvents, and PCB liquids prior to their shipment and disposal. The ensuing blaze completely destroyed the facility. Approximately 60 fire fighters responded to the scene and eventually brought the fire under control. The protective clothing or "turnout gear" worn by the fire fighters was assumed to be contaminated with PCB liquid, soot, and possibly pyrolysis products. These garments, valued at over $800 each, along with several hundred feet of fire hose, were immediately removed from service and temporarily placed in secure storage prior to disposal. A few sets of turnout gear, however, were subjected to a field decontamination procedure, which consisted of immersing the garments in a barrel containing water and detergent and then

[2]U.S. Environmental Protection Agency, Guidelines for On-Scene Coordinators, Decontamination of Personal Protection Equipment.

flushing them with clear water. Specimens were taken of these site decontaminated garments for residual PCB contamination. This incident provided the opportunity to secure turnout gear which had been exposed to a PCB fire and to conduct a limited study to evaluate removal efficiencies using a Freon[3] solvent technique.

Fabric Sampling Protocol

One major technical hurdle had to be addressed during the planning stages of the study. This issue was selecting a technique to obtain at least relative values indicating the amount of contamination on a fabric's surface. Further, it was assumed that the contaminants would not be uniformly dispersed on the fabric's outer surface, but rather that surface contamination levels would vary with location and even within specific surface areas of the garments.

A technique used to sample surfaces by wiping a measured area with solvent-soaked absorbent medium was rejected because of inherent variables, including the wiping time, pressure, and surface condition, and the difficulty of determining the exact areas necessary for comparisons. Based on a suggestion from the National Institute of Occupational Safety and Health (NIOSH), a sampling technique was devised consisting of marking an area 12.7 cm (5 in.) square and, within that area, marking off five 3.54-cm (1.0-in.) squares.[4] Four of these five smaller squares were located at the four corners of the larger square, with the fifth located at the center, producing an overall checkerboard effect. A simple template was fashioned to mark off these areas on the fabric's surface. To determine initial surface contamination, each of the five 2.54-cm squares was cut out from within the larger square and formed a composite sample. To determine a postdecontamination condition within the same area, the remaining fabric within the marked-off section was removed. This checkerboard technique was used to average out or minimize concentration variations within the area under investigation. To eliminate variations based on cutting errors and unequal areas removed, all the results were normalized by reporting contaminant concentration on the basis of the weight, not the area, of the fabric, that is, micrograms of PCB per gram of fabric. It was assumed that bulk density variations in the fabric and substrate were minor. For comparison consistency, fabric sections were removed from the same approximate location on each coat and trousers sampled.

Analytical Method

Fabric specimens so sectioned from the garment were placed in appropriate sample containers and sent to an independent laboratory for analysis. The analytical technique involved extracting the fabric (Nomex)[5] with toluene and

[3]Freon is a registered trademark of E. I. du Pont de Nemours and Co.
[4]Original measurements were in English units.
[5]Nomex is a registered trade name of E. I. du Pont de Nemours and Co.

cleaning the extract on a Florisil column. A gas chromatographic analyzer (Hewlett-Packard 5830) with electron capture detector and accessories for capillary column capabilities was used. Five percent methane in argon was used as the carrier gas. The limit of detection was 0.5 $\mu g/g$. The presence of PCB Arochlors was determined by comparison with standard specimens. No recovery studies were performed on the fabric; therefore, the reported results should be considered as the minimum amount of PCB materials present.

Evaluation of Site Contamination and Site Decontamination

Turnout gear obtained from the fire department included coats and trousers, each constructed of an outer layer of Nomex fabric. A few garment sets received had been site decontaminated using a detergent and water soak followed by a clear water rinse.

Fabric specimens were removed from approximately the same areas in (1) site-contaminated garments and (2) site-contaminated and site-decontaminated garments.

Because different garments were sampled, the results presented in Table 1 should be viewed as only indicative. Variations in site-contaminated specimens were expected because of the exposure conditions at the fire scene. It should be noted, however, that the micrograms of PCB per gram levels for the site-decontaminated fabric sections were higher than for those garments that had not been decontaminated at the site. Insufficient information regarding the actual site-decontamination events prevent a detailed analysis of the elevated PCB phenomenon.

Evaluation of Freon Solvent System Decontamination

The next step of the study involved selecting garments and marking fabric areas for sampling using the template and technique previously described.

TABLE 1—*PCB contamination of fire fighter protective clothing worn at fire site, in micrograms of PCB per gram of fabric.*

Specimen No.	Site Contaminated Fabric	Site Decontaminated Fabric (Detergent Dip and Rinse)
1-S	7.8	...
2-S	0.76	...
3-S	6.7	...
4-S	14.8	...
5-S	601	...
6-S	...	15
7-S	...	68
8-S	...	14
9-S	...	37
10-S	...	240
11-S	...	1060

From the area marked, the five smaller squares were removed for analysis. The entire garment was then placed in a rotating drum through which Freon TF was pumped for approximately 10 min. After the solvent was extracted and dried, the balance of the marked-off sampling square was removed for analysis. The procedure was repeated for several garments. The results of this decontamination procedure are presented in Table 2. The range of removal efficiencies varied from 66 to 99+%, with an average of 95%.

Since relatively low levels of PCB contamination (less than 10 000 μg PCB/g fabric) were expected as a result of site contamination, the study incorporated a step to further contaminate test areas of the garments with a range of PCB concentrations in order to evaluate removal effectiveness at elevated concentrations. Exposure to or contact with concentrated PCB liquids by fire fighters responding to a PCB spill or equipment leak is a distinct possibility.

For this test, the PCB Arochlor 1242 was diluted with a dielectric-grade mineral oil. Solutions containing approximately 1.5, 6, 25, and 75% of PCBs were applied to previously marked off sampling squares on several garments. These solutions were "brushed on" the test sampling areas. The amounts applied were sufficient to produce a nonfree-flowing yet very wet appearance, as may be encountered in an actual splash or direct contact.

After application, at least 30 min of contact time was provided to ensure surface saturation. To determine the predecontamination surface concentration, five 2.54-cm squares were cut out of each sampling area and placed in an appropriate container for laboratory analysis. The garments were then subjected to the same Freon solvent decontamination process. The remainder of the specimen squares from each garment were then removed and analyzed. Table 3 presents the predecontamination and postdecontamination results in micrograms of PCB per gram of fabric for the laboratory-contaminated test step. The results indicate 99+% removal efficiency.

TABLE 2—*PCB decontamination results*[a] *of fire-site-contaminated fire fighter protective clothing, in micrograms of PCB per gram of fabric.*

Specimen No.	Predecontamination	Postdecontamination	Percentage of Removal
1	60	5.4	91
2	480	25.0	95
3	200	1.8	99
4	56	3.5	94
5	10	1.2	88
6	35	2.4	93
7	5.3	1.8	66
8	35	4.2	88
9	45	4.3	90
Average	103	5.5	95

[a]Using Freon solvent system equipment.

TABLE 3—*PCB decontamination results*[a] *of laboratory-contaminated fire fighter protective clothing, in micrograms of PCB per gram of fabric.*

Specimen No.	Predecontamination	Postdecontamination	Percentage of Removal
10	16 000	3.9	99
12	58 000	59	99
14	180 000	62	99
15	340 000	230	99
16	940 000	180	99
17	630 000	53	99

[a]Using Freon solvent system.

The laboratory-contaminated garments were used as received from the fire site and were assumed to have site PCB contamination. No attempt was made to vary the Freon solvent decontamination process cycle during this study. All the garments processed for removal efficiency were subjected to one decontamination cycle.

Freon Solvent Decontamination Process

Figure 1 represents a basic flow diagram of the process used to remove the PCB contamination from the turnout gear. The equipment employed was a modified version of a Radkleen[6] HPS 1550 decontamination unit typically used to remove radioactive contamination from protective clothing. Modifications included the addition of an adsorber to remove PCB materials from the Freon TF solvent.

Process Description

Contaminated clothing is placed in the cleaning chamber, which consists of a sealed, rotating drum that will tumble garments through the solvent. Solvent is pumped from the solvent reservoir through a filter adsorber set (which removes particulate and organic contaminants), and then flows through the cleaning chamber. The contaminated solvent then returns to the solvent reservoir. Freon vapor and vapor pressure buildup within the sealed system are controlled by vapor recirculation across a condenser coil, with the condensate returning to the solvent reservoir. Periodically the solvent is regenerated and highly purified by a built-in distillation process. Contaminants are removed from the system as still bottoms and materials trapped on replaceable filter/adsorber cartridges.

[6]Radkleen is a registered trade name of Quadrex HPS Inc.

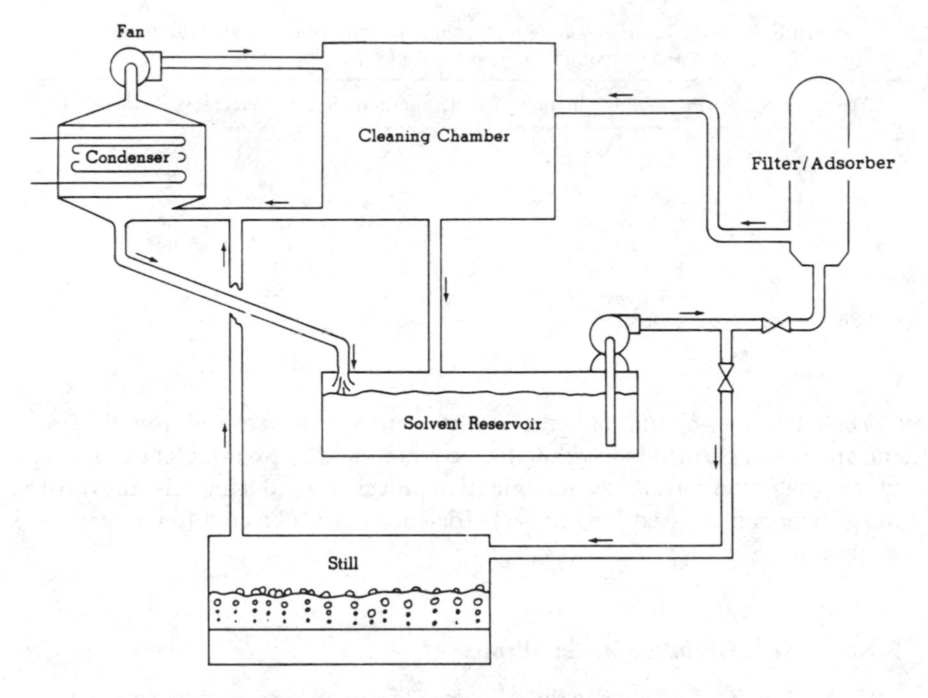

FIG. 1—*Basic Freon 113 decontamination process.*

Decontamination Techniques and Physical Damage

The fire fighter turnout gear used in this study has considerable resistance to cuts, abrasion, and gouging. The coat and trouser fabric construction is a trilayer composite consisting of outer and inner layers of Nomex fabric and a center layer of neoprene. These garments did not show any physical damage (cuts, abrasions) to the outer Nomex layer even after six repeated washings in the tumble wash action decontamination unit, Radkleen HPS 1550. The tumble wash action, however, was found to be unsuitable for at least two other types of reusable protective clothing.

Chemical resistant, total encapsulation protective clothing ensembles with supplied air connectors, as used in the chemical manufacturing industry, were obtained for decontamination evaluation. These ensembles cost more than $1000 each and are of the butyl-covered fabric construction.

Before any contaminants were applied to these ensembles, the garments were cycled in the tumble wash unit. After three cycles, certain wear points appeared on the fabric's outer surface beneath which was a fiberglass frame used to provide shape to the garment's head and face area. Further, a few small gougelike marks were noted on the fabric's outer surface.

In the latter instance, it seems that the metal snaps and fittings mounted on the suit caused the gouging during the tumble wash action. The abrasion wear spots, noted on each side of the head area, are believed to be caused by highly stressed contact and relative movement between the fabric point and the rotating drum wall of the decontamination unit. For these types of garments, tumble wash was abandoned in favor of the shower technique.

Freon Shower Decontamination Approach

In the Freon shower decontamination technique, the protective clothing ensembles were hung vertically within a Lexan[7] chamber 122 by 122 by 244 cm high (4 by 4 by 8 ft). The garments were then sprayed and flushed with Freon TF while inside the sealed chamber. Within the chamber, permanently mounted spray nozzles were used to flush both the exterior and interior surfaces of the garment. Supply pressure was 2.8 kg/cm^2 (40 psig). A pair of glove ports were installed through one of the chamber's walls to provide interior manipulation and enable one to use a high-pressure spray gun to clean specific areas. The gun nozzle pressure was varied from 56.2 to 77.3 kg/cm^2 (800 to 1100 psig) and was very effective in cleaning heavily soiled areas. After being cycled through the chamber, the garments were inspected.

None of the damage associated with the tumble wash action was found. Figure 2 presents the vertical decontamination chamber and process flow diagram.

Other than the method of applying the solvent to the garment to be decontaminated, the fundamental processes depicted in Figs. 1 and 2 are identical. The concept of using a sealed chamber in which the protective clothing ensembles are hung vertically and spray flushed with low-pressure Freon TF is very similar to an installation developed by and in use at the Kennedy Space Center (KSC). The facility at KSC is used to finish clean both interior and exterior surfaces of total encapsulating protective clothing, which are equipped with supplied air and communications connections. Prior to the solvent flush cleaning step, the garments are flushed with copious amounts of water to remove residual rocket fuel.

These garments are worn by workers when handling liquid rocket fuel. With a detailed suit maintenance procedure, these garments, each valued at several thousand dollars, have experienced a service life in excess of eight years.

Conclusion

This limited study indicates that both fire-borne and liquid PCB contamination can be removed from fire fighter turnout gear with removal efficiencies approaching 99+%.

Further, the site-decontamination technique of washing the garments in a

[7]Lexan is a registered trade name of General Electric Co.

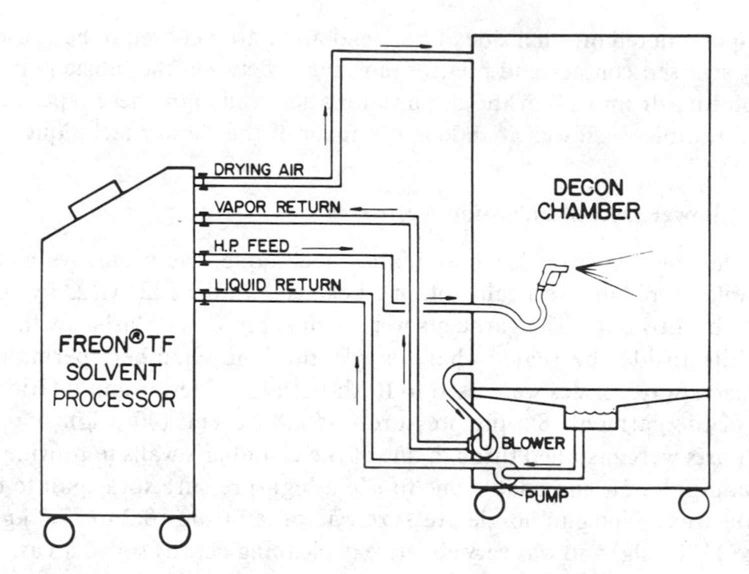

FIG. 2—*Vertical decontamination chamber and process flow diagram.*

barrel of water and detergent, followed by a water rinse, seemed to raise the average surface concentration of PCB contamination.

Although the results were not measured, one could conclude that the barrel dip wash technique used on the garments in this study would cause or increase the contamination on the garments' interior surfaces.

Freon solvent decontamination, using a tumble wash action produced by a rotating drum, did not appear to cause any observable cuts or abrasions in the turnout gear.

Tests are currently underway to determine if the solvent decontamination technique affects key fire fighter turnout gear performance criteria, that is, vapor barrier protection and flammability resistance.

This study also indicated that the tumble wash decontamination technique is unsuitable for the total encapsulation, chemical-resistant ensemble type of protective clothing. The tumbling action produced in a rotating drum caused surface abrasion and pinholes in this fabric. This destructive phenomenon was eliminated by changing the decontamination application technique. The technique of handling these garments in a sealed chamber and spray flushing the surfaces with solvent proved very satisfactory.

Future Developments

The technique of removing a section of fabric from protective clothing to determine predecontamination and postdecontamination surface contamination levels is incompatible with the practical goal of returning these garments to service. A nondestructive surface sampling apparatus and technique are required.

Another area of investigation is the extent of contamination and decontamination potential for fire fighter turnout gear that has been exposed to fires involving organic materials other than PCBs. These would include pyrolysis products occurring in industrial and commercial building fires.

Acknowledgments

The author wishes to thank individuals in the following organizations, who provided insight and assistance in the studies described: the Jacksonville, Florida, Fire Department, the International Association of Fire Fighters, the National Institute for Occupational Safety and Health, the U.S. Environmental Protection Agency, the National Aeronautical and Space Administration, Kennedy Space Center, E G & G Inc., E. I. du Pont de Nemours & Co. Inc., Quadrex HPS, Inc., and the Subcommittee on Protective Clothing Decontamination of ASTM Committee F-23 on Protective Clothing.

Laboratory Measurement of Thermal Protective Performance

Barry N. Hoschke,[1] *Barry V. Holcombe,*[1] *and*
A. Mariette Plante[1]

A Critical Appraisal of Test Methods for Thermal Protective Clothing Fabrics

REFERENCE: Hoschke, B. N., Holcombe, B. V., and Plante, A. M., **"A Critical Appraisal of Test Methods for Thermal Protective Clothing Fabrics,"** *Performance of Protective Clothing, ASTM STP 900*, R. L. Barker and G. C. Coletta, Eds., American Society for Testing and Materials, Philadelphia, 1986, pp. 311–326.

ABSTRACT: This paper discusses the merits and deficiencies of some existing and proposed test methods for the evaluation of fabrics intended for protection against radiant heat and flame. Modifications to improve the apparatus and the procedures are suggested.

The factors investigated include (1) the means used to obtain a given radiant or convective heat flux; (2) the type, design, and construction of the heat sensor; and (3) the specimen mounting.

Standard Test Methods British Standard for Clothing for Protection Against Intense Heat for Short Periods (BS 3791), ISO Test for Evaluation of Thermal Behaviour of Materials and Materials When Exposed to a Source of Radiant Heat (ISO 6942-1981) and ASTM Test for Thermal Protective Performance of Materials for Clothing by Open-Flame Method (D 4108-82) are considered, together with the radiant heat test method proposed in the Federal Aviation Administration's Proposed Flammability Standard for Commercial Transport Flight Attendant Uniforms (FAA-RD-75-176).

KEY WORDS: radiant heat, convective heat, test methods, protective clothing, heat sensor, calorimeter, specimen mounting

In assessment of the merits of various fabrics and garments intended for protection against heat, small-scale laboratory tests are more convenient, of greater general use, and less hazardous than full-scale field trials. This paper, based on experiments performed over several years, discusses the merits and shortcomings of various laboratory test methods and suggests improvements in apparatus and procedures.

[1]Principal research scientist, senior research scientist, and senior technical officer, respectively, CSIRO Division of Textile Physics, Ryde, New South Wales, 2112 Australia.

Radiant and convective heat tests are examined in turn, with consideration of the heat source, the sensor that measures heat transmitted through the specimen, and the mode of presentation of the specimen, all of which influence the measured performance of materials. Another paper [*1*] in this volume considers the interpretation of test data obtained with these methods.

Protection Against Radiant Heat

Heat Sources

Incident flux values commonly encountered in practicc range from sunlight (generally less than 1 kW/m^2) to 20 kW/m^2, which is the maximum radiation intensity at floor level in a burning building just before flashover occurs [*2*]. A person standing 1 m from a 1-m^2 fire pool also experiences a radiant heat flux of 20 kW/m^2 [*3*]. A kiln worker standing 3 m from a 0.5-m-diameter opening in a kiln heated to 1535°C is subjected to a radiant heat flux of 4.2 kW/m^2 [*4*].

The main characteristics of three radiant heat test methods are summarized in Table 1.

The International Organization for Standardization (ISO) Test for Evaluation of Thermal Behaviour of Materials and Material Assemblies When Exposed to a Source of Radiant Heat (ISO 6942-1981) uses a source consisting of six silicon carbide rods spaced one above the other with their axes horizontal. Two groups, each containing three rods in series, may be electrically connected in parallel or in series. The series connection, which results in a lower rod temperature, is normally used between tests, to prolong the life of the rods. We have performed tests with both types of connection in order to observe the influence of rod temperature on heat transfer through specimens. The desired heat flux is obtained by positioning the specimen at an appropriate distance from the source.

TABLE 1—*Main characteristics of three radiant panel test methods.*

Characteristic	ISO Test 6942-1981	BS 3791	FAA-RD-75-176
Heat source	silicon carbide heating rods	gas-fired surface combuster	sintered quartz, electrically heated
Approximate size, mm	82 by 178	300 by 300	300 by 300
Operating temperature, °C	1100 to 1220[a]	904[a]	912
Heat flux incident on specimen, kW/m^2	5, 10, 20, 40, 80	20	8.4, 16.7
Heat sensor	aluminum block calorimeter	copper disk calorimeter	heat flux meter
Measured parameters	transmission factor fabric integrity after 3 min	thermal protective index	total heat flow fabric integrity after 10 s at 16.7 kW/m^2

[a]Value not specified in the standard.

The gas-fired surface combustion heater used as the heat source in the British Standard for Clothing for Protection Against Intense Heat for Short Periods (BS 3791: 1970) produces a radiation intensity of 20 kW/m² at a point 200 mm in front of the panel. With an optical pyrometer we have measured the panel temperature as 904°C. The heat source used in the radiant heat test method proposed for commercial transport flight attendant uniforms [*5,6*] is a quartz panel electrically heated to 912°C.

Establishing that these three heat sources were equivalent for the measurement of heat transfer through fabrics at a specified incident heat flux could facilitate the adoption of an international standard test method for radiant heat protection.

To compare the two types of electrically heated panels, the procedure of ISO Test 6942-1981 was followed, except that the nature of the heat source was varied. Tests were performed with the groups of silicon carbide rods connected in parallel, with a mean surface temperature of 1220°C, and in series, resulting in a mean surface temperature of 850°C. The second heat source used was the quartz panel (Casso-Solar Heater Type C, 7.7 to 62 kW/m²), adjusted to 850°C for direct comparison with the rods.

The principle of ISO Test 6942-1981, Method B, is that the heat transmission factor, TF_x, of the test specimen at a particular incident heat flux is determined as the ratio of the transmitted and the incident heat flux measured by an aluminum block calorimeter. The subscript denotes the incident heat flux in kilowatts per square metre.

The heat transmission factor was determined for wool melton, 513 g/m², at an incident heat flux of 10 kW/m² and for aluminized asbestos, 865 g/m², at 40 kW/m²; the results are shown in Table 2.

A similar experiment has been performed in Finland [*7*] to evaluate the ISO method. In this work, the temperature of the rods was adjusted to 850 or

TABLE 2—*Comparison of heat transmission factors using different heat sources.*[a]

Material	Statistic (n = 5)	Rods at 1220°C	Rods at 850°C	Quartz Panel 850°C
		TF_{10}		
Wool melton, 513 g/m²	$\bar{X}$	0.45	0.48	0.49
	S	0.0055	0.0084	0.0045
		TF_{40}		
Aluminized asbestos, 865 g/m²	$\bar{X}$	0.041	0.045	0.046
	S	0.0040	0.0035	0.0043

[a]Key to abbreviations: $\bar{X}$ = average (arithmetic mean), S = estimated standard deviation, and n = number of observations.

1100°C by varying the power supply through an adjustable-voltage stabilizer. A summary of the results is shown in Table 3.

The thick wool melton reported in Table 2 showed a small but statistically highly significant difference between TF_{10} values obtained with the rods at 1220° and 850°C, yet there was no difference for the TF_{20} values of the thinner wool gaberdine at 1100° and 850°C, reported in Table 3. The results from different heating panels operating at the same temperature were equivalent.

The highly reflective aluminized asbestos showed considerable variability between individual measurements, but there was no statistically significant difference between the results obtained with the three heat sources.

The absolute differences between results for different sources fell well within the spread of values found between laboratories in an ISO interlaboratory trial (Committee Document ISO/TC 94/SC 9/WG1 N10 Annex 1) and are not of practical importance.

The effect of using a gas-fired radiant panel or an electrically heated panel was examined using the method described in Appendix C of the Federal Aviation Administration-Systems Research and Development Service Report FAA-RD-75-176 [5], using, alternatively, the specified panel and the BS 3791 gas-fired panel. The material used for the comparison was Whatman filter paper No. 2, exposed for 35 s with the specimen in contact with the heat flux transducer. The results are shown in Table 4. In general, the results are equivalent at the same incident heat flux, although the difference between 16.9 kW/m^2 tests for the two sources is significant at the 5% level.

Overall, it appears that in-plant quality control, or quick assessments of the practical heat transmission of a material, can be obtained with any convenient heat source. For a standard test method, it is preferable to specify the operating temperature of the radiant heat source, but the limited data reported here indicate that the nature of the source is not critical. Panels operating at the same temperature were equivalent.

Heat Sensors

The sensor used in ISO Test 6942-1981 consists of an aluminum block calorimeter containing two platinum resistance thermometers for temperature measurement and a heating coil for calibration. The large mass of this block (535 g) means that it is slow to respond to changes in heat flux, and the standard notes that the slope of the straight part of the graph of temperature rise as a function of time is accurate only after 40 s of exposure. To measure the slope over a sufficiently long part of the curve, it is necessary for the period of exposure to be greater than 60 s. In practical usage, nonmetallized materials would not normally be exposed to high-incident heat fluxes for such a long time.

The specimen is tensioned across the curved front surface of the calorimeter. The remaining calorimeter surfaces are well insulated, and this inhibits

TABLE 3—*Comparison of heat transmission factors using different source temperature* [7].[a]

Material	Statistic (n = 10)	TF_{20} Rods at 1100°C	TF_{20} Rods at 850°C
Cotton drill, 269 g/m²	$\bar{X}$	0.68	0.69
	S	0.0097	0.0127
Wool gaberdine, 310 g/m²	$\bar{X}$	0.56	0.57
	S	0.0068	0.0070

[a]Terms are as defined in Table 2.

TABLE 4—*Comparison of electrically heated and gas-fired radiant panels (material: Whatman No. 2 filter paper).*[a,b]

Heat Source	Quartz Panel			Gas-Fired Surface Combuster		
Operating temperature, °C	912			904		
Distance between panel face and specimen, mm	200	350	500	250	380	490
Incident heat flux density, kW/m²	16.9	8.9	5.9	16.9	8.9	5.9
Total heat flux, kJ/m², for 35-s contact,						
$\bar{X}$	284.0	131.3	85.1	262.9	128.5	82.6
S	11.8	5.1	4.9	14.0	4.8	4.3

[a]Terms are as defined in Table 2.
[b]The number equals five.

cooling between tests. It is common practice for users to modify the apparatus so that a portion of the insulation can be removed to accommodate a water-cooled aluminum block pressed against the calorimeter.

The amount of heat absorbed by the calorimeter is calculated from the formula

$$Q = \kappa m C_{Al}\, \Delta T$$

where

Q = amount of heat absorbed, J,
m = mass of calorimeter, kg,
C_{Al} = heat capacity of aluminum, J/kg · K
ΔT = temperature rise of calorimeter, K, and
κ = correction factor.

The correction factor, κ, is introduced to account for the presence in the calorimeter of elements other than aluminum and for the inevitable thermal losses which will vary from one piece of apparatus to another. Knowledge of the correction factor is not required for determination of the transmission factor, as κ appears in numerator and denominator, but it is necessary if the calorimeter is used to determine the incident heat flux. It is puzzling that the correction factors reported by various members of the ISO working group developing this method differ from unity by up to 20%. This large variation may be due more to the calibration procedure than to inherent differences between calorimeters. Pure aluminum blocks of equivalent size should produce the same results. We consider that further interlaboratory trials, with the block heater and calibration procedure omitted, are required before the standard can be revised. The required incident heat flux can be specified in terms of the time required for the calorimeter temperature to rise by a given amount.

The sensor specified in BS 3791 has a low mass and rapid response. It consists of a 0.4-mm-thick copper disk with a copper-constantan thermocouple, 0.4 mm in diameter, soft soldered to its rear face. The disk is cemented with silicate paint to a block of incombustible material. A spring-loaded holder pushes the calorimeter against the specimen, and the thermal protective index (radiation) is determined as the mean time in seconds for the temperature reached by the thermocouple to rise by 25 degrees Celsius.

The experience in our laboratory is that the specifications for construction of this calorimeter are not sufficiently rigid. Slight differences in technique when affixing the copper disk heat sensor to the incombustible board cause large variations in measured values of thermal protective index. Values differing by a factor of two have been observed. A modified calorimeter, incorporating an air gap behind the sensor (as for ASTM Test D 4108-82), has been found to yield consistent results independent of the manufacturer. The copper disk is set flush with the face of the insulation board, supported on a 1-mm ledge and backed by a 4-mm air gap. The time for a 25 degree Celsius temperature rise in this sensor is less than that obtained with any of the sensors made for BS 3791. As no calibration procedure for sensors exists in BS 3791, it is not possible to ascertain how the use of the new type of sensor would modify the minimum thermal protective index values originally specified for particular types of garments.

As a separate radiometer is used to adjust the initial heat flux density, the state of the surface of the copper disk (polished, oxidized, or blackened) is not critical. The calorimeter is wiped clean between tests and used as is, thus saving time lost in the ISO method by reconditioning the sensor.

The commercial heat flux meter used in the Federal Aviation Administration (FAA) test is unsuitable for testing materials, such as flame-retarded cellulosic fabrics, which deposit a tarry residue on the face of the meter. Some heat flux transducers are delicate and thus easily damaged during cleaning operations. In our experience, it is expedient to use heat flux transducers for equipment calibration, but solid metal calorimeters are more convenient for measuring heat flow through textile materials.

Specimen Mounting

In both the British and the international test methods the specimen is tensioned across the heat sensor. The specimen may therefore remain cooler and degrade less than it would in a garment exposed to the same heat flux. For this reason, ISO Test 6942-1981 has a separate test for material integrity, Method A, wherein the specimen is stretched across an open frame and exposed to the radiant heat source for 3 min. Such a long exposure, intended for aluminized materials, is irrelevant to the intended end use of many other materials, so provision exists for additional tests at shorter exposure as agreed on by the parties concerned.

Protection Against Flames

Heat Sources

Two different types of flame hazard are addressed by BS 3791 and the ASTM Test for Thermal Protective Performance of Materials for Clothing by Open-Flame Method (D 4108-82).

In the British method, burning hexane floating on water in a 140-mm-square tray, provides a soft flame which is very sensitive to draughts of air in the vicinity of the test and does not maintain contact with the vertical test specimen. A heat flux meter placed in the specimen position registered an average heat flux of about 15 kW/m^2, with peak values up to 40 kW/m^2. This level of heat flux is appropriate for garments such as civilian fire fighters' turnout coats in which occasional flame lick may occur, but the nature of the source leads to poor reproducibility. The test method discards the lowest and highest values obtained and averages the remaining three tests. This procedure, common in the subjective judging of athletic and artistic performance, is inappropriate for treating objectively measured data.

The 83 kW/m^2 heat flux employed in ASTM Test D 4108-82 corresponds to "emergency conditions" [*8*] such as exist around a crashed aircraft when fuel is burning fiercely or during flashover of a building fire. The maximum time expected for evacuation through burning aircraft fuel is 3 s [*9*]. At a heat flux of 83 kW/m^2, bare human skin will suffer a second-degree burn in 0.5 s. The Scope clause of ASTM Test D 4108-82 mentions that the method is applicable to fabrics intended for protection against a chance, short exposure to open flames.

The fuel gas used in the ASTM method may be liquid propane, Matheson B, or natural gas. Chemically pure propane is expensive, and Matheson B gas is not readily available in Australia. As part of our contribution to the development of an ISO standard based on ASTM Test D 4108-82 (and initially on Draft D-13-77-4 by ASTM Subcommittee D 13-52 on Flammability of ASTM Committee D-13 on Textiles), we have investigated the use of other fuels. Standards for fuel gases generally specify calorific value and many other parameters but do not precisely specify chemical composition.

In Australia commercial-grade liquid propane consists typically of 60 to 90% propane, 10 to 40% propylene, and 2% butane, with a calorific value of 48 850 to 50 000 kJ/kg. Commercial-grade butane is 58% *n*-butane, 40% isobutane, and 2% butylene, with a calorific value of 48 000 kJ/kg.

Natural gas, denatured with petroleum refinery by-products and air to the same calorific value as town gas, was available from mains supply and was included in our study of the effect of fuel gases on test results.

Two burners were used, one an Amal Meker burner, town gas type, modified by adding baffles and changing jets to burn propane or butane. The flat burner head had an overall diameter of 43 mm, containing a 38-mm perforated area. For natural gas, an Analite brand Meker burner, was used. This had an overall diameter of 47 mm, with a 40-mm-diameter perforated area. The diameter of the surface perforations of this burner was approximately three times greater than that of the other burner (3.1 mm compared with 1.1 mm), hence the flame cones for natural gas were fewer and longer.

The heat flux at a position 50 mm above the top of the burner was calibrated with a Medtherm 64-25-20 heat flux transducer (Medtherm Corp., Huntsville, Alabama).

The three gases all provided the 80-kW/m^2 heat flux required by the ISO draft method (ISO/TC94/SC9/WG1N38, succeeded by ISO/TC94/SC13/WG2N62). The influence of gas type on fabric performance was compared for two fabrics, an undyed aramid twill, weight 300 g/m^2, 0.59 mm thick at a pressure of 196 Pa (2 g/cm^2), and a gray wool cavalry twill, Zirpro[2] treated, 460 g/m^2, and 1.16 mm thick. In accordance with the ISO proposed method, spacers equivalent to the specimen thickness were used in conjunction with a 3-mm spacer to maintain the calorimeter face at 3 mm from the back of the specimen. Table 5 lists the results for unrestrained specimen mounting.

The chart record of the calorimeter temperature rise obtained in each test is overlaid with the standard human tissue burn tolerance overlay of ASTM Test D 4108-82. In the ISO method, the time at which the sensor response and the second-degree burn curve intersect is called the estimated burn time, EBT_{80}, at which point the subscript indicates the incident heat flux in kilowatts per square metre. This corresponds to the "thermal end point" in the ASTM method.

The ISO draft method also evaluates the influence of the fabric on human tissue tolerance to pain [*10*] as the estimated pain time, EPT_{80}.

The EPT_{80} and EBT_{80} values are listed in Table 5. While there is no statistically significant difference between the results with butane and those with modified natural gas, a significant difference at the 5% level occurs between propane and butane results for wool for both EPT_{80} and EBT_{80}. A significant difference at the 1% level was observed between propane and modified natural gas for aramid EBT_{80}. However, these differences are unlikely to be of

[2]Wool Development International Limited trademark.

TABLE 5—*Influence of fuel gas and calorimeter on the determination of EPT_{80} and EBT_{80} for an unrestrained specimen with 3-mm spacing between the specimen and sensor.*[a]

Fuel Gas	Sensors No.	Sensors Surface	Statistic (n = 3)	Aramid, 300 g/m² EPT_{80}	Aramid, 300 g/m² EBT_{80}	Zirpro Wool, 460 g/m² EPT_{80}	Zirpro Wool, 460 g/m² EBT_{80}
Commercial propane	4	oxidized	$\bar{X}$	4.3	7.3	4.2	8.6
			S	0.10	0	0	0.12
	4	blackened	$\bar{X}$	4.5	7.3	5.0	10.0
			S	0.10	0.21	0.10	0.15
	1	oxidized	$\bar{X}$	5.0	8.5	5.2	10.1
			S	0.06	0.06	0	0.10
	1	blackened	$\bar{X}$	4.5	7.3	5.2	10.2
			S	0.06	0.06	0.15	0.20
Commercial butane	1	blackened	$\bar{X}$	4.4	7.3	4.8	9.7
			S	0.21	0.26	0.06	0.15
Modified natural gas	1	blackened	$\bar{X}$	4.1	6.8	5.0	9.9
			S		0.06	0.52	0.67

[a]Initial heat flux calibrated with Medtherm heat flux transducer. The terms are as defined in Table 2.

practical importance, because at this level of incident heat flux pain is experienced at 0.2 s and second-degree burn at 0.5 s; the materials tests had EPT_{80} > 4.4 s and EBT_{80} > 6.8 s. Nevertheless, for consistency between laboratories it is preferable to specify a single gas.

No single gas appears to be universally obtainable readily and cheaply. In some countries propane is plentiful; in others butane is preferred as a commercial-grade gas. If chemically pure gases are required, testing costs will increase, and in some countries supply difficulties are experienced. At present the ISO draft method specifies commercial-grade propane as the fuel gas.

Heat Sensors

The copper disk calorimeter of ASTM Test D 4108-82 contains four thermocouples connected in parallel to average the temperature rise throughout the disk [*11*]. The fitting of four thermocouples, peened into holes drilled in the disk, is a delicate, time-consuming operation. Care is necessary to ensure that the thermocouple wires separate at the top of each hole.

The experiment reported in this paper was designed to investigate the sensitivity of using a single copper-constantan thermocouple (as opposed to the four of ASTM Test D 4108-82). The constantan wire was fixed with a minimum amount of soft solder at the center of the disk, and the copper wire was soldered near the disk perimeter. ASTM Test D 4108-82 does not directly specify the thermocouple material, although the thermocouple electromotive force examples given in the standard are for iron-constantan. It was found that both the single-thermocouple (1-TC) and four-thermocouple (4-TC) sensors, when blackened, agreed with the heat flux measured on a Medtherm heat flux transducer in direct flame exposure tests.

It was noted in the section on Heat Sensors, within the larger section on Protection Against Radiant heat, that the copper disk sensor of BS 3791 does not require special surface conditioning, and hence time is not lost in reconditioning between tests. The surface of a new copper sensor generally becomes oxidized after the first few tests. In order to determine whether or not blackening increases the measured heat flux when used in tests involving fabrics, a comparison was carried out between two calorimeters, one of which was oxidized and the other treated with an optical black agent. The results are listed in Table 5. For each test, the incident heat flux was established using a Medtherm heat flux transducer, as the oxidized sensor registered approximately 5% lower heat flux than the blackened version during direct flame exposure.

In general, the state of the copper surface has little influence on the measurement of heat transfer through these fabrics. The apparent discrepancies do not follow a consistent pattern. The 4-TC oxidized calorimeter yielded lower values for wool but not for aramid; the 1-TC oxidized calorimeter gave high values for aramid but not for wool. The blackened sensors gave equivalent results. Therefore, although blackened calorimeters require frequent reconditioning (which increases testing time and expense), they are more consistent and must be preferred, particularly if the calorimeter is used for flux measurement or in tests with an air gap.

The Influence of Specimen Mounting

In ASTM Test D 4108-82 single-layer materials are mounted either restrained to restrict heat shrinkage or relaxed to permit heat shrinkage. The procedure uses a fixed-thickness (6.4-mm) spacer, which maintains this spacing between the front (exposed) surface of the specimen and the sensor.

During development of the ISO draft method, spacers of the same thickness as the specimen were used to support the calorimeter in contact with the specimen (that is, 0-mm spacing). For tests with an air gap, an additional spacer 3 mm thick was introduced so that, no matter what thickness the material, there was a constant 3-mm gap between the specimen and the sensor.

Two further mounting procedures were investigated in a search for simplicity and relevance. In one, the specimen size was increased to a 150-mm-square, the same as the mounting plate and the sensor support, and the mass of the sensor and board was adjusted to 200 g. As the mounting plate contains a 50-mm-square exposure opening, this arrangement exerted a pressure of

98 Pa (1 g/cm^2) on the specimen. (The mass of the calorimeter and board could also be adjusted to correspond to the pressure at which thickness measurements are made. For the present study, a light pressure was chosen.)

In the second mounting procedure, the sensor was modified by introducing a curve into its front face to maintain specimen–sensor contact and eliminate variability in test results due to specimen distortion. Other test methods use curved heat sensors for the same reason. Radiant heat test ISO 6942-1981 employs a curved calorimeter, and the specimen is tensioned across this by a 200-g weight. The German Standard for Flameproof Protective Clothing for the Mining Industry: Safety Requirements and Testing (DIN 23 320, Part 1) uses the curved aluminum sensor for a convective heat exposure test with the specimen mounted horizontally. However, the mass of the aluminum block calorimeter is too great for rapid thermal response.

The new sensor was constructed by shaping the face of the soft insulation board[3] used in ASTM Test D 4108-82 into a curve with a 405-mm radius in one direction (Fig. 1). A 35-mm square of copper, 1.6 mm thick, which has approximately the same mass as the sensor in ASTM Test D 4108-82 was annealed and bent to fit the curve. A rectangular hole was machined in the center of the mounting block with a depth at the center of approximately 10 mm. A ledge left along the two flat sides parallel to the thin edges of the block served to locate the copper square, which was mounted flush with the face of the fixed mounting block with an adhesive[4] capable of withstanding high temperatures. The face of the copper calorimeter was coated with a thin layer of optically black paint.[5]

An angle bracket measuring approximately 12 by 12 by 150 mm was fixed to the rear of the mounting block 20 mm from and parallel to one of the thin edges. One end of a 150 by 300-mm specimen was clamped to this bracket while the free end was tensioned to about 2 N by a 200-g weight attached to the clamp by pulleys. This arrangement ensured intimate contact between the specimen and the sensor.

The mounting plate in ASTM Test D 4108-82 was replaced by edge supports so that the center of the specimen was 50 mm from the face of the burner.

For all mounting methods, a pneumatically driven shutter was used to increase timing accuracy. An event mark was automatically made on the recorder trace at the instant the specimen was fully exposed to the flame.

All of the results in Table 6, with the exception of those corresponding to the curved calorimeter, were obtained with a four-thermocouple calorimeter. All the copper sensors were used in the oxidized state. Values for the ASTM

[3]Marinite XL (Johns-Manville, Denver, Colorado; density 737 kg/m^3, thermal conductivity 0.166 W/m·K), was used.

[4]Dow Corning Silastic is suitable.

[5]Medtherm Optical Black Coating, Part No. 20164.

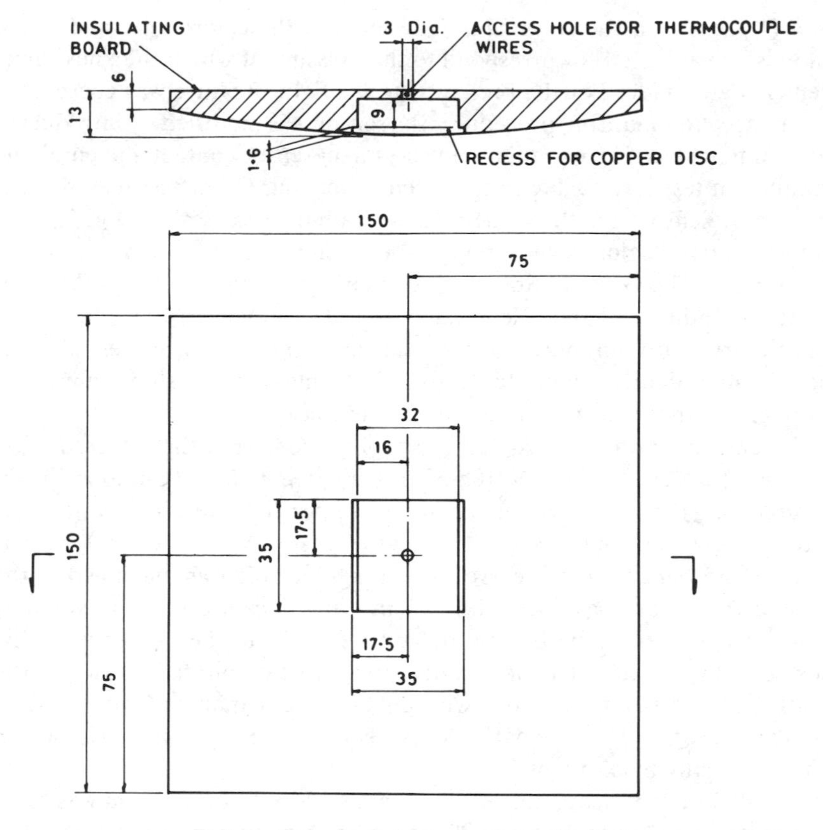

FIG. 1—*Curved calorimeter for convective heat test.*

method are also included. For this test the sensor was oxidized rather than blackened, as normally required. Initial heat flux was calibrated with a Medtherm heat flux transducer.

Representative calorimeter temperature records are shown in Fig. 2 for the four materials tested in an ISO interlaboratory trial. It is instructive to consider the shape of these curves as well as the data tabulated in Table 6. In particular, the behavior of the temperature curve in the vicinity of the standard pain and burn time curves is of interest.

When the specimen is held in intimate contact with the curved sensor, the measured pain and burn times correlate well with the fabric thickness ($r > 0.9$). Heat transfer between specimen and sensor is so efficient that specimens do not degrade before the burn threshold is reached. The chart records (Fig. 2*a*) show an initial lag corresponding to the specimen thickness, then a relatively monotonic increase in temperature.

Both procedures for flat contact—that is, 98 Pa pressure on a large specimen and using a spacer with a thickness equivalent to that of the specimen—

TABLE 6—*Influence of specimen mounting on the determination of EPT_{80} and EBT_{80} using oxidized sensors.*[a]

	Specimen Mounting			Curved Contact		Flat Contact		0-mm Restrained		3-mm Restrained		3-mm Unrestrained		6.4-mm Unrestrained	
Material	Weight, g/m^2	Thickness, mm[b]	Statistics ($n = 3$)	EPT_{80}, s	EBT_{80}, s	EPT_{80}, s	EBT_{80}, s	EPT_{80}, s	EBT_{80}, s	EPT_{80}, s	EBT_{80}, s	EPT_{80}, s	EBT_{80}, s	EPT_{80}, s	EBT_{80}, s
Aluminized glass	520	0.53	$\bar{X}$	1.5	2.2	2.8	4.2	3.0	4.2	4.9	7.1	4.4	6.9	6.1	8.7
			S	0	0.15	0.23	0.35	0.06	0.06	0.06	0.06	0.20	0.42	0.06	0.21
Zirpro wool	460	1.16	$\bar{X}$	2.4	3.9	2.9	5.2	2.8	4.9	4.9	9.6	4.2	8.6	6.1	11.4
			S	0.06	0.10	0.06	0.15	0	0.10	0.15	0.25	0	0.12	0.25	0.49
FR cotton	275	0.60	$\bar{X}$	1.1	2.0	2.1	3.6	1.9	3.7	3.7	4.0	3.7	4.0	3.7	4.3
			S	0.1	0.06	0.21	0.21	0.15	0.20	0.12	0.10	0.12	0.17	0.06	0.06
Aramid	300	0.59	$\bar{X}$	...	...	2.8	5.1	1.9	4.3	4.8	7.9	4.3	7.3	...	...
			S	...	...	0.26	0.25	0.12	0.15	0	0.10	0.10	0	...	...

[a]Terms are as defined in Table 2.

[b]Thickness is measured at a pressure of 196 Pa (2 g/cm^2).

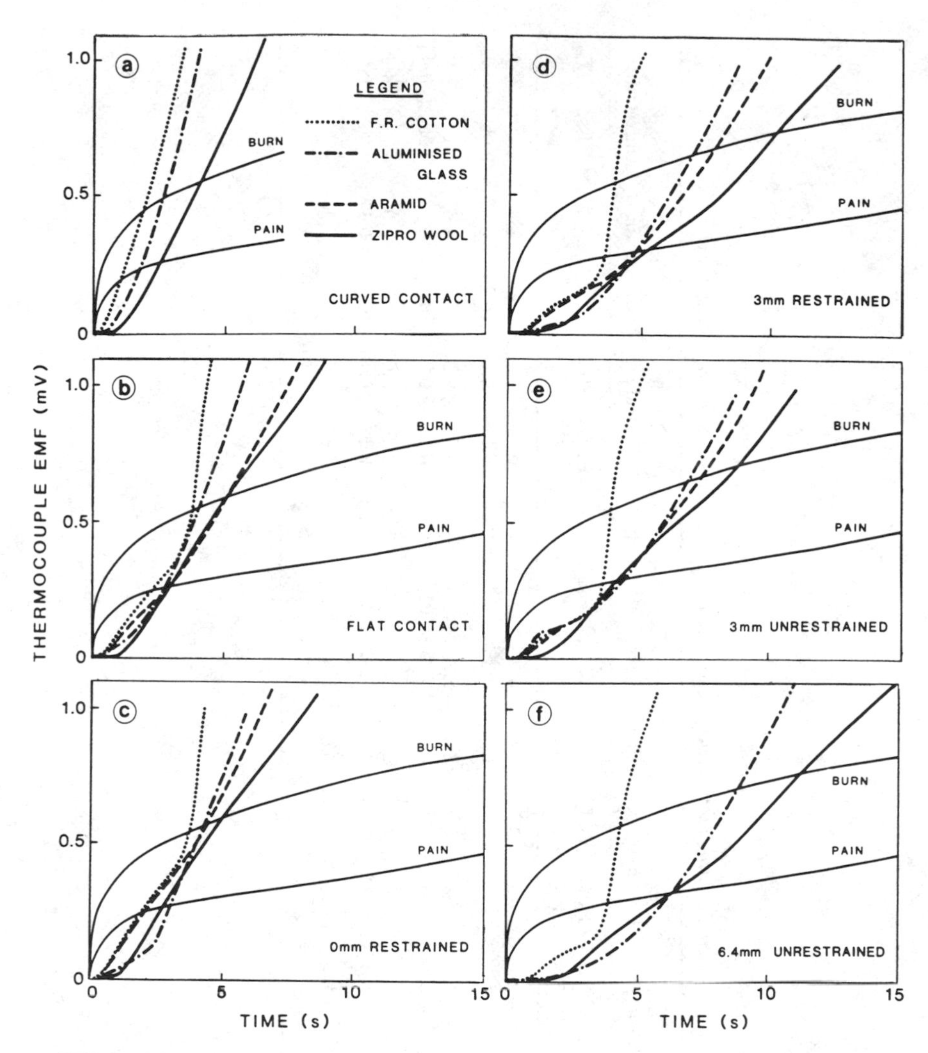

FIG. 2—*The influence of specimen mounting on convective heat transfer at 80 kW/m² incident heat flux. The ordinate is the electromotive force from a copper-constantan thermocouple.*

yielded equivalent results. The kink in the trace for aluminized glass evident in Fig. 2*c* is due to an initial bulging of the fabric away from the sensor.

The mounting plate used to locate the sensor and the test specimen contains a square aperture which acts as a physical restraint around the perimeter of the exposed specimen area, so that, as fabrics swell and buckle when the flame is applied, they tend to bulge downwards away from the calorimeter and increase the measured protection time. This effect is most evident in tests in which an air gap is used behind the specimen, as the extra insulation of the air gap causes both a more rapid temperature rise in the specimen than a contact test and a longer exposure to the flame before the burn threshold is reached. The many changes in slope of these traces are due to either a change in position of the specimen in relation to the sensor (swelling, shrinking, or buckling), a change in thickness of the specimen due to char formation (especially for wool), or the release of pyrolysis products which condense on the sensor (for example, cotton). The aramid material in Fig. 2*e* shows a knee in the curve at about 1 s as the material bulges downward and then an increase in slope as it shrinks back toward the sensor. Note that this movement has occurred before the pain time curve is reached, thus extending the time from that predicted by extrapolation of the initial rise to meet the pain curve.

Many of the protective clothing fabrics we have tested with specimen-sensor contact [*12*] distort sufficiently during tests with an air gap to make the results of doubtful relevance to the performance of protective clothing. The artificial constraint of the mounting plate has great impact on the results obtained, and materials that sag quickly upon application of heat have an unfair advantage over materials such as cotton. The flame-retardant (FR) cotton specimen did not deform during any of the tests, and hence the measured heat flow was high relative to that of materials that moved away from the sensor and increased the air gap.

Increasing the initial separation of specimen and sensor from 3 to 6.4 mm caused an increase of a couple of seconds in both pain time and burn time for the aluminized glass and the Zirpro wool. Little difference was noted for the FR cotton because the temperature rise here is more due to condensation of pyrolysis products on the sensor than to heat transfer through the material.

During garment exposure to convective heat the fabric is not artificially constrained by a small aperture, as in this test, and therefore will not bulge away from the skin and reduce the intensity of exposure. A fabric that distorts by shrinkage would probably pull onto the skin, thus increasing heat transfer and consequent burn injury. Fire-pit mannequin tests [*13*] demonstrate this effect. However, the test result does not reflect this shrinkage; it only reports a laboratory result for the apparent heat transfer.

The curved calorimeter described in this paper appears to overcome the bias of the other mounting methods considered. The measured pain and burn time values are shorter than would be expected with skin, which has a lower thermal diffusivity than the copper calorimeter; hence a margin of safety is provided in material ratings. To ensure that the material will maintain its integrity during exposure to fire, an additional test is necessary with an unbacked specimen exposed to the same heat flux intensity for a short time corresponding to the anticipated protection time requirement, for example, 3 s for aviators' clothing.

General Conclusions

Radiant heat sources consisting of silicon carbide rods, an electrically heated quartz panel, and a gas-fired radiant panel were found to be practically equivalent. The Meker burner used for convective heat tests yielded similar heat transfer results for the three fuel gases—propane, butane, and natural gas.

The copper disk calorimeter used in BS 3791 is not specified in sufficient detail to ensure reproducible construction. Incorporation of an air gap behind the disk removes the variability between calorimeters resulting from techniques of manufacture. The other sensors examined, an aluminum block calorimeter and heat flux meters, all have limitations.

Specimen mounting (for example, restrained or unrestrained, in contact with or separated from the calorimeter) has a marked influence on both measured heat transfer and fabric integrity. Maximum heat transfer occurs when the specimen is restrained in contact with a curved calorimeter, leading to a conservative assessment of protection afforded by the material.

The convective heat test methods that use horizontally mounted specimens and a restricting exposure aperture bias the results in favor of fabrics which deform during the test in a manner that is not necessarily characteristic of their behavior when worn as a garment.

References

[1] Holcombe, B. V. and Hoschke, B. N., "Do Test Methods Yield Meaningful Performance Specifications?," in this publication, pp. 327-339.

[2] Thomas, P. H., *Fire Safety Journal*, Vol. 5, 1983, pp. 181-190.

[3] Hagglund, B., *Fou-brand*, Vol. 1, 1977, p. 18.

[4] Aubertin, G. and Cornu, J.-C., *Cahiers de Notes Documentaires*, (INRS, Paris), No. 79, 2nd Trimester 1975, pp. 173-182.

[5] Braun, E., Cobble, V. B., Krasny, J. F., and Peacock, R., "Development of a Proposed Flammability Standard for Commercial Flight Attendant Uniforms," Report No. FAA-RD-75-176, U.S. Department of Transportation, Washington, DC, August 1976.

[6] Braun, E., Cobb, D., Cobble, V. B., Krasny, J. F., and Peacock, R. D., *Journal of Consumer Product Flammability*, Vol. 7, 1980, pp. 15-25.

[7] Irjala, B.-L. and Holmlund, C., "Thermal Behaviour of Textile Materials When Exposed to Radiant Heat, the ISO 6942 Method," NORDTEST-Project 354-82, Technical Research Centre of Finland, Espoo, Finland, May 1984.

[8] Abbott, N. J. and Schulman, S., *Journal of Coated Fabrics*, Vol. 6, 1976, pp. 48-64.

[9] Stoll, A. M., Chianta, M. A., and Munroe, L. R., *Journal of Heat Transfer*, Series C, Vol. 86, 1964, pp. 449-456.

[10] Stoll, A. M. and Chianta, M. A., *Aerospace Medicine*, Vol. 41, 1969, pp. 1232-1238.

[11] Behnke, W. P. and Seaman, R. E., *Modern Textiles Magazine*, Vol. 50, 1969, pp. 19-24.

[12] Holcombe, B. V., *Fire Safety Journal*, Vol. 6, 1983, pp. 129-141.

[13] Kydd, G., Askew, G. K., and Spindola, K., "Fuel Fire Tests of Selected Assemblies," Interim Report No. NADC-82121-6O, Naval Clothing and Textile Research Facility, Natick, MA, 1982.

Barry V. Holcombe[1] *and Barry N. Hoschke*[1]

Do Test Methods Yield Meaningful Performance Specifications?

REFERENCE: Holcombe, B. V., and Hoschke, B. N., **"Do Test Methods Yield Meaningful Performance Specifications?"** *Performance of Protective Clothing, ASTM STP 900*, R. L. Barker and G. C. Coletta, Eds., American Society for Testing and Materials, Philadelphia, 1986, pp. 327-339.

ABSTRACT: There is an urgent need for standards to evaluate the protection against intense heat afforded by textile materials. But there is equally a need to look carefully at where and how they should be applied. In most intense heat exposure accidents in which fires or explosions are involved, the bulk of the energy incident on the victim is in the form of radiant heat. Convective heat tests are subject to a number of problems that create uncertainty in the interpretation of test data, and in many instances, high-intensity radiation tests would provide the information required.

The performance requirements for each application should be carefully considered. In some occupations, it is more important that other risks, such as heat exhaustion, be given higher priority in the setting of garment performance requirements. Protective clothing can reduce the risk of injury but cannot guard against all hazards.

Where the test exposure exceeds about 10 kW/m^2, a mandatory pass-fail screening, consisting of a simple mechanical rigidity assessment following exposure, should be included. This eliminates materials that maintain a rigid char until the moment deformation is applied.

It is important not to misinterpret test results. Test methods are essentially a means of ranking performance, whereby the heat transfer of materials can be compared under similar conditions, and a measure of the degradation obtained under those conditions. They ought not to be regarded as duplicating real-life exposures.

KEY WORDS: thermal protection, convective heat, radiant heat, performance specifications, working environment, protective clothing

The function of thermal protective clothing is to prevent external heat energy from reaching the skin of the wearer at a rate and intensity that would cause burn injury. Evaluation of the protection afforded by any particular garment is complicated by the complexity of the relationship between the in-

[1]Senior research scientist and principal research scientist, respectively, CSIRO Division of Textile Physics, Ryde, New South Wales, 2112 Australia.

cident heat flux intensity and the rate of injury to the protected skin. The main factors that influence burn injury are the following:

(*a*) the incident heat flux intensity and the way it varies during exposure;
(*b*) the duration of exposure (including the time it takes for the temperature of the garment to fall below that which causes injury after the source is removed);
(*c*) the total insulation between source and skin, including outerwear, underwear, and the air gaps between them and the skin;
(*d*) the extent of degradation of the garment materials during exposure and the subsequent rearrangement of the clothing/air insulation; and
(*e*) condensation on the skin of any vapor or pyrolysis products released as the temperature of the fabric rises.

These factors may not be adequately considered when performance specifications are set for materials. Regulatory authorities and purchasing agencies can be faced with the situation that they would prefer a performance specification to a material specification, but no well-founded performance specification is available. Material specifications may exclude products whose performance is equal in all important respects to that of the material specified. Thus, inappropriate methods may be chosen, unrealistic values may be set, and the ramifications of such arbitrary decisions may be misunderstood. The purpose of this paper is to assist in the interpretation of test method data by describing the role of the protective garment during intense heat exposure and how current test methods relate to some typical documented hazards.

Exposure of Materials to Intense Heat

The behavior of human skin exposed to intense heat loads is well documented, and relationships between the burn injury threshold and heat flux at the skin surface (or behind the protective clothing) have been used for many years to assess how well protective clothing interposed between source and skin performs.

The behavior of textile materials subjected to intense heat depends in part on the reflectivity of the material and the nature of the exposure. The source may be purely radiant, purely convective, or a mixture of the two. Where radiant exposure is involved, the amount of heat absorbed is determined by the reflectivity of the surface because conventional textiles have very high absorption in the infrared region. This may be substantially reduced by the application of polished metal films, such as gold or aluminum, to the outer surface. When the exposure involves convective heat, the reflectivity of the surface is of little consequence.

Energy absorbed by the exposed surface of a garment raises its temperature and produces a temperature gradient through the material. Degradation such as shrinkage, swelling, charring, hole formation, or the release of pyrolysis products may occur if the exposure is sustained or intense.

The most serious garment failure for the wearer is hole formation. When the fabric remains intact, its heat flow properties do not change greatly even when the component fibers are degraded, because heat transfer is by conduction and radiation through air in the structure and by conduction through the fibers (which is relatively small). Only when these fibers melt or coalesce and displace the air, or when they bubble and form an insulating char, are heat flow properties substantially altered.

Shrinkage or expansion in the plane of the fabric does not substantially change the thermal insulation of the fabric itself. However, the spacing between fabric and skin or between garment layers may alter, with a consequent change in overall insulation. For example, if the outer layer shrinks and pulls the garment onto the body, the total insulation is reduced and the heat flow increases.

With so many unpredictable variables involved, the assessment of garment performance by laboratory simulation is a difficult task, and any test method used cannot hope to duplicate all exposure conditions.

The Nature of the Intense Heat Hazard

There arc a number of references in the literature to the typical persistence times for flame exposures. Simulated mine explosions, involving materials such as coal dust and methane, endure for 2.2 to 2.6 s and reach maximum heat flux levels varying from 130 kW/m^2 for lignite to 330 kW/m^2 for methane [*1*]; hydrogen balloon explosions last 1 s or less at ground level [*2*]; values from 3 to 10 s are quoted for escape through aircraft or vehicle crash fuel spills [*3*]; and heat flux intensities peaking between 167 and 226 kW/m^2 have been measured in mannequin tests with burning JP-4 fuel pools [*4*]. Probably the most severe environment is that of bush or wild fires, which can persist for periods from as little as 5 s for fine grass to more than 600 s for logging slash [*5*] and produce heat fluxes of the order of 100 kW/m^2 [*6*].

Using the criteria of Stoll and Chianta [*7*], which allow the time to thermal injury to human skin to be established for a given incident heat flux, the projected time to second-degree burn at a heat flux of 330 kW/m^2 is 0.07 s, while at 100 kW/m^2 it is 0.39 s. Despite the apparent rapid rate of burn development at these intense flux levels, the introduction of a material only 0.5 mm thick increases the protection time at 100 kW/m^2 from 0.39 to 2.5 s [*8*], which is adequate for a typical building flashover or mine explosion. The danger lies with parts of the body not covered by clothing, and this is confirmed by statistics which show that some 75% of all fire fighters' burn injuries in the United States are to the hands and face [*9*].

When choosing a test to evaluate the protection a garment or garment system provides against a given environment, it is necessary to ensure that the conditions simulated by the test at least parallel those in practice. Exposure intensity and duration should be such that the test result may be used as a guide to performance in practice. Studies of the composition of flames have found that fuel fires, for example, can involve as much as 80% radiation [*4*]. Without flame contact, exposure to fire sources must be considered exclusively in terms of radiant heat. Most emergency exposure situations, even when flames are present, should therefore be considered primarily in terms of radiant rather than convective heat exposure.

The Measurement of Transmitted Heat

The ideal test method would be one in which the heat flux on both the exposed face and the rear of the fabric is monitored without influencing the performance of the fabric. If a heat sensor or heat flux meter is placed in contact with the rear face of the specimen, heat will be conducted away by the sensor and reduce the rate of fabric temperature rise and damage. If there is an air gap between the specimen and the sensor, the air acts as an additional insulator, and the rate at which the sensor receives heat is then dependent on the size of the gap.

There are a number of tests, either available or under development, that can be used to evaluate performance in extreme heat conditions. The International Organization for Standardization (ISO) Test for the Thermal Behaviour of Materials Exposed to a Source of Radiant Heat (ISO 6942-1981) was prepared as a means of evaluating aluminized materials. This does not preclude its use for the evaluation of conventional materials at very high flux intensities. The equipment is capable of providing incident fluxes as high as 100 kW/m^2 if required.

If a predominantly convective exposure is anticipated, the ASTM Test for Thermal Protective Performance of Materials for Clothing by Open-Flame Method (D 4108-82) employs a Meker burner as the flame source, with an incident flux of 83 kW/m^2. The International Organization for Standardization is also currently investigating a similar test, with some differences in detail.

Despite the use of the term *convective*, approximately 25% of the energy released by the Meker burner used in both ASTM Test D 4108-82 and the ISO draft method[2] has been found to be in the form of radiant heat [*10*]. A comparison of the results obtained using both convective and radiant heat sources under identical conditions indicates that the influence of the source is generally small for nonaluminized materials (see Table 1). The convective source was the Meker burner of ASTM Test D 4108-82, and the radiant source was the panel of silicon carbide rods of ISO Method 6942-81. The heat

[2]Proposed draft International Organization for Standardization test method for protective clothing against flames (predominantly convective heat), ISO/TC94/SC13/WG2-N62, 1983.

TABLE 1—*Comparison of radiant and convective heat sources.*

Fabric	Thickness, mm at 196 Pa	Burn Threshold, s	
		Radiant	Convective
Aluminized glass	0.53	>30	2.6
Flame-retardant cotton	0.72	2.2	2.4
Aramid	0.97	2.7	3.1
Zirpro wool	1.16	3.1	4.1
Wool melton	3.64	6.7	8.8
Aramid + cotton interlock	1.77	5.0	5.5
Zirpro wool + cotton interlock	1.96	4.3	6.8
Bare skin		0.5	0.5

flux in both instances was 80 kW/m^2. The heat sensor was similar to the calorimetric device used for both the ASTM and ISO methods but with the front face curved, rather than flat, to ensure intimate contact. Specimen orientation was vertical in each case.

The wool materials are possible exceptions, as they tend to produce an insulating char more readily with convective exposure than with radiant. A similar char was observed during the radiant exposure, but it did not appear until after the burn threshold had been reached. Char formation, provided the fabric retains sufficient mechanical integrity, is a useful and positive material property that will be discernible under any appropriate test procedure.

It has been demonstrated [*11*] that mixed-mode tests [*12*] give no more information than a radiant test alone. It would seem that a high-intensity radiant exposure test such as that of ISO Method 6942-1981 should give adequate prediction of flame contact performance for most end uses. When the anticipated hazard is known to involve a significant proportion of convective heat, or the product is known to produce an insulating char under flame lick conditions, an additional test for intense convective exposure would be required.

A further reason for encouraging the use of radiant heat tests is that both the ASTM and the ISO convective heat test methods are subject to problems associated with the method of mounting, which make the relevance of their results questionable. This is discussed in detail elsewhere [*13*]. As an example, Fig. 1 illustrates the behavior of a typical aramid fabric tested with a 3-mm air gap in both restrained and unrestrained conditions. The distortion in the unrestrained state is sufficient to move the fabric several millimetres away from the sensor, whereas when it was restrained by taping to the support plate, it barely moved at all. The movement of the fabric substantially increases the apparent protection time. Thus, slight variations of the same test method yield quite different results.

Figure 2 shows one of the most extreme examples of distortion we have observed to date with this test method. The material is a neoprene-backed wool, and it appears that the neoprene has expanded considerably, causing

FIG. 1—*Convective heat exposure test specimens showing the influence of different mounting methods:* (a *and* b) *unrestrained;* (c *and* d) *restrained.*

FIG. 2—*An example of exaggerated specimen distortion in the convective heat test.*

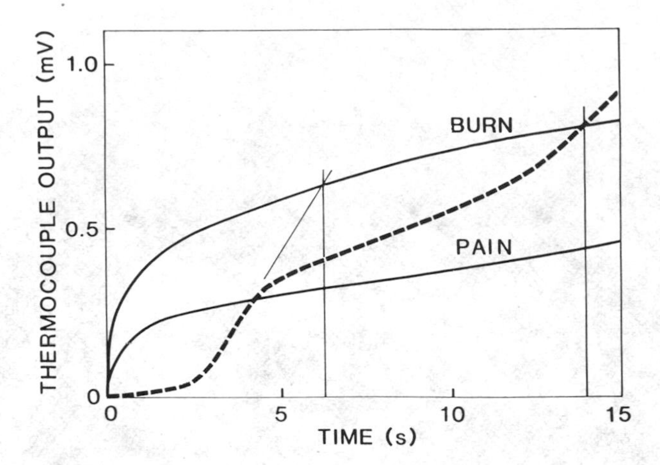

FIG. 3—*The sensor response derived from the specimen shown in Fig. 2.*

the center of the exposed area to bulge outward. The air gap is now several times greater than the thickness of the fabric itself. Figure 3 shows the corresponding calorimeter thermocouple trace plotted against time. It is particularly striking that in this test, the change in the position of this material in relation to the sensor has extended the burn injury threshold from 6 s (which is obtained by extrapolation of the linear portion of the graph and which would have resulted if contact had been maintained) to 14 s. This would not occur with the outer layer of a garment, for which the lack of restraint would reduce the fabric's ability to lift away from the body.

This raises doubts as to the validity of applying the results obtained from this test to general garment performance specifications. If the result is so dependent on the mounting method, and the mounting method is not directly applicable to the intended application, then how is the result to be interpreted?

Interpretation of Burn Time and Pain Time Data

Both ASTM Test D 4108-82 and the ISO draft method employ a common form of assessment; that is, the heat flux at the rear of the fabric is measured during exposure, and this is used to determine the time it would take for human skin in the position of the sensor to reach a particular burn threshold. Burn injuries are only partly determined by the total energy absorbed by the skin. The ability of the body to remove heat by conduction and through blood flow means that the formation of burns depends on both input heat intensity and duration.

The concept of using burn injury threshold for assessing garment protection was first applied by Stoll and Chianta [7]. They derived a relationship between incident heat flux and thermal damage from animal experiments in which pig skin was exposed to fixed levels of heat flux for specific periods. They emphasized that in order to apply their data, the heat pulse reaching the skin must be rectangular, because any variation from this shape invalidates the results.

In laboratory tests in which a calorimeter positioned behind the exposed specimen is used, the measured heat flux must be constant for the criteria to be validly applied. Smissaert [*14*] observed that the pulse measured behind the fabric is not rectangular; it starts at zero, rises slowly to a level determined by the thermal resistance of the fabric, then remains relatively steady until degradation of one form or another occurs. Factors such as hole formation and distortion or movement of the specimen cause the measured heat pulse to be far from rectangular.

Figure 4 shows the instantaneous heat flux measured by a transducer in contact with the rear face of the specimen for materials of different thickness exposed to a Meker burner source at 80 kW/m^2. Each trace continues only until the burn threshold. In no case does the pulse approach the required rectangular profile. In the extreme example, no heat flux appears behind the fabric until after 1 s has elapsed, and the maximum heat flux is barely reached at the threshold. Thus, the direct application of heat flux/time to burn data introduces an error into the determination which cannot be readily assessed and will vary depending on the heat flux/time behavior of the fabric. Where air gaps are used in the test, distortion of the fabric from its initial plane adds further complication and renders the use of burn criteria less justifiable.

While the second-degree burn threshold assessment is misleading in some respects, it is preferable to alternatives such as the time to a specific temperature rise. It is essential when using burn criteria for the test method to be performed in such a way that the measured heat flux remains relatively con-

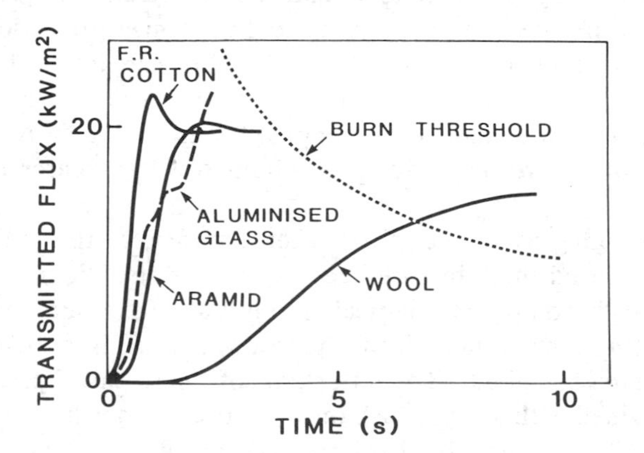

FIG. 4—*Typical convective heat exposure sensor response curves, each continued to the appropriate burn threshold.*

stant. Thus, good specimen-sensor contact should be maintained to maximize heat transfer. This will ensure the most rapid rise in transmitted heat flux and give the best approximation to a rectangular heat pulse.

Since the burn threshold determined by this technique can deviate substantially from that obtained under ideal conditions, it is essential that the results be presented in such a way that they can in no way be interpreted as duplication of real exposure conditions. Further, it would seem that the subsequent determination in the ISO draft of factors such as pain alarm time and residual heat transfer after exposure can only be considered even more unrealistic, and they therefore contribute no additional information of value.

We have given some consideration to the use of skin simulant sensors [*15*] in place of the calorimetric sensor. These sensors consist of a fine thermocouple embedded at a precise depth beneath the surface of a material whose thermal properties approximate those of human skin. From the time/temperature history of the thermocouple, it is possible to determine the rate of development of burn injury in accordance with the work of Henriques [*16*] and Stoll and Greene [*17*]. This takes into account nonrectangular application of the heat flux at the skin surface and is clearly more realistic than the calorimetric sensor.

Evaluation of data obtained by this method to determine the time at which the injury threshold is reached is complex and requires computational facilities to perform the necessary calculation. While it is possible to couple the sensor directly to a microprocessor that can provide the result directly, the added cost and sophistication puts the method beyond the realm of a standard test intended for quality control and comparison purposes. As technology in computing changes, this situation may change.

Fabric Integrity

Tests using an air gap yield longer protection times than those measured in contact tests. It has been our experience that the longer protection time does not eventuate in practice, as most lightweight fabrics are totally degraded. All that remains over the exposed area is a brittle char, and even the slightest body movement would result in the exposed area falling away or breaking open. This would give no useful protection to a live wearer in a similar situation.

Mechanical integrity of the fabric after exposure to the burn threshold should be measured in all intense heat exposure tests. We have found that some materials have little mechanical strength after exposure, and, in some instances, lifting the specimen from the test apparatus is sufficient to cause the exposed area to fall out. The latter part of the so-called protection time has been obtained with a fragile ash or char. It is of no value for a wearer to think that his clothing affords 20 s of protection against the convective heat of a given flux in a laboratory test if the last 5 s of that protection are attributable to a fragile char which disintegrates with the slightest body movement. This is dangerously misleading.

Disintegration is not a significant problem with tests in which intimate specimen-sensor contact is maintained, as the second-degree burn times for materials under these conditions are generally less than the time required to cause fabric degradation. This arrangement provides the most conservative approximation to the burn time for fabric against the skin. Where multiple layers are involved, the extra insulation provides longer exposure times to the burn threshold, and degradation of the outer layer is more severe. The important criterion is that the layer closest to the skin remains intact, even if the outer layer or layers break open.

A simple test for retained integrity would involve folding a specimen, previously exposed to the known burn threshold, about a rod 3 mm in diameter. If the inner layer of the exposed area cracks or breaks open, the material fails the test. The relevance of so-called break-open tests, in which the specimen is subjected to convective heat, typically for 30 s at 84 kW/m^2, and is subsequently examined for hole formation, is doubtful for all but self-contained fire entry suits. Bare skin cannot endure these conditions for more than 0.5 s.

Relevance of Test Results

Intense levels of heat flux are normally generated only in explosions, flashovers, or during fireball envelopment, and as has been stated earlier, these are predominantly radiant heat sources. The amount of radiation contained in the current convective heat tests differs substantially from that experienced in real exposures. It would seem that radiant heat tests are more relevant for clothing specifications for such environments.

The importance of these tests should be seen in its proper perspective. Legislators and designers of protective clothing tend to be aroused by the unpleasant aspects of heat exposure accidents and do not pay sufficient attention to the less newsworthy but far more prevalent injuries associated with heat stress and heart strain. Requirements such as the ability of turnout coat materials to withstand 300 s of exposure to 260°C temperatures [*18*] are unrealistic and suggest, as has been indicated by others [*19*], that the intention is to exclude all but highly specialized and expensive materials when a whole range of other products are satisfactory.

Of all the injuries experienced by fire fighters in the United States [*9*], 7.6% of those reported in 1981 were burns and only 25% of these were to parts of the body covered by clothing. The 1982 U.S. fire fighter death statistics, on the other hand, indicate that only 2.6% of the total were due to burns alone, whereas 46.1% were the result of heart attacks [*20*]. While not all heart attack deaths can be attributed solely to thermal stress, the numbers serve to demonstrate the relative importance of burn protection.

A fire fighter enveloped in a wildfire for periods of more than a few seconds would need very thick and bulky protective clothing to survive, and if the exposure was for tens of seconds, self-contained breathing and cooling systems would also be required. Since it is not uncommon for the period of duty to extend into many hours on the fire line during an Australian summer (which is itself a substantial heat load), the chances of a fully protected fire fighter surviving heat exhaustion for periods long enough to be effective in fighting fires in Australia are remote.

There is a need to introduce some form of assessment into protective clothing standards to take account of the nature of the working environment and the level of wearer activity. This would oblige designers to give active consideration to comfort aspects such as ventilation and water shedding capabilities. The metabolic penalty of such things as impermeable vapor barriers in fire fighters' turnout clothing has already been demonstrated [*21*], and it would seem that more attention should be devoted to the incorporation of adequate means of dissipating metabolic heat.

Conclusion

More careful attention needs to be given to standards for evaluating the protection afforded by textile materials from exposure to intense heat and to performance specifications based on these test methods. Test results are of no value unless they can be interpreted in relation to the end use of the product. Current standards do not pay sufficient attention to the nature of the working environment, nor do they properly relate to the nature of the risks involved. It is perhaps regrettable that a test designed to assess a specific hazard, that of escaping from a burning aircraft, and published as a useful research and development tool is becoming widely applied to materials intended for other end uses. Consideration should be given to the proper development of a test to simulate intense radiant heat exposure, which is far more applicable to most flame exposure incidents than current convective heat tests.

In setting performance requirements for protective clothing, it is essential that the relative risks are properly balanced and that undue emphasis is not placed on aspects that injury statistics suggest are not warranted. Most explosions and flashovers are of very short duration, and adequate protection can be provided by a complete covering of lightweight materials. At the opposite extreme, persons caught in burning forests or trapped inside burning buildings may need protection for minutes against heat flux intensities of the order of hundreds of kilowatts per square metre. No amount of protection will enable them to survive, and more emphasis should be placed on training people to avoid these situations.

It is imperative that the protective system be able to dissipate the anticipated metabolic load of the wearer during normal usage (that is, other than emergency exposure), as the combination of heavy exertion and a substantial external heat load can lead to heat stress and impaired judgment, and thence to more serious consequences.

Unduly restrictive specifications may result in considerable overexpenditure on protective clothing without any improvement in safety. The use of a parameter such as second-degree burn time and pain time in technical marketing may unwittingly mislead consumers who lack a proper appreciation of the test method.

References

[1] Hausman, A., "Vetements Antiflamme pour Sauveteurs," *Annales des Mines de Belgique*, Part 5, 1978, pp. 537-580.
[2] Hoschke, B. N., Moulen, A. W., Holcombe, B. V., Grubits, S. J., and Madden, J. J., *Journal of Consumer Product Flammability*, Vol. 7, 1980, pp. 48-58.
[3] Stanton, R. M., "Heat Transfer and Flammability of Fibrous Materials," Technical Report AFML-TR-70-238, Air Force Materials Laboratory, Wright-Patterson Air Force Base, OH, Feb. 1971.
[4] Elkins, W. and Thompson, J. G., "Instrumented Thermal Mannikin," NTIS Report No. AD-781 176, Air Force Systems Command, Wright-Patterson Air Force Base, OH, 1973.
[5] McArthur, A. G. and Cheney, N. P., *Australian Forest Research*, Vol. 2, No. 3, 1966, pp. 36-45.
[6] Packham, D. and Pompe, A., *Australian Forest Research*, Vol. 5, No. 3, 1971, pp. 1-8.
[7] Stoll, A. M. and Chianta, M. A., *Aerospace Medicine*, Vol. 41, 1969, pp. 1232-1238.
[8] Holcombe, B. V., *Fire Safety Journal*, Vol. 6, 1983, pp. 129-141.
[9] Veghte, J. H., *Fire Service To-Day*, Aug. 1983, pp. 16-19.
[10] Holcombe, B. V. and Plante, A. M., "Results for an Interlaboratory Trial of a Method for Determining the Resistance of Protective Clothing Materials to a High Intensity Convective Heat Source," Report N50 to ISO/TC94/SC9/WG1, International Organization for Standardization Meeting, Zurich, Switzerland, April 1983.
[11] Braun, E., Cobble, V. B., Krasny, J. F., and Peacock, R., "Development of a Proposed Flammability Standard for Commercial Transport Flight Attendant Uniforms," U.S. Department of Transportation, Federal Aviation Administration, Report No. FAA-RD-75-176, Washington, DC, 1976.
[12] Behnke, W. P., *Fire Technology*, Vol. 13, No. 1, Feb. 1977, pp. 6-12.
[13] Hoschke, B. N., Holcombe, B. V., and Plante, A. M., "A Critical Appraisal of Test Methods for Thermal Protective Clothing Fabrics," in this publication, pp. 311-326.
[14] Smissaert, L., *Proceedings*, International Conference on Protective Clothing Systems, Stockholm, Sweden, 1981, pp. 267-280.
[15] Maggio, R. C., "A Moulded Skin Simulant Material with Thermal and Optical Constants Approximating Those of Human Skin," Laboratory Project 5046-3, Part 105, Naval Material Laboratory, Brooklyn, NY, 1956.
[16] Henriques, F. C., *Archives of Pathology*, Vol. 43, 1947, pp. 489-502.
[17] Stoll, A. M. and Greene, L. C., *Journal of Applied Physiology*, Vol. 14, 1959, pp. 373-382.
[18] "Protective Clothing for Structural Fire Fighting 1975," NFPA No. 1971-1975, National Fire Protection Association, Boston, MA, Nov. 1975.
[19] Utech, H. P., "Are Federal Standards Adequate, Occupations at High Burn Risk: A Summary Report," NIOSH Contract No. 210-77-0218, National Institute of Occupational Safety and Health, Miami, FL, April 1977.
[20] Washburn, A. E., Harlow, D. W., and Fahy, R. F., *Fire Service To-Day*, May 1983, pp. 15-20.
[21] Reischl, U. and Stransky, A., *Textile Research Journal*, Vol. 50, 1980, pp. 643-647.

Meredith M. Schoppee,[1] *Judith M. Welsford,*[1] *and Norman J. Abbott*[1]

Protection Offered by Lightweight Clothing Materials to the Heat of a Fire

REFERENCE: Schoppe, M. M., Welsford, J. M., and Abbott, N. J., **"Protection Offered by Lightweight Clothing Materials to the Heat of a Fire,"** *Performance of Protective Clothing, ASTM STP 900*, R. L. Barker and G. C. Coletta, Eds., American Society for Testing and Materials, Philadelphia, 1986, pp. 340–357.

ABSTRACT: Factors affecting the thermal performance of protective clothing are discussed in a general way, and a laboratory method of achieving heat absorption rates typical of those occurring during exposure to a large fire is described. Using this method, the strength retention of fabrics during short-term exposure at high heat flux levels has been found to depend on the temperature achieved at a given instant during exposure and to be independent of the mechanism of heat absorption. A comparison of the duration of exposure to high heat flux levels that causes various polymeric fabrics to lose most of their original strength or to autoignite predicts that such high-temperature materials as polybenzimidazole (PBI) and Nomex/Kevlar can provide a few extra seconds of protection against the extreme heat of a large fire.

Some difficulties associated with the use of an instrumented skin-simulant device for determining the rate of conductive heat transfer through fabrics of various kinds during standard gas flame impingement tests are also discussed. There is, at present, no accurate way to translate temperatures measured at a depth of 500 μm in a fabric-covered skin simulant to temperatures appropriate for a depth of 80 μm, the skin depth to which clinical data regarding tissue damage are related. Until the conductive heat flow equation can be revised, ranking of fabrics by the maximum temperature achieved in the skin simulant remains the only reliable method of using data from this device.

KEY WORDS: protective clothing, fire, radiant heat, strength retention, ignition time, heat conduction, flame impingement, skin simulant

Much has been written about the protective performance of various clothing materials in situations in which they are called upon to protect the wearer from the heat of a fire, and many thermal properties of the fabrics from which

[1]Senior research mathematician, research assistant, and associate director, respectively, Albany International Research Co., Dedham, MA 02026.

such clothing is made have been measured in the laboratory. However, since the degree of protection offered depends on a complex combination of properties of both the thermal environment and the material exposed, the total risk situation must be taken into account when assessing the overall protection potential of lightweight clothing items. The three factors that control performance are the total amount of thermal energy absorbed by the outer surface of a protective garment, the increase in temperature that results from this energy absorption and its effect on the properties of the material from which the garment is made, and the rate at which heat is transferred from the outer absorbing layer to the inner surface of the garment and from there to underlying tissue.

The net thermal energy absorbed by a clothing layer per unit of area and per unit of time depends primarily on the level of heat flux incident on it from the thermal source and on the respective magnitudes of the radiative and convective components. The absorption of radiant energy as a fraction of the incident flux is determined by the radiative properties of the surface layer—its absorptance, reflectance, and transmittance; each of these properties, in turn, is affected by the temperature and surface condition of the exposed layer and by the wavelength distribution of radiation from the source. Convective heat transfer from a moving gas stream, such as the turbulent flow within and around a large fire, depends on the temperature, pressure, and velocity of the moving gases and on the type of boundary layer present at the garment surface.

Only energy that is absorbed can cause the temperature of an exposed material to increase; this increase occurs at a rate determined by the net rate of heat absorption and by the specific heat capacity of the material. The ultimate temperature achieved and its effect on the molecular structure of the material controls such thermal behavior as loss of strength and integrity, melting, charring, and ignition.

Heat is transferred from the outer, exposed layer of a protective garment to its inner surface principally by conduction at a rate that is directly dependent on the rate at which heat is absorbed; the rate of transfer is also governed by the thickness, density, thermal conductivity, and specific heat capacity of the clothing layer. The mechanism and rate of subsequent heat transfer to the body depends on the extent of clothing contact with the skin. If air layers that separate clothing layers from each other or from the body are not diminished by fabric shrinkage or body movement during exposure, then radiative interchange, governed by the temperature and emissivity of the inner fabric surface, becomes important in determining the amount of heat transferred to the skin. If, on the other hand, skin and clothing are in intimate contact, heat conduction proceeds at a rate influenced by the thermal properties of body tissue.

It is obviously important that any attempt to model exposure to a fire and its effect on protective clothing on a small scale must be capable of inducing energy absorption at the same rate that would occur during an actual fire. The mechanism of absorption is not important, however; only the rate at which absorption occurs determines both the rate of temperature increase and the rate of subsequent conductive transfer.

This paper describes a laboratory method of determining the effects of net energy absorption at a known rate on material behavior, including strength loss during exposure, melting, and ignition, and it further discusses shortcomings in the methods used to translate heat transfer data into estimates of burn injury potential. The heat source used to examine the effects of energy absorption on material properties is a radiant source capable of producing the extreme heat flux levels associated with a large fire while also duplicating the spectral distribution of this energy source fairly well. Heat transfer data were collected from standard gas flame impingement tests using an instrumented skin-simulant material as the sensing element. Those fabric materials studied include polybenzimidazole (PBI), Nomex/Kevlar (aramid), Kynol (phenolic), untreated cotton, and polyester.

Modeling the Thermal Environment of a Large Fire

Large fueled fires behave essentially as blackbody radiators with average internal temperatures in the range of 860 to 1150°C, corresponding to radiant heat flux levels between 90 and 230 kW/m^2 (2.2 to 5.6 cal/cm^2/s) [*1*]. See Fig. 1 for a comparison with other heat sources. The dominant wavelength of radiation emitted, as computed from Planck's law, is between 2 and 2.5 μm, a region of the spectrum in which polymeric materials are highly absorbing. It is estimated that a person caught in such a fire has between 3 and 10 s to escape before he succumbs [*2*].

Laboratory heat sources capable of duplicating these high flux levels, such as quartz or tungsten lamps, generally must operate at considerably higher temperatures than those in a fire because they emit less efficiently than blackbodies. Because of this, the dominant emitted wavelength is shifted consider-

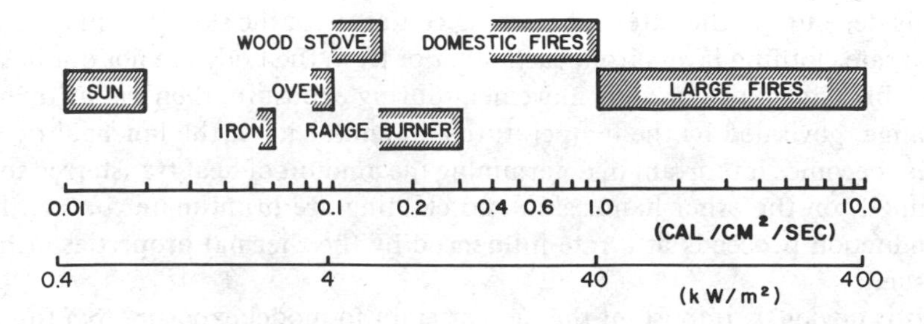

FIG. 1—*Radiant heat flux emitted by various sources.*

ably from the infrared to the visible region of the spectrum, a region in which radiative absorption by polymeric materials is both low and very wavelength dependent [*1,3*]. However, it is possible to attain the appropriate flux levels and wavelength distribution in laboratory exposures by using two quartz-faced heater panels as a facing pair, as illustrated in Fig. 2; although neither surface, used singly, emits with the efficiency of a blackbody, the pair establish a thermal environment between them that can successfully duplicate the radiant heat environment of a large fire. The bilateral configuration has the further advantage of simplifying the heat flow equation by eliminating heat losses from the exposed specimen to the surroundings, since both sides are irradiated equally; in addition, the equilibrium temperature is well defined by the wall temperature of the surrounding heater panels. As a result, the *net* heat absorbed by a fabric specimen and its rate of temperature rise can be well defined during exposure, and measured properties can be related to actual thermal conditions within the exposed material.

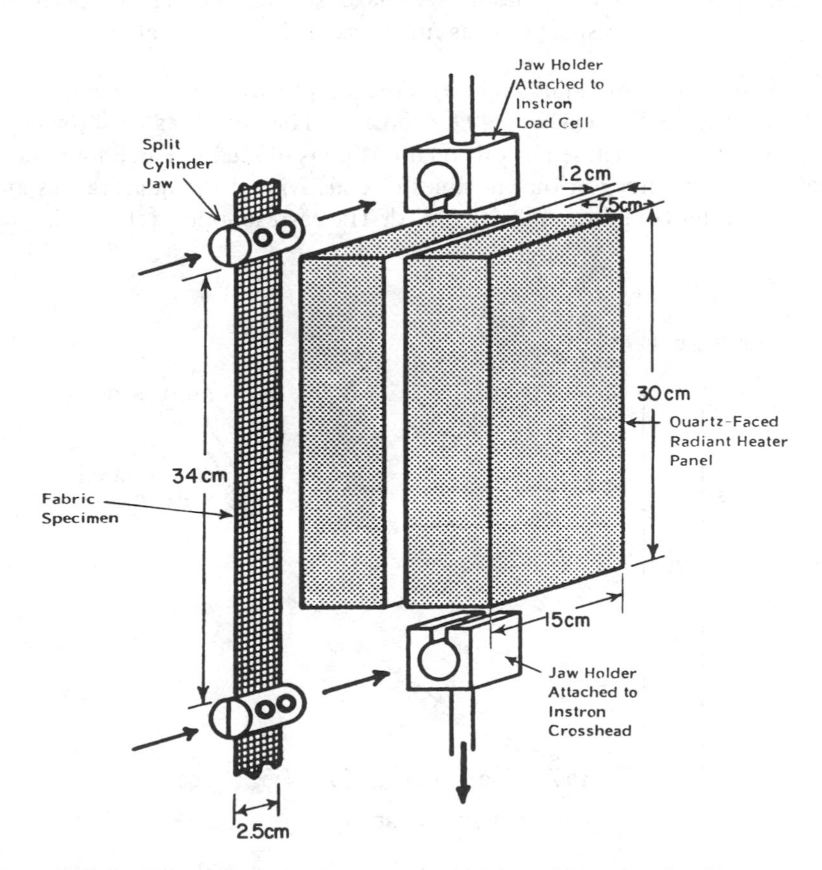

FIG. 2—*Test configuration for exposure of fabrics to bilateral radiant heat.*

The bilateral radiation interchange between the heater surfaces at equilibrium temperature and the exposed fabric strip is described by the following equation [*4,5*]

$$Q = \frac{2\sigma(T_1^4 - T_2^4)}{\dfrac{1}{\epsilon_1(T_1)} + \dfrac{1}{\epsilon_2(T_2)} - 1} \tag{1}$$

where

Q = amount of heat exchanged per unit of time and per unit of area,
σ = the Stefan-Boltzmann constant,
T_1, T_2 = temperature, degrees K, of the quartz faces and specimen, respectively, and
$\epsilon_1(T_1)$, $\epsilon_2(T_2)$ = emissivities of the quartz surfaces and of the specimen, respectively, as functions of their temperature.

It is obvious from this equation that as the temperature of the specimen, T_2, increases, the net heat exchange, Q, decreases. The *initial* heat absorbed by a fabric specimen as defined by this relationship is plotted in Fig. 3 for various heater temperatures for known values of emissivity of the quartz faces and literature values of fabric emissivity [*1,3*]. The corresponding relationship for

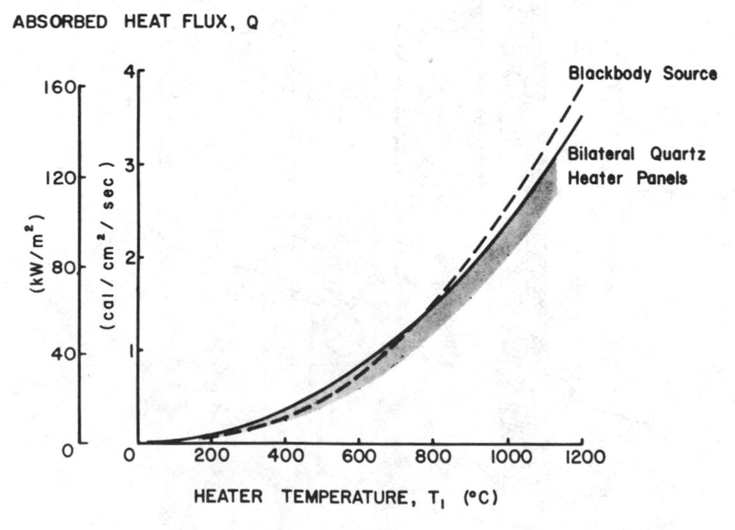

FIG. 3— *Variation of the initial heat flux absorbed by polymeric fabrics with the temperature of the radiant source.*

a blackbody enclosure is also shown in Fig. 3 to illustrate the general level of similarity between the laboratory situation and that of a large fire. The dominant wavelengths emitted in the two environments at different temperatures [*4,6*] are compared in Fig. 4. The variation of the absorptive efficiency of polymeric fabrics with temperature in general is shown in Fig. 5 [*1,3*].

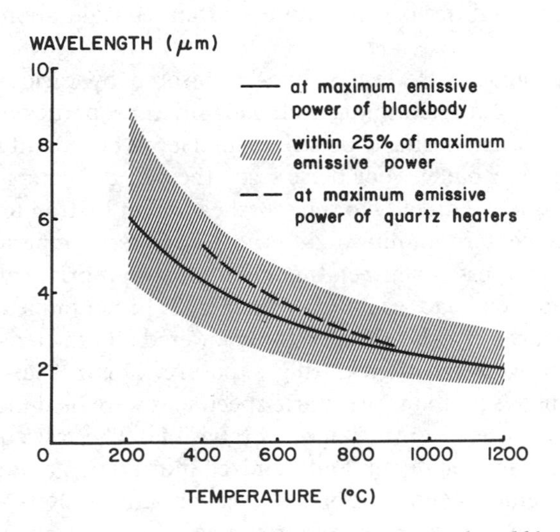

FIG. 4—*Dominant wavelength of radiation emitted by quartz panels and blackbody radiator.*

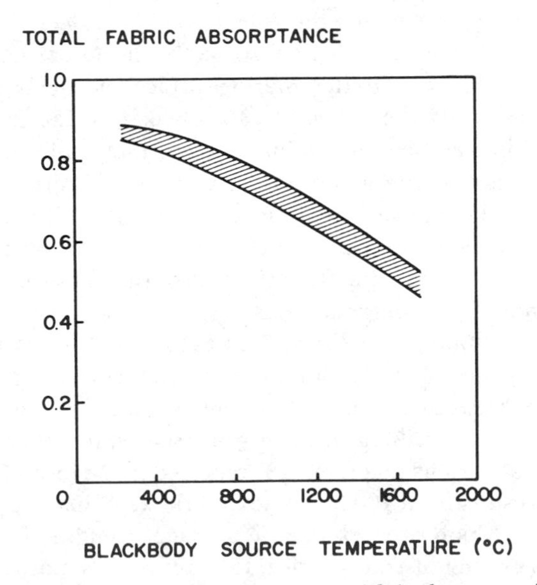

FIG. 5—*Effect of the source temperature on fabric absorptance [*1,3*].*

Strength Retention and Ignition During Exposure

While running to escape the vicinity of a fire, an active wearer imposes mechanical stresses and strains on the fabric of his clothing. If the fabric becomes so weakened during exposure that it loses its integrity under slight stressing, it can no longer provide the cover necessary to protect the wearer from the heat of the flame. Therefore, the effect of increased temperature on the strength of the fabrics becomes an important consideration in its protective ability.

During the course of several studies sponsored by various government agencies, the strength retention as well as ignition properties of many fabric types were determined during exposure to high levels of radiant heat provided by the quartz heater panels. For these tests, the pair of heaters was mounted in a chamber attached directly to the crosshead of an Instron tensile-test machine. With the heaters equilibrated to the desired test temperature, a fabric strip was quickly inserted between them and into appropriate attachments to the load cell and crosshead, as indicated in Fig. 2. Either immediately or after a specified time, the crosshead was rapidly lowered and the tensile properties of the specimen were recorded during exposure. For measurements of the time to spontaneous ignition, the fabric specimens were held in place with no motion of the crosshead until flames were visible. The test equipment and methods are described in detail, and itemized information obtained from the strength measurements and ignition tests are contained in Refs *4, 7, 8,* and *9*; only general highlights will be summarized here.

During our initial experiments, it quickly became apparent that even such heat-resistant materials as PBI, Kynol, and Nomex ignite instantaneously at *absorbed* flux levels approaching those *incident* in a large fire (80 kW/m^2; 2 $cal/cm^2/s$). In a fire some of the heat absorbed at the outer surface of a clothing layer is conducted away, to the interior surface and the body, which acts as a heat sink; as a result, the *net* energy absorbed at the fabric surface is only a fraction of the incident energy. While one advantage of our experimental setup is that it permits definition of property changes in terms of *net* absorption by eliminating heat losses from the fabric, it also necessitates looking at the effect of lower incident flux levels on fabric mechanical properties and the tendency to ignite in order to maintain the appropriate absorption rate and to be able to differentiate among materials.

The strength retention properties and times to ignition of various fabrics of interest have been documented at initial absorbed flux levels ranging between 7 and 54 kW/m^2 (between 0.2 and 1.3 $cal/cm^2/s$) for times of exposure from a few seconds to 1 or 2 min. Measurements of strength retention as a function of exposure time at various flux levels are reported in the format shown in Fig. 6, in which the results for a 100% Nomex fabric weighing 136 g/m^2 (4.0 oz/yd^2) are given; each point represents the average strength of several broken specimens. This graph illustrates typical features representative of the behav-

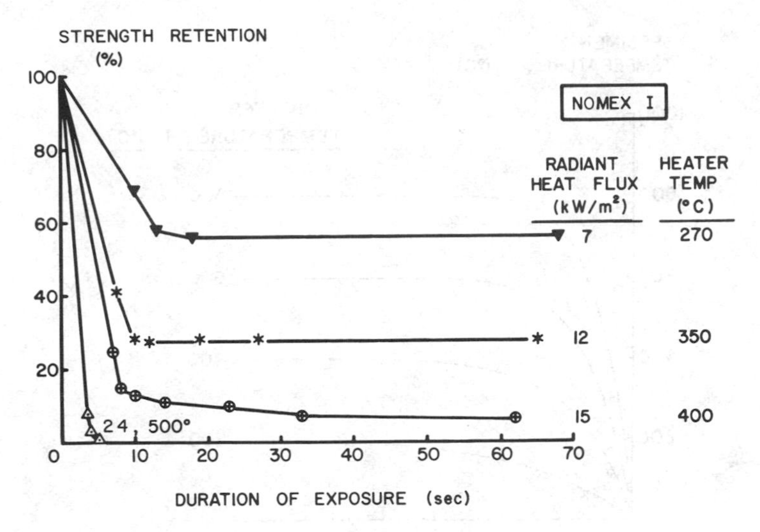

FIG. 6—*Strength retention of Nomex fabric (136 g/m²) at various times during radiant exposure.*

ior of many of the materials tested, such as the rapid initial strength loss during the first few seconds of exposure followed by a transition region (at the lower flux levels) to a more or less steady level of strength retention for exposure times up to a minute. The shape of these strength-retention curves is consistent with the shape of temperature rise-time curves for the exposed specimen, as calculated from the following relationship plotted in Fig. 7

$$\frac{dT_2}{dt} = \frac{Q(T_1, T_2)}{(W/A)Cp(T_2)} \tag{2}$$

where

T_2 = specimen temperature,
t = exposure time,
$Q(T_1, T_2)$ = absorbed heat flux as a function of heater and specimen temperature, respectively,
W/A = weight per unit of area of the exposed specimen, and
$Cp(T_2)$ = the specific heat capacity of the exposed material as a function of its temperature.

In fact, experiments carried out in a circulating hot-air oven with a variety of materials showed generally excellent agreement, as documented in Fig. 8, between the strength retained in the oven and that retained during radiant exposure when compared at equilibrium temperature; such differences as do exist are probably related to the greater availability of oxygen during the radi-

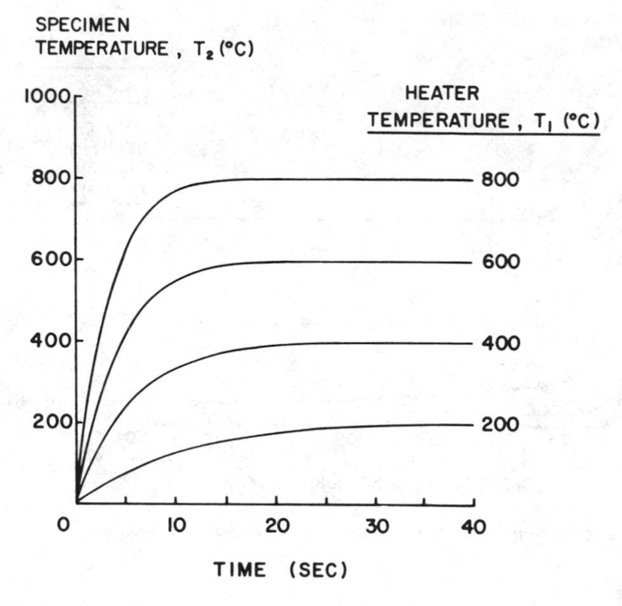

FIG. 7—*Calculated specimen temperature as a function of time during radiant exposure (fabric weight, 200 g/m²).*

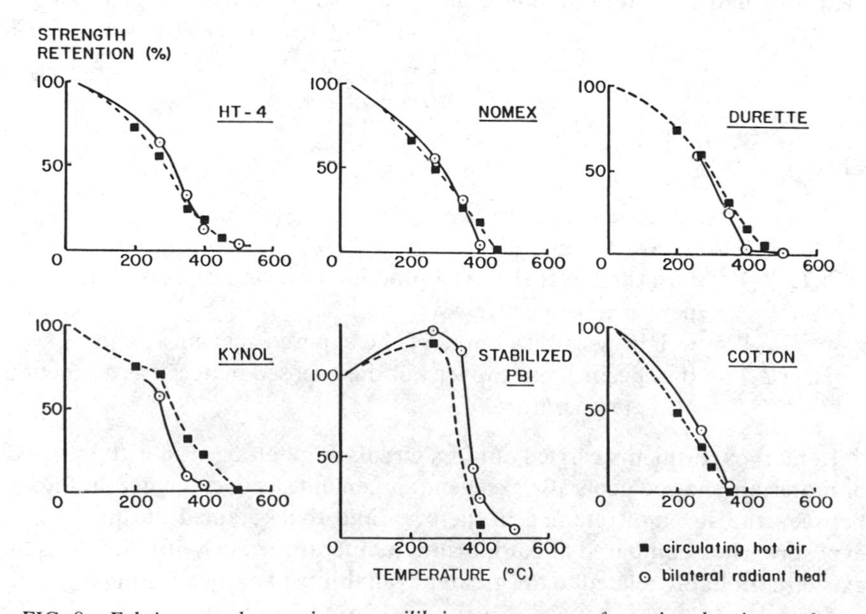

FIG. 8—*Fabric strength retention at equilibrium temperature for various heating modes.*

ant heat exposure. (A natural convective air stream flows between the panels [4].) In addition, when the measured values of strength retention of a PBI and Nomex/Kevlar fabric determined at various times of radiant exposure from a few seconds to 1 min [7] are plotted, as in Fig. 9, at the temperatures achieved at the time of the test (calculated on the basis of Eq 2), the close grouping of data points confirms the correspondence between strength retention and specimen temperature during exposure even under nonequilibrium conditions. It seems reasonable to conclude from this evidence that if the temperature of a fabric is known or can be calculated reasonably well at any moment during its term of exposure, then its strength at that time can be closely predicted from known strength-temperature profiles such as those shown in Fig. 10. We may also conclude that those aspects of material behavior that are primarily temperature dependent are independent of the mechanism of heat absorption; only the amount of heat absorbed is important.

Strength loss during exposure to extreme heat, or any other temperature-related property change that is detrimental to the performance of a protective fabric, can be delayed by increasing the fabric weight per unit of area, be-

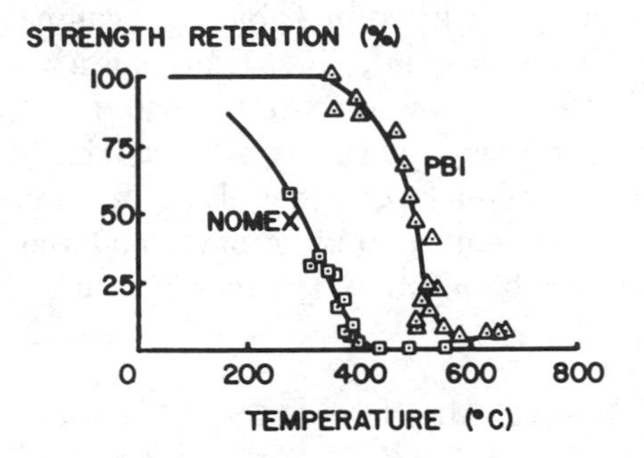

FIG. 9—*Strength retention of Nomex and PBI fabrics at calculated transient temperatures during radiant exposure.*

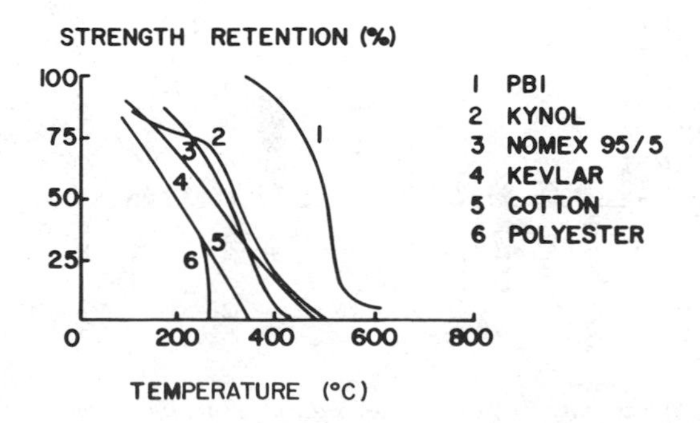

FIG. 10—*Strength retention of various fabrics at various temperatures.*

cause the rate of temperature rise is inversely proportional to weight (see Eq 2). Any comparison of properties on a time basis that hopes to show differences in the intrinsic behavior of different materials ideally should be made with fabrics of the same weight; however, since this is not always feasible, some adjustments can be made to the time scale against which results are measured. For example, in Fig. 11 the times required for fabrics of several different material types to lose 90% of their original strength are compared at two absorbed heat flux levels. The actual weights of these particular fabrics, given in Table 1, range from 150 to 220 g/m^2 (4.5 to 6.5 oz/yd^2), but the results given in Fig. 11 are normalized to a fabric weight of 200 g/m^2 (6.0 oz/yd^2). This normalization is carried out by multiplying the average measured times to 90% strength loss by the ratio of 200 to the actual weight of the fabric in g/m^2. The results in Fig. 11 show that PBI fabric retains some mechanical strength for considerably longer times at flux levels of 25 and 30 kW/m^2 (0.6 and 0.7 cal/cm^2/s) than Nomex/Kevlar, Kynol, cotton, or polyester. However, at higher fluxes, about 50 kW/m^2 (1.2 cal/cm^2/s), even PBI loses all strength within 5 s [7].

During our testing with the radiant heater panels, the loss of all mechanical strength was usually followed closely by ignition of the specimen. Times to ignition for those fabric types listed in Table 1 are compared in Fig. 12, also normalized to a weight of 200 g/m^2. Since ignition is a path-dependent event that depends not only on the temperature reached during exposure but also on the rate of heating and the extent of oxygen availability, the actual time to ignition given in Fig. 12 should be considered specific to the particular testing conditions. However, the relative rankings of the materials are probably generally valid and show that both PBI and Nomex/Kevlar resist ignition for con-

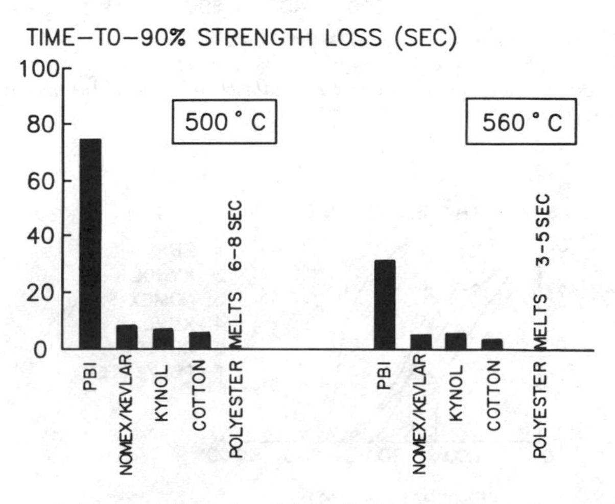

FIG. 11—*Strength loss times for various fabrics during radiant exposure: (a) 25 kW/m^2; (b) 30 kW/m^2.*

TABLE 1—*Description of the fabrics compared in Figs. 10, 11, and 12.*

Material	Weight g/m^2	Weight oz/yd^2	Weave Type
PBI	153	4.5	plain
Kynol	159	4.7	2 by 1 twill
95/5 Nomex/Kevlar	180	5.3	cloque
Cotton	217	6.4	oxford
Polyester	220	6.5	plain

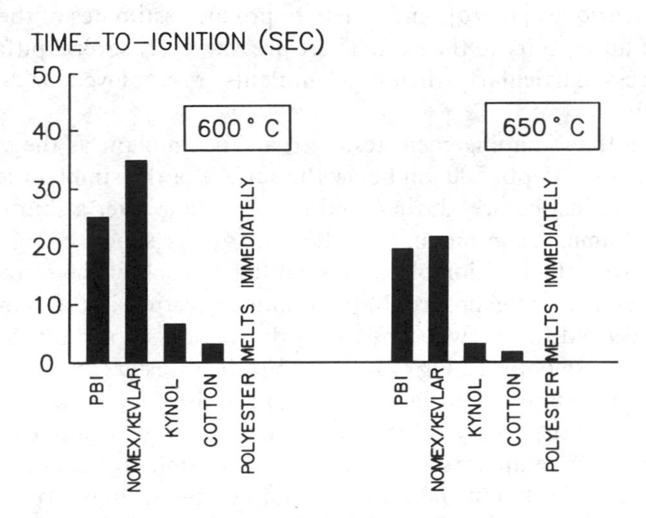

FIG. 12—*Ignition times for various fabrics during radiant exposure: (a) 34 kW/m^2; (b) 42 kW/m^2.*

siderably longer than either Kynol or cotton at fluxes to 42 kW/m^2 (1.0 cal/cm^2/s); polyester fabric melts immediately at this heat flux.

No lightweight polymeric fabric may be considered to be nonflammable; all will ignite if the rate of heat absorption is great enough. But clothing made from high-temperature-resistant materials such as PBI and Nomex/Kevlar can provide a few extra precious seconds of protection time to escape from the fire vicinity.

Heat Transfer and Estimates of Burn Injury

During the period of exposure before integrity is lost or ignition occurs, the most important aspect of protection from burn injury is the ability of clothing to retard the conductive transfer of heat inward to the wearer's body. In this aspect of performance, the high-temperature polymeric materials have nothing out-of-the ordinary to offer. Since neither the absorption rate, the thermal conductivity, nor the specific heat differs greatly from fabric to fabric, protection from conducted heat must rely principally on providing sufficient thickness to retard heat flow to the body.

Heat transfer characteristics of candidate fabrics for protective clothing are commonly measured in the laboratory during impingement by a gas flame using either a calorimeter [ASTM Test for Thermal Protective Performance of Materials for Clothing by Open-Flame Method (D 4108-82)] or an instrumented skin-simulant device [*10,11*] to record the passage of heat through the test specimen. The results from such tests are then generally used in conjunction with tissue damage rate information gathered by Derkson [*12*], Stoll [*13,14*], Henriques [*15,16*], and others to produce estimates of the potential severity of burn injury to the skin. There are, however, several pitfalls in this latter process, particularly when skin simulants are employed, as discussed in the following.

During a flame impingement test with a skin simulant as the sensor, the temperature at a depth 500 μm below the surface of the simulant is recorded for the period just before, during, and for some time after a shuttered exposure to the flame. A continuous record, such as those shown in Fig. 13, *a* and *b*, is produced; the position of the dashed line in each of these traces represents the cessation of exposure. Maximum temperature in the skin simulant is always reached some few seconds after the flame is turned off. Such traces are not always smooth, as Fig. 13*b* for a 50/50 polyester/cotton fabric illustrates; the momentary delay in the temperature rise curve in this case probably results from the highly endothermic melting of the polyester fraction.

The maximum temperatures reached in a skin-simulant device located behind a variety of different fabric types and outerwear/underwear combinations have been measured during flame impingements of 3 and 6-s duration at an incident flux of 92 kW/m^2 (2.2 cal/cm^2/s) [*8*]. The relationship between these measured results in terms of temperature increase and fabric thickness is illustrated in Fig. 14, *a* and *b*. These data represent worst-case situations, since the fabric specimens were held tightly against the skin simulant during exposure. For this group of fabrics, which ranges in total weight between 100 and 460 g/m^2 (3 and 13.5 oz/yd^2) and in thickness between 2 and 15 mm, the lowest temperature rise recorded during the 3-s exposures was 11.8°C, 18.3°C during the 6-s exposures. Tissue damage accrues all of the time the basal layer of skin, 80 to 100 μm below the surface, is at a temperature greater than 44°C, 11.5 degrees Celsius above the normal skin temperature of 32.5°C. The damage rate increases logarithmically, so that at a skin temperature of 72°C, 39.5 degrees Celsius above normal, tissue destruction occurs instantaneously [*14*]. The severity of the laboratory flame exposure can be judged by the fact that, within the group of fabrics tested, all transmitted sufficient heat, even during a 3-s exposure, to raise the temperature in the skin simulant above the damage temperature; in more than half of the 6-s exposures, the temperatures in the simulant increased beyond the level at which instantaneous destruction occurs. These increases were measured in

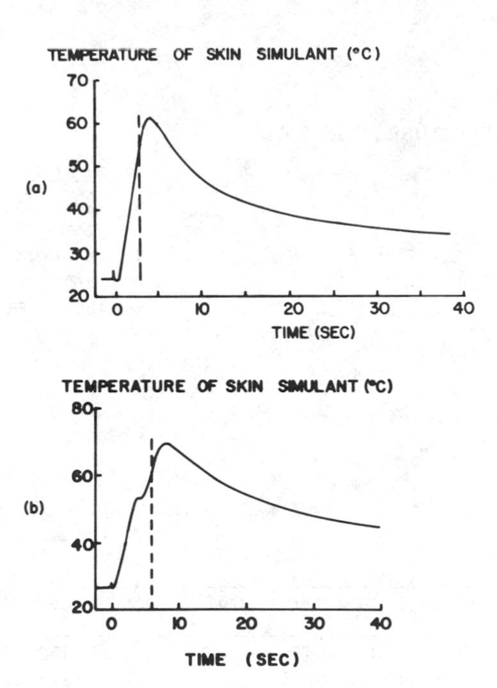

FIG. 13—*Temperature in the fabric-covered skin simulant during gas flame exposure at 92 kW/m²: (*a*) Nomex/Kevlar fabric, 156 g/m², at a 3-s exposure; (*b*) 50/50 polyester/cotton fabric, 234 g/m², at a 6-s exposure.*

the skin simulant well beyond the 80 to 100 μm depth to which the clinical data are keyed, a depth at which temperatures would have been considerably higher than those measured at 500 μm.

Stoll [*17*] suggests that for exposures from 5 to 20 s, a constant ratio of about 0.8 or 0.9 between the temperature measured at 500 μm in the skin simulant and the temperature of the basal layer of skin at 80 μm may be used to convert skin-simulant temperatures to skin temperatures for burn injury estimates. However, using her heat flow equation derived for a contiguous two-layer system [*14*] to calculate the relationship between temperature rise at 80 μm and temperature rise at 500 μm in the skin-simulant material for covering layers of various thicknesses and thermal diffusivities representative of the range of real fabrics properties, it quickly becomes apparent that no single conversion factor is appropriate. For example, as the results of these calculations show in Fig. 15, at a high thermal diffusivity the ratio between temperature at 500 μm and that at 80 μm ranges between 0.05 and 0.4, depending on the thickness of the covering layer, and at a low thermal diffusivity, it ranges between about 0.08 and 0.5. Thus, even for the ideal conditions implicit in the heat flow equation, the ratio of temperature at 500 μm to temperature at 80 μm may differ by nearly an order of magnitude, depending on the nature of the fabric layer covering the skin simulant.

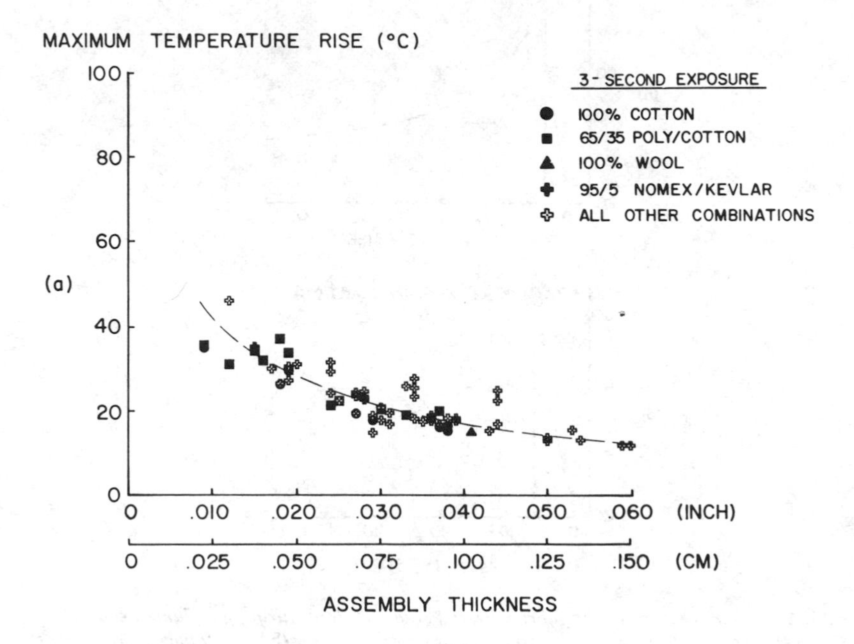

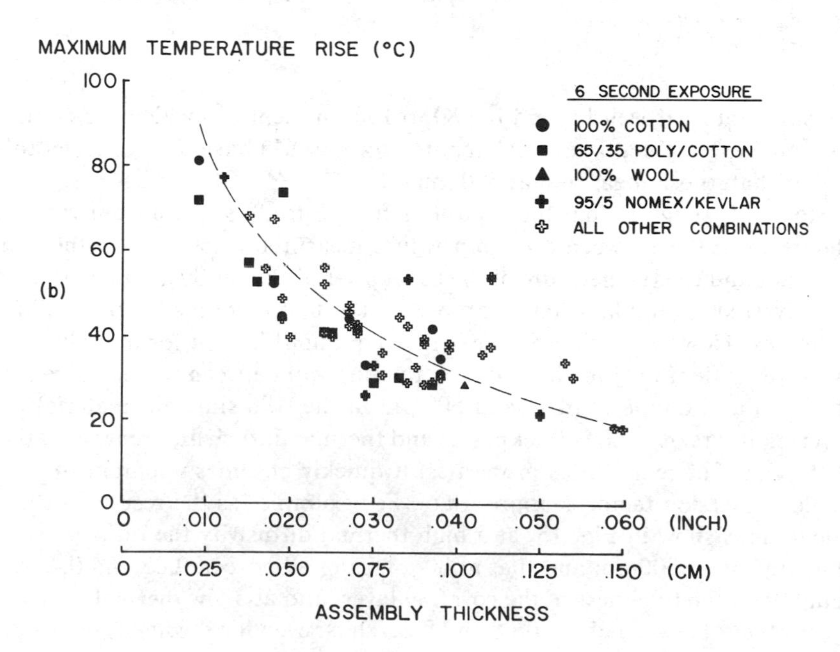

FIG. 14—*Effect of assembly thickness on temperature rise in the fabric-covered skin simulant during gas flame exposure at 92 kW/m²*.

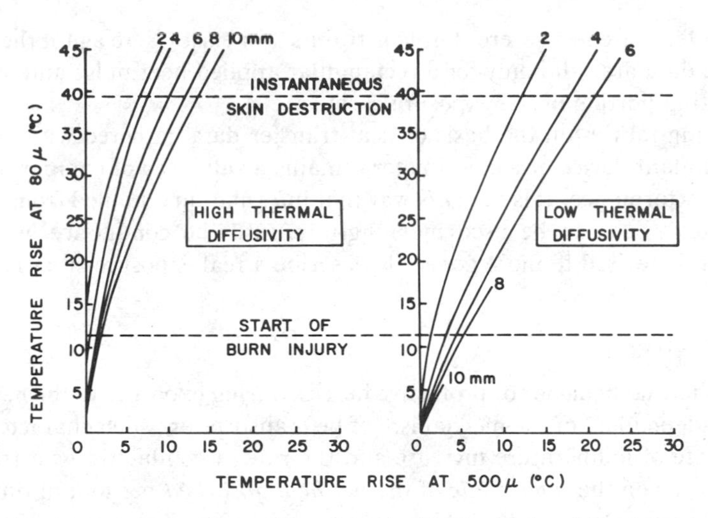

FIG. 15—*Calculated temperature rise at depths in the skin-simulant device for 6-s exposure at 92 kW/m*^{2}.

The use of an instrumented skin simulant to help determine the extent of potential burn injury with various fabric coverings is attractive for the following reasons: the temperature record produced is independent of the shape of the heat pulse; heat transfer during both the heating and the cooling phases of exposure is recorded; and this method holds promise for linking directly with clinical data relating burn injury to skin temperature. Unfortunately, however, there is no way, including the use of Stoll's heat flow equation, to translate accurately the temperature measured in the simulant at 500 μm to the temperature at 80 μm. This equation is valid only for a contiguous two-layer system with components whose properties do not change with increasing temperature and whose surface absorbs heat at a constant rate. In actuality, the density, thermal conductivity, and specific heat capacity of most fabrics do change somewhat with temperature, particularly if a large thermoplastic fraction is included; absorption proceeds at a rate which decreases as the exposed surface temperature increases, or increases markedly if exothermic reactions occur in the fabric as it degrades. In addition, the equation is applicable only while heat is being applied and not during the equally important period after exposure ceases but during which maximum temperature is reached and damage continues to accumulate. Attempts to remedy this latter deficiency by theoretically superimposing a negative pulse on the initial positive pulse at a time equivalent to cessation of exposure are inappropriate because they fail to take into account new boundary conditions at the start of the negative pulse; as a result, this method vastly overestimates the maximum temperature achieved [*8,18*].

If a calorimeter rather than a skin-simulant device is used to monitor the rate of heat transfer through covering fabric layers during flame exposure, the problems are less severe, but constraints still exist. Here again the tissue damage data are valid only for a rectangular applied heat pulse and only for the heating portion of the exposure cycle.

Ranking fabrics on the basis of heat transfer data gathered with either a skin-simulant device or a calorimeter remains a valid way of comparing protective performance. It is the *only* way that information obtained from a skin-simulant device can be used meaningfully until the conductive heat flow equation is revised to more generally describe a real exposure situation.

Conclusions

The thermal behavior of protective fabrics during exposure to the heat of a fire is independent of the mechanism of heat absorption; such characteristics as the rate of temperature increase and the rate of conductive heat transfer depend only on the *magnitude* of the *net heat absorbed* per unit of time and per unit of area. As a consequence, when modeling a fire environment in the laboratory, one need only duplicate the rate of absorption of thermal energy to determine the effect of exposures of different degrees of severity on such fabric properties as strength loss, ignition, and the rate of heat transfer.

Using a pair of quartz-faced radiant heater panels capable of achieving absorbed flux levels to 120 kW/m^2 (3 $cal/cm^2/s$), the short-term strength retention of a variety of fabric types has been shown to depend principally on the temperature reached in the heated material. Fabrics composed of materials that degrade slowly at high temperatures, such as PBI, retain strength and resist ignition longer at high fluxes than those materials, such as cotton or polyester, that degrade rapidly at lower temperatures.

Maximum temperatures measured in a fabric-covered skin-simulant device during gas flame impingement can be used successfully to rank protective fabric resistance to heat transfer and to show the dependence of this property on fabric thickness. However, such temperature measurements cannot be used with any degree of accuracy to predict the extent of burn injury using the Stoll criteria because the measurements made at a depth of 500 μm below the surface of the simulant cannot be translated into temperatures at a depth of 80 μm, the depth in the skin at which blisters form. Significant revision of the conductive heat-flow equation given by Stoll would be necessary before such calculations could be performed accurately.

References

[1] Morse, H. L., Thompson, J. G., Clark, K. J., Green, K. A., and Moyer, C. B., "Analysis of the Thermal Response of Protective Fabrics," Technical Report AFML-TR-73-17, Air Force Materials Laboratory, Dayton, OH, Jan. 1973.

[2] Stanton, R. M., "Heat Transfer and Flammability of Fibrous Materials," Technical Report AFML-TR-70-238, Air Force Materials Laboratory, Dayton, OH, Feb. 1971.

[3] Quintierre, J., "Radiation Characteristics of Fire-Fighters' Coat Fabrics," *Fire Technology*, May 1974, pp. 153-161.

[4] Schoppee, M. M., Skelton, J., and Abbott, N. J., "The Transient Thermomechanical Response of Protective Fabrics to Radiant Heat," Technical Report AFML-TR-77-72, Air Force Materials Laboratory, Dayton, OH, May 1977.

[5] Howell, J. R. and Siegel, R., *Thermal Radiation Heat Transfer*, Vol. II, NASA SP-164, National Aeronautics and Space Administration, Washington, DC, 1969, pp. 73-74.

[6] Sparrow, E. M. and Cess, R. D., *Radiation Heat Transfer*, Brooks/Cole, Belmont, CA, 1966, pp. 6-7.

[7] Schoppee, M. M., "Comparative Performance of T456 Nomex and PBI Fabrics at Extreme Heat Flux Levels," Unpublished report prepared under Contract DAAK60-79-M-2475, U.S. Army Natick Research and Development Command, Natick, MA, Oct. 1979.

[8] Schoppee, M. M., Welsford, J. M., and Abbott, N. J., "Resistance of Navy Shipboard Work Clothing Materials to Extreme Heat," Technical Report No. 148, Navy Clothing and Textile Research Facility, Natick, MA, Oct. 1982.

[9] Schoppee, M. M., Welsford, J. M., and Abbott, N. J., "Resistance to Navy Shipboard Outerwear Garments and Fire-Resistant Fabrics to Extreme Heat," Technical Report No. 153, Navy Clothing and Textile Research Facility, Natick, MA, Dec. 1983.

[10] Stoll, A. M., Chianta, M. A., and Monroe, L. R., "Flame-Contact Studies," *Journal of Heat Transfer*, Paper No. 63-WA-121, 1963, pp. 1-8.

[11] Derksen, W. L., Delhery, G. P., and Monahan, T. I., "Thermal and Optical Properties of the NML Skin Simulant," Report No. DASA 1169, Naval Material Laboratory, Brooklyn, NY, Jan. 1960.

[12] Derksen, W. L., Monahan, T. I., and Delhery, G. P., "The Temperature Associated with Radiant Energy Skin Burns," Vol. 3, Part 3, *Biology and Medicine*, Reinhold, London, 1961, Chapter 16, pp. 171-175.

[13] Stoll, A. M. and Greene, L. C., "Relationship Between Pain and Tissue Damage Due to Thermal Radiation," *Journal of Applied Physiology*, No. 14, 1959, pp. 373-382.

[14] Stoll, A. M. and Chianta, M. A., "Heat Transfer Through Fabrics as Related to Thermal Injury," *Transactions of the New York Academy of Sciences*, Vol. 33, No. 7, Nov. 1971, pp. 649-670.

[15] Henriques, F. C. and Moritz, A. R., "Studies of Thermal Injury, II. The Relative Importance of Time and Surface Temperature in the Causation of Cutaneous Burns," *American Journal of Pathology*, Vol. 23, 1947, p. 695.

[16] Henriques, F. C., "Studies of Thermal Injury: V. The Predictability and Significance of Thermally Induced Rate Processes Leading to Irreversible Epidermal Injury," *Archives of Pathology*, Vol. 43, 1947, p. 489.

[17] Stoll, A. M. and Chianta, M. A., "Method and Rating System for Evaluation of Thermal Protection," *Aerospace Medicine*, Nov. 1969, pp. 1232-1238.

[18] Claus, W. D., Jr., "Heat Conduction in a Two-Layer System with Application to Heat Transfer to the Skin," *Journal of Fire and Flammability*, Vol. 4, Jan. 1973, pp. 52-55.

Itzhak Shalev[1] *and Roger L. Barker*[2]

Predicting the Thermal Protective Performance of Heat-Protective Fabrics from Basic Properties

REFERENCE: Shalev, I. and Barker, R. L., **"Predicting the Thermal Protective Performance of Heat-Protective Fabrics from Basic Properties,"** *Performance of Protective Clothing, ASTM STP 900,* R. L. Barker and G. C. Coletta, Eds., American Society for Testing and Materials, Philadelphia, 1986, pp. 358–375.

ABSTRACT: Novel experimental techniques were developed to measure changes in the weight, thickness, density, heat capacity, heat conductivity, and infrared (IR) transmission of protective fabrics occurring during a thermal protective performance (TPP) test. Comparisons are made between polybenzimidazole (PBI), aramid, a PBI/aramid blend fabric, and flame-retardant (FR) cotton fabrics in the 250 g/m^2 (7.5-oz/yd^2) weight range. This research analyzes changes in fabric heat transfer properties produced through mechanisms of pyrolysis, char formation, and shrinkage. Fiber character is shown to play a decisive role in determining the direction and extent of change in thermophysical properties. Retention of air volume is found to be critical to prolonged thermal protection performance. Experimental data indicate that air and fiber conduction dominate in intense exposures to a mixture of radiant and convective thermal energy; direct radiant transmission is not an important contributor to the total heat transferred in these exposures. The ability of fabrics to maintain surface fibers is thought to have significant impact in blocking convective heat transmission. The degradation behaviors of different materials are compared and related to their thermal protective performance.

KEY WORDS: thermal protection, thermal protection performance, protective clothing, heat transfer, thermal injury, high performance fabrics

One of the most important properties of fabrics intended for use in protective garments is the ability to block heat transfer and reduce burn injury potential. Protection against radiant and convective exposures can be measured in the laboratory using a thermal protective performance (TPP) test. These tests use Stoll criteria to determine a protective index from heat flux measurements [*1,2*]. However, routine analyses by these methods, while highly useful

[1]Assistant professor, Shenkar College of Textile Technology and Fashion, Ramat-Gan 52526, Israel.

[2]Associate professor, School of Textiles, North Carolina State University, Raleigh, NC 27650.

for comparative studies, provide limited information on the specialized role of a particular fiber or flame-retardant (FR) finish. There is a need, therefore, to develop a more fundamental understanding of TPP tests as they are used to evaluate protective fabrics. Such knowledge would be of immediate practical value since it would permit a clearer interpretation of standard test results and make the TPP procedure a more powerful tool for laboratory analysis.

Protective fabrics, made with different fibers or FR finishes, display characteristic nonlinear calorimetric traces in TPP tests [*3*]. The slope of the temperature–time curve varies during the exposure because the heat exposure produces fundamental changes in fabric thermal and spatial properties through mechanisms of pyrolysis, char formation, and shrinkage. This research measures these changes in fabric properties by stopping a TPP test at predetermined time intervals and characterizing the exposed specimens. The goal is to investigate the properties controlling heat transfer (for example, unit mass, thickness, density, and optical and thermal properties) and to relate the retention of these properties to performance in a TPP test. A deeper understanding of how fabrics degrade in a TPP exposure is of considerable practical interest. Such information identifies the relative contribution of fiber and fabric variables and indicates how fabrics can be engineered to improve thermal protective performance.

Experimental Procedures

Materials

The fabrics tested and their initial properties are listed in Table 1. Test fabrics were chosen to represent several of the commercially available state-of-the-art high-performance, thermally stable woven fabrics. An effort was made to minimize variables by selecting fabrics of similar initial unit mass, thickness, and construction. A heavier FR cotton fabric was included because of its widespread use in industry. Two common FR finishes for cotton were represented. These are the condensed-phase active ammonia-cured THPOH

TABLE 1—*Initial properties of the test fabrics.*

Fabric Composition	Fabric Weight, g/m^2	Fabric Construction	Fabric Thickness, mm	Fabric Density, g/cm^3
PBI	303	twill	0.83	0.36
Kevlar	269	twill	1.01	0.26
40/60 PBI/Kevlar	246	twill	0.92	0.27
FR cotton No. 68[a]	525	denim	0.85	0.60
FR cotton No. 34[b]	271	twill	0.44	0.62

[a]Halogenated finish.
[b]Phosphonium finish.

phosphonium salt-based treatment and the vapor-phase active decabromobiphenyloxide modified with antimony oxide finish.

Method

The difficulty in continuously monitoring changes in the thermal and physical properties of fabrics under real or simulated intense thermal exposure conditions is obvious. A recently developed TPP tester (Fig. 1) allows precise control of thermal flux level and exposure duration. The tester employed two meker burners burning 95% pure methane gas. The burners converged on the specimen at an angle of 35° to the horizontal. A bank of nine electrically heated quartz tubes supplied a radiant flux. A pneumatic shutter controlled by a digital timer allowed control of the exposure time down to 0.25 s. A nitrogen gas quenching device on the tester allowed immediate cool-down and arrested further degradation resulting from accumulated heat and smoldering. The distance from the back of the fabric to the calorimeter face was kept at 6.3 mm using a stainless steel spacer. The exposed fabric area was 100 cm^2.

The total incident heat flux was kept at 2 $cal/cm^2 \cdot s$ (8.4 W/cm^2). The combined radiant component emanating from the gas flames and the quartz panel was 1 $cal/cm^2 \cdot s$, as determined by a Hy-Cal Model R-8015 water-cooled, quartz-window radiometer, placed at the fabric surface plane. The total flux was determined using the copper calorimeter extensively described in the ASTM Test for Thermal Protective Performance of Materials for Clothing by Open-Flame Method (D 4108-82).

An interrupted exposure approach was used to obtain series of test fabrics exposed for incrementally (0.5 s) increasing periods of time. These fabrics

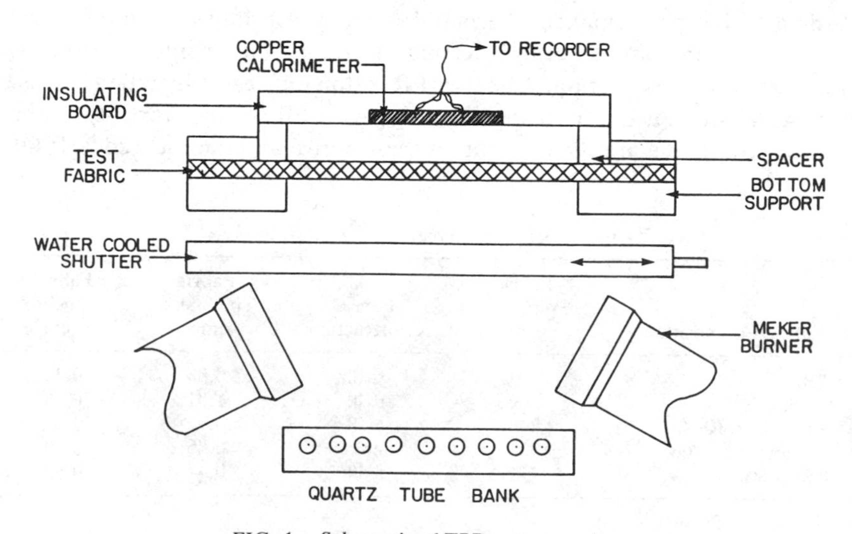

FIG. 1—*Schematic of TPP test apparatus.*

were then analyzed off line using various techniques for thermal and physical properties. At least five specimens of each fabric type were exposed at each time increment. Temperature profiles for the fabric flame and back sides were determined by multiple measurements with fine-gage thermocouples placed on the fabric surfaces. Following exposure, the fabrics were stored in a conditioned atmosphere (21°C, 65% relative humidity).

Weight loss as a function of exposure time was determined by weighing the fabrics immediately prior to and after the exposure sequence. The thickness of the fabrics was determined at a low pressure since it was suspected that surface fibers play a significant role in the effective thermal thickness of the fabrics. Reproducible, low-pressure thickness measurements were facilitated by a highly sensitive compressometer. The plunger on this instrument is lowered mechanically at a fixed rate of 0.02 mm/s. The presser foot area is 2 cm^2. A continuous pressure versus thickness profile is plotted automatically. Thicknesses were taken at 2 gf/cm^2 (196 N/m^2). This is the minimal pressure at which good reproducibility is obtained.

The momentary bulk densities for each exposure time were computed using the average values of momentary unit mass and thickness. In addition, the fiber densities before and after exposures were determined using a Beckman air compressor pycnometer. The fabric and fiber densities were used to compute air volume fractions. Fabric transmission in the infrared (IR) range was determined using a Perkins-Elmer IR spectrophotometer Model 281-B. A novel null method, which allows rapid determination of the insulative disadvantage of a fabric layer as opposed to a stagnant air layer, was used. A differential scanning calorimeter was used for this purpose. The temperature gradient between the specimen holders and ambient atmosphere was kept at 200°C.

Detailed descriptions of all the analytical test procedures used in this study are given in Ref *4*.

Results

The TPP traces of the test fabrics are shown in Fig. 2. The tolerance times (TT) according to the Stoll criteria and TPP ratings are given in Table 2. The initial and final values of the fabric air volume fractions are given in Table 3.

The TPP trace for the 40/60 polybenzimidazole (PBI)/Kevlar blend falls between the traces of the pure fabrics. The 100% PBI fabric exhibited significantly lower heat transfer rates. Back-side ignition of Cotton No. 34 is clearly indicated by a sharp increase in the heat transfer rate at 3.5 s, which subsequently decreases as charring proceeds. Evidence for concurrent flame and back-side ignition of Cotton No. 34 can be observed visually. Extensive charring occurs on both fabric sides simultaneously. Flame-side ignition for this fabric was determined to occur at 3 s. The surface temperature at ignition for Cotton No. 34 was 311°C; the back-side temperature was 128°C.

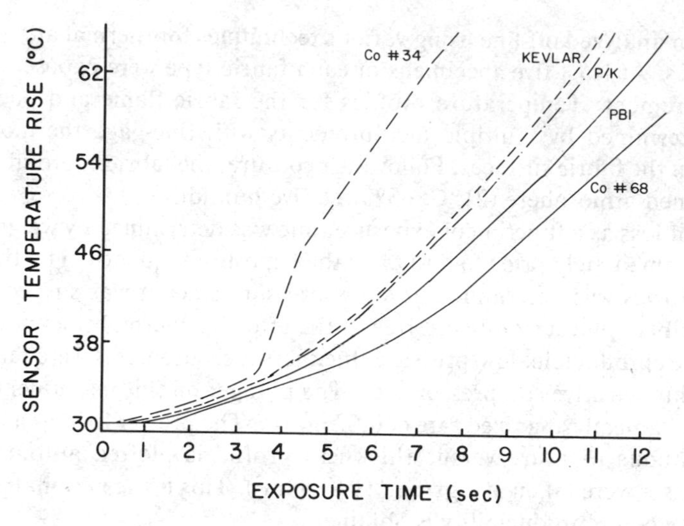

FIG. 2—*TPP traces of the test fabrics in a 2-cal/cm² · s, 50/50 radiant/convective exposure.*

TABLE 2—*TT and TPP values for the test fabrics.*

Fabric	Tolerance Time, s	TPP, cal/cm²
PBI	8.1	16.2
Kevlar	7.3	14.6
PBI/Kevlar	7.6	15.2
FR Cotton No. 68	9.1	18.2
FR Cotton No. 34	4.7	9.4

TABLE 3—*Initial and final air volume fractions of the test fabrics.*

	Initial Fractions			Final Fractions, 9 s.		
Fabric	ρ_{fab}, g/cm³	ρ_{fib}, g/cm³	$\frac{V_a}{(V_a + V_f)}$, %[a]	ρ_{fab}, g/cm³	ρ_{fib}, g/cm³	$\frac{V_a}{(V_a + V_f)}$, %
PBI	0.36	1.44	76	0.25	1.46	83
Kevlar	0.25	1.54	84	0.28	1.31	78
PBI/Kevlar	0.27	1.45	81	0.26	1.43	82
FR Cotton No. 68	0.60	1.72	65	0.51	1.52	66
FR Cotton No. 34	0.62	1.33	53	0.40	1.16	65

[a] V_a = air volume; V_f = fiber volume.

The Cotton No. 68 fabric displayed a gradual, two-stage trace with a distinct change in slope at 5 s. This exposure time coincides with the observed flame-side ignition time. The surface temperature at ignition was 490°C and the back-side temperature was 140°C. A marked difference was observed in the extent of charring on the flame and back-side of Cotton No. 68 as a function of exposure time. No ignition was observed for any of the other test fabrics.

Table 3 illustrates the extent of change in the fiber and fabric density due to thermal exposure. PBI and Cotton No. 34 exhibited an increase in air volume fraction. Kevlar exhibited a drop in air volume fraction. The limited change in fiber density of the PBI fabric compared with that of the other test fabrics, all of which underwent a significant drop, underscores its enhanced thermal stability. The fabric densities of the cotton fabrics and the PBI fabric dropped. The fabric density of the Kevlar fabric increased. The PBI/Kevlar blend displayed a moderate drop. Despite extensive change in fiber and fabric densities, the computed air volume fraction of Cotton No. 68 changed little.

Mass Retention

Figures 3 and 4 show the percentage of mass retention as a function of exposure time of the test fabrics. PBI exhibited a 20% drop in unit mass at 9 s. The moisture content of PBI at standard conditions is above 13%, so a large portion of the mass loss must be attributed to moisture evaporation. Morse [5] has reported a 10.7% mass loss at a temperature of 455°C. This is

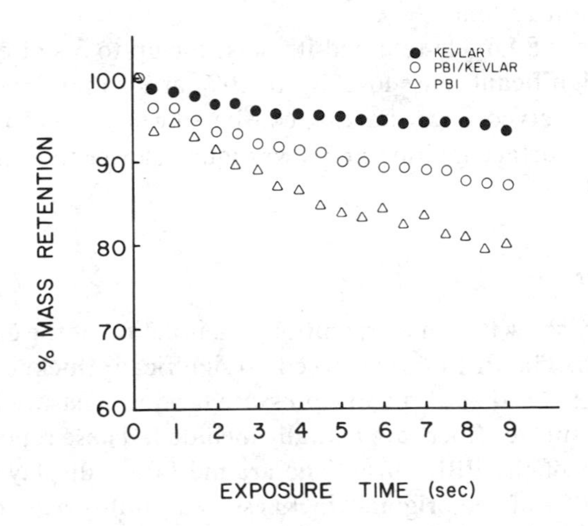

FIG. 3—*Mass retention of PBI, Kevlar, and blend in TPP exposure.*

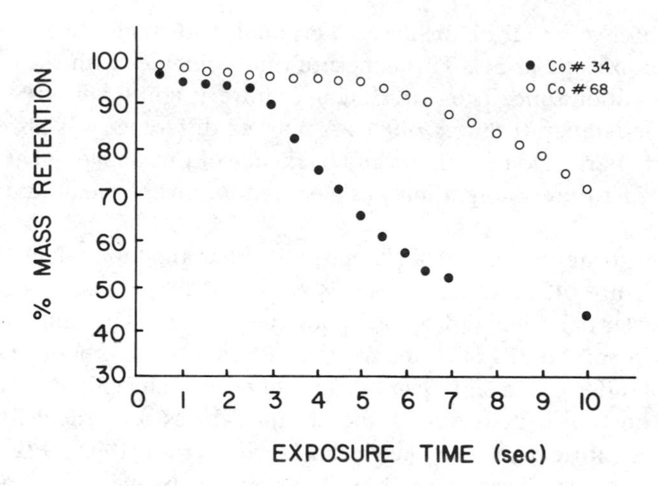

FIG. 4—*Mass retention of FR cotton fabrics in TPP exposure.*

the average temperature attained in the present study at 6 s. The mass loss at this exposure time was 16%.

Kevlar displayed minimal unit mass loss. A 6.5% loss was recorded at 9 s. The PBI/Kevlar blend displayed an intermediate mass loss rate, attaining a 13% loss at 9 s.

The fabric Cotton No. 34 showed a three-stage mass loss profile similar to its TPP trace. A small mass loss was observed at the start of the exposure because of moisture loss. At 3.5 s, coinciding with ignition time, a drastic mass loss occurred. This leveled off as charring was completed. Only 45% of the original unit mass was retained at 9 s. This indicates a complete failure of the flame-retardant system.

The Cotton No. 68 largely retained its mass for up to 5 s of exposure and then exhibited significant mass loss, up to 20% at 9 s and 35% at 12 s. No leveling off was observed since charring continued well beyond 12 s. The rate of mass loss after surface ignition at 5 s was much slower than that observed for Cotton No. 34.

Thickness Retention

Figures 5 and 6 show thickness retention as a function of the exposure time of the test fabrics. The PBI fabric showed no significant thickness change. It should be appreciated that at the low pressure used to measure fabric thickness, protruding surface fibers are actually included. These remained largely intact in the case of the PBI fabric. The aramid fabric displayed a stepped drop, down to 80% of the original thickness. Variability was initially large because of the irregular surface of the fabrics and the high fiber modulus of Kevlar. The variability decreased considerably after surface fibers were

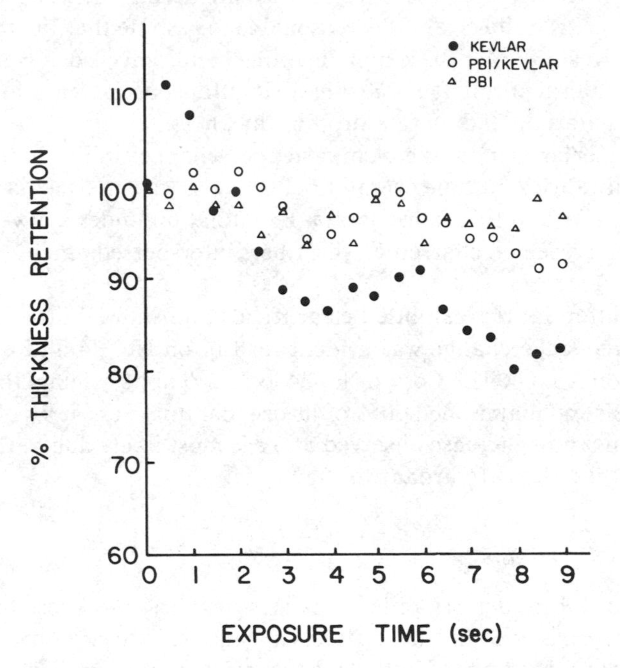

FIG. 5—*Thickness retention of PBI, Kevlar, and blend in TPP exposure.*

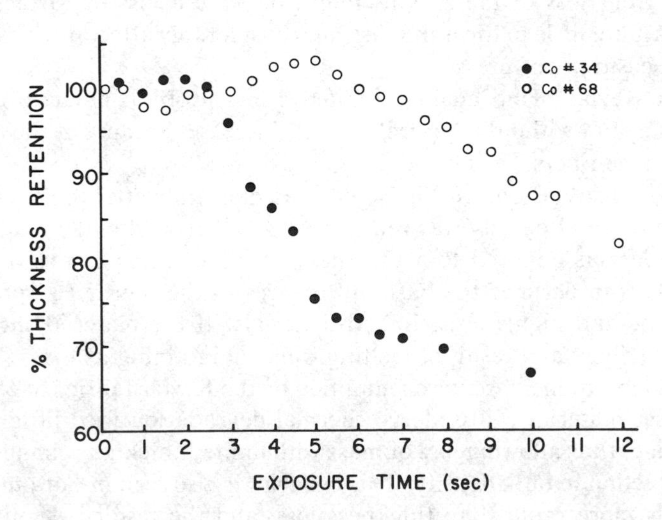

FIG. 6—*Thickness retention of FR cotton fabrics in TPP exposure.*

charred significantly and the surface was smoothed out. The downward trend, however, is quite clear. It is reasonable to assume that the momentary thickness of Kevlar under these high heat flux conditions is determined by an interplay of lateral shrinkage, causing a simultaneous increase in thickness and loss of material, leading to a drop in thickness.

The PBI/Kevlar blend underwent a stepped change similar to that for the 100% Kevlar fabrics, but the overall thickness loss was intermediate between the 100% Kevlar and PBI fabric losses. Examination under a low-power stereomicroscope revealed charred Kevlar fibers interspersed among uncharred PBI fibers.

The FR cotton fabrics exhibited clear trends and showed little variability. Again, a three-stage change was evident for Cotton No. 34 and a two-stage one for Cotton No. 68. The Cotton No. 34 lost 32% of its original thickness at 9 s. Cotton No. 68 maintained 90% of its original thickness at 9 s of exposure time. The thickness increase observed at 3 s is most likely due to the binders used to affix the flame retardant to the fabric.

Bulk Density Retention

Changes in fabric density reflect changes in fabric thickness and weight caused by pyrolysis and shrinkage effects during the TPP exposure. Figures 7 and 8 show the percentage of bulk density retention as a function of exposure time for the test fabrics. A clear drop is observed for PBI largely because of moisture loss. More than 83% of the original bulk density was retained at 9 s. The bulk density of the Kevlar fabric increased to 118% of the original density at 8 s. This is caused directly by a decrease in thickness without significant concurrent loss of mass. This might indicate a loss of surface fibers which contribute little to the unit mass but considerably affect its thickness at 2 gf/cm^2 measurement pressure.

The PBI/Kevlar blend again exhibited a loss profile similar to that for 100% Kevlar, but without an overall density increase, because of the remaining PBI surface fibers.

The bulk density of Cotton No. 34 decreased steadily after 3.5 s, and only 65% of the original density was retained at 9 s. This is directly related to a large mass loss as opposed to a thickness loss, indicating extensive loss of material. This underlines the basic differences in the thermal responses of thermoplastic and highly cross-linked materials. If the material shrinks or loses surface fibers as a result of melting while maintaining its mass, its bulk density will tend to increase, as exemplified by the Kevlar fabric. In contrast, a theromoset material will undergo thermal degradation with little dimensional change, thus allowing loss of mass with limited thickness change, leading to a reduction in bulk density. This behavior is also seen in Cotton No. 68. Mass loss is more rapid than thickness loss, and the result is significantly reduced fabric density at the end of the thermal exposure. Since density is a

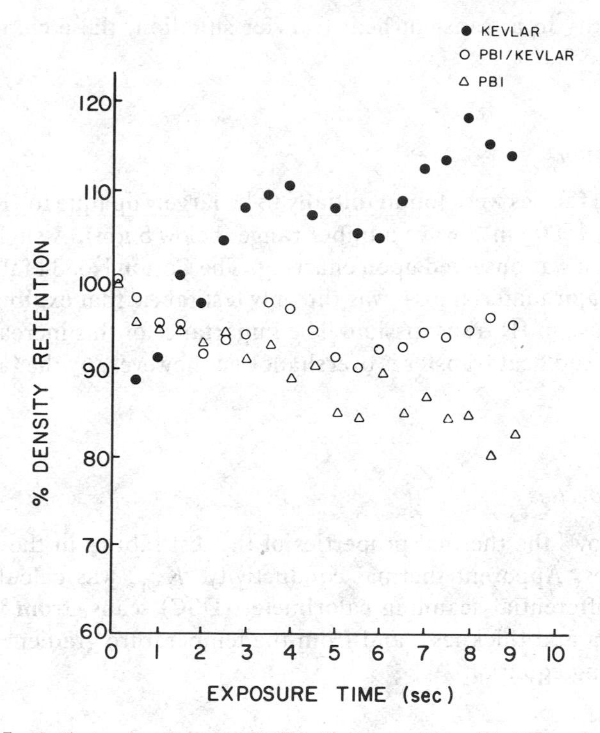

FIG. 7—*Bulk density retention of PBI, Kevlar, and blend in TPP exposure.*

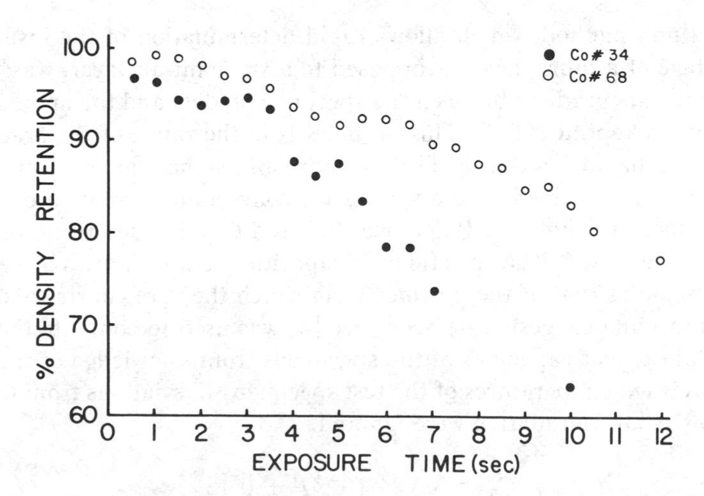

FIG. 8—*Bulk density retention of FR cotton fabrics in TPP exposure.*

critical quantity in a transient heat transfer situation, these changes are of great interest.

IR Transmission

Most of the fabrics were found initially to be largely opaque to IR radiation in the 4000 to 1500-cm^{-1} wave number range (below 5 μm). A slight increase in transmission was observed upon charring. The Cotton No. 34 fabric, which underwent major material loss, was the only test fabric that exhibited a measurable increase in IR transmission. The importance of this increase as it relates to enhanced heat transfer is overshadowed, however, by the fabric backside ignition.

Thermal Properties

Table 4 shows the thermal properties of the test fabrics in the virgin and exposed states. Apparent thermal conductivity, K_{app}, was calculated from isothermal differential scanning calorimeter (DSC) scans, from the known specimen area and thickness, and from the temperature gradient according to the following equation

$$K_{app} = \frac{\Delta cal/s \cdot thickness}{area \cdot \Delta T} \quad (1)$$

where Δcal/s is the measured difference in the rate of heat loss between the fabric layer and stagnant air, and ΔT is the temperature gradient across the fabric.

A novel null method, which allows rapid determination of the insulative disadvantage of a fabric layer as opposed to a stagnant air layer, was used. The temperature gradient between the specimen holders and the ambient atmosphere was kept at 200°C. This gradient is in the range of the gradients measured on the fabrics during TPP exposure and is the maximum gradient that could be maintained before significant convection currents are set up above the specimen holders. Reference *4* gives a detailed description of the DSC techniques used. The specific heat capacities, C_p, of fabrics are essentially the same as that of the polymer from which they are constructed. An empirical formula suggested by Schoppee [*6*] was used to compute the momentary fabric heat capacities of the specimens from knowledge of the momentary average temperatures of the test specimens. Deviations from linearity due to partial degradation were ignored.

$$C_p = 5.3 \times 10^{-4} T + 0.15 \quad (2)$$

TABLE 4—*Initial and final thermal properties of the test fabrics.*

Fabric	Exposure Time, s	ℓ, mm	T_{avg}, °C	ΔT, °C/mm	K_{app}, (cal/cm · s · K) × 10^5	K/ℓ, (cal/cm^2 · s · K) × 10^4	ρ, g/cm^3	Heat Capacity, C_p, cal/g · K	$K/\rho C_p$, (cm^2/s) × 10^4
PBI	0	0.83	30	0	9.31	11.21	0.36	0.31	8.34
	9	0.81	560	320	9.12	11.26	0.25	0.59	6.18
Kevlar	0	1.01	30	0	9.69	9.53	0.26	0.31	12.02
	9	0.83	605	301	9.35	11.26	0.28	0.61	5.47
PBI/	0	0.92	30	0	9.34	10.15	0.27	0.31	11.15
Kevlar	9	0.84	577	327	9.24	11.00	0.26	0.60	5.92
Cotton No. 34	0	0.44	30	0	8.47	19.25	0.62	0.31	4.54
	9	0.30	615	431	8.14	27.13	0.39	0.62	3.34
Cotton No. 68	0	0.86	30	0	9.57	11.26	0.60	0.31	5.06
	9	0.80	363	265	9.22	11.58	0.51	0.49	3.66

The following equation was used to compute the average temperature, T_{avg}

$$T^t_{avg} = \frac{(T^t_{front} - T^t_{back})}{2} \quad (3)$$

where T^t = temperature at time, t.

The fabrics exhibited a decrease in apparent conductivity after exposure. The Kevlar fabric underwent a large increase in thermal conductance, and the PBI and PBI/Kevlar blend changed little. The Cotton No. 34 fabric showed a large increase while Cotton No. 68 exhibited a small increase after 9 s. The heat capacity, C_p, of all the fabrics increased as the result of the TPP exposure. Changes in inherent bulk thermal conductivity are obviously related to heat-induced modification of the air volume and fibrous configuration within the fabric. Unfortunately, quantitative analysis of the transient heat effects on the thermal properties of fibrous materials is quite involved. However, we have been successful in gaining analytical insights by using semiempirical models to predict how transient changes in thermal properties affect the heating-up rate of these materials. The results of this analysis, though beyond the scope of this paper, are described in some detail in Ref *4*.

Discussion

FR Cotton Fabrics

A combination of FR finish mechanism and low mass and thickness contribute to the early failure of Cotton No. 34. Despite the reduced volume of combustible products, obtained by the phosphonium finish, Cotton No. 34 eventually evolves enough combustible volatiles to exceed the lower flammability limit, and ignition occurs almost simultaneously on both sides of the fabric. This is aided by the relatively small temperature gradient across the fabric (128 to 311°C) at ignition because of its low mass and thickness. This castastrophic ignition is attested by the rapid drop in density retention (Fig. 8) at 3 s of exposure time, concurrent with the sharp rise in the TPP trace (Fig. 2).

The Cotton No. 68 fabric ignites visibly on its flame side long before ignition, and significant charring is observed on the back side. This is due to the large momentary temperature gradient across the fabric (140 to 490°C at ignition) and the continuously evolving halide radical scavenger species in the air gap [*7*]. A relatively high air gap temperature is required to exceed the lower flammability limit, and back-side flaming is largely suppressed. Subsequently, failure is not as sharp as in Cotton No. 34 (Fig. 2). The rate of heat transfer increases at surface ignition time (5 s) as an internal heat source is introduced. It then remains largely constant through to TT. This suggests that ignition in the gap is not the rate-controlling factor. Constant flux output conditions are set up after surface ignition occurs as heat crosses the fabric mainly by conduction. Diathermancy remains minimal throughout, as is shown in the IR transmittance scans [*4*].

The Cotton No. 34 fabric undergoes a dramatic increase in diathermancy concurrent with a large loss of material at 3 s, as predicted in the plots of mass, thickness, and density retention. This change may also affect the rapid rate of transfer after surface ignition, but its significance is overshadowed by the complete failure of the material on ignition as its flaming back surface becomes the major heat source to the sensor. Constant flux output is achieved in Cotton No. 34 after charring is complete at 4.5 s as conduction mechanisms take over. The rate of temperature rise decreases significantly (6.0 degrees Celsius per second as opposed to 12.5 degrees Celsius per second). It is, however, still greater than the postignition Cotton No. 68 (3.7 degrees Celsius per second) because of a smaller residual thickness and decreased opacity. The property retention plots mirror the three-stage process occurring in Cotton No. 34 rather than the two-stage process occurring in Cotton No. 68.

The minimization of temperature in the air gap is clearly important in preventing fabric back-side ignition, which leads to immediate TPP failure. This may be achieved by increased thickness and weight. Conversely, the same end may be achieved by chemical prevention of ignition at any air/fuel combination, as in the halogen-finished cottons. Back-side ignition does not seem to occur even for lightweight halogen-finished fabrics, as is shown by the characteristic halogenated finish TPP traces previously obtained for these fabrics [*3*].

PBI and Kevlar Fabrics

TPP exposure caused pronounced changes in the spatial and thermal properties of both the PBI and Kevlar specimens. The degradation response of the PBI fabric is characterized by rapid loss of mass during exposure, a result of heat-induced evolution of moisture contained by PBI fibers. It seems reasonable to speculate that, in a convective test, the higher moisture contained in PBI not only acts to increase heat capacity but also provides an ablative effect beneficial to thermal protective performance. We also noticed that, while Kevlar shows a rapid loss of initial thickness, PBI maintained fabric thickness. PBI specimens actually increased in air fraction after the TPP exposure (Table 3). The maintenance of the air/polymer fraction can be attributed to the thermal stability of PBI and the resulting ability to maintain surface fibers. The importance of surface fibers cannot be overstressed. In a convective exposure, the efficiency of heat transfer to the fabric bulk is dependent on the thickness of the laminar flow boundary air layer on the fabric surface. A profusion of surface fibers serves as a baffle and extends the still air layer, that is, the effective thermal thickness of the fabric. The ability of the fabric to maintain these fibers at high temperatures, as exemplified by the PBI fabrics, impacts directly on the retention of its heat blocking properties. Akaoui [8] has calculated that these fibers heat up 14 times faster than the fabric bulk, making this a formidable material requirement.

If these fabrics are exposed to a purely radiant source at the same total flux level, the distinctions between the fabrics can be expected to change. The shading provided by surface fibers in this case is minimal, and the surface effect is lost. Aluminization of the fabric surface has also been shown to be detrimental in a convective exposure for the same reason [3]. The role of the moisture in the fabric may also be quite different in a radiant test.

Predicting TPP from Initial Properties

These experiments show that no single fiber or fabric variable solely controls heat transfer in a TPP exposure. The thermal diffusivity property, $k/\rho C_p$, which combines the effects of fabric density, conductivity, and heat capacity does predict the rate of temperature equalization in the fabrics (Table 4). The observation that the initial diffusivity of the PBI specimen is lower than that of the Kevlar fabric is explained by the difference in fabric density. Differences in initial density and thermal diffusivity are a direct result of differences in fabric thickness and weight and are related only incidentally to differences between the material properties of Kevlar and PBI fibers. On the other hand, changes in fabric thermal diffusivity during the TPP test are controlled by the thermal stability of component fibers, in the sense that the reaction of the polymer to heat determines the rate and direction of changes in the fabric spatial properties. Changes in fabric bulk, through shrinkage or loss of surface fibers, affects the all-important air/polymer fraction. This is indicated by observed changes in fabric heat capacity, by changes in fabric thickness, and by fluctuation in fabric density associated with mass loss. The ablative effect of mass loss due to evolution of heated moisture is a significant but unexplained factor which undoubtedly accounts for many differences in the characteristic response of PBI and Kevlar.

If the fabric ignites in the TPP test, as is the case with FR cotton systems, then all these considerations change. In these cases, the TPP response is indicative of ignition-dominated behavior, especially ignition on the back side of the fabric, which produces heating of the air gap between the fabric and heat sensor. The FR mechanisms that discourage ignition in the fabric/skin air gap limit the flame to the fabric surface, and fabric thermal properties become important again. This was clearly observed in the TPP of Cotton No. 68. Flame-side ignition occurs at 5 s, but TPP failure occurs only at 9.1 s.

The profound effect of thickness and density variations on transfer rate, coupled with the high opacity of the test fabrics, points to an air and fiber conduction-dominated situation. This has been shown to be true in comfort-level heat transfer studies [9] and seems to hold true under these extreme conditions also. The intuitive assumption that hot gases are blown through the fabric under these turbulent conditions seems unfounded in light of these data and the data previously reviewed. The total lack of correlation of TPP values with the initial measured air permeability obtained in a previous study [*3*] further corroborates this finding.

Conclusions

This research has demonstrated novel and useful techniques for in-depth analysis of fabric response in the TPP test environment. The interrupted exposure approach has allowed critical fabric thermal and physical properties to be monitored and related to the actual thermal output at the fabric/sensor interface. The importance of this method lies not in the characterization of specific candidate fabric specimens but rather in the fundamental insight it provides into FR fabric thermal response as it relates to thermal protection. Identification of rate-controlling processes is the key to improved design and full utilization of material properties.

The specific conclusions of this study may be summarized succinctly as follows:

1. Basic thermophysical properties *do* change extensively during a TPP exposure. Therefore, simple correlations with initial properties are not to be expected. Charring alone is not indicative of a protective disadvantage.
2. The polymer-to-air ratio emerges as an important variable predicting TPP. Fabrics that maintain effective thermal thickness during intense heat exposures can be expected to prolong thermal insulation in convective exposures. The mechanism by which thermal insulation is maintained depends on the fiber or FR finish. Given no back-side ignition, charring is not detrimental to insulative performance (in some FR cotton fabrics) as long as mechanical integrity is maintained (we have no way to judge retention of mechanical strength in a static TPP test). In fact, decreased density resulting from charring tends to reduce the rate of thermal transfer, thus improving heat-blocking capacity. This is why the slope of the TPP trace decreases with time as charring progresses in FR cotton fabrics. On the other hand, PBI and PBI/Kevlar blend fabrics maintain air volume by maintaining effective thermal thickness while the fabric mass is decreasing. The characteristic TPP of PBI fabrics (and the distinctive differences between the thermal response of PBI and Kevlar fabrics) is indicated by mass loss as a result of moisture evolution from PBI during the test. We suspect that the ablative effect of moisture loss, along with retention of air volume fraction, contributes to improving the thermal protective performance of materials in a convective exposure.
3. Air and fiber conduction dominate in a nominal 50/50 radiant/convective TPP test. Direct radiant transmission is not an important contributor to the total heat transfer in this exposure. This means that a competitive advan-

tage could be maximized in some applications by using low-density structures or fabrics that incorporate a layer of surface fibers (for example, flocked or napped fabrics). We suspect that the ability of a fabric to maintain its surface fibers has a significant impact on flame heat transfer. A profusion of surface fibers serves as a baffle and extends the still thermal thickness of the fabric. When fabrics are exposed to a purely radiant source at the same flux level, distinctions between competitive fabrics will be less pronounced. The shading provided by surface fibers in this case is minimal, and the surface effect is lost.

4. In a TPP test, surface ignition may be tolerated for several seconds if back-side ignition does not occur. This is why vapor-phase active flame retardants, which suppress air gap ignition when treated cotton fabrics are tested, perform better in a TPP exposure than solid-phase retardants.

Therefore, the TPP of specific fabrics is better understood through monitored changes in fabric properties during exposure. Certainly, much more remains to be learned about fabric response to intense heat, especially about the role of moisture in controlling the rate of heat transfer through the fabric. At the same time, more needs to be known about factors affecting TPP when the heat exposure is rich in radiant energy. Research is continuing at North Carolina State University to answer fundamental questions on the relationship between the transient thermophysical properties of fabrics and thermal protective performance.

Authors' Note

Care must be taken in deriving conclusions concerning safety benefits from these data. These data describe the properties of fabrics in response to the controlled laboratory exposure and conditions that are specified. They are not presented to predict actual field conditions, in which the nature of the thermal exposure can be physically complicated and unqualified. The authors wish to emphasize that it is not their intention to recommend, exclude, or predict the suitability of any commercial product for a particular end use. Neither do they claim that the test specimens are unqualifiably representative of given fabric types or that they represent the best fabrics that have been developed for protective applications. The PBI, PBI/Kevlar, and Kevlar specimens were prototypes and produced in laboratory lots. They are not necessarily representative, in their construction or finish, of fabrics sold into commercial applications.

Acknowledgments

Portions of this research were funded by Celanese Fibers Operations. Other portions were funded by Cotton Inc. We are grateful for advice and material support received from technical personnel at both organizations, especially

W. S. Harmon and N. Byars of Celanese and W. F. Baitinger of Cotton Inc. We also acknowledge the contribution of Celanese in providing the specimens of PBI and Kevlar for analysis.

References

[*1*] Stoll, A. M., Chianta, M. A., and Piergallini, J. R., "Heat Transfer Measurements of Safety Apparel Fabrics," Technical Report, NADC-78209-60, Naval Air Development Center, Warminster, PA, 1978.

[*2*] Stoll, A. M. and Chianta, M. A., "Method and Rating System for Evaluation of Thermal Protection," *Aerospace Medicine*, Vol. 40, 1969, pp. 1232–1238.

[*3*] Shalev, I. and Barker, R. L., "Analysis of the Heat Transfer Characteristics of Fabrics in an Open Flame Exposure," *Textile Research Journal*, Vol. 53, 1983, pp. 475–483.

[*4*] Shalev, I., "Transient Thermophysical Properties of Thermally Degrading Fabrics and Their Effect on Thermal Protection," Ph.D. thesis, School of Textiles, North Carolina State University, Raleigh, NC, 1984.

[*5*] Morse, H. L., Thompson, J. G., Clark, K. J., Green, K. A., and Moyer, C. B., "Analysis of the Thermal Response of Protective Fabrics," Technical Report AFML-TR-73-17, Air Force Materials Laboratory, Wright-Patterson Air Force Base, OH; NTIS-AD-759 525, National Technical Information Service, Springfield, VA, 1973.

[*6*] Schoppee, M. M., Skelton, J., Abbott, N. J., and Donovan, J. G., "The Transient Thermomechanical Response of Protective Fabrics to Radiant Heat," Technical Report AFML TR-77-72, Air Force Materials Laboratory, Wright-Patterson Air Force Base, OH, 1976.

[*7*] Pitts, J. J., "Antimony-Halogen Synergistic Reactions in Fire Retardants," *Journal of Fire and Flammability*, Vol. 33, 1972, pp. 51–84.

[*8*] Akaoui, S. E., "A Heat and Mass Transfer Study—Analysis of Continuous Processes for the Coloration of Polyester Fabric. II. Modelling and Characteristics of Fabric Heating," *Journal of Applied Polymer Science*, Vol. 27, 1983, pp. 4713–4733.

[*9*] Verschoor, J. D. and Greebler, P., "Heat Transfer by Gas Conduction and Radiation in Fibrous Insulations," *American Society of Mechanical Engineers, Transactions*, Vol. 74, 1952, pp. 961–968.

Arne Schleimann-Jensen[1] *and Krister Forsberg*[2]

New Test Method for Determination of Emissivity and Reflection Properties of Protective Materials Exposed to Radiant Heat

REFERENCE: Schleimann-Jensen, A. and Forsberg, K., **"New Test Method for Determination of Emissivity and Reflection Properties of Protective Materials Exposed to Radiant Heat,"** *Performance of Protective Clothing, ASTM STP 900,* R. L. Barker and G. C. Coletta, Eds., American Society for Testing and Materials, Philadelphia, 1986, pp. 376–386.

ABSTRACT: A new test method has been developed for rapid testing of protective materials' behavior when exposed to infrared IR radiation. The test equipment uses an IR thermometer, which, without contact with the test specimen, calculates the surface temperatures by measuring the IR radiation emitted from the object. The thermometer is updated every 1/4 s.

Several protective materials have been studied, and emissivity has been measured from 0.1 to 1.0. Reflection has been calculated from 0.0 to 0.9. Dynamic temperature measurements have been performed on the "cool" side of the protective material when the other side was suddenly exposed to radiant heat. The results show great differences for various protective materials.

The authors conclude that the method has a capacity that makes it possible to select radiant heat protective clothing materials. The method is also useful for dynamic temperature measurements, showing the time required to heat and cool the material. Four types of dynamic heat behavior were observed when the whole heating and cooling phase was measured.

This method should be discussed as a standard method by ASTM Committee F-23 on Protective Clothing.

KEY WORDS: emissivity, infrared thermometer, radiant heat, radiant heat protective clothing materials, reflection index, transmission, protective clothing

When selecting radiant heat protective clothing materials for jobs with a high risk of flame or metal splash or both, there are high demands on the

[1]Staff member, Scandiaconsult AB, S-102 65 Stockholm, Sweden.

[2]Researcher, Department of Work Science, The Royal Institute of Technology, S-100 44 Stockholm, Sweden.

materials used. Examples of suitable textile fibers are wool, flame-resistant cotton, and viscose. Nomex and Kelvar are nonflammable materials.

When working in environments with high heat radiation, such as the steel industry, forge shops, melting shops, fire fighting, and power stations, the materials most often used are protective clothing made from heat-reflective materials, which must be able to stand open fires, metal splash, and radiant heat. The clothing must cover most of the human body.

An aluminum layer is used together with woven, felt, or nonwoven materials made from cotton, wool, viscose, glass fiber, carbon fiber, aramid fibers, and others. The application techniques vary, and among the reflective materials used are foil, vacuum metal, and different plastic treatments of the reflecting layers, such as Mylar.

This study is part of a project with the objective of developing protective clothing with a capacity not found in the normal "8-h clothing." The work to study different materials' heat protective properties was started with an infrared (IR) camera (AGA Thermovision System 750). The results led to certain modifications of the method. A new method was developed, and this was used for a second test in an experimental study.

The new method for the study of the protective material in connection with radiant heat reflection, heat emissivity, and heat transmission was developed by Scandiaconsult.

IR Radiation Theory

All objects emit more or less IR radiation. The IR heat, which is an electromagnetic radiation, is found in the wavelength range of 0.7 to 1000 μm (1 μm = 0.000 001 m).

When an object is exposed to IR radiation, part of the radiation will be absorbed, some will be reflected, and some will be transmitted. The following symbols—α, ρ, and τ—for the different parts are valid and

$$\alpha + \rho + \tau = 1$$

An object absorbing all incoming radiation is called black. This is symbolized as

$$\alpha = 1 \qquad \rho = \tau = 0$$

No object is totally black in this respect.

As mentioned before, all objects emit IR radiation. A black body radiates, or emits, as much energy as the object absorbs. The symbol for emissivity is ϵ. Then you get

$$\epsilon = \alpha \qquad \text{(Kirchhoff's law)}$$

The emissivity ϵ, equals the ratio of the radiant power from an object to that from a black body at the same temperature, as follows

$$\epsilon = \frac{W \text{ object}}{W \text{ black body}} \qquad (0 \leq \epsilon \leq 1)$$

where W = emissive power (in watts per metre square). The emissivity, ϵ, for some materials can be found in Table 1.

The relationship of emissivity to the reflection in the black body can be achieved from the formula

$$\alpha + \rho + \tau = 1$$

when

$$\alpha = \epsilon_1 \qquad \tau = 0$$

$$\longrightarrow \epsilon + \rho = 1 \qquad \longrightarrow \epsilon = 1 - \rho$$

or when

$$\tau \neq 0, \qquad \alpha = \epsilon$$

$$\longrightarrow \epsilon = 1 - \rho - \tau$$

By using an infrared camera or a special IR thermometer, it is possible to determine the emissivity of different materials. Materials with a reflective surface layer create a problem in connection with measuring the emissivity of the material. Because of reflections of the surroundings against the reflective side, the IR thermometer cannot "see" the material, but instead sees the reflections of the surroundings. By using special methods, described later, one can eliminate these problems.

TABLE 1—*Emissivities of some materials at certain temperatures.*

Material	Temperature, °C	Treatment	Emissivity, ϵ
Aluminum	100	polished	0.05
Brass	100	highly polished	0.03
Copper	100	polished	0.05
Gold	100	highly polished	0.02
Silver	100	polished	0.03
Wool cloth	20	...	0.91
Natural silk cloth	20	...	0.91
Cotton cloth	20	...	0.91
Human skin	30	...	0.90
Black body	...	...	1.00

The radiation from a surface is in direct proportion to thc cmissivity. The radiated heat from a surface can be calculated by the following formula

$$P = A_0 \times \epsilon_0 \times C_s \times (T_0^4 - T_a^4)$$

where

P = radiated heat effect, W,
A_0 = the surface of the object, m^2,
ϵ_0 = the emissivity of the object,
C_s = the Stefan-Bolzmann constant (5.67×10^{-8} W · s/m^2 × K^4),
T_0 = the temperature of the object, K, and
T_a = the ambient temperature, K.

Apparatus

During the tests, the following apparatus was used:

(*a*) IR thermometer (AGA Thermopoint 80),
(*b*) digital contact and air thermometer (Jenway 8501),
(*c*) IR radiator (Philips IR 250 R), and
(*d*) a stand for apparatus mounting (test rig).

IR Thermometer—The IR thermometer is an instrument developed for measuring surface temperatures without contact with the surface. The thermometer measures continuously the IR radiation from an object. The temperature is shown directly on a display. The instrument can be programmed to take into consideration the emissivity of the object as well as the ambient temperature around it. These two functions make it possible to determine the emissivity of a material.

The IR thermometer is very easy to work with, and the results from these measurements are reproducible. The wavelength range is 8 to 14 μm. The resolution of the IR thermometer is 1°C.

Digital Contact and Air Thermometer—The contact thermometer transmitter must not have too large a contact area or too much thermal mass in relation to the test specimen. If it does, this results in inaccurate measurements of the materials.

The digital thermometer used in this test has a thermoelement transmitter of nickel-chromium with low volume (≈1 mm^3). The transmitter is mounted with spring force against the materials during the measurements. The resolution has been within 0.1 degree Celsius of the temperature range of interest.

IR Radiator—The radiation source used was a Philips IR 250 R heating lamp with a power of 250 W. The lamp has a red filter to reduce the optical radiation within the wavelength range 400 to 750 nm. The visible light is in the range of 400 to 800 nm. The maximum heat flux was 130 mW/cm^2.

The IR thermometer measures in the center of the test body, where the radiation intensity from the lamp is most intense. In this case the maximum heat flux on the test object was 130 mW/cm^2 from the IR lamp. The diameter of the surface measured, in this case, is 20 mm. The IR radiator used here simulates the shortwave IR radiation (0.2 to 5 μm) that sources such as melted metal, glass, or fire emit to the environment. The IR lamp also radiates in the wavelength range of 8 to 14 μm.

Test Rig—Figure 1 shows the test rig. The test specimen is mounted in the test frame (25 by 25 cm). The shield tube mounted on the IR thermometer is needed to eliminate reflections of possible external heating sources (such as the operator) when the reflective surface is turned toward the IR thermometer. Without this shielding, it was not possible to obtain reproducible measuring results of the reflective surface of the test specimen.

The insulated shield between the test frame and the IR radiator is used for total shielding of the test specimen from the IR radiator, which is needed for some of the tests.

Procedure

Determination of Emissivity—Reflection Side

The test specimen is mounted with the reflection side toward the IR thermometer. After this, the back of the test specimen is heated by the IR radiator, and the surrounding temperature is programmed into the IR thermometer. With the digital contact thermometer the temperature of the heated reflection side of the cloth is measured. Different values have been obtained by testing different emissivities in the IR thermometer. When the temperature of the IR thermometer is in line with the temperature of the digital thermometer, the emissivity is read for the reflection surface.

Determination of Emissivity—Cloth Side

The emissivity of the test specimen (cloth side) is determined as outlined in the preceding section. The values obtained, in most cases, are the same for all materials ($\epsilon \approx 0.9$).

Dynamic Temperature Measurements

The test specimen was mounted in the test frame with the reflection side toward the IR radiator (a distance of 350 mm).

The temperature of the back of the test specimen was regularly measured

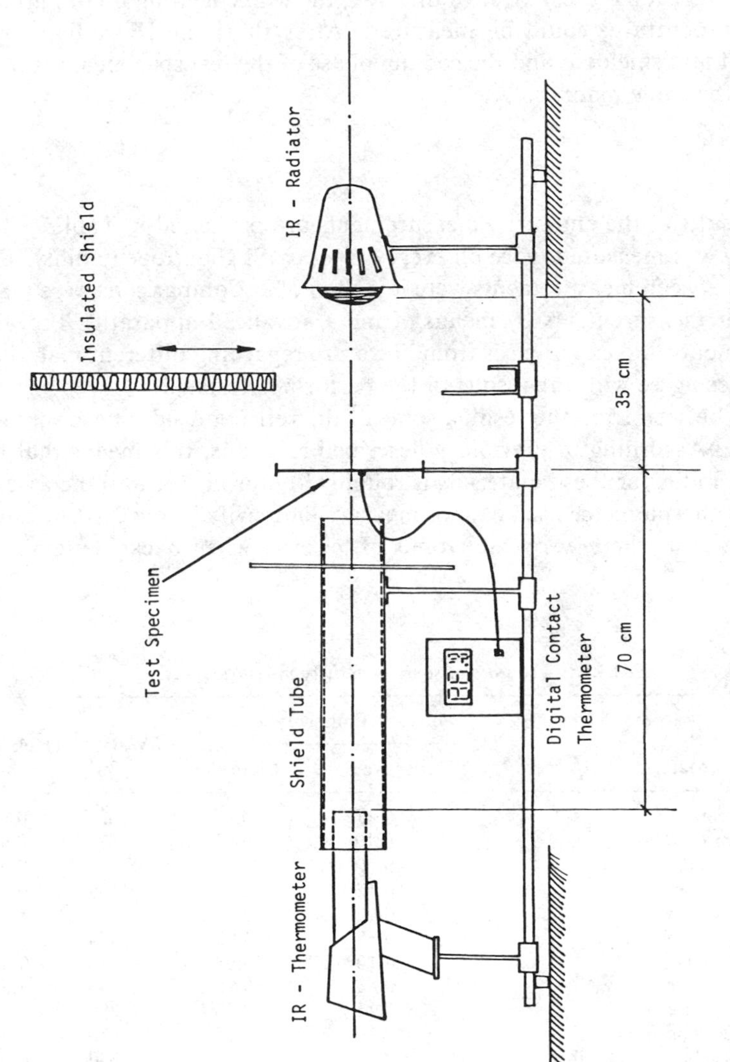

FIG. 1—*Test equipment for determination of protective materials' infrared reflection and radiant properties.*

by the IR thermometer. A value was registered every 10 s. The IR thermometer was programmed for $\epsilon = 0.95$ and the ambient temperature, $T_a = 23°C$.

At first, the temperature of the test specimen was measured without the IR radiator turned on. After that, the IR radiator was turned on and the temperature was registered every 10 s. In this way, the whole heating phase, up to a stable temperature, could be measured. After that, the IR radiator was turned off and shielded, and the cooling phase of the test specimen was measured in the same manner.

Results

The results of the emissivity measurements are presented in Table 2. The emissivity was measured twice on every object, with good repeatability. Differences between measurements were less than 5%. Comparisons were made with other measurements by means of more advanced apparatus. Previous measurements and experiences from literature regarding different materials' emissivities agree with the results of the performed tests.

As can be seen from the results, some of the reflection sides have very low emissivity. According to previously described relations, this means that the reflection index for these materials is roughly 0.9, or higher in some cases.

The IR thermometer used cannot measure emissivity below 0.10. In some cases, however, there were indications of lower ϵ, when backside tempera-

TABLE 2—*Results of some emissivity measurements.*[a]

Material	Emissivity, ϵ		Weight, g/m²	Thickness, mm
	Reflective Side	Cloth Side		
Aluminum-cotton	0.11	0.95	424	0.66
Double aluminum-viscose	0.10	0.95	528	0.99
Aluminum-viscose	0.20	0.95	551	0.92
Aluminum-viscose	0.12	0.95	549	0.90
Aluminum-Nomex	0.12	0.95	511	1.14
Aluminum-Nomex (yellow)	0.85	0.95	454	1.05
Aluminum-Kevlar	0.10	0.95	≈350	≈0.50
Aluminum-Kevlar (yellow)	0.85	0.95	443	2.52
Aluminum-viscose	0.14	0.95	788	3.80
Aluminum-carbon-fiber	<0.10	0.95	656	0.98
Aluminum-carbon/amide-fiber	<0.10	0.95	540	0.98
Aluminum-wool	0.10	0.95	543	1.44
Aluminum-fiber-glass	0.15	0.95	518	0.44
Aluminum-painted silicon-fiber-glass	1.00	0.95	611	0.48
Aluminum-polypropylene (nonwoven)	0.75	0.95	75	0.50

[a]All test specimens were taken from nonused materials. All material temperatures were in the range of 23 to 65°C. The ambient temperature was 23°C.

tures of the exposed materials differed and the reflection properties for the reflective sides were the same.

In other cases the reflection index has stayed low, even though there is a visible reflection layer on the test specimen. A reason for this could be that the visible, see-through surface layer in front of the reflection layer has been heat absorbent and, in this way, has conducted the heat into the cloth through the reflection layer.

The result of the dynamic temperature measurements is shown in Figs. 2, 3, and 4, which represent different types of results. Several measurements were made—some before and some after the test materials were perforated (65 holes/cm^2).

Four types of dynamic heat behavior were observed in our complete study. In Table 2 materials with Types 1 and 2 behavior are presented.

Type 1

A reflecting layer with a high reflection index is present in materials with Type 1 behavior. Very thin cloth is behind. A small thermal mass is present, with rapid heating and fast cooling off. The temperature is on the cool side.

Type 2

Type 2 material behavior is like Type 1, but with thicker cloth accumulating more heat. Most of the heat is reflected away. A big thermal mass is pre-

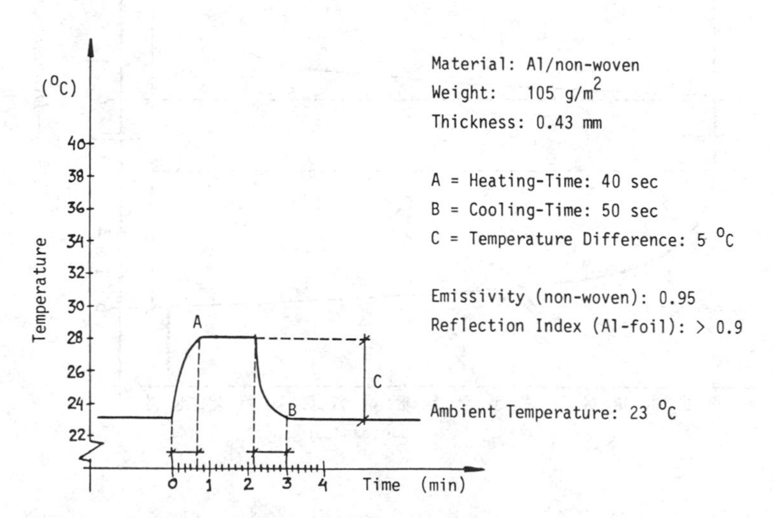

FIG. 2—*Dynamic temperature measurements for nonwoven aluminum.*

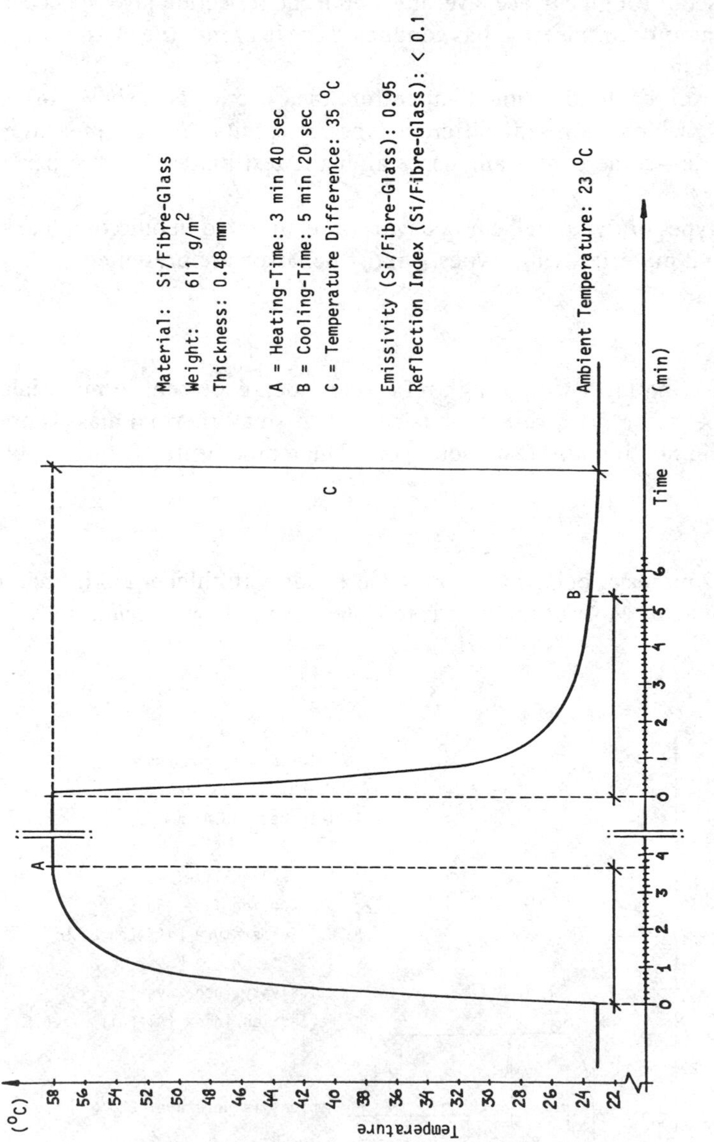

FIG. 3—*Dynamic temperature measurements for silicon/fiberglass.*

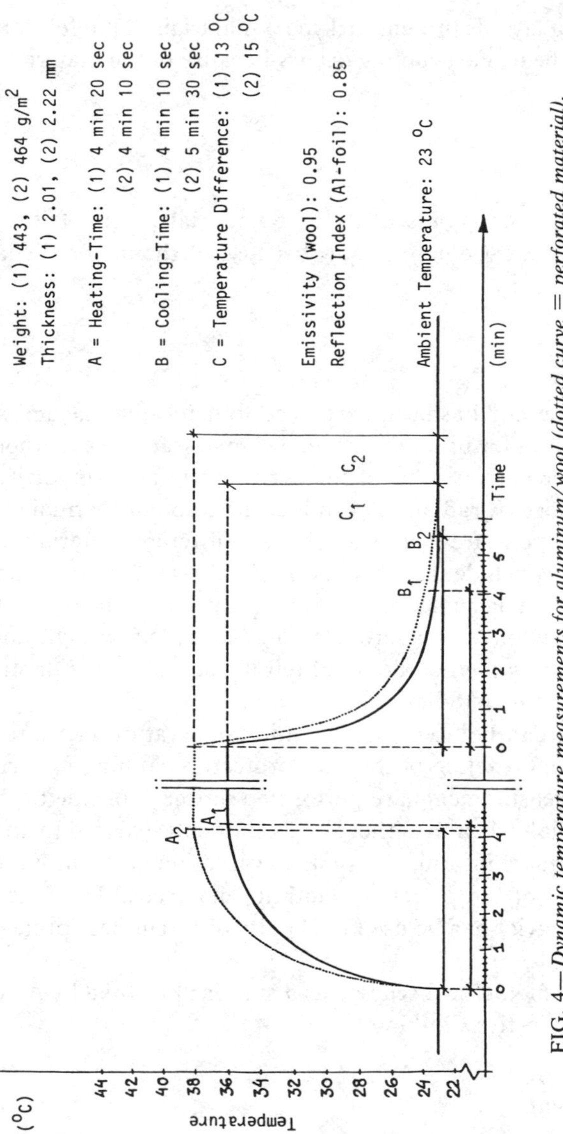

FIG. 4—*Dynamic temperature measurements for aluminum/wool (dotted curve = perforated material).*

sent, with slower heating and cooling off. The highest temperature is the same as for Type 1.

Type 3

No reflecting layer is present in Type 3 materials. Thin felt rapidly absorbs and transmits heat. Fast cooling occurs because of the material's small thermal mass.

Type 4

No reflecting layer is present in these materials. Thick felt slowly absorbs and transmits heat. Cooling is slow because of the material's large thermal mass.

Discussion

A new test method has been developed to determine the emissivity and reflection properties for different protective materials. The method has proven to be very simple and fast when it comes to determining emissivity, the protective effect against IR radiation, and identification of thermal characteristics at dynamic temperature measurements for different materials.

The authors conclude that this new method makes it possible to differentiate accurately various materials' thermal properties. The method seems to be suitable for selection tests before selecting takes place at high-intensity radiation tests. The new method does not tell if the material is flammable or will melt at high-intensity radiation.

The method can be used for dynamic temperature measurement, which gives the thermal reaction of the heat protective clothing material. Dynamic temperature measurements are performed on the cool side of the protective clothing material, when the other side is suddenly exposed to an IR radiator. This gives adequate information on the evaluation of the properties from the "wearer's side" of the protective clothing materials. The four types of dynamic heat behavior can be used to classify different heat protective clothing materials.

The method should be discussed as a standard method by ASTM Committee F-23 on Protective Clothing.

Acknowledgments

This study has been supported by the Swedish Work Environment Fund under Contract ASF 82-0277.

Evaluating Materials for Thermal Protective Clothing

Randall E. Bouchillon[1]

Protective Performance of Polybenzimidazole-Blend Fabrics

REFERENCE: Bouchillon R. E., "**Protective Performance of Polybenzimidazole-Blend Fabrics,**" *Performance of Protective Clothing, ASTM STP 900*, R. L. Barker and G. C. Coletta, Eds., American Society for Testing and Materials, Philadelphia, 1986, pp. 389–404.

ABSTRACT: The paper describes the physical, thermal, and fire-protective characteristics of fabrics prepared from Celanese polybenzimidazole (PBI) fiber blended with other fibers (for example, flame-retardant rayon, aramids). Fabric engineering design criteria, production techniques, testing, and correlations with applications requirements are presented. Computer statistical analyses were used to define optimum performance regimens and are discussed in relation to laboratory data and end-use performance evaluations. Data from various standard tests are interpreted in light of PBI fiber performance contribution.

KEY WORDS: protective clothing, high-performance fibers, polybenzimidazole (PBI), aramid, Nomex, Kevlar, permanently flame-resistant (PFR) rayon

The introduction of polybenzimidazole (PBI) fiber, with commercial production begun in early 1983, offers a new dimension in engineering options for high-performance fiber products. In pure form or in blends with other high-performance fibers, PBI contributes unique thermal, chemical, and textile properties that are useful in a variety of critical performance applications.

The potential of PBI in protective apparel has been reviewed previously [*1,2*]. First used in various National Aeronautics and Space Administration (NASA) applications, PBI is now successfully entering commercial areas such as use in protective apparel, for asbestos replacement, and with other high-temperature fabrics.

This paper reviews the performance of PBI fiber in general terms, as well as how it specifically contributes to improving industrial protective apparel.

[1]End-use manager, Celanese Fibers Operation, Charlotte, NC 28232.

Fiber Properties

General

PBI is the acronym for poly[2,2′-(*m*-phenylene)-5,5′-bibenzimidazole]. The basic chemical reaction for PBI production is shown in Fig. 1, and the acid derivation step is outlined in Fig. 2. It is marketed in fiber and other polymer forms by the Celanese Corp. Various forms, such as films, foams, adhesives, papers, and fibrids, are being developed by Celanese Corp. for nontextile applications needing the thermal and chemical resistance of PBI polymer.

PBI fiber has the following unique combination of thermal, chemical, and textile properties:

(*a*) does not burn in air,
(*b*) remains dimensionally stable at high temperatures,
(*c*) retains suppleness even when charred,
(*d*) emits little smoke on exposure to heat or flame,
(*e*) does not melt or drip,
(*f*) has excellent resistance to chemicals and solvents,
(*g*) has excellent comfort because of high moisture regain and low modulus, and
(*h*) is processed well on conventional textile equipment.

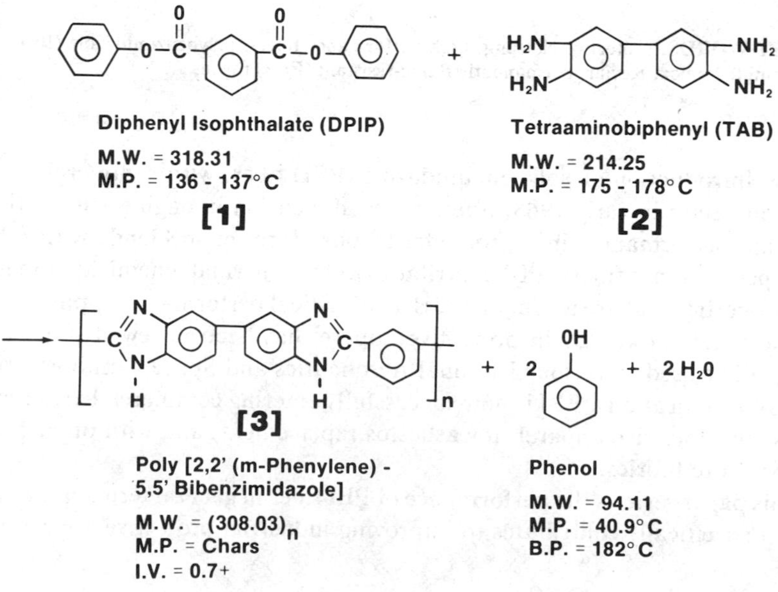

Diphenyl Isophthalate (DPIP)
M.W. = 318.31
M.P. = 136 - 137°C
[1]

+

H_2N NH_2 H_2N NH_2
Tetraaminobiphenyl (TAB)
M.W. = 214.25
M.P. = 175 - 178°C
[2]

→ [3] + 2 Phenol (OH) + 2 H_2O

Poly [2,2′ (m-Phenylene) - 5,5′ Bibenzimidazole]
M.W. = $(308.03)_n$
M.P. = Chars
I.V. = 0.7+

Phenol
M.W. = 94.11
M.P. = 40.9°C
B.P. = 182°C

FIG. 1—*PBI fiber chemical reaction.*

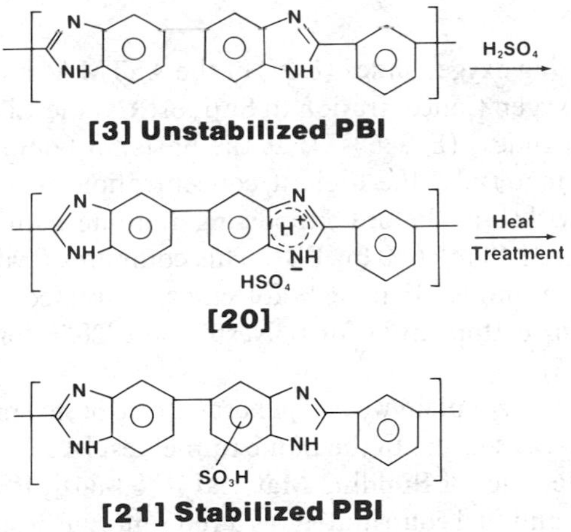

H_2SO_4

[3] Unstabilized PBI

Heat Treatment

HSO_4^-

[20]

SO_3H

[21] Stabilized PBI

FIG. 2—*PBI fiber acid derivation reaction.*

The fiber's physical properties are summarized in Table 1. These fiber properties allow production of fabrics that are durable and comfortable, while providing excellent protection from heat and flame. PBI fiber can be readily formed into woven, knitted, needle-punched, and other nonwoven fabrics for a wide variety of industrial and apparel applications. PBI has been core-spun on Dref spinning equipment to create composite yarns with special characteristics. In blends with other high-temperature fibers, PBI's unique properties contribute to a significantly improved protective fabric [*3*].

TABLE 1—*Physical properties of Celanese PBI fiber.*

Property	Values (Metric Units in Parentheses)
Denier per filament	1.5 denier (1.7 denier tex[a])
Tenacity	2.7 g/denier (2.4 dN/tex)
Breaking elongation	28%
Initial modulus	34 g/denier (30 dN/tex)
Crimps per unit length	12/in. (0.5/mm)
Percent crimp	30%
Density	1.43 g/cm^3
Moisture regain, at 25°C, 65% RH	15%
Boiling-water shrinkage	<1%
Hot-air shrinkage, at 204°C	<1%
Specific heat	~0.3 cal/g°C
Limiting Oxygen Index	41%
Surface area resistivity, at 25°C, 65% RH	0.7 × 10^{10} Ω/m^2

[a]1 tex = 1 g/100 m.

Flame Resistance

Using the limiting oxygen index (LOI) of the ASTM Method for Measuring the Minimum Oxygen Concentration to Support Candle-Like Combustion of Plastics (Oxygen Index) (D 2863-77) as the basis for comparing the relative flammability of materials, the highest concentration of oxygen that fabric specimens can withstand before supporting flaming combustion has been found to range from 40 to 46% for PBI. This compares favorably with values of 25 to 31% for aramids, 35 to 36% for chloride-treated aramids, 23% for nylon 66, 20% for cotton, 21% for polyester, and 28% for flame-retardant (FR) rayon [*1,4,5*].

PBI fabrics (175 g/m^2 plain weave) generate little or no smoke when heated or exposed to a flame source. In the flame tunnel test [ASTM Test for Surface Burning Characteristics of Building Material (E 84-84)], PBI was rated 0 for smoke generated and fuel contributed; as a reference, asbestos is rated 0, and red oak tests at 100 [*6*]. In smoke generation measurements [ASTM Test for Specific Optical Density of Smoke Generated by Solid Materials (E 662-83), nonflaming mode], PBI smoke density reached a specific optical density, D_s, of 2.0 on decomposition. Other fabrics tested produced D_s values of 8.0 for regular aramids, 3.0 for high-strength aramids, and 36.0 for FR rayon. In the federal test method (FTSM 191-5903) for vertical burn testing, PBI showed no afterflame and a 0.3-cm char length after 10-s exposure to methane flame. The char lengths were 5.8 cm for regular aramids, 1.5 cm for high-strength aramids, and 7.9 cm for FR rayon [*7*].

PBI maintains integrity and dimensional stability on exposure to flame or heat. Fabric shrinkage for PBI fabric (175 g/m^2 plain weave) was 1% at 315°C and 8% at 425°C for 10-min exposures, approximately 35% less than that of aramid fabric; also, the aramid fabric was badly charred and curled at 425°C, while the PBI fabric remained supple [*8*]. PBI fabric will carbonize (or char) when exposed to extreme temperatures. The char forms predominantly on the surface, allowing PBI fabrics to retain their integrity, pliability, and protective performance [*2*].

Thermal Stability

PBI fiber has excellent thermal stability. Over a wide range of temperatures and environments, PBI strength and weight retention are excellent [*3,5*].

Thermogravimetry (TG) measures the weight loss of a material exposed to elevated temperatures. Figure 3 shows the weight loss of PBI fiber in air and nitrogen as a function of temperature using a programmed heating rate of 20 degrees Celsius per minute. Rapid degradation starts at about 450°C in air, with completion of loss reached at approximately 640°C. Weight loss of aramids begins at 430°C, with completion of loss at 595°C. The amount of

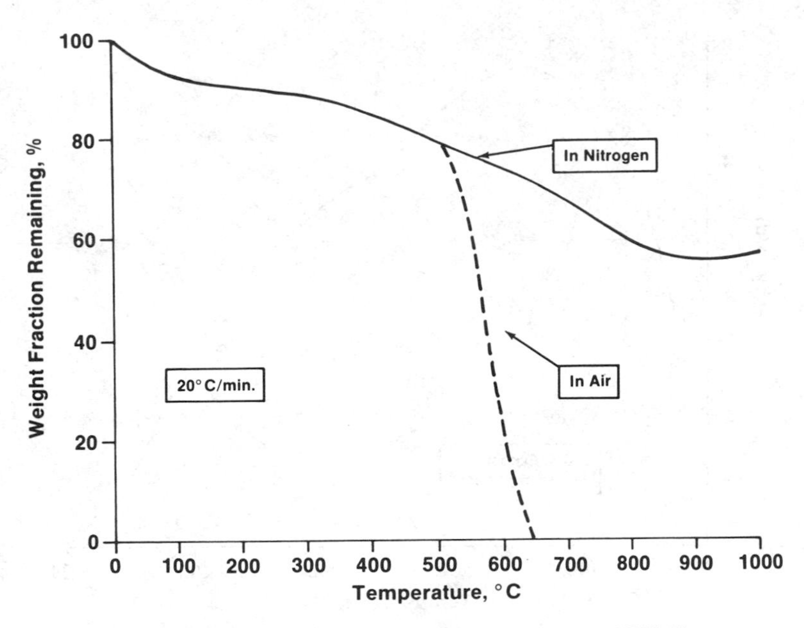

FIG. 3—*Programmed temperature thermogravimetry of PBI fiber.*

degradation for PBI is considerably less in nitrogen, with more than half of the specimen weight remaining at 975°C [*1,9*].

PBI strength retention is also excellent. Figure 4 shows the fiber strength at short-term (2-min) exposures. For longer exposures, PBI maintains 100% of its strength up to 400°C for 30 min, and 50% of its strength after more than 300 h at 300°C. The useful life of PBI fibers at elevated temperatures depends on the amount of oxygen in the environment. In an inert atmosphere or in vacuum, PBI maintains 100% of its initial strength even at 350°C for 300 h [*6*]. Regular and high-strength aramids had lost 57 and 86% of their respective initial strengths at 300°C after 300 h. PBI fiber also offers excellent resistance to steam hydrolysis: after 75 h of exposure to steam at 150°C, PBI fabric retains 96% of its original breaking strength [*3,5*].

Chemical Resistance

In addition to its thermal properties, PBI fiber exhibits outstanding resistance to many chemicals. These include strong acids, bases, and organic chemicals. Table 2 gives the tensile strength retention values for the fiber after exposure to acids. In these conditions, the tensile strength of PBI is largely unaffected, whereas regular aramids are often destroyed under these conditions [5]. Table 3 shows similar data for PBI in solutions of strong inorganic bases. These results show PBI's chemical resistance as a significant attribute for both gaseous and liquid filtration [*1,3*].

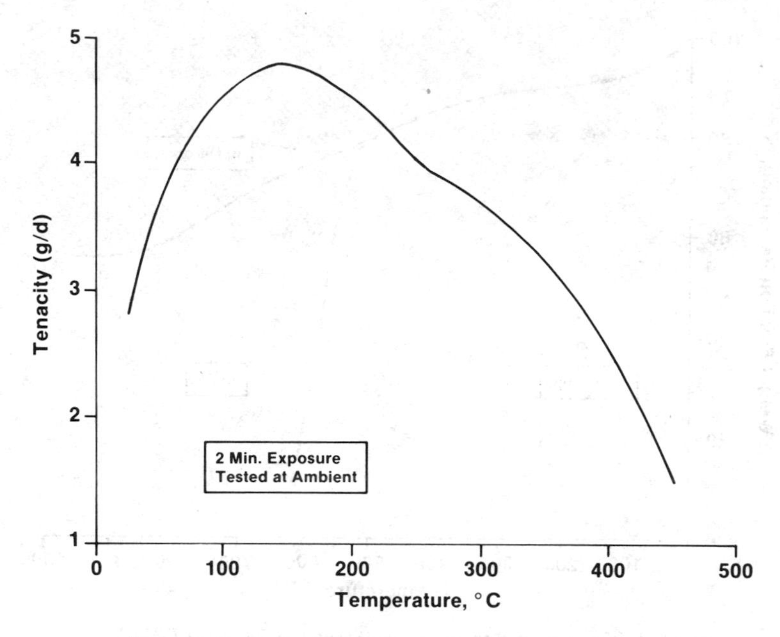

FIG. 4—*Strength of PBI fiber at an elevated temperature.*

TABLE 2—*Tensile strength of PBI after immersion in inorganic acids.*

Acid	Concentration, %	Temperature, °C	Time, h	Tensile Strength Retained, %
Sulfuric acid	50	30	144	90
	50	70	24	90
Hydrochloric acid	35	30	144	95
	10	70	24	90
Nitric acid	70	30	144	100
	10	70	48	90

TABLE 3—*Tensile strength of PBI after immersion in inorganic bases.*

Base	Concentration, %	Temperature, °C	Time, h	Tensile Strength Retained, %
Sodium hydroxide	10	30	144	95
Sodium hydroxide	10	93	2	65
Potassium hydroxide	10	25	24	88

Resistance to organic chemicals is another valuable attribute of PBI. The chemicals listed in Table 4 have no effect on the tensile properties of the fiber after one week of exposure [*1,3*].

Comfort

Fabrics made from PBI fiber have high moisture regain and soft "hand," both of which would be expected to contribute to wearer comfort. Comfort is a desirable characteristic for protective garments. The user will be more likely to wear a garment in marginal exposure situations if it is comfortable. In severe exposures, a more comfortable garment will allow longer work times before fatigue or discomfort occurs.

As part of a large comfort development program, PBI was evaluated against competitive fibers by the Gillette Research Institute. In the study, wearers were asked to evaluate the garments on a five-point scale (5 = total comfort) for various descriptors. The test subjects rated the garments' comfort while exercising (Exercycle, 10 min at 13 to 20 km/h) and while at rest under various environmental conditions (17 to 35°C, 20 to 75% relative humidity). The garments were laundered (10 min at 49°C, tumble dried) twice before each exposure condition. Table 5 shows comparative results. In the study, PBI was rated as comfortable as cotton and rayon [*10*]. In military evaluations of flight suits, PBI was the preferred fiber for comfort [*11–13*]. It has also been proposed that the high moisture regain of PBI contributes to improved thermal protective performance: PBI fabrics require more heat to dissipate the additional moisture [*2*].

TABLE 4—*Tensile strength of PBI after immersion in organic chemicals.*

Compound	Temperature, °C	Time, h	Tensile Strength Retained, %
Acetic acid	30	168	100
Methanol	30	168	100
Perchloroethylene	30	168	100
Dimethylacetamide	30	168	100
Dimethylformamide	30	168	100
Dimethyl sulfoxide	30	168	100
Kerosene	30	168	100
Acetone	30	168	100
Gasoline	30	168	100
Heptane	22	336	100
Toluene	22	336	100
Oil No. 1	22	336	90
Oil No. 3	22	336	100

TABLE 5—*Comfort ratings of PBI versus cotton garments.*

Comfort Descriptor	Comfort of PBI versus Cotton
Snug	PBI more comfortable
Heavy	PBI more comfortable
Loose	PBI more comfortable
Lightweight	PBI more comfortable
Stiff	PBI more comfortable
Sticky	no significant difference
Nonabsorbent	no significant difference
Clammy	no significant difference
Damp	no significant difference
Clingy	no significant difference
Picky	no significant difference
Rough	no significant difference
Scratchy	no significant difference

Engineering Protective Fabrics with PBI

General

The most significant engineering option for PBI fabrics is determining the optimum PBI fiber blend percentages with other high-performance fibers. PBI is easily processible on all conventional textile equipment and can be readily formed into woven, knit, and nonwoven fabrics. PBI's excellent dimensional stability and nonembrittlement characteristics allow fabrics of PBI to maintain their integrity even after exposure to extreme conditions. In fabric form, PBI offers improved thermal and flame resistance, durability, chemical resistance, dimensional stability, and comfort in comparison with other high-performance fibers. By blending PBI with other high-performance fibers, the design engineer can usually improve the performance of currently available protective apparel.

Figures 5 through 7 and Table 6 illustrate PBI performance in blends with high-strength aramids. The PBI blend effect was determined by statistically smoothing (by least-squares linear regression) the performance of four to nine blend fabric specimens. PBI offers improved flammability resistance, durability, softness, and retained strength after exposure to heat; the high-strength aramid significantly improves the initial fabric strength. Blend trials with knit fabrics show similar results. A 40/60 PBI/aramid blend ratio has been determined as optimal for overall fabric performance.

Figure 8 best illustrates PBI's impact in blends with FR rayon. A 20 to 50% PBI blend with FR rayon has been determined as the best overall blend range.

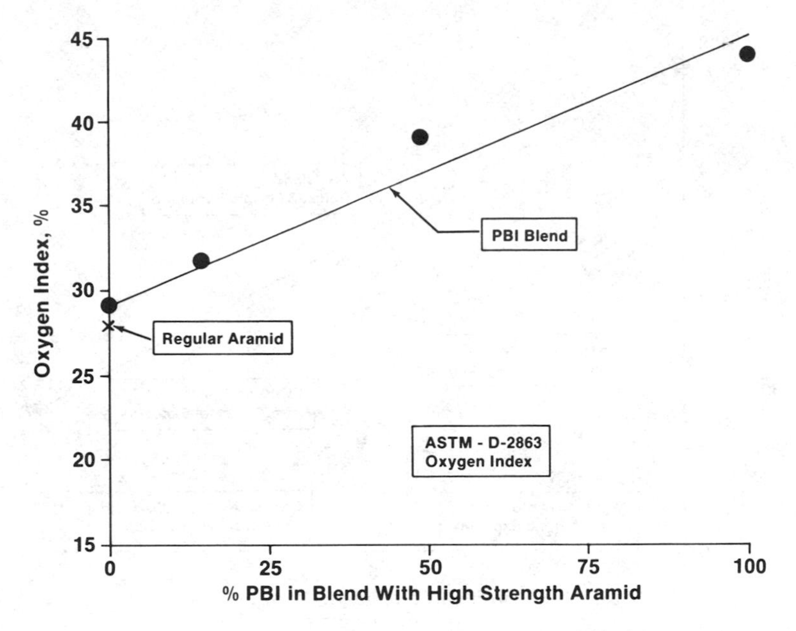

FIG. 5—*PBI effect on the limiting oxygen index (190 g/m² plain weave).*

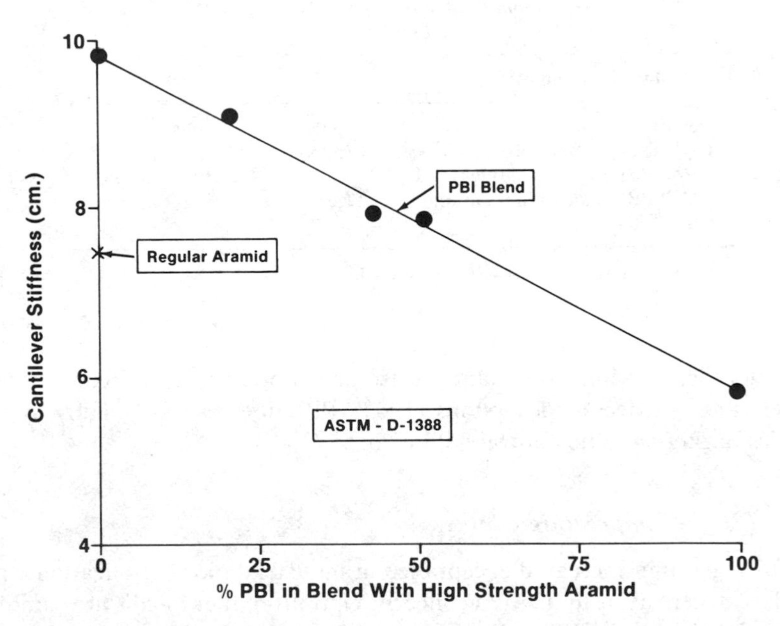

FIG. 6—*PBI effect on fabric stiffness (ASTM Tests for Stiffness of Fabrics [D 1388-64 (1975)]).*

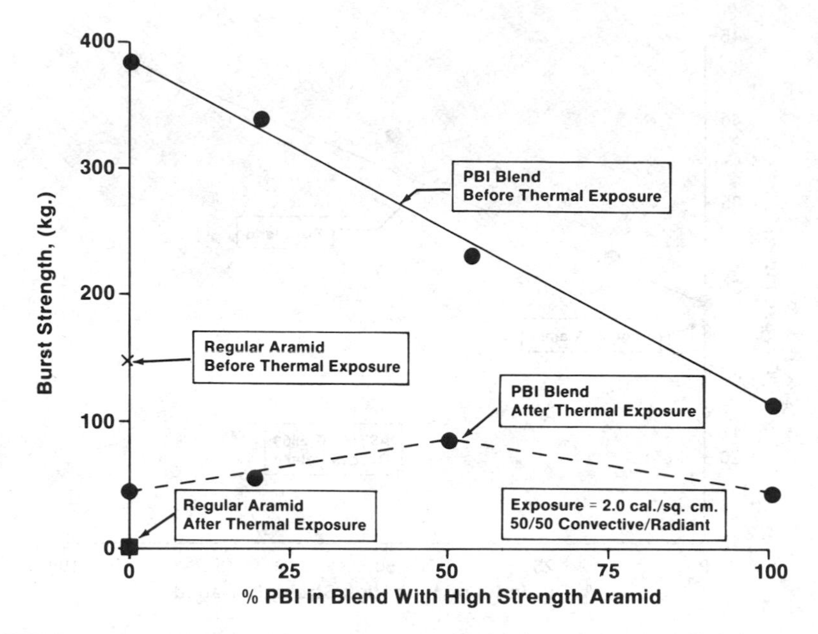

FIG. 7—*PBI effect on fabric burst strength (190 g/m² plain weave) [ASTM Test for Bursting Strength of Knitted Goods: Constant-Rate-of-Traverse (D 3787-80a)].*

TABLE 6—*Random tumble pilling results (ASTM D-3512-82) after 30 and 60 min.*[a]

Material Composition	30 min	60 min
Regular aramid	1 (pills)	1 (pills)
High-strength aramid	1 (pills)	1 (pills)
20/80 PBI/high-strength aramid	3 (fuzz)	3 (fuzz and pills)
50/50 PBI/high-strength aramid	4 (fuzz)	4 (fuzz)
PBI	4	4

[a]Rating system against visual standards: 5 = best, 1 = worst.

PBI significantly improves flame resistance, while the rayon contributes a wider range of dye shade options (100% PBI fiber can be readily dyed to shades darker than its natural gold color).

PBI in High-Temperature Gloves

PBI is gaining increased acceptance in industrial glove applications, particularly where asbestos use is a concern. Optimum fiber blends are combinations of PBI/glass and PBI/Kevlar/glass, for use in the 480 to 815°C contact heat temperature range [*14*].

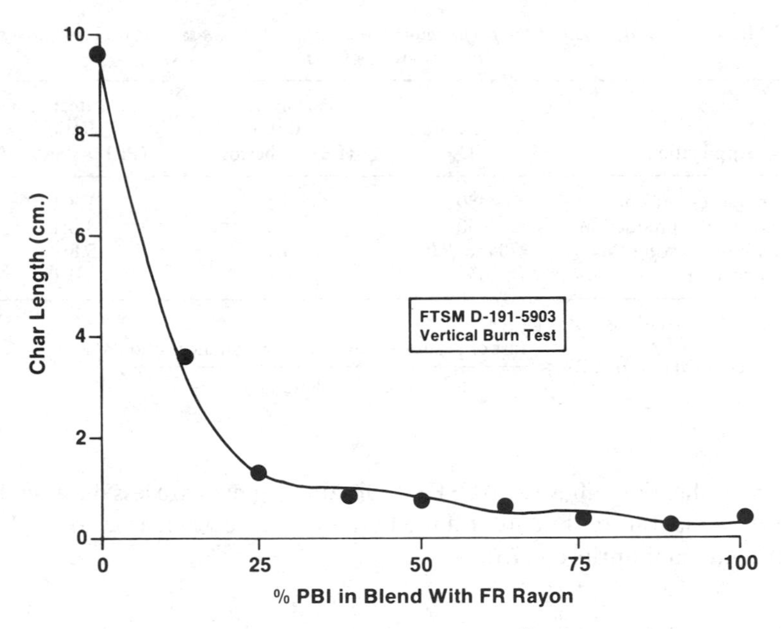

FIG. 8—*PBI effect on fabric char length (150 g/m² Swiss pique knit).*

Use of a glove fabric with high contact heat resistance is crucial to worker safety. Traditionally made of asbestos, these gloves are often subjected to severe abrasion and short-duration temperatures up to 815°C. Because of these harsh conditions, asbestos gloves deteriorate rapidly. Gloves containing PBI fiber can stand up to more exposures than gloves containing asbestos. Gloves properly engineered using PBI fiber are softer and more supple, offering the worker greater mobility, comfort, and protection, even if the fabric becomes charred [*15*].

Field trials have demonstrated the cost-effectiveness of gloves made from a PBI/high-strength aramid/glass fabric blend in comparison with asbestos gloves. Although initially more expensive, PBI-blend gloves lasted up to eight times longer than asbestos gloves. Results of the trials are shown in Table 7. In the aluminum extrusion application, highly abrasive, sharp-edged rails were guided manually by operators. In fabricating welded tubes, operators removed the jagged waste metal, called the scarf, from the welded seam at about 650°C. In aluminum and steel casting, workers fabricated molds that were cured at 650 to 760°C. The hot molds, very rough and abrasive on the surface, were removed from the furnace by hand and then filled with molten metal [*5,14*].

These uses demonstrate that PBI gloves are superior to asbestos gloves. They exceed asbestos gloves in abrasion resistance, mobility, and molten-metal splash resistance; and they are more economical because of longer wear

TABLE 7—*Comparison of the performance and cost-effectiveness of PBI blend gloves and asbestos gloves.*

Application	Temperature, °C	Performance Ratio (PBI/Asbestos)[a]	Cost-Effectiveness Ratio (PBI/Asbestos)[b]
Aluminum extrusion	480	3:1	1.8:1
Welded tube production	650	6:1	3.6:1
Aluminum casting	704 to 760	6:1	5.8:1
Steel casting	650	8:1	5.0:1

[a]Life of PBI versus asbestos gloves.

[b]Cost-effectiveness of PBI $= \dfrac{\text{cost of asbestos gloves} \times \text{performance ratio}}{\text{cost of PBI blend gloves}}$.

life. In applications above 815°C, PBI-containing gloves are less durable than at lower temperatures but are still the best alternative when asbestos replacement is the main priority [*15*].

PBI in Foundry and Fire Proximity

Aluminized PBI knit and woven fabrics compare favorably with heavier asbestos fabrics in applications in which workers are exposed to large radiant heat fluxes. In foundry applications, workers report longer exposure times with lightweight (270-g/m^2) aluminized PBI/FR rayon fabrics than with conventional aluminized glass fabrics (640 g/m^2).

In evaluations for molten metal splash, aluminized PBI fabrics offer better thermal protection than heavier aluminized asbestos fabrics. Test results are shown in Table 8. The test procedure is similar to that being proposed as the standard for evaluating molten metal splash protection [*16–19*]. The procedure consists of pouring 0.91 kg of molten iron (1510°C) onto fabric specimens attached to a supporting board at an angle of 70°C from the horizontal and from a height of 460 mm [*20*]. The test is schematically illustrated in Fig. 9. Maximum heat flux and back-side temperature rise are measured after 30 s, and the fabric appearance is subjectively rated for charring, shrinkage, perforation, and metal adherence.

PBI in Firemen's Protective Apparel

PBI fabrics are rapidly gaining acceptance in firemen's protective apparel. Compared with the conventional aramid shell fabric, PBI offers significant improvement in protection.

In the outer shell fabric for the turnout coats, a blend of 40/60 PBI/high-strength aramid was chosen as the optimum blend. The PBI/high-strength

TABLE 8—*Molten iron splash protection of aluminized fabrics.*

Conditions and Appearance	Fabric			
	220-g/m² Knit 50/50 PBI/PFR	220-g/m² Knit 100% PBI	240-g/m² Woven 100% PBI	353-g/m² Woven Asbestos
Fabric back-side thermal conditions				
Maximum heat flux cal/cm²/s	0.9	1.1	1.1	1.1
Maximum temperature rise, °C	10	15	10	19
Appearance after splash[a]				
Char	3	3	2	2
Shrinkage	1	1	1	1
Perforation	2	1	1	2
Adherence	2	2	1/2	2

[a]1 = excellent; 5 = poor.

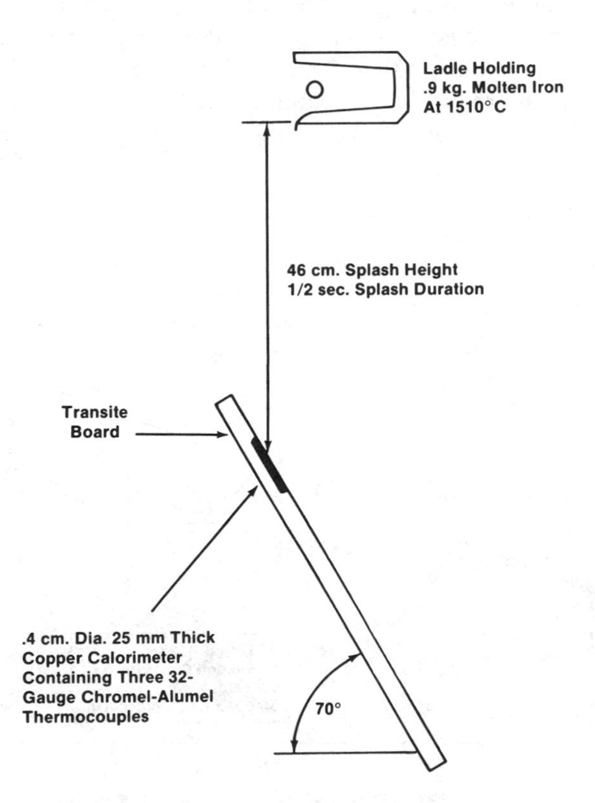

FIG. 9—*Molten metal test apparatus.*

aramid fabric offers the following significant performance improvement over that of regular aramids:

(*a*) lower char lengths in flammability tests,
(*b*) no afterflame on FTMS 191-5905 vertical flammability,
(*c*) a superior thermal protective performance [ASTM Test for Thermal Protective Performance of Materials for Clothing by Open-Flame Method (D 4108-82)] rating (Fig. 10),
(*d*) superior LOI (see Fig. 5),
(*e*) no embrittlement after thermal exposure,
(*f*) superior fabric burst strength (Fig. 7),
(*g*) superior contact heat resistance (Table 9), and
(*h*) superior pilling performance [ASTM Test for Pilling Resistance and Other Related Surface Changes of Textile Fabrics: Random Tumble Pilling Tester Method (D-3512-82)].

To illustrate the nonembrittlement characteristics of PBI fabrics, the fabric burst strength was measured after severe thermal exposures. The high flux exposures were generated using TPP test apparatus. The tester is used to simulate protection time against severe burns. Protective fabrics can be tested under various flame and radiant heat conditions. Figure 7 shows that the PBI/high-strength aramid fabric maintains 30% of its burst strength after thermal exposure, while the regular aramid fabric has no strength remaining

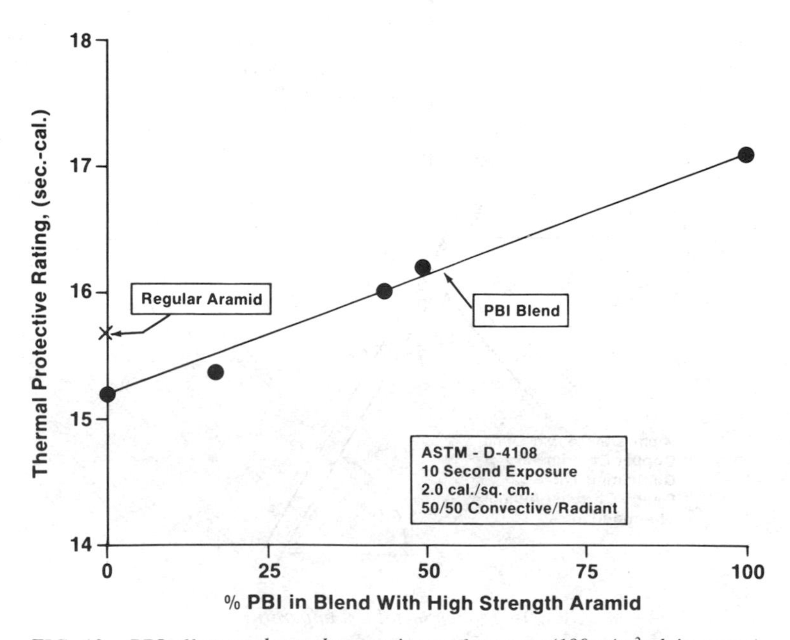

FIG. 10—*PBI effect on thermal protection performance (190 g/m² plain weave).*

TABLE 9—*Contact heat testing time to ignition at 790 and 860° C.*[a]

Fabric Composition	At 790°C, s	At 860°C, s
Regular aramid	5.7	0.5
High-strength aramid	no ignition	1.2
20/80 PBI/high-strength aramid	no ignition	3.5
50/50 PBI/high-strength aramid	no ignition	4.5
PBI	no ignition	no ignition

[a]Procedure involves heating a 0.41-kg iron block to the test temperature in a forced air oven for 1 h. The block is then applied to the test fabric for 15 s.

after exposure. This test illustrates the nonembrittlement properties of PBI under test conditions simulating actual use by firemen.

In firemen's head hoods (sock hoods), a 20/80 PBI/FR rayon knit fabric is becoming preferred over aramid hoods because of its improved comfort and reduced flammability (char lengths 1.3 cm versus 5.4 cm).

Conclusions

PBI fiber marks a significant step forward in engineering options for protective apparel. Originally designed for NASA applications, PBI is now available for various commercial protective apparel end uses. The improved flammability, thermal and chemical resistance, comfort, and textile processability of PBI should allow new levels of safety and productivity for industrial workers.

Acknowledgments

Many people were involved in developing the information presented in this paper; however, W. S. Harmon and T. E. Schmitt must be singled out for their contributions. The author also acknowledges the support and encouragement of the management of Celanese Research Co. and Celanese Fibers Operations.

References

[1] Coffin, D. R., Serad, G. A., Hicks, H. L., and Montgomery, R. T., "Properties and Applications of Celanese PBI—Polybenzimidazole Fiber," *Textile Research Journal*, Vol. 52, 1982, pp. 466-472.

[2] Jackson, R. H., "PBI Fiber and Fabric—Properties and Performance," *Textile Research Journal*, Vol. 48, 1978, pp. 314-319.

[3] "Celanese PBI Polybenzimidazole High-Performance Fiber," Technical bulletin, 2nd Ed., Celanese Corp., Charlotte, NC, 1983.

[4] Barker, T. L. and Brewster, E. P., "Evaluating the Flammability and Shrinkage of Some Protective Fabrics," *Journal of Industrial Fabrics*, Vol. 51, 1981, pp. 7-17.

[*5*] Buckley, A., Steutz, D. E., and Serad, G. A., "PBI—A Unique High Performance Fiber," *Textile Research Journal*, in press.
[*6*] Brown, J. R. and Browne, N. McM., "Actinic Degradation of High Performance Fiber," Report AD A032858, Australian Defense Department, Melbourne, Australia, 1976.
[*7*] Harmon, W. S. and McFadden, T. D., "New End Use Characterization of PBI Blend Fabrics," Celanese Technical Report, Celanese Corp., Charlotte, NC, May 1983.
[*8*] Freeston, W. D., Jr., "Flammability and Heat Transfer Characteristics of Cotton, Nomex, and PBI Fabric," *Journal of Fire and Flammability*, Vol. 8, 1971, pp. 57-75.
[*9*] Dees, J. R., "TGA and DSC Analyses of PBI, Nomex, and Durette Samples," Technical Report to Celanese Corp., Charlotte, NC, 28 Feb. 1984.
[*10*] DeMartino, R. N., "Comfort Properties of Polybenzimidazole Fiber," *Textile Research Journal*, Vol. 54, 1984, pp. 517-521.
[*11*] Stanton, R. M., "Heat Transfer and Flammability of Fibrous Materials," AFML-TR-70-238, Air Force Materials Laboratory, Wright-Patterson Air Force Base, OH, 1971.
[*12*] Stanton, R. M., "The Protective Characteristics of PBI and Nomex Coveralls in JP-4 Fuel Fires," AFML-TR-73-27, Air Force Materials Laboratory, Wright-Patterson Air Force Base, OH, 1973, pp. 11, 12, 13, 23, 41, 42.
[*13*] Stanton, R. M., Schulman, S., and Ross, J. H., "Evaluation of PBI and Nomex II for Air Force Flight Suites," AFML-TR-73-28, Air Force Materials Laboratory, Wright-Patterson Air Force Base, OH, 1973 pp. 5, 6, 32, 63, 72, 73, 84, 105-108.
[*14*] "PBI in High-Temperature Protective Gloves," Applications Bulletin HPF-P1, Celanese Corp., Charlotte, NC, 1983.
[*15*] Slapak, M. J., "Asbestos Replacements—PBI," Presentation to the National Industrial Glove Distributors Association, San Diego, CA, 14 March 1984.
[*16*] ASTM, Standard Test for Evaluating Heat Transfer Through Materials Upon Impact of Molten Substances, Draft No. 2, ASTM F23-20 on Physical Properties, Subcommittee of ASTM Committee F-23 on Protective Clothing, 8 Oct. 1982.
[*17*] Barker, R. L. and Yener, M., "Evaluating the Resistance of Some Protective Fabrics to Molten Iron," National Institute for Occupational Safety and Health, Washington, DC, NIOSH Grant No. R010H00910, 1982.
[*18*] Benisek, L. and Edmondson, G. K., "Protective Clothing Fabrics, Part I. Against Molten Metal Hazards," *Textile Research Journal*, Vol. 51, 1981, pp. 182-190.
[*19*] Benisek, L., Edmondson, G. K., and Phillips, W. A., "Protective Clothing—Evaluation of Wool and Other Fabrics," *Textile Research Journal*, Vol. 49, 1979, pp. 212-221.
[*20*] "Response of Fabrics to Molten Iron Impact," SORI-EAS-82-1008, Southern Research Institute, Birmingham, AL, 1982.

Ludo Benisek,[1] *G. Keith Edmondson,*[1] *Parvez Mehta,*[2] *and W. Alden Phillips*[1]

The Contribution of Wool to Improving the Safety of Workers Against Flames and Molten Metal Hazards

REFERENCE: Benisek, L., Edmondson, G. K., Mehta, P., and Phillips, W. A., **"The Contribution of Wool to Improving the Safety of Workers Against Flames and Molten Metal Hazards,"** *Performance of Protective Clothing, ASTM STP 900,* R. L. Barker and G. C. Coletta, Eds., American Society for Testing and Materials, Philadelphia, 1986, pp. 405-420.

ABSTRACT: The type of fiber and flame-retardant compound plays an important role in the transfer of heat through clothing when exposed to flames or molten metals. The fiber/ flame retardant should preferably form a well-developed char on exposure to heat, without softening or melting, to retain low thermal conductivity of the fabric and also prevent adhesion of molten metal onto the fabric, thereby avoiding excess heat transfer through clothing. Wool performed extremely well in practice and in the draft International Organization for Standardization (ISO) tests evaluated, because of its good char-forming property, which is further enhanced by the Zirpro flame-retardant treatments. Fabric construction parameters—weight, thickness, and density—are important for a given fiber type with regard to protection against flames and molten metal. Also, accumulation of moisture in clothing has a significant effect; in multilayer clothing assemblies, the accumulation of moisture in outer fabric layers as a result of exposure to water increases protection against flame by increasing fabric heat capacity, whereas the accumulation of moisture (body perspiration) in the inner layer when wearing clothing containing a vapor barrier significantly decreases protection against flames by increasing the fabric thermal conductivity. High moisture permeability of wool clothing demonstrably reduces wearer heat stress. Interlaboratory evaluations of two draft ISO test methods for molten metal splashes showed superior repeatability and reproducibility of the test method based on a polyvinyl chloride (PVC) skin simulant as an end-point detection system in comparison with a calorimeter heat sensor. The response and sensitivity of calorimeter heat sensors were significantly affected by the calorimeter construction, contamination, and damage during testing and by the location of the heat sensor *vis-á-vis* the poured metal.

[1]Director, Textile Technology, and senior development officers, respectively, International Wool Secretariat, Ilkley, W. Yorkshire, England LS29 8PB.

[2]Manager, Apparel Services, The Wool Bureau, Inc., Woodbury, NY 11797.

KEY WORDS: protective clothing, hazards, flames, molten metals, wool, comparisons, laboratory test methods, heat stress, skin simulant

Previous work on the selection of fabrics for protective clothing against heat hazards showed that it is not sufficient to rely solely on the flame resistance properties of the materials used but also necessary to consider the performance of the fabric assembly, as it would exist in clothing, against various types of heat exposures [*1-5*], for example, convective (flames), conductive (molten metal splashes), and radiation sources.

There are some conflicting reports on the effect of fiber type and fabric construction on heat transfer from various sources. In the case of exposure to radiant heat, there is strong agreement that fabric thickness plays a predominant role [*1,2,6,7*]. However, for exposures to flames and molten metals there is less agreement between workers. Some have emphasized the importance of fiber type [*1-5,7-9*] while others have suggested that fabric construction, particularly thickness, is the important parameter [*6,10*].

The purpose of this work was to evaluate various flame-retardant (FR) fabrics for heat transfer properties from convective and molten metal sources in order to establish the contribution of fiber type and fabric construction on heat hazards.

The test methods employed were based on draft International Organization for Standardization (ISO) methods, and their repeatability and reproducibility, based on ISO/TC92/SC13/WG2 (Protective Clothing Against Heat and Flames) interlaboratory evaluations, are reported.

In the case of molten metal splashes, two test methods were compared as end-point detection systems, one based on a simulated polyvinyl chloride (PVC) skin and the other using a heat sensor, to assess the advantages and disadvantages of each detection system.

Finally, the contribution of moisture transport properties in alleviating heat stress as reported by other workers in field trials is reiterated [*11-14*], and the role of moisture in modifying the protective performance of clothing to heat and flames is demonstrated.

Experimental Procedure

Fabrics

The commercial fabrics evaluated are described in Tables 1 through 6. All the fabrics were flame-resistant according to Federal Aviation Regulation (FAR) 25.853b, a vertical flame test given in the *Federal Register* (Vol. 37, 1971, p. 3971) and similar to Federal Test Method Standard 191, Method 5903.

Test Methods

Fabric Thickness—Fabric thickness was measured according to British Standard (BS) 2544:1967 using a 50-cm^2 area foot plate and a load of 50 g.

Heat Transfer from Flame Exposure—A proposed draft ISO test method for protective clothing against flames (predominantly convective heat), ISO/TC94/SC13/WG2:N62, which is similar to the ASTM Test for Thermal Protective Performance Materials for Clothing by Open-Flame Method (D 4108-82) was used. A horizontally oriented specimen was subjected to the flames from a gas burner and was placed beneath it, producing an incident heat flux of 80 kW/m^2 (75% convective, 25% radiant heat), and the amount of heat passing through the specimen was measured by means of a small copper calorimeter, placed behind the fabric. The time/temperature relationship of the calorimeter was used to determine the following threshold times, in accordance with specified burn criteria [*15*]:

EPT_{80} = estimated pain time at 80 kW/m^2,
EBT_{80} = estimated burn time at 80 kW/m^2, and
PAT_{80} = pain alarm time—the difference in time, in seconds, between EBT_{80} and EPT_{80}.

Heat Transfer from Molten Metal Splashes—Two test methods were evaluated in the ISO/TC92/SC13/WG2 interlaboratory evaluations. In both cases the molten metal pouring equipment as specified in BS 6357 was used. In this test, small quantities of molten metal were poured, using an automatic pouring apparatus, onto the specimen supported at an angle to the horizontal by a small pin frame.

BS 6357—The performance of the specimen was judged by the amount of heat transferred through the test assembly by placing a PVC skin simulant directly behind the specimen and noting the damage to the skin simulant after pouring. Any adherence of the metal to the specimen surface was also noted. The test was repeated using a greater or smaller mass (in grams) of metal, until the minimum mass (in grams) of molten metal poured that would cause damage to the skin simulant—the molten metal splash index (MMI)—was obtained.

French Experimental Standard S74-107—Alternatively, 100 g of molten metal was poured, using the preceding pouring assembly, onto the specimen supported by a noncombustible board with four embedded thermocouples and held at a 60° angle to the horizontal. The temperature rise was continuously recorded and covered a period of at least 60 s after the termination of pouring. Depending on the temperature rise, the materials were rated GPI-1 ($<$40°C rise) or GPI-2 ($\geq$40°C rise), as recommended in French Experimental Standard S74-107.

Results

Flame Exposure

Repeatability and Reproducibility (ISO Interlaboratory Evaluation—Under the supervision of the ISO working group, four fabrics were evaluated by each of the six laboratories participating in this trial using the draft ISO method. The results are summarized in Table 1.

The within-laboratory repeatability was excellent, the coefficient of variation (CV) being less than 5% in the majority of cases. The main exception was the pain alarm time values for FR cotton, caused by the very small differences between EPT_{80} and EBT_{80} values for this fabric.

The CV values for between-laboratory reproducibility were 10% or less in the majority of cases. This good reproducibility was confirmed by a statistical analysis of variance. Pain alarm time values for FR cotton were an exception. The CV values for estimated pain time, EPT_{80}, for wool also exceeded 10%. This is possibly associated with the inherent voluminous char formation of Zirpro wool on exposure to flames, which at the initial stage of the exposure involves some variability in the rate of growth of the char. The other fibers produced a flat, more reproducible char and consequently lower CV values.

Effect of Air Gap (Separation) Between Fabric and Sensor—In actual clothing there is a variable air gap because of variable separation between the fabric and the human skin, depending on the design of the garment and the location of the clothing vis-á-vis the human body. To evaluate the effect of the air gap, single-layer and two-layer assemblies were tested by the draft ISO flame exposure test, with an air separation between the innermost fabric and heat sensor varying between 0 and 6 mm (Table 2).

As expected, the threshold times increased with increasing air gap, because of the beneficial effect of the trapped air layer. In the case of the two-layer assemblies, the threshold time were significantly longer than for the single-layer fabrics. Once again, the increased thickness of the assembly, with more insulating air, was responsible for this.

The differences in pain and second-degree burn threshold times between the single-layer outerwear fabrics were relatively small, particularly with the 0-mm air separation. In the two-layer assemblies evaluated, using the same innerwear fabric, the EBT_{80} results for each of the outerwear fabrics, at a given air separation, were very different. This could not be explained by the slight differences in the construction of the outerwear fabrics. The effect of fiber type and flame-retardant treatment becomes predominant in this case. In the assembly test, the order of merit of the three fiber types evaluated was Zirpro wool $\gg$ aramid $>$ FR cotton.

Effects of Fabric Construction and Fiber Type—Comparison of four Zirpro wool fabrics in denim, twill, serge, and melton constructions indicated that the time to a given skin damage criterion increased with an increase in fabric

TABLE 1—*Repeatability and reproducibility of draft ISO flame exposure test (sensor/fabric separation—3 mm, specimen restrained).*[a]

Fiber Composition	Fabric Construction	Weight, g/cm^2	Thickness, mm	Density, g/cm^3	Skin Damage Criteria		Repeatability						Reproducibility
							Lab 1	Lab 2	Lab 3	Lab 4	Lab 5	Lab 6	
Aluminized glass fiber	herringbone	520	0.59	0.88	EPT_{80}	mean	4.83	6.06	...	5.00	4.77	4.83	5.10
						CV[b]	6.3	1.9		0.0	1.5	1.4	10.68
					EBT_{80}	mean	6.83	8.58	...	7.20	6.87	7.02	7.30
						CV	5.9	3.6		1.4	0.8	0.8	10.00
					PAT_{80}	mean	2.00	2.52	...	2.20	2.10	2.19	2.20
						CV	5.0	8.6		4.5	0.0	0.8	8.87
Aramid	2 by 1 twill	292	0.59	0.49	EPT_{80}	mean	4.83	5.57	4.66	4.36	4.50	4.80	4.79
						CV	3.2	5.1	3.3	4.8	0.0	0.4	8.84
					EBT_{80}	mean	8.00	9.02	7.83	7.13	7.10	7.89	7.83
						CV	2.5	2.1	2.7	3.2	1.4	1.2	8.98
					PAT_{80}	mean	3.16	3.45	3.16	2.77	2.63	3.09	3.04
						CV	1.8	7.3	1.8	2.1	5.8	3.0	9.75
FR cotton	2 by 1 reverse twill	275	0.75	0.37	EPT_{80}	mean	3.26	3.98	3.80	3.60	3.40	3.62	3.61
						CV	1.8	1.1	2.6	2.8	2.9	3.2	7.22
					EBT_{80}	mean	3.63	4.36	4.23	4.16	3.77	4.00	4.02
						CV	1.6	3.1	3.6	1.4	1.5	2.8	6.97
					PAT_{80}	mean	0.36	0.38	0.43	0.56	0.37	0.38	0.41
						CV	15.7	28.8	13.3	20.4	15.7	9.5	18.34
Zirpro wool	whipcord	460	1.11	0.41	EPT_{80}	mean	5.13	7.48	5.43	5.63	4.83	4.93	5.57
						CV	9.8	4.4	4.6	3.7	3.2	3.3	17.62
					EBT_{80}	mean	9.10	11.9	9.46	9.86	8.70	9.62	9.77
						CV	4.0	3.0	3.4	2.1	3.0	2.9	11.45
					PAT_{80}	mean	4.03	4.62	4.03	4.23	3.80	4.69	4.23
						CV	1.4	9.6	5.7	6.8	9.5	3.9	8.37

[a]FR = flame-retardant; EPT_{80} = estimated pain time, at 80 kW/m^2; EBT_{80} = estimated burn time (blister), at 80 kW/m^2; and PAT_{80} = pain alarm time, at 80 kW/m^2.

[b]The mean is the arithmetic mean of three results.

TABLE 2—*The effect of air separation between fabric and heat sensor on the protection time of single and two-layer assemblies.*

Fabric Description	Weight, g/m² (Thickness, mm)	Fabric Sensor Separation, mm											
		0, s			2, s			4, s			6, s		
		EPT_{80}	EBT_{80}	PAT_{80}	EPT_{80}	EBT_{80}	PAT_{80}	EPT_{80}	EBT_{80}	PAT_{80}	EPT_{80}	EBT_{80}	PAT_{80}
SINGLE LAYERS													
Zirpro wool, 2 by 1 twill	290 (0.80)	2.2	3.6	1.4	2.5	5.0	2.5	3.5	7.9	4.4	4.9	10.7	5.8
Aramid, 3 by 1 twill	259 (0.80)	2.1	3.1	1.0	2.4	3.8	1.4	3.6	6.1	2.6	4.7	7.6	2.9
FR cotton, 3 by 1 twill	315 (0.90)	3.3	4.4	1.1	3.8	4.9	1.1	4.3	5.4	1.1	5.0	6.4	1.4
TWO-LAYER ASSEMBLIES[a]													
Zirpro wool, 2 by 1 twill	290 (0.80)	6.5	18.4	11.9	6.8	22.1	15.3	8.4	23.5	15.1	9.4	30.5	21.1
Aramid, 3 by 1 twill	259 (0.80)	6.7	10.5	3.8	7.6	12.1	4.5	7.8	12.0	4.2	9.4	14.6	5.2
FR cotton, 3 by 1 twill	315 (0.90)	5.3	7.0	1.7	5.1	6.8	1.7	5.7	7.9	2.2	6.1	10.9	4.8

[a]Innerwear fabric consisted of Zirpro wool, 2 by 2 rib, 230 g/m², 2.40 mm thick.

thickness *and weight* (Table 3). Thickness alone, however, was not a sufficient criterion, since although the five melton fabrics had a steady increase in thickness, the protection, measured as EBT_{80}, did not improve. This is probably because weight did not increase correspondingly with an increase in fabric thickness to maintain a constant density.

Similarly, as the thickness and weight of three FR cotton fabrics (windowpane, sateen, and drill) treated with a nonphosphorous flame-retardant compound increased, the protection, as measured by EPT_{80} and EBT_{80}, also increased. However, heavier and thicker duck and whipcord fabrics produced shorter and consequently poorer results than the windowpane and sateen fabrics. The duck and whipcord fabrics were treated with organophosphorous-based flame retardants, whereas the others had a nonphosphorous-based flame-retardant treatment, suggesting that the type of flame-retardant chemical may also play a part in the protection offered.

The results for the aramid, FR cotton, and Zirpro wool single-layer fabrics (Table 3) clearly indicate the importance of fiber type, mentioned previously. Although the three fabrics had exactly the same thickness, there were differences in their EBT_{80} values. These could not be explained by either fabric

TABLE 3—*Effect of fabric construction for a given fiber type on the skin damage criteria time (fabric sensor separation—0 mm).*[a]

Fiber Type	Fabric Construction	Weight, g/m^2	Thickness, mm	Density, g/cm^3	Skin Damage Criteria, s		
					EPT_{80}	EBT_{80}	PAT_{80}
			TWO-LAYER ASSEMBLIES				
Zirpro wool	denim	246	0.70	0.35	7.2	13.5	6.3
	twill	244	0.74	0.33	7.3	16.1	8.8
	serge	377	1.09	0.35	7.7	18.3	10.6
	melton	585	2.07	0.28	10.1	21.3	11.2
	melton	575	2.10	0.27	8.9	17.5	8.6
	melton	592	2.23	0.27	8.9	18.1	9.2
	melton	534	2.24	0.24	9.0	18.3	9.3
	melton	565	2.55	0.22	9.7	17.3	7.6
FR cotton	windowpane[b]	228	0.54	0.42	5.8	9.9	4.1
	sateen[b]	401	0.70	0.57	7.2	11.4	4.2
	duck[c]	454	0.97	0.47	5.7	6.6	0.9
	drill[b]	448	1.10	0.41	8.0	12.1	4.0
	whipcord[c]	426	1.12	0.38	5.5	6.5	0.9
			SINGLE LAYERS				
Aramid	twill	236	0.70	0.34	6.3	10.1	3.8
FR cotton	sateen	401	0.70	0.57	7.2	11.4	4.2
Zirpro wool	denim	246	0.70	0.35	7.2	13.5	6.2

[a]All the evaluations except the last three were made in the two-layer assemblies with 100% Zirpro wool, 254 g/m^2, and 3.01-mm knitted innerwear fabric (2 by 2 rib with tucks).
[b]Nonphosphorus flame retardant.
[c]Phosphorus flame retardant.

weight or density as the heaviest and densest FR cotton fabric produced a shorter EBT_{80} time than the much lighter and less dense Zirpro wool fabric of the same thickness.

Effect of Moisture—In the past, a known procedure adopted by many fire fighters when single-layer tunics were popular was to wet the tunic before entering a burning building to increase the thermal capacity and delay the onset of heat transfer to cause skin damage.

In modern times, however, multilayer tunics have been developed and, often, the layers are separated by a coated fabric to act as a vapor barrier. Although this creates a desirable situation wherein external water is prevented from reaching the inner clothing, it also results in body perspiration being prevented from escaping by diffusion through the fabric layers to the environment and thus accumulating in the inner layers of clothing.

The skin burn hazards associated with condensed moisture in the inner layer are demonstrated by the results given in Table 4. The second-degree burn time of an unwetted commercial aramid fire service tunic with a vapor barrier was 30 s. When the inner layer was wetted by 7% to simulate condensation of body perspiration by placing the assembly over a water container maintained at 40°C for 2 h, the burn time was drastically reduced by some 65%, to only 13 s.

In the case of the wool tunic, without vapor barrier, second degree burn time was increased from 30 to 46 s after the simulated perspiration test, mainly because of the increased thermal capacity of the perspiration water contained inside the fabrics and distributed through all the layers. These results suggest possible serious safety hazards for vapor barrier clothing exposed to flame.

When the outer layer was also wetted, by about 10% in the case of the water-repellent-treated aramid tunic, by subjecting it to a shower for 1 min to

TABLE 4—*The effect of moisture in fabrics on protection against flame.*

Specimen Preparation	Fiber	Pain Sensation Time, s	Second-Degree Burn Time, s	Pain Alarm Time, s
Normal laboratory conditioning	wool[a]	19.8	30.0	10.2
	aramid[b]	19.7	27.2	7.5
After 2 h in perspiration test	wool	29.6	46.0	16.5
	aramid	9.5	12.9	3.4
After 2 h in perspiration test and 1 min in rain test	wool	23.2	38.2	15.0
	aramid	43.1	46.7	3.6

[a]Three-layer wool tunic *without* vapor barrier, total weight 3.8 kg.
[b]Three-layer aramid tunic *with* vapor barrier, total weight 4.0 kg.

simulate external water hazard, the burn time increased because of the extra thermal inertia of the water delaying onset of skin damage. However, the pain alarm time was unaffected and very short.

In the case of wool, a better water-repellent treatment further enhanced the natural water repellency of the outer fabric, as an alternative to the vapor barrier. Thus, the absorption of rain was less than that for the aramid tunic, less than 2%; hence, the delay in onset of pain was not as great as it was with the aramid tunic, but the pain alarm time was unaffected and greater than that for the aramid tunic. Nevertheless, the wool tunic subjected to a simulated perspiration and rain test offered a far better protection against heat transfer from convective heat exposure, without a vapor barrier, than the same dry, conditioned wool assembly.

Thus, the wool tunic provides the optimum protection by allowing perspiration to escape because of the absence of a vapor/water barrier and stops external water penetration with its water-repellent finish.

Molten Metal Splash

Reproducibility (ISO Interlaboratory Evaluation)—Five different fabrics were evaluated by four laboratories according to BS 6357, using a simulated PVC skin to detect the end point, and French Test S74-107, using thermocouples embedded in a noncombustible board to determine temperature rise (Table 5).

The coefficient of variation for the BS 6357 test was significantly lower than that for the French test, and the same applies to the percentage spread of the results. The coefficient of variation for the latter test was particularly high when the metal adhered to the specimen (Fabrics 1 and 5).

Because the metal pouring equipment for both tests was the same, these differences in reproducibility are associated with the end-point detection system—PVC simulated skin (in BS 6357) and thermocouples (in the French test).

The French standard divides fabrics into two classes: GPI-1 (temperature rise <40 degrees Celsius) and GPI-2 (temperature rise ≥ 40 degrees Celsius). According to this classification, Fabrics 1 through 4 would fall into the same class, when, in fact, according to BS 6357 there are large differences in the performances of these fabrics, as seen by their molten metal indices, which range between 51.3 and 373 g.

Effect of Fabric Construction and Fiber Type (BS 6357)—The results for the Zirpro wool fabrics (Fabrics 1 through 6) and the FR cotton fabrics (Fabrics 7 through 10) (Table 6) indicate that for a given fiber type and flame-resistance treatment the results are dependent on fabric construction, for example, fabric weight, density, and thickness. In most cases, with an increase in these three parameters the MMI increases. For similar fabric weights, an increase in fabric density can be beneficial (Fabrics 1 and 2).

TABLE 5—*Reproducibility of molten metal test methods (BS 6357, French method) in ISO trial using molten aluminum 780° C ± 20° C.*

	BS 6357 Method											French Method[a]						
	Lab 1, g		Lab 2, g		Lab 3, g		Lab 4, g		Average Molten Metal Index, $\bar{x}$ g	% Spread[c]	Coefficient of Variation, %	Lab 1, 60°	Lab 2, 60°	Lab 3, 60°	Lab 4, 60°	Average Temperature Rise, $\bar{x}$ °C	% Spread[c]	Coefficient of Variation, %
Material	60°[b]	45°	60°	45°	60°	45°	60°	45°										
Aluminized glass fiber, 520 g/m², 0.59 mm, 0.88 g/cm³	40	...	69	...	38	...	58	...	51.3	60	29	16.5	23.5	79.3	10.3	32.4	213	97.8
Wool melton, 588 g/m², 1.78 mm, 0.3 g/cm³	>286	>300	>435	>390	>464	...	>200	...				12.0	16.0	15.9	7.1	12.8	70	32.8
Wool cavalry twill, 460 g/m², 1.12 mm, 0.41 g/cm³	>285	>300	368	329	>452	417	>200	...	373[d]	24[d]	16.9[d]	16.1	23.3	14.3	5.7	14.9	118	48.3
Wool gaberdine, 277 g/m², 0.80 mm, 0.35 g/cm³	182	...	168	...	165	...	>200	...	172[e]	>8.9	5.3	20.7	32.6	23.4	7.0	20.9	122	50.7
FR cotton, 427 g/m², 0.80 mm, 0.53 g/cm³	<25	...	<42	...	<27	...	<25	...				47.2	106.0	387	46.0	147	232	110.9

[a]Temperature rise in degrees Celsius after 100 g molten aluminum poured at 780°C ± 20°C.
[b]Angle of test specimen to horizontal.
[c]Percentage spread = (highest value in row − lowest value in row) × 100.
[d]Results of Laboratories 2 and 3, at 45° angle.
[e]Average of the results of Laboratories 1, 2, and 3.

TABLE 6—*Effect of fabric construction and fiber type (BS 6357:1978) in the molten metal test.*

Fabric No.	Fabric	Weight, g^{-2}	Thickness, mm	Density, $g \cdot cm^{-3}$	MMI Iron[a]	MMI Aluminum[b]	Weight of Metal Adhered, g	
							Iron	Aluminum
1	Zirpro wool gaberdine	277	0.65	0.42	50	122	0	0
2	Zirpro wool denim	254	0.55	0.46	...	165	...	0
3	Zirpro wool twill	330	0.80	0.41	76	312	0	0
4	Zirpro wool twill	407	0.88	0.46	...	420	...	0
5	Zirpro wool melton	530	1.44	0.37	117	>519	0	0
6	Zirpro wool melton	616	1.96	0.31	127	...	0	...
7	FR cotton twill[c]	358	0.88	0.41	<25	<25	<0.1	1.9
8	FR cotton moleskin[c]	620	1.38	0.45	86	40	<0.1	0.75
9	FR cotton twill[d]	307	0.65	0.47	42	<25	<0.1	0.35
10	FR cotton moleskin[d]	498	1.50	0.33	106	38	<0.1	0.9
11	Aramid twill	265	0.63	0.42	<25	<25	0.2	1.0
12	Glass fiber twill	343	0.43	0.80	<25	<25	2.05	5.9
13	Asbestos plain weave	844	1.45	0.58	<25	<25	0.49	2.4
14	Novoloid plain weave	395	0.97	0.41	62	<25	<0.1	1.8
15	Carbon fiber (plain weave)	339	0.82	0.41	<25	<25	0.1	1.1

[a]Molten cast iron—1400°C ± 20°C at 60°C angle to the horizontal.
[b]Molten aluminum—780°C ± 20°C at 45°C angle to the horizontal.
[c]Organophosphorus flame retardant Type 1.
[d]Organophosphorus flame retardant Type 2.

Fiber type plays a major role in the performance of the fabric against molten aluminum and iron. Fabrics 11 through 13 and 15 (aramid, glass fiber, asbestos, and carbon fiber gave extremely low MMI values, despite their relatively high fabric weight and thickness (for example, Fabric 13—844 g/m^2, 1.45 mm), in comparison with Zirpro wool fabrics of similar or lower weights and thicknesses. Also, adhesion of molten aluminum and iron to the fabrics occurred with Fabrics 7 through 15, while this was not the case with Fabrics 1 through 6. This is also associated with the physical and chemical properties of the fiber substance.

FR cotton and novoloid produced very low MMI results for molten aluminum, but in the case of molten iron they were only slightly inferior to Zirpro wool of comparable weight and thickness.

Discussion

The tests carried out on various fabric types available commercially to evaluate the suitability for protective clothing where heat, flame, and molten metal are likely hazards have clearly emphasized that although all fabrics meet the most stringent flame-resistant requirement, many could pose serious safety hazards to the wearer if used for protective clothing. It is shown to be necessary to evaluate their performance by other appropriate tests measuring heat transfer.

Test Methods

The preceding recognition has led to the development of new standards in many countries, and in some cases, interlaboratory evaluations have been conducted under the auspices of the ISO groups comparing different national tests. Much of the work reported here is based on the data generated in our laboratories working for the British Standards Institute and ISO. Some of the data presented here with the permission of these organizations show that the flame exposure test ISO/TC94/SC13/WG2:N62 has high repeatability and reproducibility in the interlaboratory evaluations. It is expected that this draft method will be adopted in due course as an ISO standard.

The interlaboratory evaluations of the British and the French tests on the evaluation of fabrics against molten metal hazards have indicated that the use of a PVC skin simulant to detect end point has merits over thermocouple or calorimeter detection systems. The following observed disadvantages of the use of a calorimeter were also suggested by Proctor [*16*]:

1. The design of the heat sensor is critical. The heat flow measured is influenced by the variations in thickness of the copper disk and the mass of the solder used in attaching the lead wires. This requires initial and periodic calibration of the unit.

2. The presence of an inadvertent air gap between the specimen and the sensor occurring because of nonprecision mounting of the calorimeter in the base plate significantly alters the reading obtained.
3. A slight variation in the air gap between the copper sensor and the noncombustible board can have a significant effect on the sensitivity of the heat sensor.
4. A buildup of charred material on the calorimeter occurring with each test alters the reading obtained.
5. Penetration of molten metal through test assembly, if it occurs, damages the heat sensor, possibly irreparably.
6. The restricted size of the heat sensor requires the molten metal poured on the test specimen to be precise in the area of the specimen. In practice, this is almost impossible to achieve because of the distortion of the crucible by heat, contamination from previous tests, occurrence of a skin of semisolidified material on the molten metal surface, differences in the viscosities of different metals, and other factors.
7. The high thermal conductivity and heat capacity of copper extracts significant heat very rapidly through the specimen during testing. These heat sink properties are associated with the high thermal inertia of copper compared with human skin. The thermal inertia of the PVC skin more closely simulates that of human skin.

The claimed disadvantages of the skin simulant, such as consistency and reliability of supply over a long period of time, have been obviated by a standard test method to assess the thermal properties of the skin simulant, described in BS 6357.

Importance of Fiber Type and Fabric Construction

Flame Exposures—The results shown in this paper emphasize the primary importance of the fiber type and flame-resistance treatment used. The fabric should be made from a nonthermoplastic fiber so that it does not melt away or soften when in contact with flames and so that it has a low thermal conductivity. For example, the poor protection against flames for glass and asbestos fiber fabrics in spite of their excellent fabric integrity in flames is due to the high thermal conductivity of these fibers [*17,18*]. The fiber type required for protection against flames is the one producing a well-developed insulating char of high strength on exposure to flames. Zirpro wool fabrics have shown excellent protection against flame exposure largely because of metal complex flame retardants increasing char production in wool [*19-21*]. When wool is treated with an organobromine compound as a flame retardant, it does not increase the char of treated wool when exposed to flame, and consequently such treated wools offer inferior protection to that of Zirpro wool [*2,22*]. A similar effect of flame-retardant type is observed on cotton fabrics. It is amply clear from these evaluations that a fabric's protective properties cannot be predicted solely from its thickness, as suggested by some workers [*6*].

A multilayer approach to garment design—tightly woven outerwear fabric with a high integrity against flames, and a bulky, low-density, thick knitted innerwear fabric, both made from char-forming fibers such as Zirpro wool—offers additional insulation against flame exposure, associated with the air trapped in the knitted fabric.

As condensation of moisture in the fabric increases both its thermal capacity and its thermal conductivity, it is easy to visualize why a wet outer and a dry inner fabric offer the best protection. In this context, it is surprising to find that garments designed to encourage moisture condensation in the inner layers, by using vapor-impermeable barriers, are officially approved for use in fire hazards.

Molten Metal Splashes—The arguments used in the preceding section are also valid for protection against molten metal splash in terms of the need for a nonmelting fiber type. Aramid fibers are known to soften at around 316°C [*23*], and this causes trapping of molten metal and subsequent excessive heat transfer.

Untreated cotton fabric offers good protection against molten aluminum, but the application of some organophosphorous flame-retardant compounds make molten aluminum adhere to the fabric, with resultant excessive heat transfer [*24*]. This, however, is not the case with some flame retardants based on organobromine compounds [*4*].

Untreated and Zirpro-treated wool offer excellent protection against most metals because metal sheds readily from wool's surface, and the fabrics have low thermal conductivity values. To prevent molten metal penetrating the fabric at the point of impact, a relatively heavy and tightly constructed outer fabric is required. In this sense, contrary to the case with exposure to flames, a multilayer approach for garment design is not suitable for protection against molten metal hazards [*7*].

Physiological Stress

The hazards of work in impermeable clothing have been documented in the past, and there are now at least three papers covering field studies [*11-13*] demonstrating the extent of physiological stress produced in reducing either the worker efficiency or time of working. During rest, most of the surplus body heat is lost by conduction or radiation, whereas during exercising or high physical activity, the dominant means of losing excess body heat is by the evaporation of perspiration. It is therefore not surprising that researchers in Holland [*12*] found that the duration of arduous work decreased linearly with a decrease in fabric water vapor permeability. It is also not surprising that Atterbom and Mossman [*13*] showed that the maximum performance of a subject wearing clothing with a vapor barrier was some 60% less than that of the same subject wearing the same clothing but without a vapor barrier.

Even in two sets of clothing varying in water vapor permeability by a small amount, the differences in worker performance have been significant. A two-piece wool coke oven worker's garment which had higher vapor permeability was preferred over a garment of similar style in FR cotton [*14*].

Conclusions

To summarize, it is important to have realistic and fair test methods to assess the heat transfer properties of fabrics intended for protective clothing against flames and heat. These should be made of a suitable fiber type and fabric construction to offer maximum protection against heat transfer from external heat sources and at the same time maximize comfort properties, such as effective moisture and internal heat transfer from metabolic heat. The vapor barrier concept can lead to decreased protection against convective heat exposure, particularly when the inner layer is wet from perspiration, a common situation in the case of fire fighters.

References

[*1*] Benisek, L., Edmondson, G. K., and Phillips, W. A., *Textile Research Journal*, Vol. 49, 1979, pp. 212-221.
[*2*] Benisek, L., Edmondson, G. K., and Phillips, W. A., *Fire and Materials*, Vol. 3, 1979, pp. 156-166.
[*3*] Benisek, L. and Phillips, W. A., *Clothing Research Journal*, Vol. 7, 1979, pp. 2-20.
[*4*] Benisek, L. and Edmondson, G. K., *Textile Research Journal*, Vol. 51, 1981, pp. 182-190.
[*5*] Benisek, L. and Phillips, W. A., *Textile Research Journal*, Vol. 51, 1981, pp. 191-196.
[*6*] Holcombe, B. V., *Fire Safety Journal*, Vol. 6, 1983, pp. 129-141.
[*7*] Mehta, P. N., *Textile Research Journal*, Vol. 50, 1980, pp. 185-193.
[*8*] Shalev, I. and Barker, R. L., *Textile Research Journal*, Vol. 53, 1983, pp. 475-482.
[*9*] Krasny, J. F., Singleton, R. W., and Pettingill, J., *Fire Technology*, Vol. 18, 1982, pp. 309-318.
[*10*] Barker, R. L. and Yener, M., *Textile Research Journal*, Vol. 51, 1981, pp. 533-541.
[*11*] Mehta, P. N. and Norman, D. L., *Textile Research Journal*, Vol. 53, 1983, pp. 153-159.
[*12*] Linde, P. J. G. van de and Lotens, W. A., "Restraint by Clothing upon Firefighters Performance," in *Proceedings*, International Conference on Protective Clothing Systems, Stockholm, Sweden, 1981.
[*13*] Atterbom, H. A. and Mossman, P. B., "Effect of Impermeable Clothing and Respirator on Work Performance," *Proceedings*, 19th International Congress on Occupational Health, Dubrovnik, Yugoslavia, R. Plestina, Ed., *Institute of Medical Health and Occupational Health*, Vol. 3, 1980, pp. 1301-1322.
[*14*] Davies, D. G., "The Development of Protective Clothing for Coke Oven Workers," *Proceedings*, Ergonomics Society Conference on Clothing and the Working Man, Loughborough University, Loughborough, England, 1979, pp. 57-81.
[*15*] Stoll, A. M. and Chianta, M. A., *Aerospace Medicine*, Vol. 41, 1969, pp. 1232-1238.
[*16*] Proctor, T. C., "Research into Equipment for Protection Against Molten Metal Splashes," *Proceedings*, Second International Conference on Protective Equipment, Torremolinos, Spain, 1982, pp. 653-678.
[*17*] Kaye, G. W. C. and Laby, T. H., *Table of Physical and Chemical Constants*, Longman, London, 1973, pp. 60-61.

[*18*] Morton, W. E. and Hearle, J. W. S., *Physical Properties of Textile Fibers*, Textile Institute, London, 1975, p. 590.
[*19*] Benisek, L., *Journal of the Textile Institute*, Vol. 65, 1974, pp. 102–108.
[*20*] Benisek, L., *Journal of the Textile Institute*, Vol. 65, 1974, pp. 140–145.
[*21*] Gordon, P. G., McMahon, D. T. N., and Stephens, L. J., *Textile Research Journal*, Vol. 47, 1977, pp. 699–711.
[*22*] Benisek, L. and Craven, P. C., *Textile Research Journal*, Vol. 53, 1983, pp. 438–442.
[*23*] Stepniczka, H. E., *Industrial and Engineering Chemistry Products Research and Development*, Vol. 12, 1973, pp. 29–41.
[*24*] Baitinger, W. F., *Proceedings of Symposium on Textile Flammability*, LeBlanc Research Corp., RI, 1978, p. 168.

William F. Baitinger[1] *and Ludmilla Konopasek*[2]

Thermal Insulative Performance of Single-Layer and Multiple-Layer Fabric Assemblies

REFERENCE: Baitinger, W. F. and Konopasek, L., "**Thermal Insulative Performance of Single-Layer and Multiple-Layer Fabric Assemblies,**" *Performance of Protective Clothing, ASTM STP 900,* R. L. Barker and G. C. Coletta, Eds., American Society for Testing and Materials, Philadelphia, 1986, pp. 421–437.

ABSTRACT: A variety of cotton fabrics treated to confer flame resistance characteristics were examined with the use of highly specialized laboratory equipment and procedures to predict their potential for thermal protection. Thermal sources based on convective (flame) and radiant exposures that approximate possible high-risk situations were employed. At 1.0 and 2.0-cal/cm^2/s heat flux levels of exposure, fabric weight or thickness was found to be directly related to the amount of thermal energy passed through to the sensor.

Several investigators are currently examining the properties of wool, Nomex, flame resistant (FR) cotton, and other protective fabrics by similar approaches. Our work emphasizes the state of the art with flame resistant cotton fabrics. Very little has been published previously on multiple-layer assemblies and the effect of air spaces in such assemblies, so only a few model systems have been reported. Air spacing in multiple-layer systems has been found to be a significant contributor to insulation.

KEY WORDS: cotton fabrics, flame resistance, thermal protection, convective heat flux, radiant heat flux, protection time, single-layer assemblies, multiple-layer assemblies, insulative performance, protective clothing

The degree to which an apparel fabric will delay the formation of a second-degree burn to the wearer when he is exposed to an external heat source is the most singularly significant piece of information required to predict its protective capacity. Simple laboratory tests do not suffice to reveal information on the subject. Logically, we put on a shirt to protect ourselves from overexposure to the sun (a pure radiant heat source) and add layers of clothing for insulation against cold (the opposite of heat exposure); this suggests there is a

[1]Vice-president, Westex, Inc., Chicago, IL 60632.
[2]Albany International Research Co., Dedham, MA 02026.

relationship between the heat source and the body which has protective significance.

In the 1960s, Stoll and Chianta [1] developed sophisticated tests to measure burn potential from external thermal insult related to military requirements for postcrash egress from aircraft and protection from nuclear flash burns. Ratings for different fabrics' performances were developed. More recently, Behnke [2] published an equipment modification. Pintauro and Barker [3] have studied and published on fabric performance characteristics as well. A variation of the methodology was most recently published as ASTM Test for Thermal Protective Performance of Materials for Clothing by Open-Flame Method (D 4108-82).

As the 100% cotton flame resistant (FR) fabrics for industrial safety apparel were developed at Cotton Incorporated, studies were conducted and reported [4] on the insulating capacity of fabrics of various weight by exposing them to different types of heat sources. The present study examining 28 different fabrics was carried out on a piece of equipment developed for Cotton Incorporated by Custom Scientific. In addition to an examination of single-layer fabrics, a study was conducted on multiple-layer systems of flame resistant shirting and pant fabrics in combination with 100% cotton knit underwear fabric not treated for flame resistance.

Experimental Procedures

Materials

Fabrics of various weights, weaves, and colors, all 100% cotton, were selected for this study. See Table 1 for a general description. All of the fabrics were treated with a phosphonium salt-based flame retardant cured with ammonia gas. Generically this system is recognized as the ammonia cure FR process. When tested by a standard test method [5], all had vertical char lengths of 15.2 cm (6 in.) or less initially and after washing, thus classifying the fabrics as self-extinguishing.

Instrumental Methods

The thermal protective performance (TPP) tester provides a thermal flux from two heat sources. A pair of Meker burners set at 45° angles provide convective and radiant heat from gas-fed flames controlled by flow rate. A bank of nine quartz tubes (radiant tubes mounted in a heat-resistant enclosure and controlled by three powerstats mounted on a common shaft) controlled by a rheostat provides supplemental radiant heat. Diagrams of this setup are published elsewhere [2,6]. The test specimen is positioned horizontally to the heat source and is held in a folding specimen frame, as described for ASTM Test D 4108-82. The height of the specimen above the heat source determines the incident heat flux. The TPP tester has an electrically timed, water-cooled shutter that is employed to produce the square-wave heat pulse necessary for the sensitive measurements. The temperature rise behind the fabric specimens is measured by a copper calorimeter containing carefully embedded thermocouples (ASTM Test D 4108-82). Heat flux is calculated with the following equation

$$q = \frac{(m)(Cp)(\Delta T)}{(A)(\Delta\theta)} \qquad \text{units} = \frac{\text{cal}}{\text{cm}^2/\text{s}}$$

where

m = mass of copper disk, g,
Cp = heat capacity of copper, cal/g(°C),
T = temperature rise, °C,
A = area of copper disk, cm^2, and
$\Delta\theta$ = time, s.

For the sensor employed for this work, this simplifies as

$$q = \frac{(17.4)(0.0942)(\Delta T_{10})}{(3.14 \times 4)(10)} = 0.013\ (\Delta T)$$

For the 2-cal/cm^2/s exposures, the specimens were exposed to a gas flow sufficient to produce the desired exposure flux when the sensor was positioned 8.9 cm above the unit base plate and the powerstat controlling the radiant bank was set at 38.[3] To obtain a 1.0-cal/cm^2/s exposure, the preceding conditions were maintained and the specimen platform was raised to 14.7 cm above the unit base. Frequent calibrations were run to verify that the heat flux was stable during the course of the specimen exposures.

For the 0.3-cal/cm^2/s radiant (only) exposure, the specimen holder was positioned at 8.9 cm from the brass base plate of the unit, and the powerstat was adjusted to a setting appropriate for the desired flux level (usually 69 to 70).

For the 1.0 and 2.0-cal/cm^2/s tests, the specimen exposure time was 20 s. For the 0.3-cal/cm^2/s tests, the exposure time was 90 s.

All the fabrics were exposed with the fabric and sensor in direct contact (results are in Table 1). For a few specimens, a 0.64-cm spacer was placed between the sensor and the fabric to illustrate the effect of air space (Table 2). Each value reported represents the average of at least three tests. Variations of spacing were employed in the multiple-layer studies, and these will be discussed later.

[3]These conditions were determinations carried out by Barker and Behnke at the du Pont laboratories and forwarded as a private communication. By measurement, they produce a 50/50 radiant/convective mix.

TABLE 1—*Fabric description and protection times for single-layer assemblies.*

Sample No.	Fabric Type	Color	Weight, g/m^2 (oz/yd^2)	Thickness, mm (in.)	Protection Time, s		
					Radiant Heat, 0.3-cal/cm^2/s Contact	Convective and Radiant Heat, 1.0-cal/cm^2/s Contact	Convective and Radiant Heat, 2.0-cal/cm^2/s Contact
1	flannel	print	142.5(4.2)	0.406(0.016)	41.5	6.2	3.4
2	Westex—breeze tone	blue	190.0(5.6)	0.330(0.013)	50.5	6.5	3.7
3	Graniteville—plain jean shirting	blue	203.5(6.0)	0.330(0.013)	30.1	6.4	3.7
4	knit interlock	navy	203.5(6.0)	0.508(0.020)	58.3	7.5	4.5
5	Ameritex—flannel	print	223.9(6.6)	0.508(0.020)	49.7	10.0	4.8
6	Graniteville—supershirting	green	230.6(6.8)	0.457(0.018)	54.0	7.4	3.9
7	Westex—shirting	navy	244.2(7.2)	0.483(0.019)	49.8	7.4	4.2
8	Graniteville—supershirting	blue	244.2(7.2)	0.406(0.016)	50.0	7.3	4.0
9	Graniteville—supershirting	navy	247.6(7.3)	0.457(0.018)	59.7	6.7	4.1
10	Westex—Style 65	blue	264.6(7.8)	0.508(0.020)	...	...	4.3
11	double knit	navy	274.8(8.1)	0.660(0.026)	67.2	8.2	5.1

12	Westex—2/1 twill	blue	298.5(8.8)	0.457(0.018)	...	...	4.6
13	Graniteville—supertwill	green	312.1(9.2)	0.533(0.021)	65.3	10.2	4.9
14	Graniteville—supertwill	yellow	315.4(9.3)	0.533(0.021)	51.7	9.3	4.9
15	Westex—sateen	gray	318.8(9.4)	0.584(0.023)	63.0	9.8	5.4
16	Graniteville—supertwill	navy	325.6(9.6)	0.508(0.020)	74.5	10.3	5.3
17	Westex—Lot 308	yellow	329.0(9.7)	0.610(0.024)	65.2	10.3	5.6
18	Westex—sateen	navy	339.2(10.0)	0.610(0.024)	76.5	9.6	5.8
19	Westex—sateen	green	342.6(10.1)	0.635(0.025)	70.5	10.3	6.0
20	Westex—sateen	cream	352.8(10.4)	0.660(0.026)	62.7	10.4	5.8
21	Westex—twill	green	373.1(11.0)	0.711(0.028)	88.7	10.6	6.3
22	Westex—denim	navy	417.2(12.3)	0.762(0.030)	73.0	10.8	7.0
23	Graniteville—bull denim	green	417.2(12.3)	0.711(0.028)	65.7	11.4	6.5
24	Westex—whipcord	green	424.0(12.5)	0.686(0.027)	77.0	10.8	6.6
25	Graniteville—mallard duck	black	407.0(12.0)	0.686(0.027)	47.4	11.7	6.1
26	Graniteville—mallard duck	white	410.4(12.1)	0.737(0.029)	56.8	...	...
27	Graniteville—mallard duck	yellow	430.8(12.7)	0.711(0.028)	61.6	11.9	6.4
28	Graniteville—mallard duck	red	464.7(13.7)	0.737(0.029)	54.6	...	...

TABLE 2—*Comparison of contact versus spaced fabric/sensor configuration.*

Specimen No.	Fabric	Weight, g/m^2(oz/yd^2)	Thickness, mm (in.)	Protection Times, s					
				Radiant Heat, 0.3 cal/cm^2/s		Convective Heat			
						1.0 cal/cm^2/s		2.0 cal/cm^{2}s	
				Spaced	Contact	Spaced	Contact	Spaced	Contact
2	breezetone	190.0(5.6)	0.330(0.013)	61.2	50.5	7.3	6.5	4.0	3.7
19	sateen	333.2(10.0)	0.610(0.024)	91.8	76.5	10.4	9.6	6.1	5.8
23	denim	417.2(12.3)	0.762(0.030)	113.3	73.0	12.5	10.8	7.1	7.0
25	whipcord	424.0(12.5)	0.686(0.027)	124.3	77.0	12.2	10.8	6.9	6.6

A least-squares analysis was performed at each exposure level for protection time, determined by use of an overlay of the Stoll data [*1, 7*], versus both fabric weight and thickness by using a Tektronix computer and graph plotter with an available program.

Results and Discussion

For the 28 fabrics, arranged generally in order of ascending weights, the protection times (the intersection of the injury curve with the temperature rise curve) are reported in Table 1. The results are summarized for each level as follows:

1. *0.3 Cal/cm²/s Radiant Heat Only*—Protection times range from 41.5 to 88.7 s. A typical curve for this series of runs is illustrated in Fig. 1, and the least-squares analyses are presented in Figs. 2 and 3. For radiant heat, the temperature rise curve is essentially linear through the first minute of exposure, then it rises slightly more steeply. For this level of radiant heat exposure, considerable scatter of the results is the consequence of the contribution of variable fabric parameters such as construction, weave, color, and finish in addition to weight and thickness. An indication of the correlation with increasing fabric weight and thickness is suggested from the data.

2. *1.0 Cal/cm²/s Convective/Radiant Exposure*—Protection times for this series range from 6.2 to 11.9 s. A typical curve for the 1.0 cal/cm²/s exposure is presented in Fig. 4. This curve differs in shape from the previous temperature rise curve (in Fig. 1) in that after a linear rise of about 10 s, a very steep temperature rise occurs, most likely accompanying the rapid loss of water from the cellulosic polymer of cotton that accompanies char formation catalyzed by the flame retardant treatment. Although some scatter is still apparent, the correlations of protection time with weight (Fig. 5) and thickness (Fig. 6) are better than those for the previous example. Unlike the specimens exposed at 0.3 cal/cm²/s, which were not visibly damaged, these specimens show a fully charred portion of the fabric that is exposed to the thermal flux.

3. *2.0 Cal/cm²/s Convective/Radiant Exposure*—Protection times for this series range from 3.4 to 6.6 s, and the results generally parallel those at 1.0 cal/cm²/s. The model curve (Fig. 7) is essentially identical. The statistical data are presented in Figs. 8 and 9 and show a high degree of correlation with both fabric weight and thickness.

A composite summary for all data is presented in Figs. 10 and 11. The effect of a 0.64-cm air gap between the sensor and the fabric in providing extended protection times, particularly at lower heat flux levels, is illustrated for selected fabrics over the weight range in Table 2.

Interesting effects of color on radiant heat exposure of flame resistant cotton duck fabrics of identical construction are summarized in Table 3. For this

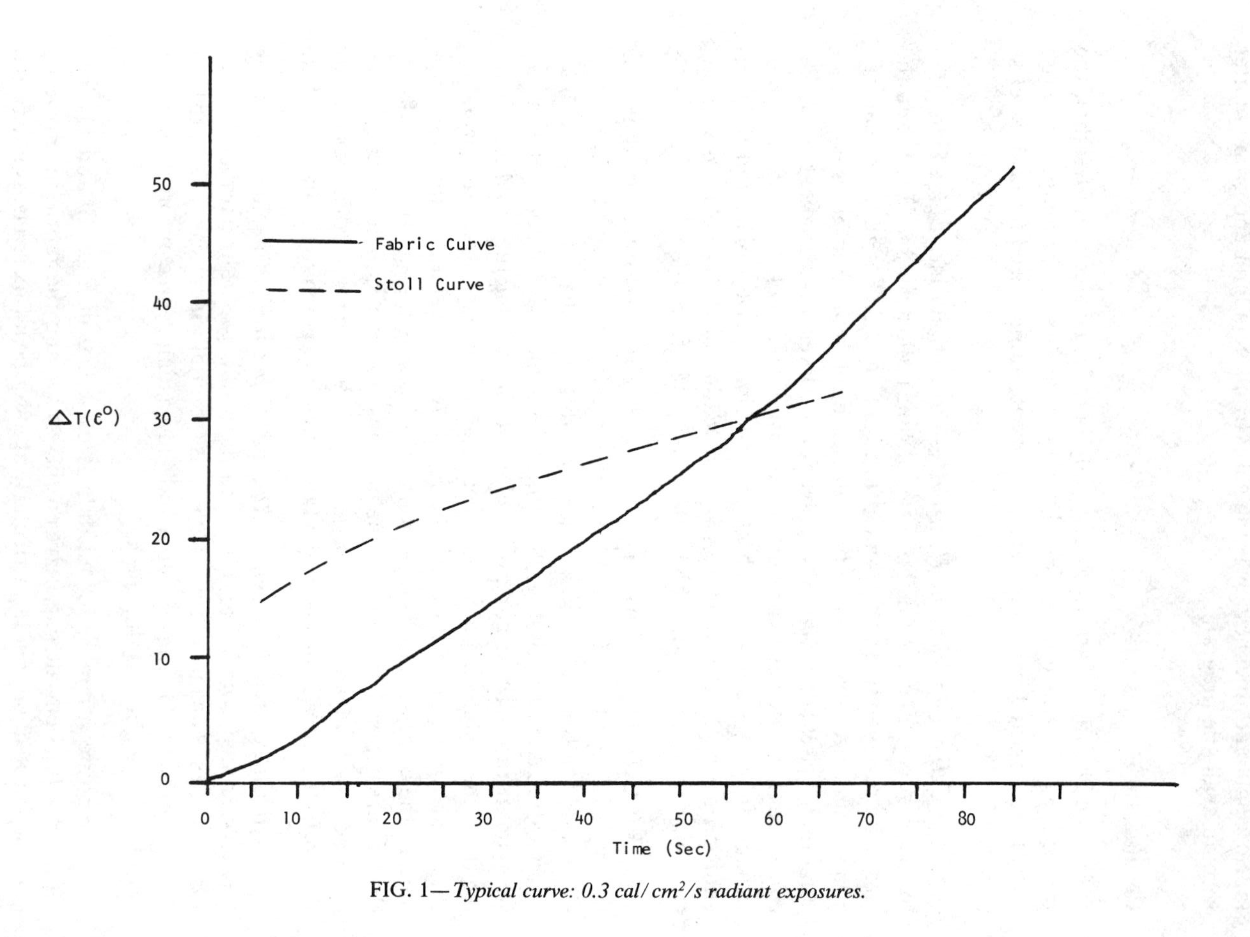

FIG. 1—*Typical curve: 0.3 cal/cm²/s radiant exposures.*

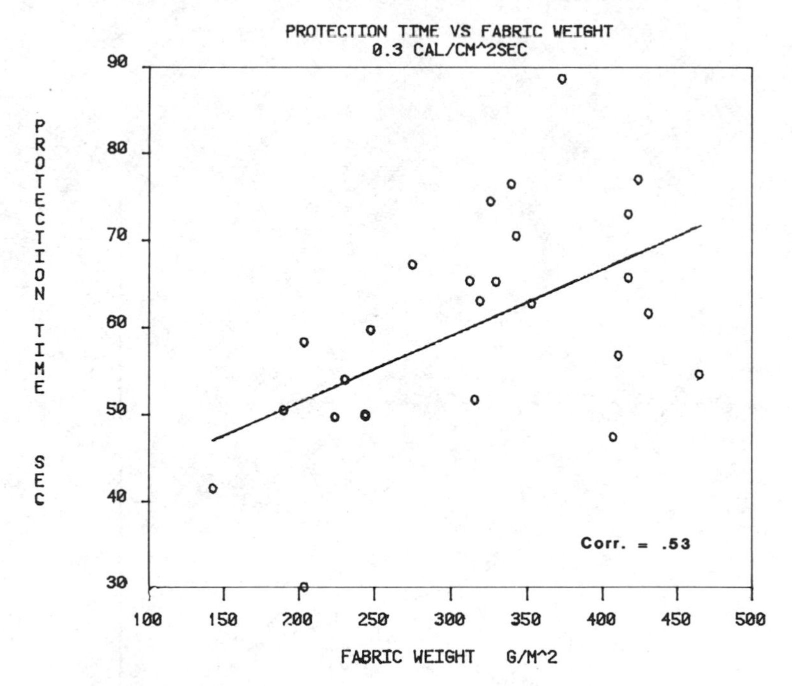

FIG. 2—*Protection time versus fabric weight correlation (0.3 cal/cm²/s).*

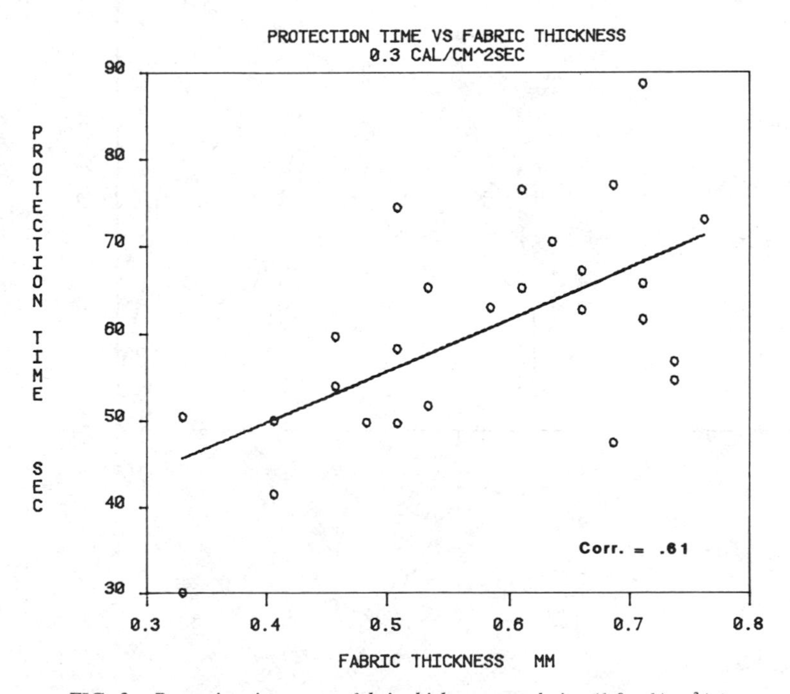

FIG. 3—*Protection time versus fabric thickness correlation (1.0 cal/cm²/s).*

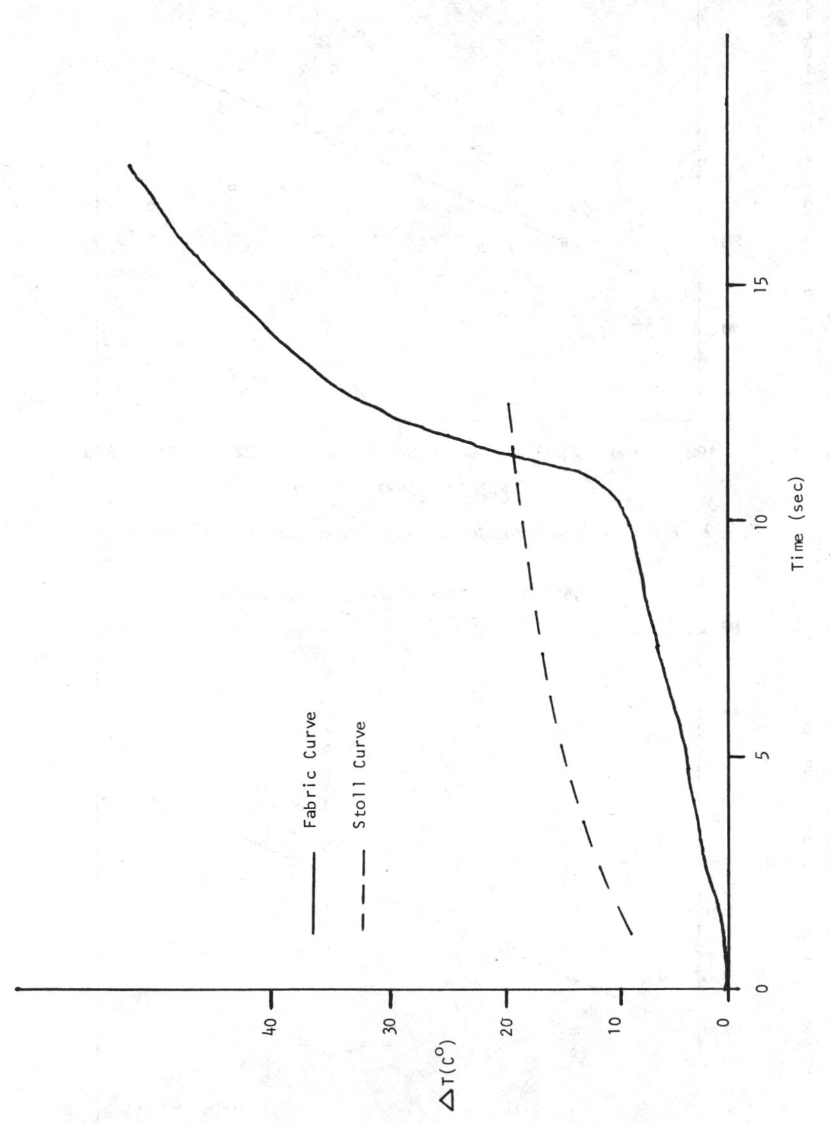

FIG. 4—*Typical curve: 1.0 cal/cm²/s convective/radiant exposure.*

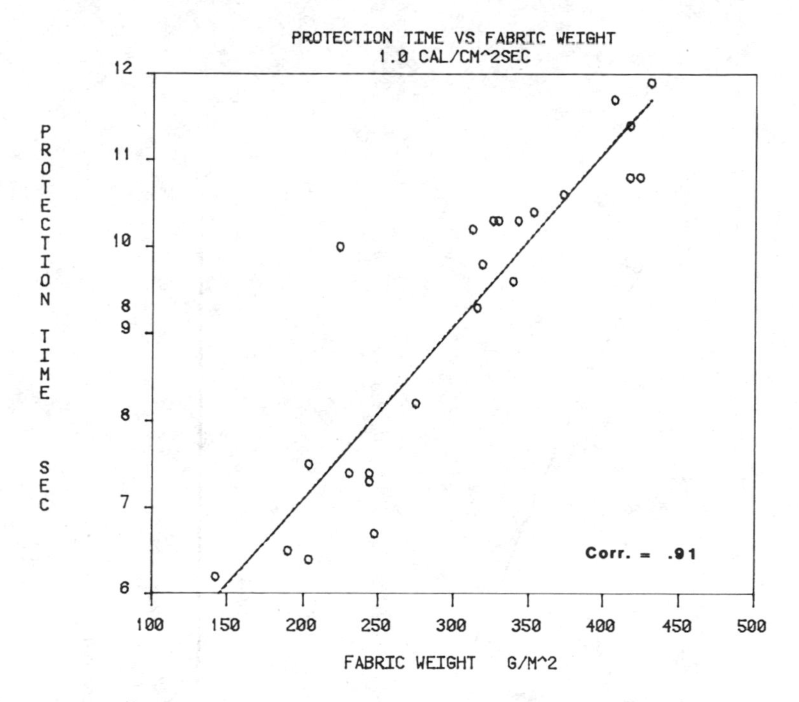

FIG. 5—*Protection time versus fabric weight correlation (1.0 cal/cm²/s).*

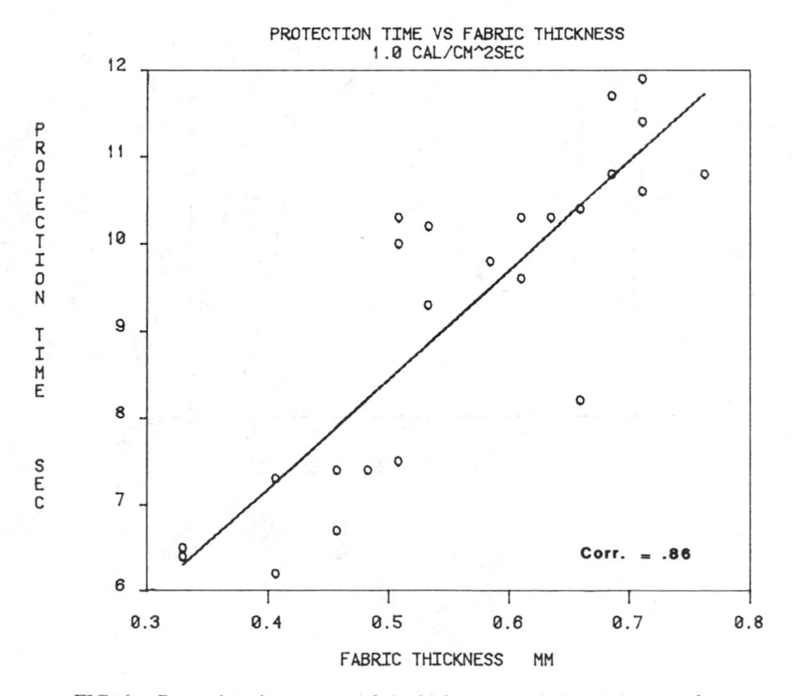

FIG. 6—*Protection time versus fabric thickness correlation (1.0 cal/cm²/s).*

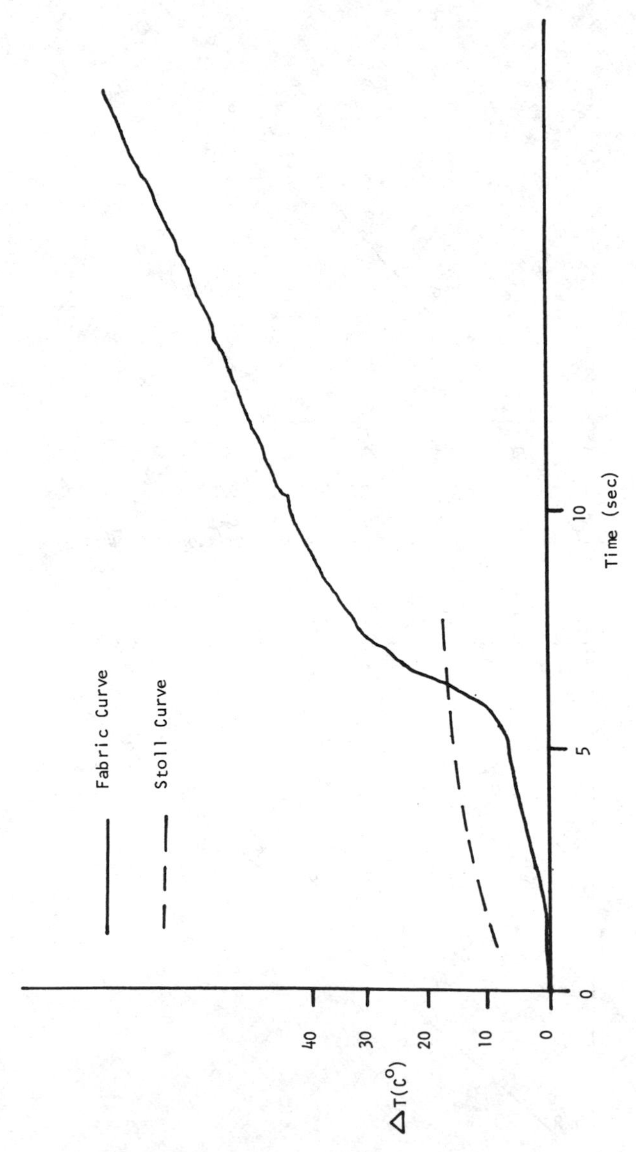

FIG. 7—*Typical curve: 2.0 cal/cm^2/s convective/radiant exposure.*

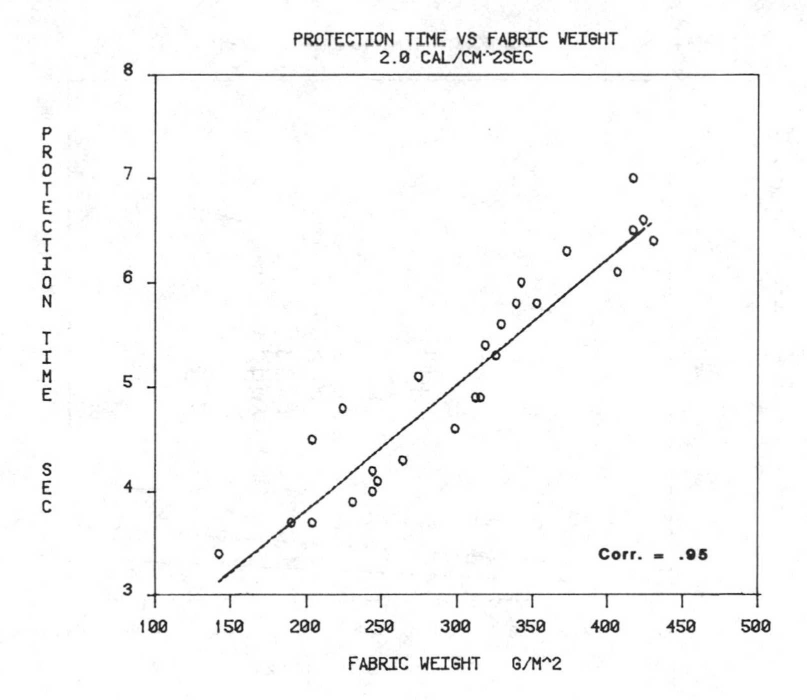

FIG. 8—*Protective time versus fabric weight correlation (2.0 cal/cm²/s).*

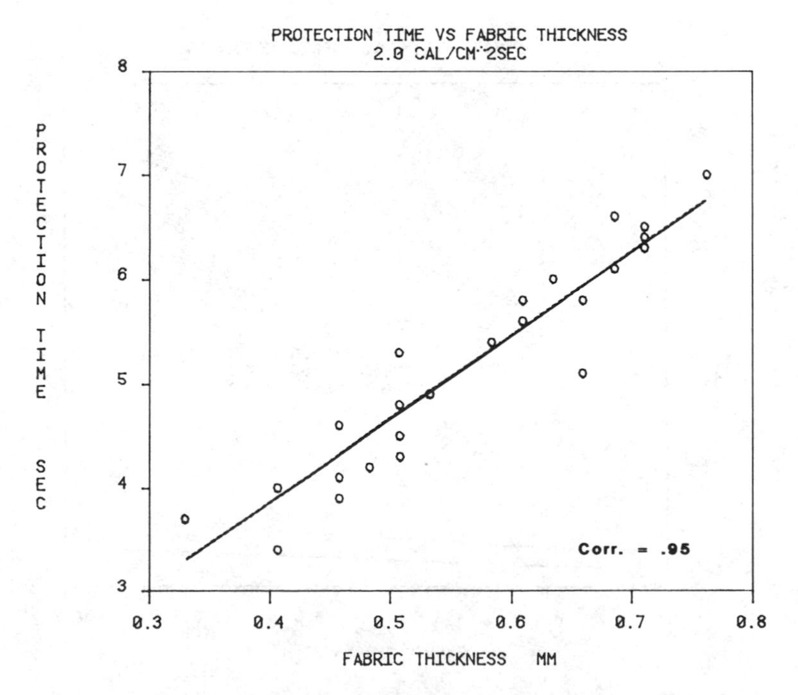

FIG. 9—*Protection time versus fabric thickness correlation (2.0 cal/cm²/s).*

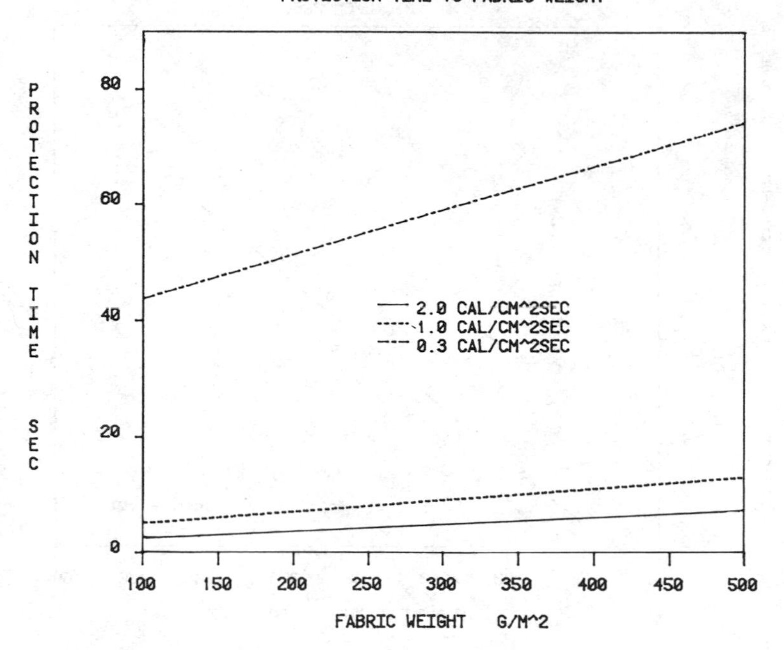

FIG. 10—*Protection time versus fabric weight—summary.*

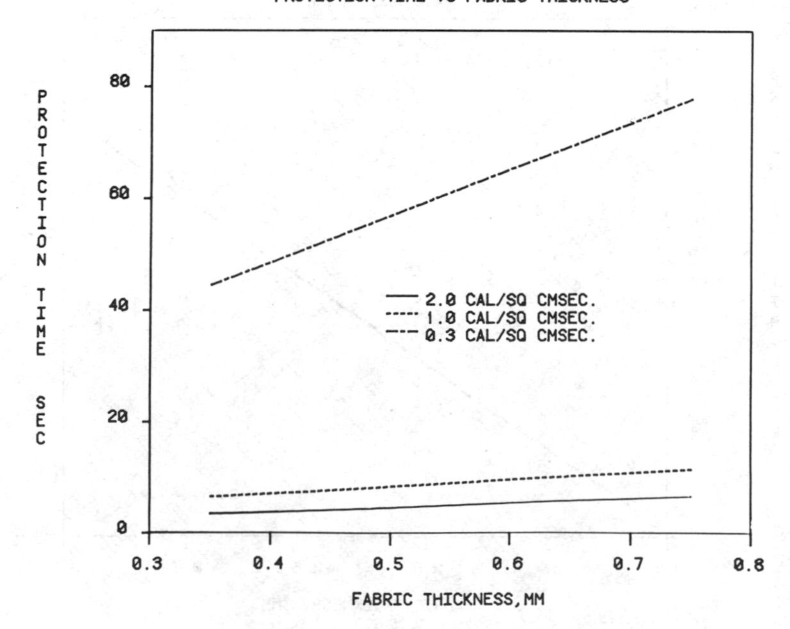

FIG. 11—*Protection time versus fabric thickness—summary.*

TABLE 3—*Effect of color on radiant heat protection for duck fabrics.*

Specimen No.	Color	Weight, g/m²(oz/yd²)	Thickness, mm (in.)	Count, yarns/in., Warp and Fill	Protection Time, s
26	black	407.0(12.0)	0.686(0.027)	84 and 26	47.4
27	white	410.4(12.1)	0.737(0.029)	84 and 27	56.8
28	yellow	430.8(12.7)	0.711(0.028)	85 and 26	61.6
29	red	464.7(13.7)	0.737(0.029)	84 and 27	54.6

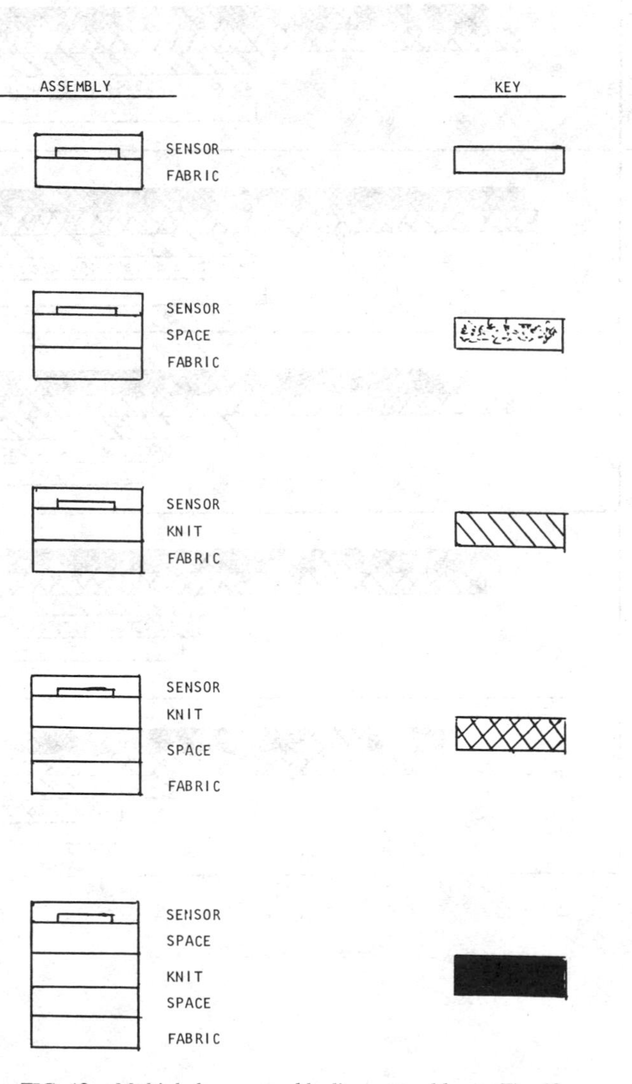

FIG. 12—*Multiple-layer assembly diagram and key to Fig. 13.*

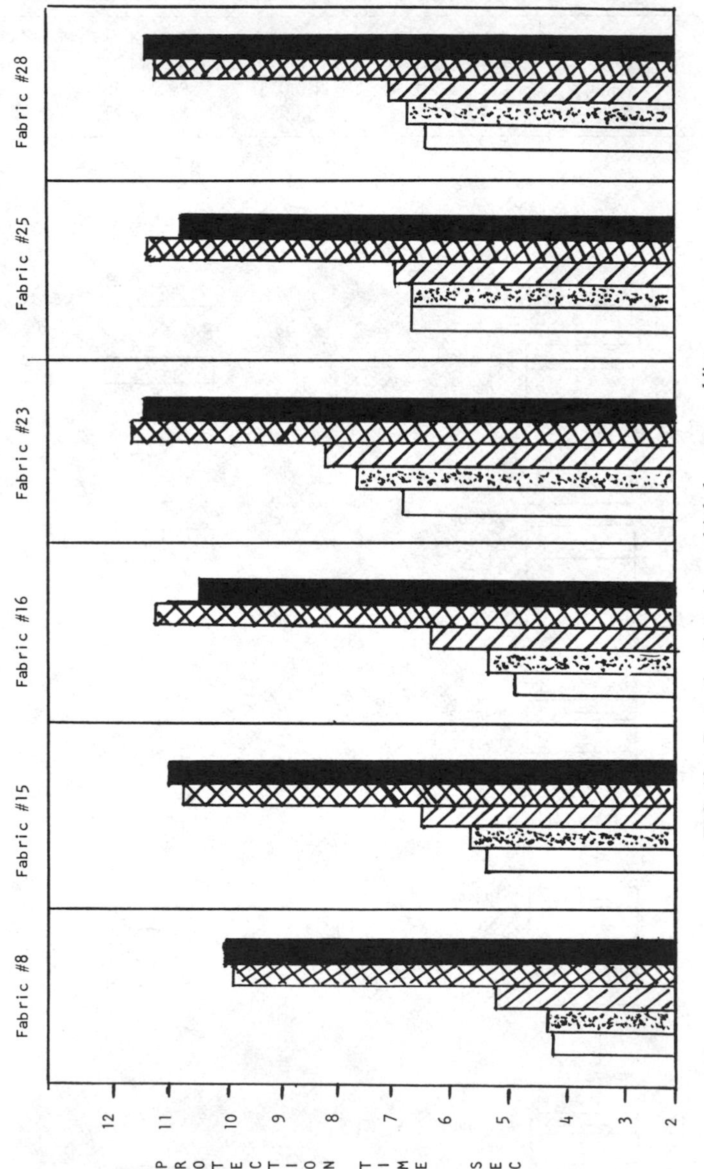

FIG. 13—*Protection times for multiple-layer assemblies.*

series of fabrics (Nos. 26 through 29), the black provides the lowest protection level, the yellow provides the highest level of protection, and red and white dyed fabrics are intermediate.

In addition to the work covered extensively in single-layer assemblies, we were interested in the effect of layering in enhancing insulative protection. The benefit of multiple layers, in addition to providing added fabric thickness, is the inclusion of air space; clothing assemblies are not at all in intimate contact with the skin, and a small 0.64-cm (1/4-in.) air layer can provide increased insulation. Generally speaking, insulation is improved significantly by the use of multiple layers of clothing.

The models of layered systems studied are illustrated in Fig. 12, along with a bar graph key for the actual data obtained (Fig. 13) when specimens were exposed to a 2.0-cal/cm^2/s heat flux, as described previously.

As can be seen in the bar graph results, the dramatic increase in protection time occurs with the combination of outer layer fabric, air spacing, and knit underwear; two to three times the protective capacity of single-layer counterparts is obtained.

Conclusions

By using sophisticated experimental procedures, the heat transferred through single and multiple-layer fabric assemblies can be accurately measured and interpreted in terms of burn potential. Generally, protection increases with increasing fabric weight or fabric thickness. A dramatic increase is observed for multiple-layer air-spaced assemblies. By using this information, safety directors can develop protective clothing systems that maximize personal protection.

References

[*1*] Stoll, A. M. and Chianta, M. A., "Method and Rating System for Evaluation of Thermal Protection," *Aerospace Medicine*, Vol. 40, 1969, pp. 1232-1238.

[*2*] Behnke, W. P., "Thermal Protective Performance Test for Clothing," *Fire Technology*, Vol. 13, 1977, pp. 6-12.

[*3*] Pintauro, B. E. and Barker, R. L., "A Summary of Research on Heat Resistant Fabrics for Protective Clothing," *American Industrial Hygiene Association Journal*, Vol. 44, 1983, pp. 123-130.

[*4*] Baitinger, W. F., "Product Engineering of Safety Apparel Fabrics: Insulation Characteristics of Fire Retardant Cottons," *Textile Research Journal*, Vol. 49, 1979, pp. 221-225.

[*5*] "Flame Resistance of Cloth; Vertical," Method 5903, U.S. Federal Test Method Standard No. 191, Textile Test Methods, 1979.

[*6*] Stoll, A. M., Chianta, M. A., and Piergallini, J. R., "Heat Transfer Measurements of Safety Apparel Fabrics," Technical Report, NADC-78209-60, Naval Air Development Center, Warminster, PA, 1978.

[*7*] Derksen, W. L., Monahan, T. I., and De Lhery, G. P., "The Temperature Associated with Radiant Energy Skin Burns" in *Temperature—Its Measurement and Control in Science and Industry*, C. M. Herzfeld, Ed., Vol. 3, Part III, Reinhold, New York, 1961, pp. 171-175.

Vladimir Mischutin[1] *and David Brown*[2]

Advances in Flame-Retardant Safety Apparel

REFERENCE: Mischutin, V. and Brown, D., **"Advances in Flame-Retardant Safety Apparel,"** *Performance of Protective Clothing, ASTM STP 900*, R. L. Barker and G. C. Coletta, Eds., American Society for Testing and Materials, Philadelphia, 1986, pp. 438–445.

ABSTRACT: The paper discusses recent advances in fabrics used for flame-retardant work wear in those occupations in which people are exposed to open flames, flammable materials, hot molten substances, and high intensity currents.

Some of these work wear needs are currently satisfied by fabrics made of inherently flame-resistant fibers or topically treated 100% cotton or wool. However, these fabrics do not fully satisfy the requirements of some critical professions.

New technology has now been developed to economically flame retard 100% cotton fabrics suitable for use by foundry workers and polyester/cotton fabrics for use in other occupations.

In addition to fire retardancy, fabrics can also be made water-repellent and oil-repellent with good properties for easy care and shrinkage resistance, which makes them attractive to some fashion-conscious professions.

KEY WORDS: protective clothing, flame-retardant fabrics, 100% cotton fabrics, thermal protection, physical properties, molten metals

Flame-retardant work wear has been available in the United States and throughout the world for many years. These protective garments are used by personnel engaged in hazardous occupations, such as fire fighters, foundry workers, and utility workers. Fire-retardant safety apparel is made from two kinds of fabrics: inherently flame-retardant (FR) and topically FR treated fabrics.

Inherently flame-retardant fabrics are produced from Nomex, modacrylic, FR polyester, FR rayon, and other specialty fibers. These fibers have a multi-

[1]Director of flame retardants, White Chemical Corp., Newark, NJ 07114.
[2]Riegel Textile Corp., Greenville, SC.

tude of problems, such as high cost, thermoplastic properties, difficulties in weaving and dyeing, and poor shrinkage properties, which have prevented them from gaining universal acceptance in all industries.

Topically FR-treated fabrics are composed of natural fibers, such as cotton or wool, which have been chemically treated with fire-retardant agents. The most popular treatment for cotton is the so-called ammonia cure process. This process employs a chemical, tetrakishydroxymethyl phosphonium chloride or sulfate (THPC or THPS), which is applied to fabrics by padding, using a precondensate of this chemical with urea; the fabrics are then dried to approximately 20% moisture and passed through a chamber containing anhydrous ammonia gas. After ammoniation, the fabrics are oxydized with hydrogen peroxide and dried.

The ammonia-cure-treated fabrics have gained worldwide acceptance, particularly in the steel industry, since these fabrics have excellent hand, are reasonably priced, and retain the desirable properties of cotton.

Wool is mostly treated by a proprietary process developed by the International Wool Secretariat called the Zirpro process.

Zirpro-treated wool is not as widely used as the FR cotton in secondary protective clothing because the cost of wool is high, its thermal properties make FR wool garments uncomfortable for some people to wear, and above all, wool garments must be dry-cleaned; however, it is widely used for primary protection.

In view of this, the metals industry has relied almost exclusively on the use of cotton garments treated with the ammonia cure process. Then, some ten years ago, it was discovered that ammonia-cure-treated fabrics promoted the sticking of molten aluminum, making these fabrics unusable for workers in the aluminum trade. The reason for this phenomenon is not yet clearly understood, but it is believed that there is some kind of interaction between phosphorus and the molten aluminum. A similar finding was made with other nonferrous metals.

This finding caused concern throughout the aluminum industry, which, after the steel industry, is the largest employer in the metals field. As a result, the Aluminum Association began an intensive research program to find suitable substitutes for the ammonia-cure-treated fabrics. During this program dozens of flame-retarded fabrics were evaluated using the molten aluminum pour test. After all the tests were completed, all the fabrics were rated according to their metal sticking and thermal resistant properties.

Two types of fabrics emerged as the forerunners in this series of tests: Zirpro-treated wool and cotton fabrics treated with the White Chemical Corporation's Caliban flame retardant. The problems inherent with 100% wool fabrics were mentioned before, and in view of them the general consensus in the industry was to adopt the Caliban treatment as the flame-retardant treatment of choice for garments used in the aluminum trade.

Discussion

The Caliban finish for textiles is an aqueous dispersion of two chemicals, decabromodiphenyloxide

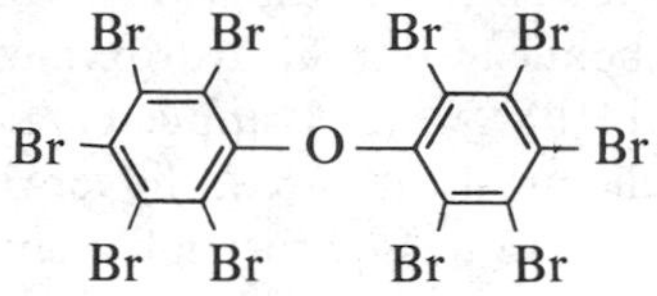

and antimony trioxide (Sb_2O_3), both materials having been reduced to a fine particle size of 1 μm or less and dispersed in water at a high concentration of over 65% by weight. Both ingredients in the Caliban flame retardant are totally inert and have no reactive sites that would allow them to attach themselves to the cotton fibers. In view of this, it is necessary to use some kind of binder that will glue the particles of flame retardant to the fibers and thus make the material durable. Numerous binder systems can be used with Caliban; however, since the fabrics must retain as much as possible of their desirable properties, such as softness of hand, durability in washings, air and water vapor permeabilities, and water absorbency, the selection of the right binder is very critical. Acrylic latex binders, and more specifically those that contain butyl acrylate as the major component, are known to have the softest hand with outstanding wash durability. However, the butyl acrylate polymers are hydrophobic in nature and, when applied to fabrics, impair water vapor and air permeabilities. Therefore, a totally new butyl acrylate binder had to be developed, which would be hydrophilic and have improved permeability properties without loss of softness or durability.

This binder, which is called Caliban Binder 145, is commercially available and has been mill tested on a commercial scale. The initial evaluations of Caliban-treated fabrics were performed on laboratory-treated specimens.

To satisfy increasing industry demand for commercial quantities of treated fabrics, it was necessary to enlist the help of a textile company, preferably a vertical operation to ensure close control of the greige fabrics. A major textile manufacturer, Riegel Textile Corp. of Greenville, South Carolina, undertook the task of scaling up the Caliban process and bringing it to commercial reality.

The road from the laboratory specimens to commercial-scale production was not an easy one. It took a considerable number of trials to produce acceptable-quality fabrics reproducibly and consistently on a commercial scale. At this point, all the trial work has been completed, and Caliban-treated fabrics are available on demand.

Fabric Evaluation

Fabric properties that are of significance for primary protective clothing for foundry workers include the following:

(*a*) physical properties, such as, strength, stability, and ease of care,
(*b*) flame resistance, and
(*c*) thermal protection on contact with molten metals.

All the physical property and flame resistance tests were performed by the fabric finisher. Protective properties against molten metal contact were evaluated at a major aluminum manufacturer and were limited to aluminum metal.

It is worth noting that the physical properties of the untreated denim (see Table 1) were determined on regular finished fabrics, which are normally sold to the apparel market. Prior to application of the Caliban finish, these fabrics were mill washed to remove all sizing and then flame-retardant treated. The treated fabrics, when tested on the molten metal pour tester, gave a temperature rise well below the blister line with no metal sticking or fabric ignition and minimal scorching (Fig. 1).

The shirting fabric also exhibited excellent thermal protection properties when tested using the molten aluminum pour test with no metal sticking and minimal fabric scorching (Fig. 2).

TABLE 1—*Physical properties of 100% cotton denim.*

Property	Untreated	Treated
Style	TSH[a]	DFR[b]
Weight, g/m^2	380	460
Construction (W by F), ends/cm[c]	27 by 17	27 by 17
Tensile strength (W by F), g	64 by 41	82 by 47
Tear strength (W by F), g	4100 by 3200	2400 by 1800
Flex abrasion (W by F), cycles	1000 by 1000	>1000 by >1000
Shrinkage (W by F), %		
After 3 washes at 60°C	3 by 3	3 by 0.7
After 25 washes at 60°C	NA[d]	5.4 by 0.7
Appearance rating after 25 washes at 60°C	NA	4+
Air permeability, m^3/min × cm^2	NA	11
Stiffness (Chatillon test), kg	5.9	3.3
Char length (W by F), cm		
Initial	BEL[e]	6.6 by 7.4
After 30 washes at 60°C	BEL	9.6 by 8.1

[a]TSH =
[b]DFR =
[c]W by F = warp by fill.
[d]NA = not available.
[e]BEL = burned entire length.

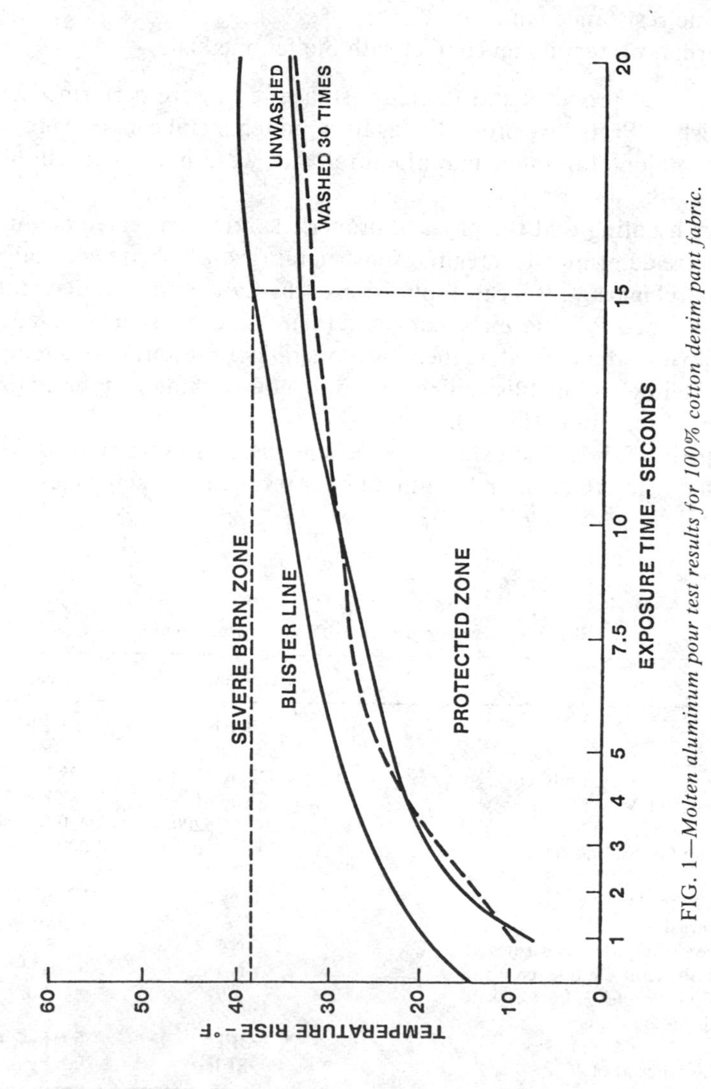

FIG. 1—*Molten aluminum pour test results for 100% cotton denim pant fabric.*

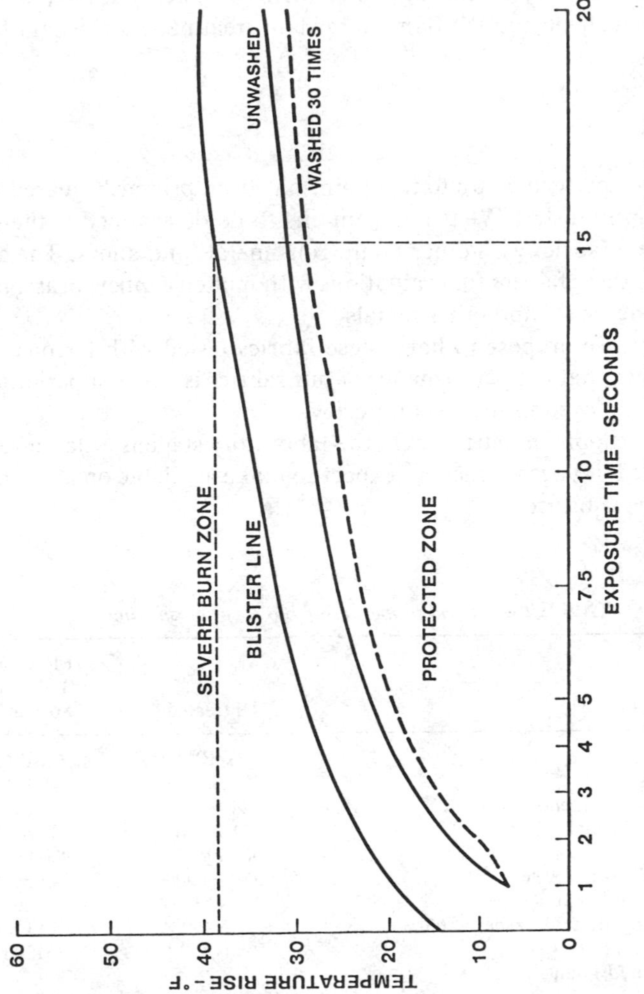

FIG. 2—*Molten aluminum pour test results for 100% cotton shirting fabric.*

As can be seen from Tables 1 and 2, both treated fabrics had excellent flame retardancy, which remained practically unchanged through 30 washings. Two important properties that make the Caliban-treated fabrics superior to the ammonia-cure-treated fabrics are the low shrinkage and the excellent wash-and-wear appearance ratings of the fabrics after multiple washings.

Other important factors that should be mentioned are that the fabrics contain no formaldehyde and the flame-retardant treatment is resistant to chlorine bleaches.

Future Goals

All the work performed up to this point has been primarily geared toward the aluminum industry. What now remains to be determined is the performance of these fabrics with other nonferrous metals and alloys. The authors have targeted these fabrics for evaluations with molten copper, brass, bronze, magnesia, zinc, lead, and other metals.

In addition, we propose to have these fabrics tested with ferrous metals, since we believe that the performance of our fabrics is vastly superior to what is currently employed in the steel industry.

Another development currently in the laboratory stage is a flame-retarded flannel usable for gloves, which we expect to make available on a commercial scale in the near future.

TABLE 2—*Physical properties of 100% cotton shirting.*

Property	Untreated	FR Treated (Laboratory Specimens)
Style	PWBW[a]	SFR[b]
Weight, g/m^2	204	282
Construction (W by F), ends/in.[c]	24 by 24	24 by 24
Tensile strength (W by F), kg	21 by 18	37 by 24
Tear strength (W by F), g	1600 by 1200	900 by 750
Flex abrasion (W by F), cycles	350 by 300	342 by 300
Air permeability, m^3/min × cm^2	NA[d]	26
Appearance rating after 30 washes at 60°C	NA	4.0+
Stiffness (Chatillon test), g	NA	1300
Char length (W by F), cm		
Initial	BEL[e]	9.4 by 11.4
After 30 washes at 60°C	BEL	12.2 by 12.0

[a]PWBW =
[b]SFR =
[c]W by F = warp by fill; 1 in. = 2.54 cm.
[d]NA = not available.
[e]BEL = burned entire length.

Acknowledgments

The authors wish to thank Louis Stone from Riegel Textile Corp., Ware Shoals Division, for his support in the laboratory and in the plant. Particularly we wish to express our appreciation to Gerry Fleming from Kaiser Aluminum and Chemical Corp., Oakland, California, without whose help and support this program could not have been completed successfully.

Bal Dixit[1]

Performance of Protective Clothing: Development and Testing of Asbestos Substitutes

REFERENCE: Dixit, B., **"Performance of Protective Clothing: Development and Testing of Asbestos Substitutes,"** *Performance of Protective Clothing, ASTM STP 900*, R. L. Barker and G. C. Coletta, Eds., American Society for Testing and Materials, Philadelphia, 1986, pp. 446–460.

ABSTRACT: The origin of asbestos use is traced and its unique properties are discussed. A comparison of various substitutes available to replace asbestos shows that there is no single substitute. Development of an effective substitute from fiberglass is described. Lack of proper test methods for measuring the effectiveness of the substitutes in safety clothing required the development of new test methods, which are described. These tests show that Zetex, made from highly texturized fiberglass, offers insulation and strength properties superior to those of asbestos. Results of molten metal splash resistance tests on asbestos and Zetex and the effects of various treatments are discussed.

KEY WORDS: protective clothing, asbestos, Zetex, asbestos substitutes, protective clothing evaluation, glove evaluation, texturized fiberglass

To develop a substitute for asbestos, it is necessary to understand its nature, properties, and diverse applications. Asbestos has been known for more than 2000 years. Archelogical studies in Finland have shown evidence of asbestos fibers in pottery as far back as 2500 B.C. In the fifth century B.C. asbestos cloth was used for cremation. In the first century A.D. Romans used asbestos in lamps in some of the Greek temples, and in 15th century asbestos found its way into body armor. Asbestos textiles, gloves, socks, and handbags were manufactured in Russia around 1720. The commercial production of asbestos began in Italy around 1850 with the manufacture of paper and cloth. With the Industrial Revolution and the development of asbestos mining in Canada and South Africa in 1880, the manufacture of various asbestos products grew rapidly.

[1]President, Newtex Industries, Inc., Victor, NY 14564.

Asbestos is the common name for a group of fibrous silicates that occur in different forms in many countries throughout the world. The word *asbestos*, derived from the ancient Greek language, means unquenchable. Some 30 or more minerals of fibrous structure comprise the asbestiform group, but only 6 have economic significance. Out of the 6, chrysotile is the most commonly used asbestos and accounts for 95% of the total world production of natural mineral fibers.

During the last 20 years various products with different trade names have been offered as replacements for asbestos, but generally they are made from the following types of basic raw materials:

(*a*) fiberglass,
(*b*) aramid,
(*c*) high silica or leached fiberglass,
(*d*) ceramic, and
(*e*) heat-stabilized carbon fiber.

These fibers offer more high temperature resistance than cotton, polyamide, or polyester fibers and have been considered suitable substitutes for asbestos. Table 1 compares the properties of asbestos and substitute fibers. It is evident from the properties shown in Table 1 that no single substitute fiber material possesses all the desired qualities to the same degree as asbestos. Some of the properties are better for substitutes—for example, fiberglass and aramid fibers have better tensile strength than asbestos. Fiberglass and high silica fibers offer better continuous operating temperature than asbestos, but both products lack the cut resistance of asbestos. Aramid fibers have excellent cut resistance but lack the high temperature and thermal insulation properties of asbestos. The thermal insulation properties of carbon and high silica fabrics are not as good as those of asbestos, and the cost of substitute material, except for fiberglass, is higher. Ceramic fibers are reinforced with a carrier fiber just like asbestos and generally exhibit similar properties in textile form but are weaker and very expensive. Continuous-filament ceramic yarns are available, but they are very expensive and additional work is necessary to improve the yarn handling characteristics.

Another important consideration in the selection of a suitable substitute is the effect of temperature. Since asbestos is used in applications involving high temperature, many of the organic fibers may not be suitable as they are likely to give off toxic gases such as hydrogen cyanide and carbon monoxide.

However, for any particular application a substitute fiber can often be employed to achieve the necessary combination of properties. This paper details the extensive amount of work that has gone into developing an economical asbestos-free substitute from fiberglass. In order to understand the process, let us first examine the unique properties of asbestos, which are the following:

(*a*) resistance to heat,
(*b*) strength,

TABLE 1—*Comparison of properties of asbestos and substitute materials.*

Property[a]	Asbestos	Fiberglass	Aramid	Leached Fiberglass	Carbon (Heat-Stabilized Polyacrylonitrile)	Ceramic
Specific gravity	2.50	2.54	1.44	2.54	1.40	2.73
Fiber strength, psi						
At 72°F	100×10^3	500×10^3	200×10^3	100×10^3	100×10^3	250×10^3
At 1000°F	0	250×10^3	0	. . .	0	N/A
Continuous operating temperature, °F	750	1100	600	1800	600	2300
Melting point, °F	2770 (fusion point)	2000	800 (carbonize)	3000	5000	3200
Resists molten steel[b]	yes	yes	no	yes	yes	yes
Thermal insulation	good	excellent	fair	fair	fair	excellent
Toxic off-gasing at 1000°C						
Hydrogen cyanide (HCN), ppm	trace	trace	2 000	4.5[c]	yes[d]	trace
Carbon monoxide (CO), ppm	trace	trace	10 000	650[c]	yes[d]	trace
Die cut ability	good	good	poor	excellent	good	good
Ability to be sewn	good	good	excellent	good	good	good
Seam strength	good	good	excellent	poor	good	good
Product handling ability	good	fair	excellent	fair	good	fair
Abrasion resistance	fair	fair	excellent	poor	poor	fair
Cut resistance	good	fair	excellent	poor	fair	fair
Weight loss at 1000°F, %	15 to 35	2 to 4	100	2 to 4	up to 100	up to 25
Fiber diameter, μm	0.02	6	12	9	9	1 to 12
Price per square yard, $	5 to 14	3 to 12	8 to 28	8 to 25	10 to 30	80 to 135

[a]Metric conversion factors:
1 psi = 0.07 kg/cm^2.
°C = 5/9 (°F − 32).
1 yd^2 = 0.8361 m^2.
[b]Ability to withstand at least 6.35-mm (1/4-in.)-thick A-36 hot-rolled steel plate cut 203.2 mm (8 in.) directly above the fabric.
[c]For a 4.18-m^2 (5-yd^2) specimen.
[d]Quantities unknown.

(*c*) thermal insulation,
(*d*) durability,
(*e*) corrosion and rot resistance,
(*f*) ability to withstand molten metal,
(*g*) ability for the product to be cut, sewn, and handled, and
(*h*) good electrical properties.

In addition to possessing these properties, asbestos is very economical. In 1977–78, a major effort was launched to find a suitable material for asbestos replacement. After studying various alternatives, the author decided to concentrate on fiberglass, since fiberglass exhibited the following properties:

(*a*) resistance to heat,
(*b*) strength,
(*c*) corrosion and rot resistance, and
(*d*) outstanding electrical properties.

In addition, fiberglass is inorganic and also economical. However, the continuous-filament fiberglass did not have the bulk and insulation properties, had poor abrasion resistance compared with that of asbestos, and had very poor seam strength, which is very important in fabricating safety clothing. In order to improve the thermal insulation properties, the author decided to texturize fiberglass highly in order to trap more air in the yarn and fabric structure. The texturizing also allowed several strands of fiberglass filament yarn to be combined into one heavy textured yarn without twisting. The most commonly used asbestos fabric is either a commercial or Underwriters' Laboratories-grade asbestos weighing 1220 to 1350 g/m^2 (36 to 40 oz/yd^2). By analyzing the construction, the proper size of texturized fiberglass yarn and fabric construction were chosen to duplicate the bulk, thickness, appearance, hand, and feel of asbestos fabric. This was accomplished by using a 10 by 8 construction with 1.75-cut to 5.00-cut yarn. The properties of the resulting fabric in comparison with those of asbestos, are shown in Table 2.

The data in Table 2 show that it was possible to achieve the same bulk, hand, feel, and thickness as are found in asbestos fabric by using the heavy textured yarns. In fact, the 1200 g/m^2 (35-oz) fabric that was offered as a substitute for 1350 g/m^2 (40-oz) asbestos fabric shows better strength and thickness, even though it was 12.5% lighter than asbestos. The 1200 g/m^2 (35-oz) fabric was considerably stronger than asbestos and was twice as good an insulator as asbestos. The seam strength and abrasion resistance also improved and were comparable to those of asbestos.

In order to improve the ability of the fabric to be cut, sewn, and handled, a nonhalogen type of proprietary treatment was developed. This treatment was applied to one side of the fabric in order to retain the soft hand and feel of the fabric on the other side. The treatment provides excellent cutting, sewing, and handling properties and also improves the abrasion resistance of the fab-

TABLE 2—*Comparison of properties of asbestos and texturized fiberglass (Zetex).*

Property[a]	Asbestos, Chrysotile	Zetex 1200, G Filament	Zetex 1100, DE Filament
Weight			
oz/yd^2	40	35	32
g/m^2	1350	1200	1100
Thickness			
in.	0.075	0.080	0.070
mm	1.905	2.032	1.778
Construction, ends and picks/in.	20 × 10	10 × 8	10 × 9
Breaking strength, lb			
Warp	160	346	442
Fill	80	423	428
Abrasion testing			
Taber—cycles to failure, (CS10 wheel, 500 g/wheel, 70% vacuum level)	230 to 824	493[b]	871
Thermal properties			
"*K*" value, Btu/h/in./ft^2/°F	0.8190	0.3385	0.3399
Leachable chlorides, ppm	100+	0.5	0.5

[a]Metric conversion factors:
1 in. = 25.4 mm.
1 lb = 0.4516 kg.
1 Btu/h/in./ft^2/°F = 0.81 J/h/cm/cm^2/°C.
[b]CS17 wheel, 1000 g/wheel.

ric. There was some improvement in the irritation factor as the treatment reduced the loose fibers that could become airborne during fabrication.

Development of Safety Clothing

All the work on developing safety clothing was done with Zetex, a trademark name (Newtex Industries, Inc.) for a highly texturized fiberglass fabric. A considerable number of problems were experienced in 1978 and 1979 in manufacturing safety clothing from Zetex. These are briefly described here, along with actions taken to resolve some of the problems:

1. Severe irritation problems were experienced in the beginning. The highly texturized fiberglass yarns were made from G filament, which is about 9 μm in diameter. This filament size was chosen for two reasons. First, it was considerably larger than the carcinogenic fibers such as asbestos, and, second, it was more economical. Because of the considerable difficulty experienced in working with G filament, an intensive effort was launched to manufacture the highly texturized yarns from DE filament, which is 6 μm in diameter. This substantially improved the ability to handle the product and reduced irritation significantly. The DE texturized yarn also improved the abrasion resistance, as shown in Table 2.

2. The conventional design for asbestos gloves and mittens did not work out satisfactorily and had to be modified to allow more seam.

3. The design of the glove had to be changed from an inset thumb to a wing thumb design, which reduced the failure of the glove at the base of the thumb.

4. Even though the abrasion resistance of Zetex fabric was comparable to that of asbestos, the cut resistance of the fabric was inferior. The Zetex gloves and mittens had to be reinforced with leather or lightweight Kevlar in the palm, fingers, and thumb area to improve the cut resistance.

5. Other changes that had to be made in order to improve the performance were in the number of stitches per inch and the type of thread used for sewing the gloves. Surprisingly, the most commonly used thread for sewing asbestos gloves is cotton. We found that the 70 Tex Spun Kevlar (registered trademark of E. I. du Pont de Nemours and Co.) thread was more suitable for manufacturing gloves from Zetex.

Tests were conducted to compare the effect of temperature on single-palm gloves manufactured using commercial-grade asbestos weighing 1350 g/m^2 (40 oz/yd^2) and Zetex weighing 1100 g/m^2 (32 oz/yd^2). All the gloves used during the test had the same type of wool lining, weighing 339 g/m^2 (10 oz/yd^2). Three pipes, each weighing 1.70 kg (3.75 lb) and 76 mm (3 in.) in diameter, were used for test purposes. During the first test, the pipes were heated to 316°C (600°F) in a muffle furnace. Using the gloves to be tested, these heated pipes were held in the hand, as shown in Fig. 1. The rise in temperature was noted every 5 s using a thermocouple during the 1st and 10th cycles. The time, in seconds, for which the pipe could be held for the 1st and 20th cycles for each of the gloves was also noted. The results are tabulated in Tables 3 and 4.

It is obvious from the data shown in Table 3 that, even though Zetex fabric

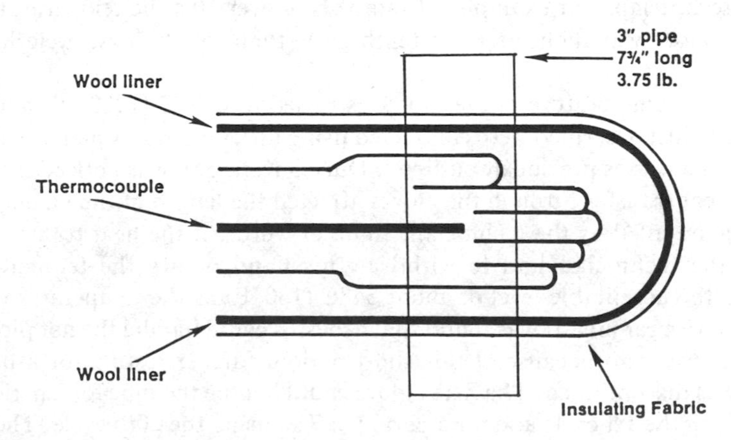

FIG. 1—*Schematic of the testing glove.*

TABLE 3—*Rise in temperature of the glove at 316°C (600°F) for the first and the tenth cycles, in degrees Farenheit.*[a]

	Zetex		Asbestos	
Time, s	Cycle 1	Cycle 10	Cycle 1	Cycle 10
0	77	77	77	77
5	86	86	86	86
10	89	96	95	98
15	99	110	112	115
20	112	121	123	128
25	123	133	131	137
30	133	143	136	145
35	140	...	142	...
40	...	...	...	...

[a]°C = 5/9 (°F − 32).

TABLE 4—*Time for which pipe at 316°C (600°F) could be held during 1st and 20th cycles, in seconds.*

Cycle No.	Zetex	Asbestos
1	32	26
20	26	21

was 20% lighter than asbestos fabric, Zetex provides better insulation than asbestos; for example, at intervals of 10, 15, 20, 25, and 30 s, Zetex shows a lower temperature than asbestos. It is also obvious that the temperature inside the glove rises more rapidly with an increasing number of cycles and, as the data in Table 4 show, reduces the length of time the pipe can be handled by the technician. For example, at intervals of over 10 s the temperature inside the glove was higher for the tenth cycle than for the first cycle for all gloves.

Next, the temperature of the pipe was raised to 649°C (1200°F), and the same tests just described were conducted using different gloves manufactured in the same way as previously outlined. During testing, it was noticed that the rate of heat transfer through the gloves affected the length of time a hot pipe could be handled by the technician. In other words, if the heat transfer was rapid, the technician had to withdraw his hand before the temperature reached the acceptable level of about 54°C (130°F) as the temperature continued to rise rapidly. It was found that asbestos could handle the hot pipe for less than 6 s, and because of this short period, further testing for asbestos gloves was discontinued. The Zetex glove could handle the pipe for a period of 14 s during the 1st cycle and for a period of 7 s during the 20th cycle. The rise in temperature for both materials is tabulated in Table 5.

TABLE 5—*Rise in temperature of the glove at 649°C (1200°F), in degrees Fahrenheit.* [a]

Time, s	Zetex	Asbestos
0	80	80
5	85	105
10	92	135
15	110	...
20	135	...
25	...	...
30	...	...
35	...	...
40	...	...

[a] °C = 5/9 (°F − 32).

The temperature of the pipe was further raised to 871°C (1600°F). Using the Zetex glove, the pipe could be held in the hand during the first cycle for a period of 10 s and during the tenth cycle for 7 s. The palm of the glove in contact with pipe was flexible after the tenth cycle. Further testing could not be continued because of the short period for which the hot pipe could be handled. Asbestos gloves could not be tested at this temperature for the same reason. Single-palm gloves, either asbestos or Zetex, are not recommended at this temperature. This test was conducted to see the effect of temperature on the insulating properties of the glove. In reality, double-palm gloves are used in industry for handling these temperatures.

The Zetex single-palm gloves were then tested for the effect of the weight of the heated object being handled on the insulating properties of the glove. This test was conducted as follows: three pieces of 19 mm (3/4-in.) steel plate weighing 0.453 kg (1 lb), 2.269 kg (5 lb), and 4.537 kg (10 lb) were heated in a muffle furnace to 649°C (1200°F). Figure 2 illustrates the way this test was

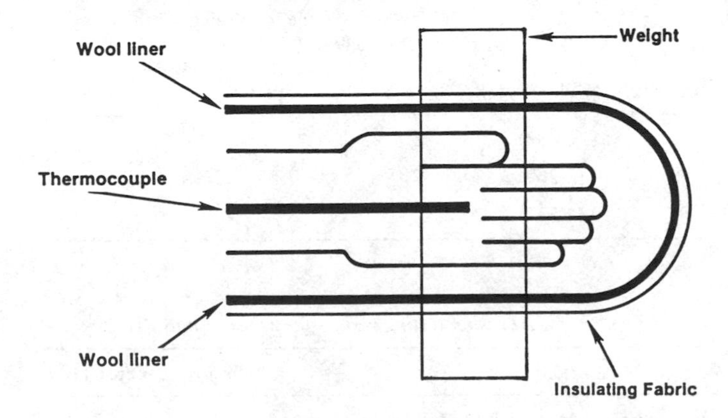

FIG. 2—*Schematic of the testing glove with various weights.*

conducted. The time for which each of these pieces could be held in the hand using Zetex gloves was noted. As expected, the data in Table 6 show that the insulating properties of the glove are adversely affected by the weight of the steel plate. In other words, the insulating properties of Zetex are inversely proportional to the weight of the object. It is a common practice in industry to use double-palm gloves and mittens and ceramic-reinforced double-palm mittens to handle hot and heavy objects.

The Zetex 1100 and the asbestos fabrics used in making gloves were tested for abrasion according to ASTM Test for Abrasion Resistance of Textile Fabrics (Rotary Platform, Double-Head Method) (D 3884-80). The Taber abrasion tester was equipped with CS-10 wheels and 500-g/wheel weight. Using a 70% vacuum, cycles to failure were determined; the data are presented in Table 7.

Although the average numbers of cycles to failure for asbestos and Zetex are not significantly different, it should be noted that the readings for abrasion for asbestos are higher than those that would be obtained under normal testing because the asbestos fabric was coated on both sides while the Zetex was coated on one side only. Also, Zetex 1100 is 20% lighter than asbestos.

Molten Metal Testing

In order to determine the molten metal splash resistance, various Zetex and asbestos fabrics were sent to an independent laboratory where these fabrics were tested and compared for their ability to provide thermal protection against controlled splashes of molten iron at 1510°C (2750°F). The test procedure consisted in pouring about 0.907 kg (2 lb) of molten iron at 1510°C (2750°F) onto fabric specimens attached to a board held at an angle of 70°

TABLE 6—*Effect of weight on the insulating properties of gloves.*

Weight of Plate, lb (kg)	Time for Which Plate at 649°C (1200°F) Could Be Held Using a Zetex Single-Palm Glove, s
1 (0.453)	27
5 (2.269)	15
10 (4.537)	10

TABLE 7—*Abrasion testing results for asbestos and Zetex.*

	Asbestos	Zetex 1100
Average number of cycles to failure	824	871

from the horizontal and from a height of 30.5 cm (12 in.). The test apparatus is shown schematically in Fig. 3. Each fabric was placed on a transite board and held in place with clips along the upper edge. The preheated ladle was filled with molten iron. The weight of molten iron was maintained at 0.907 ± 0.057 kg (2 lb ± 2 oz). The filled ladle was placed in a ladle holder. A fixed delay of 20 s after the start of the furnace pour was used to ensure a consistent temperature. The molten metal was then dumped onto the fabrics, and the results were assessed. The total splash was maintained at 0.5 ± 0.1 s.

Each of the fabrics was examined for (1) visual appearance and (2) heat transfer through the fabric. The visual appearance of each of the outer layers after molten metal impact was rated subjectively in the following four categories;

(*a*) charring,
(*b*) shrinkage,
(*c*) metal adherence, and
(*d*) perforation.

The rating system uses numbers from 1 to 5 in each category, with 1 representing good and 5 representing poor. Table 8 shows the visual ratings of

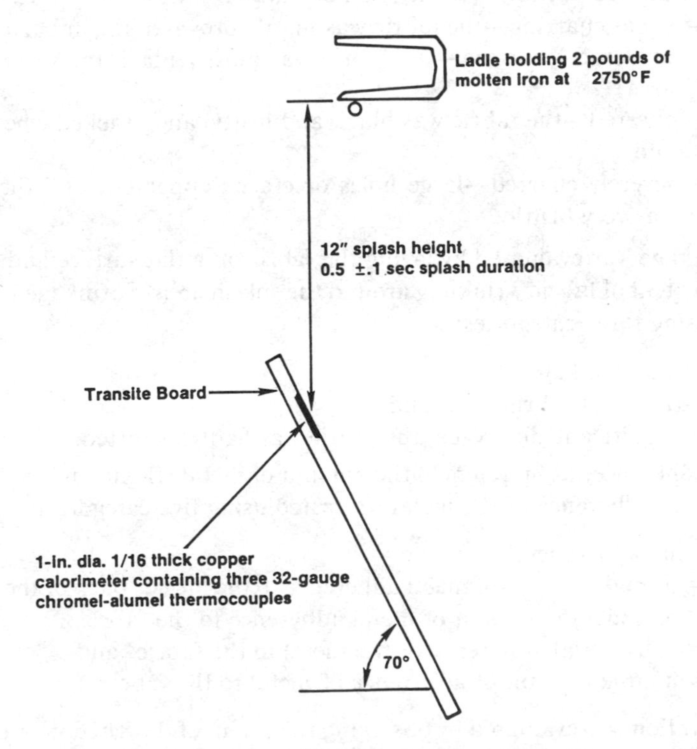

FIG. 3—*Schematic of the molten metal impact apparatus.*

TABLE 8— *Visual ratings of fabrics used in a single layer over a tee-shirt fabric lay-up exposed to molten iron.*

Material Number	Material Designation	Rating of Outer (Impacted) Layer			
		Charring	Shrinkage	Perforation	Adherence
1	asbestos	2	1	4	4
2	Zetex 1200	2	1	4	3
3	Zetex 1200 silicone	2	1	1	1
4	Zetex 1100 with Coating A	3	1	1	3
5	Zetex 1100 with Coating B	2	1	1	3
6	Zetex 800 aluminized	4	1	4	1
7	Zetex 600 aluminized	3	1	3	1

various fabrics exposed to the molten iron test. An outline of the rating system in detail follows.

Grading System Used to Evaluate Fabric Damage—The fabric specimens were evaluated visually for charring, shrinkage, and perforation to provide an indication of the extent of damage to the outer impacted layer. Five grades were used in evaluating the extent of charring:

1 = slight scorching—the fabric had small brown areas;
2 = slight charring—the fabric was mostly brown in the impacted area;
3 = moderate charring—the fabric was mostly black in the impacted area;
4 = charred—the fabric was black and brittle and cracked when bent; and
5 = severely charred—large holes or cracks appeared, and the fabric was very brittle.

Shrinkage was evaluated by laying the fabric on a flat surface and observing the extent of fabric wrinkling around the splash area. Shrinkage was evaluated using three categories:

1 = no shrinkage,
3 = moderate shrinkage, and
5 = significant shrinkage; the fabric was badly distorted.

The adherence rating refers to the amount of metal sticking to the front of the fabric. Adherence of the metal was rated using five categories:

1 = no adherence,
2 = a small amount of metal adherence to the face or back of the fabric,
3 = a moderate amount of metal adherence to the fabric,
4 = substantial adherence of the metal to the fabric, and
5 = a large amount of adherence of metal to the fabric.

Perforation was evaluated by observing the extent of destruction of the fabric, usually by holding it up to a light. Five grades were used in evaluating perforation:

1 = none;
2 = slight, with small holes in the impacted area;
3 = moderate, with holes in the fabric;
4 = metal penetration through the fabric, with some metal retained on the fabric; and
5 = heavy perforation; the fabric exhibited gaping holes or large cracks or substantial metal penetration to the back side.

The heat transfer data for various fabrics attached to the transite board are obtained by a 25.4-mm (1-in. diameter) and 1.6-mm (1/16-in.-thick) copper calorimeter located under the point of molten metal impact. Table 9 shows the average calorimeter temperature rise and the maximum rate of heat flow in molten iron splash evaluation. After comparing the results tabulated in Tables 8 and 9, the following conclusions can be derived:

1. Comparing the readings for asbestos 1350 g/m^2 (40-oz/yd^2 fabric) with those for Zetex 1200, 1200 g/m^2 (35-oz/yd^2 fabric), it can be seen that, although the visual ratings in Table 8 for both are almost identical, the maximum temperature rise as well as the maximum rate of heat flow from Table 9 for asbestos are much higher than those for Zetex 1200 fabric. This difference in the readings for the two fabrics can be attributed to the better insulating property of Zetex, in comparison with asbestos.

2. The coating on the fabric does play an important role in the temperature rise and maximum rate of heat flow during the molten iron splash test. This is shown by Materials 2 and 3. In the case of these fabrics, although the basic fabrics were the same, the coatings were different. The silicone coating does not allow molten metal to adhere to the fabric. This results in less of a temperature rise as well as in a lower rate of heat flow. The importance of the type of coating used is also shown by the results for Fabrics 4 and 5. The basic fabric was the same for both Fabrics 4 and 5, but the coatings were different. Even though the visual ratings of both fabrics are almost identical, the values for

TABLE 9—*Average calorimeter temperature rise and maximum rate of heat flow in molten iron splash evaluations for a single layer over tee-shirt fabric lay-up.*

Material Number	Material Designation	Maximum Temperature Rise, °F[a]		Maximum Rate of Heat Flow, cal/cm^2/s
		5 s	40 s	
1	asbestos	86.8	115.2	3.74
2	Zetex 1200	24.2	68.9	0.43
3	Zetex 1200 silicone	3.9	16.8	0.09
4	Zetex 1100 with Coating A	13.4	37.0	0.30
5	Zetex 1100 with Coating B	22.9	71.4	0.43
6	Zetex 800 aluminized	14.9	26.8	0.27
7	Zetex 600 aluminized	21.6	29.9	0.49

[a]°C = 5/9 (°F − 32).

both the maximum rise in temperature and the maximum rate of heat flow vary significantly.

Figure 4 compares the data for aluminized Zetex 800, aluminized fiberglass, and aluminized asbestos in a different series of molten metal impact tests from a height of 457.2 mm (18 in.). It is obvious from Fig. 4 that the rise in temperature for aluminized asbestos at any given time during the test period is always much higher than that for aluminized Zetex.

Figure 5 compares the results for a molten metal splash height of 304.8 mm (12 in.) with those for a height of 457.2 mm (18 in.) for aluminized Zetex 800. The data from Figs. 4 and 5 are presented in Table 10. It is interesting to note from these data that the height of the molten metal splash has a significant effect on the maximum temperature rise for aluminized Zetex. The maximum temperature after 5 s increases from 1.5 to 14.9°C when the splash height is decreased from 457.2 to 304.8 mm (18 to 12 in.).

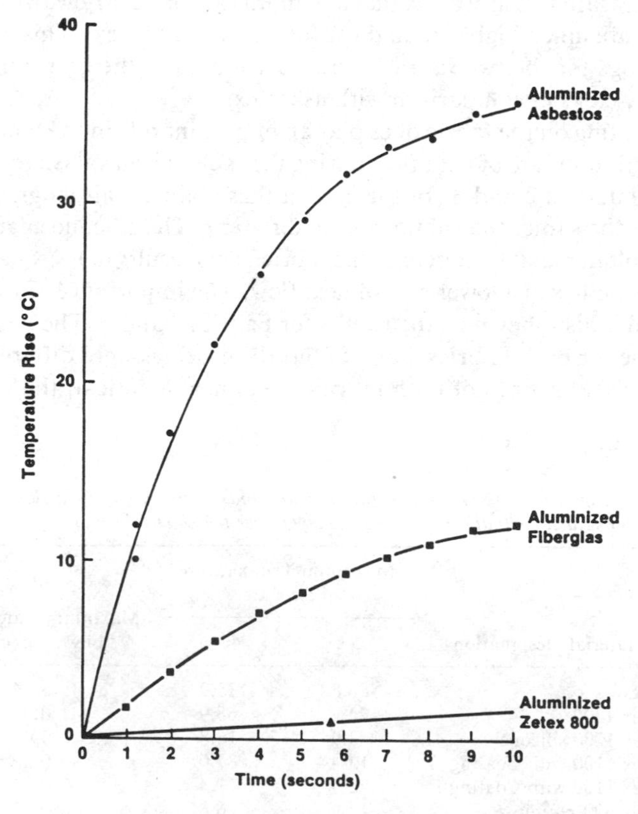

FIG. 4—*Calorimeter temperature rise under several fabrics during and after molten metal impact from a height of 457.2 mm (18 in.).*

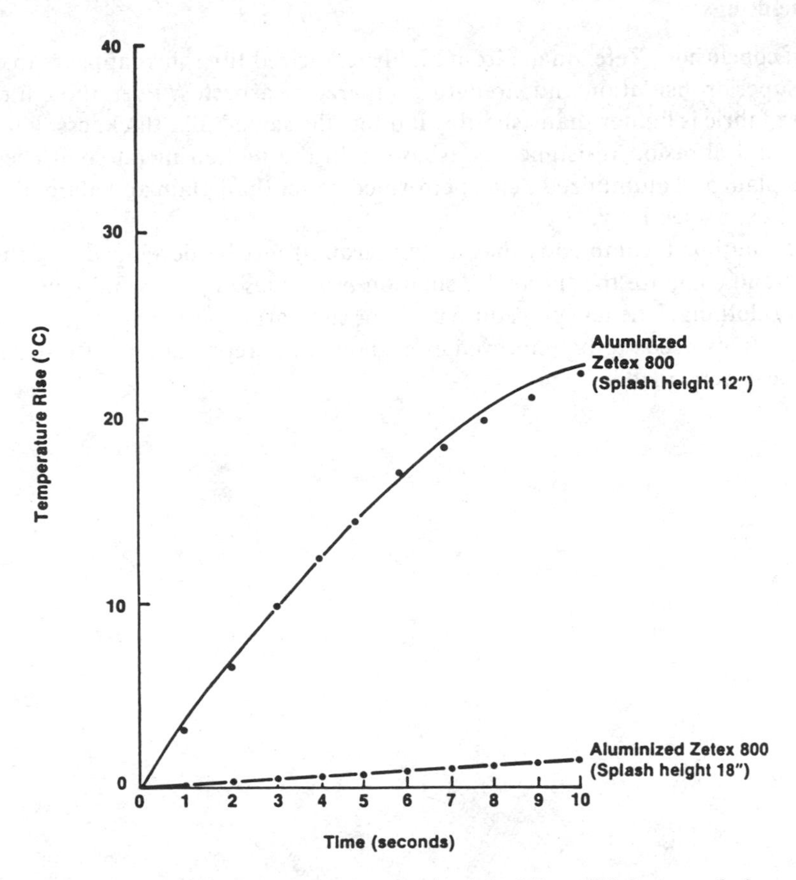

FIG. 5—*Calorimeter temperature rise under aluminized Zetex 800 during and after molten metal impact from heights of 304.8 and 457.2 mm (12 and 18 in.).*

TABLE 10—*Average calorimeter temperature rise in molten iron splash evaluations for aluminized asbestos, aluminized Zetex 800, and aluminized fiberglass.*

	Maximum Temperature Rise, °C, at a Splash Height of 12 in.		Maximum Temperature Rise, °C, at a Splash Height of 18 in.	
Fabric	5 s	10 s	5 s	10 s
Aluminized asbestos	...	...	28.8	36.0
Aluminized Zetex 800	14.9	22.8	1.5	3.0
Aluminized fiberglass	...	...	8.0	12.0

Conclusions

In conclusion, Zetex made from highly texturized fiberglass appears to offer superior insulation and strength compared to asbestos. Even though the Zetex fabric is lighter than asbestos, it offers the same bulk, thickness, hand, feel, and abrasion resistance as asbestos. In the molten metal splash test, both plain and aluminized Zetex performed better than plain and aluminized asbestos, respectively.

The author recommends that a standardized test be developed to determine and compare the thermal insulation properties of gloves, mittens, and safety clothing. The test procedure used for comparing performance of gloves and mittens needs to be improved even though the repeatability of test data has been very good.

Clothing Systems for Industrial and Fire-Fighting Applications

John F. Krasny[1]

Some Characteristics of Fabrics for Heat Protective Garments

REFERENCE: Krasny, J. F., **"Some Characteristics of Fabrics for Heat Protective Garments,"** *Performance of Protective Clothing, ASTM STP 900*, R. L. Barker and G. C. Coletta, Eds., American Society for Testing and Materials, Philadelphia, 1986, pp. 463–474.

ABSTRACT: Some principles of protection afforded by clothing in fire situations are reviewed briefly. Several examples of measurements of heat protective properties are given. The materials covered are single layers of fabrics appropriate for work uniforms, the same type of fabric combined with four popular underwear fabrics, and typical fire fighters' turnout coat assemblies, consisting of a shell fabric, vapor barrier, and thermal barrier. Such protective garment assemblies should not burn; they should retain their integrity, that is, not shrink, melt, or form brittle chars that break during the stresses induced by movement of the wearer; and they should provide as much insulation against heat on the fire scene as possible without interfering with the wearer's activities.

KEY WORDS: burn injuries, clothing, fire fighters, garments, protective clothing, thermal insulation

This paper summarizes some principles of protection by clothing in fire situations [*1–4*] and gives examples of test results on materials used in heat protective garments. In these examples, two types of heat protective garments are considered: turnout coats for structural fire fighters and work garments to be worn in areas where accidental fire may occur, for example by flight attendants and engine room personnel of ships.

Heat exposure in fire situations may consist primarily of radiation, but convective and conductive heat (if, for example, molten metal or hot paint falls on a garment) may be encountered. Under any of these conditions, the garments should not ignite; they should remain intact, that is, not shrink, melt, or form brittle chars which may break open and expose the wearer; and they

[1]Textile technologist, Center for Fire Research, National Bureau of Standards, Gaithersburg, MD 20899.

should provide as much insulation against heat (heat protection) as is consistent with not diminishing the wearer's ability to perform.

Fire fighters fighting a room fire have been reported to be exposed to up to 12.5 kW/m^2 and up to 300°C temperatures for a few minutes [2]. Greater exposures can occur in emergencies, such as a falling burning roof. For such cases, one can only hope that the turnout gear will preserve life and minimize injury.

For the purpose of protecting the wearer from pain and burn injury, several garment characteristics are important. The major protective property is thermal resistance, which, in the first approximation, is related to fabric thickness. For a given thickness, the lower the density, the greater the resistance [1]. Moisture content reduces the resistance. The resistance can be reduced by high temperature, especially if the fibers shrink, melt, or otherwise deteriorate. Furthermore, the thermal resistance decreases with curvature, requiring more thickness to protect fingers than large body areas [1].

Thermal inertia (the product of density times heat capacity times thermal conductivity) is another factor in reducing heat flux through garments when the exposure is of high intensity and short duration [1]. For most organic fibers, this is proportional to the fabric weight. Because garments retain heat, they should be removed as soon as possible after exposure and the skin cooled, especially if pain is already felt; otherwise burn injury may result. The time interval between occurrence of pain and a second-degree burn can be very short [5].

Clean, reflective surfaces are very effective in providing heat protection. Thus, surface temperatures of fabrics exposed to radiation are reported to be reduced to about one half in still air by the use of aluminized surfaces with 90% reflectivity and considerably more in moving air [1]. However, aluminized surfaces lose much of their effectiveness when dirty [6]. For nonreflective fabrics differing widely in construction, including coated fabrics, total reflectance was found to vary from 0.06 to 0.13 for a blackbody source temperature of 500 K, and from 0.07 to 0.35 for 1500 K [6].

Moisture present in a heat protective garment cools the garment, but it may also reduce its thermal resistance and increase the heat stored in it [1]. If the garment gets hot enough, steam may form on the inside and cause burn injury. In most fire fighters' turnout coats in the United States, a vapor barrier is used either on the outside or between the outer shell and the inner liner. This prevents moisture and many corrosive liquids from penetrating to the inside, but, on the other hand, it interferes with the escape of moisture from perspiration and increases the heat stress. Some European fire departments omit vapor barriers.

Another mechanism of heat transfer from a garment to the skin is by the tar and gases formed during fabric decomposition; this has been discussed in some detail in Refs *7* and *8*.

Methods

Ignition Resistance and Flammability

Ignition resistance of fabrics used in heat protective clothing can be measured in various ways. One popular method involves exposing the bottom edge of a vertical specimen to a Bunsen burner flame for 12 s (Method 5903, Flame Resistance of Cloth, Federal Test Method Standard 191). The National Fire Protection Association (NFPA) Standard 1971, Protective Clothing for Structural Fire Fighting (1981), specifies that, in using this method, the length of the charred material must not exceed 100 mm and flaming must cease within 2 s after the flame is removed. This test, in one way, is severe because both sides of the specimen are exposed to the flame. On the other hand, it does not account for preheating and drying of a garment exposed in a fire situation. Actual fire conditions are more closely simulated in tests in which the specimen is exposed to radiation and the time to ignition, with or without pilot ignition, is reported [*9,10*]. The limiting oxygen index (LOI) test is also used to characterize flammability of heat protective fabrics [*2*]. It measures the minimum oxygen concentration at which materials continue to burn [*2,3*]. Normally, materials with an LOI > 21 do not burn in air whereas those with an LOI < 21, do. Most fiber materials have LOIs of around 20%; materials with LOIs of 25% are not easily ignited in air [*2*].

Heat Protective Properties

Heat protective properties of garments can be measured, in the first approximation, by the thickness. All methods to measure actual heat protection employ a radiant or flame heat source, a specimen holder, and a heat sensor placed on the side opposite the heat source. One such method is the ASTM Test for Thermal Protective Performance of Materials for Clothing by Open-Flame Method (D 4108-82). A gas flame impinges on the lower surface of the horizontal specimen. The heat sensor is a copper disk with four thermocouples embedded. When the flame is properly adjusted, the sensor reading 50 mm above the burner rim, without a specimen, corresponds to 84 kW/m^2.[2]

This apparatus has been modified to allow a mixture of radiant and convective heat. This improved version will probably be specified in future issues of the NFPA Standard 1971. Use of this method would replace the present requirement addressing heat protective properties—that is, that the turnout coat assembly be at least 4.5-mm (0.175-in.) thick.

Another apparatus consists of a radiant quartz panel, a vertical specimen holder, and a water-cooled heat flux gage behind the specimen [*10–14*]. To

[2] 84 kW/m^2 = 20 kcal/m^{2}s = 84 kJ/m^{2}s = 7.5 BTU/ft^{2}s.

obtain a specified radiative/convective heat ratio, air can be blown through slots on the periphery of the quartz panel toward the specimen.

Several other methods have been reported in the literature. For example, the military used a JP-4 burner as a laboratory test [*15*] and uses Thermoman, a mannequin with about 110 thermosensors which is pulled through jet fuel pool fires, for full-scale testing [*2*]. Another interesting development is the simulated skin thermosensor, which has thermal properties closer to those of skin than copper disks or heat flux gages [*16*]. The temperature rise during a given time period is used to compare fabrics.

Results and Discussion

Most methods for comparing the heat protective properties of garments rely on the work by Stoll [*5,17*] or Derksen [*18*]. Figure 1 shows the injury curve based on Derksen's results (extrapolated from 30 to 60 s) and a typical time–heat curve measured behind a fabric in the test described in Refs *10* and *11*. As long as the time–heat curve is below the Derksen curve, no burn injury would be expected; above it, at least an incipient second-degree burn could occur. The time at which the time–heat curve crosses the Derksen curve will be called the time to injury in this paper.

Many heat protection tests are carried out with the specimen in contact with the heat sensor (contact mode) or with a predetermined distance between specimen and sensor (distance mode). In real life, the garment–wearer dis-

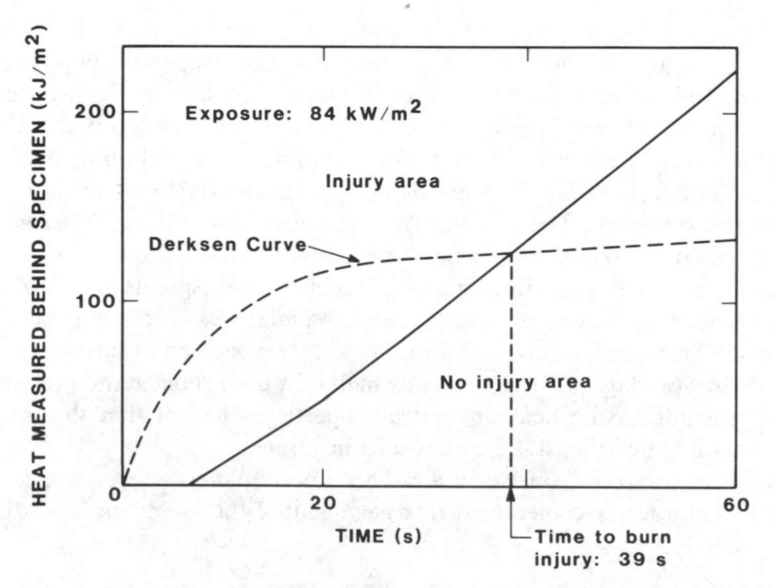

FIG. 1—*Time–heat relationship behind an exposed turnout coat specimen [*10*].*

tance may vary as the wearer moves. The lowest time to injury results are almost always obtained in the contact tests; however, the equally important tendencies of fabrics to shrink, melt, or form brittle chars are more readily observed in the distance mode because the heat sensor does not act as a heat sink. A fabric evaluation should include both modes.

Rank ordering of different fabrics determined from heat flux measurements and analyzed in accordance with the Derksen curve can be expected to predict the rank ordering in actual fires. A recent attempt to verify this is described in Ref *11*. However, the computed times to injury may not necessarily be good measures of the actual times before the skin is injured. The data for the Derksen curve was obtained by exposure of human and porcine skin to square wave radiation, which would not be the real-life exposure mode. Also, skin thickness and thus susceptibility to burn injury varies widely between persons and over the body of any one person.

Three replicates were usually tested in the three studies described here; the scatter of the results was generally about 10%.

Single Layers of Fabrics

Figure 2 shows the effect of radiant and flame exposure on work uniform fabrics containing aramid, flame-retardant (FR) cotton, modacrylic, polyes-

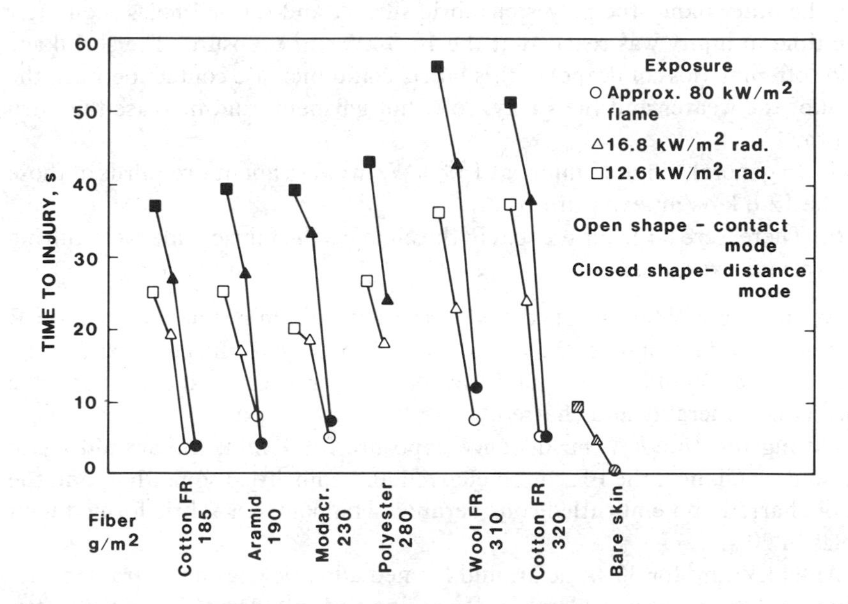

FIG. 2—*Time to injury for various work uniform fabrics from radiant heat and flame exposures.*

ter, and wool fibers [*12–14*]. All these fibers pass the vertical flammability test. Radiant exposures were 12.6 and 16.8 kW/m^2, and the flame exposure was 84 kW/m^2. In each case the specimens were tested in contact with the heat sensor (contact mode) and at some distance from it (distance mode). The distance was 13 mm in the radiation exposure and 3.2 mm in the flame exposure.

The results can be summarized as follows:

1. A single layer of these medium to heavy-weight fabrics (as used in work clothing) extended the time to burn injury by 15 to 45 s in the radiant exposures, in comparison with naked skin exposure. In the flame exposure, this additional time was 3 to 12 s.
2. The time to injury increased with the fabric weight. This can be seen by comparing the two FR cotton fabrics.
3. The wool fabric had longer times to injury (that is, greater insulation) than can be accounted for by its weight. The reasons are probably the greater thickness of wool fabrics than of comparable-weight fabrics made with other fibers. Furthermore, in the contact mode, the hairy surface of wool reduced the intensity of contact with the heat sensor (and with skin, which accounts, to a large degree, for wool's reputation of being comfortable). Both the relatively high thickness and low contact can be ascribed primarily to the crimp and irregular fiber length of wool. Similar modest, beneficial effects of high-crimp weaves have been found in another study of heat protective fabrics [*11*]. On the other hand, the polyester fabric shrank and formed holes soon after the time to injury was reached at the 16.8 kW/m^2 exposure. The shrinkage and softening (loss of drape) of this fabric could increase contact between the skin of the wearer and the sticky, soft, hot garment, and increase the burn depth.
4. In general, times to injury at 16.8 kW/m^2 were about two thirds of those at the 12.8 kW/m^2 exposure.
5. There were large differences in the changes the fabric underwent during the various exposures.

At the 12.6-kW/m^2 distance mode exposure for 2 min, the aramid and FR cotton showed no change; the wool distorted somewhat; the polyester formed melt holes at points of stress and shrank in other areas; and the modacrylic shrank considerably and charred.

During the 16.8-kW/m^2 distance exposure for 1 min, the aramid again showed no change, the FR cotton charred and embrittled somewhat, and the wool charred and embrittled considerably. The polyester fabric formed melt holes in 20 s.

At 84 kW/m^2 for 30 s, the aramid formed a brittle char and shrank somewhat, and there was considerable off-gassing and smoldering in the FR cotton but no shrinkage. The wool fabric shrank and formed a brittle char. The modacrylic shrank into a charred ball.

Outerwear/Underwear Combinations

Figure 3 shows the time to injury for two outerwear fabrics, each exposed alone and in combination with four underwear fabrics [*10*]. The outerwear fabrics were a shirt-weight aramid and a somewhat heavier FR cotton fabric. The underwear fabrics were 100% cotton and 65/35 polyester/cotton, each in woven and knitted form (Table 1). They were submitted to four radiant heat levels, as well as to flame exposure in ASTM Test D 4108-82. However, the flame temperature was adjusted to that of JP-4 fuel fires (1040°C) by mixing nitrogen with the methane. This would be a considerably less severe exposure than in the ASTM test.

Exposed without underwear, the heavier FR cotton gave more protection than the aramid fabric at a relatively low exposure, 8.4 kW/m^2. In another laboratory, the FR cotton also produced a lower temperature rise in a simulated skin in 3 and 6-s flame impingement exposures [*9*]. In the presence of underwear, the fabrics performed similarly at this level.

With increasing heat level, the aramid fabric resulted in somewhat longer times to injury and less charring of the underwear. Presumably off-gassing

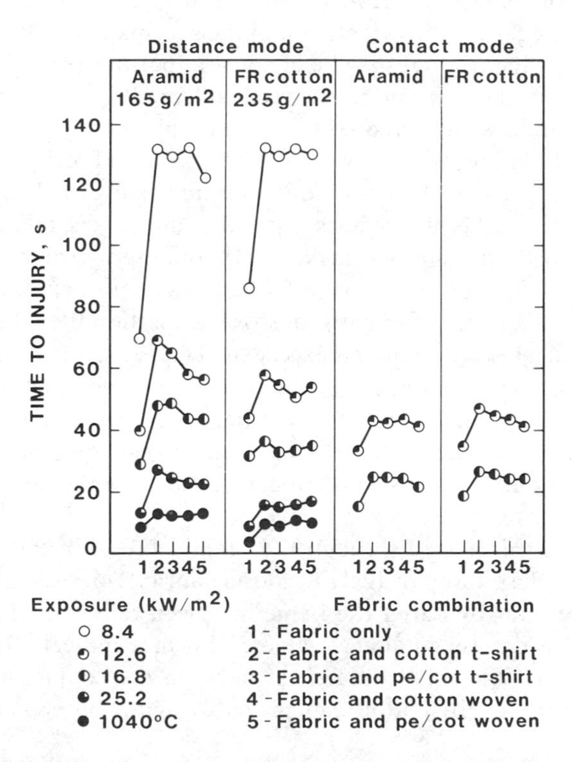

FIG. 3—*Time to injury for uniform/underwear combinations for various radiant heat and flame exposures.*

TABLE 1—*Fabric specifications.*

Fiber Content	Weight,[a] g/m²	Thickness, mm	Color
Outerwear			
100% Nomex aramid[b]	165	3.8	olive green
95/5 Nomex/Kevlar aramid[c]	165	3.8	olive green
100% FR cotton	235	4.6	dark blue
Underwear			
100% cotton tee-shirt knit	140	4.2	white
65/35 polyester/cotton tee-shirt knit	115	3.4	white
100% cotton woven	100	3.0	white
65/35 polyester/cotton woven	100	3.0	white

[a]Areal density.
[b]This fabric was used in the 8.4, 12.6, and 16.8-kW/m^2 exposures.
[c]This fabric was used in the 16.8, 25.2, and 84-kW/m^2 exposures; the results in the 16.8-kW/m^2 exposure were the same as those obtained with the other aramid fabric.

from the FR cotton increased the heat delivered to the sensor beyond that level which would be expected from a fabric of this weight. This effect may be maximized by the horizontal orientation of the specimen with the heat sensor above it in the flame test but may be of some importance in garments, even though the fabrics there are primarily vertical [*7,8*].

All types of underwear increased the time to injury, with little systematic differences between the types. However, the knit cotton fabric (the densest of the underwear fabrics) ignited behind the aramid fabric at the 25.2 kW/m^2 radiant exposure in all replicate tests. The other underwear fabrics ignited in some replicate tests and not in others, with both outerwear fabrics.

The total heat transmitted through fabrics at various exposure levels can also be used to compare their heat protective properties after the time to incipient second-degree burn injury is exceeded. This is useful in comparing the anticipated severity of burn injury behind the fabrics. Figure 4 compares the total heat transmitted after 10 and 20 s of flame exposure, and 10 s after the flame had been removed, for the outerwear/underwear combinations. Unlike in the previous figures, a low result indicates a fabric with good heat protective properties.

Less heat was transferred through the aramid fabric in the first 10 s of flame exposure than through the FR cotton fabric. However, after 20 s of flame exposure, and 10 s after the flame had been removed, the FR cotton generally produced lower total heat values. This may indicate the protective value of the char. The small differences between the contact and distance mode results for the FR cotton can probably again be explained by the off-gassing.

All the specimens were badly charred after exposure. The polyester/cotton underwear stuck to the aramid but not to the FR cotton fabrics.

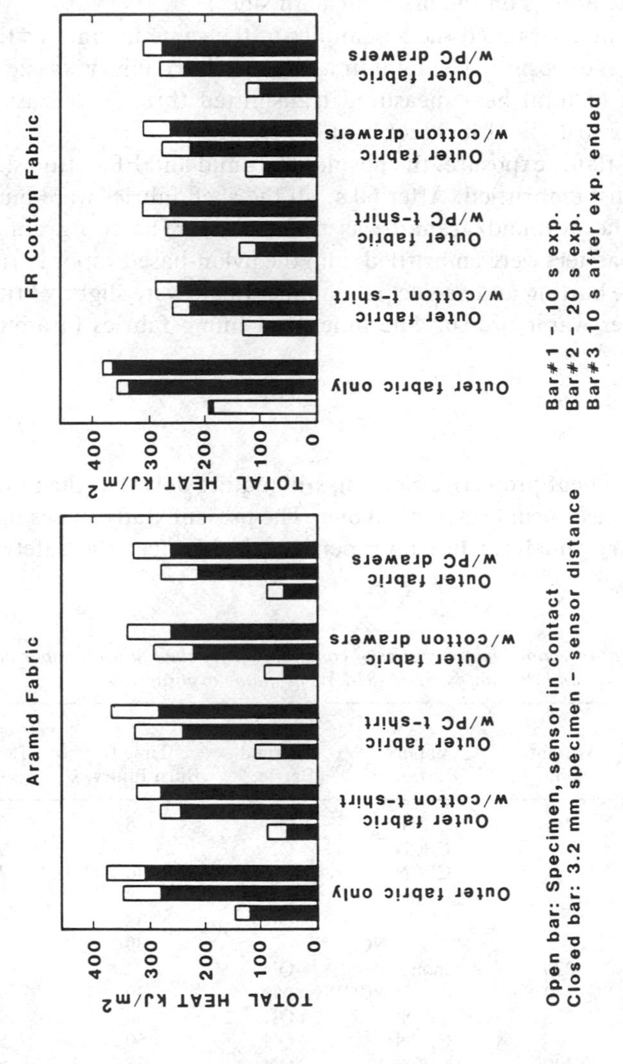

FIG. 4—*Total heat transmitted through uniform fabric/underwear combinations during flame exposure.*

Fire Fighters' Turnout Coat Assemblies

Table 2 shows the range of heat protection properties during 84-kW/m^2 exposure of commercial turnout coats for structural fire fighters [*19*]. Such coats generally consist of an assembly of an outer shell, a vapor barrier, and a thermal barrier. The latter is frequently a batting or a needle punch construction with fabric linings on the inside or both sides.

The time to injury ranked the assemblies in the same manner as the total heat after 60 s of exposure. The total time range to burn injury was 28 to 50 s, and the range of total heat measured transmitted through the assemblies ranged from 12 to 19% of incident heat.

After 10 s of flame exposure, the novoloid/aramid and FR cotton shell fabrics seemed most embrittled. After 60 s, all the shell fabrics were much embrittled, but the novoloid/aramid was the least so. The two aramid-base-coated vapor barriers were embrittled, and the nylon-based vapor barrier had melt holes. The batting and needle punch inner liners were slightly brittle; the wool inner liner was more so. The innermost lining fabrics (aramid) were unchanged.

Conclusions

Some users of heat protective clothing still require only that the fabric pass a flammability test, usually a vertical one. The present study shows that such fabrics can vary widely in other properties which affect the safety of the

TABLE 2—*Heat protection characteristics of combinations of shell fabrics, vapor barriers, and thermal barriers (84-kW/m^2 flame exposure).*

Shell Fabric Fiber Content	Weight, g/m^2	Vapor Barrier[a]	Thermal Barrier[b]	Time to Burn Injury, s	Total Heat after 60 s, kJ/m^2
Aramid	245	NCN	QN	28	925
		CPCN	QN	31	900
		CPCN	NPN	40	780
		CPCN	FW	38	740
		GN	QN	32	900
		NCNPN		39	745
Coated aramid	320	none	QN	34	850
Novoloid/aramid	340	NCNPN		30	980
Aramid blend[c]	255	NCN	QN	30	940
		CPCN	QN	50	590
FR cotton	440	CPCN	QN	38	855

[a]NCN = neoprene-coated aramid fabric, 225 g/m^2; CPCN = coated aramid pajama check fabric, 265 g/m^2; GN = Gore-Tex-coated nylon-base fabric.

[b]NCNPN = neoprene-coated aramid needle punch fabric, 735 g/m^2; QN = aramid quilt, consisting of aramid batting with aramid pajama check fabric attached, 245 g/m^2; NPN = aramid needle punch fabric, 245 g/m^2; FW = felt, wool, 430 g/m^2.

[c]50/50 Nomex/Kevlar.

wearer, for example, heat protection of the skin and loss of integrity (shrinkage, melting, forming a brittle char). Test methods that measure these properties are outlined, and typical results are shown. The relative ranking of protective aramid and FR cotton fabrics can be reversed when the heat exposure is made more severe. Modacrylic and wool fabrics charred at lower heat exposure than FR cotton and aramid fabrics.

Acknowledgments

This paper is a review of work by National Bureau of Standards personnel performed for the Federal Aviation Administration, the Navy Clothing and Textile Research Facility, and by a U.S. Department of Agriculture research associate, R. M. Perkins, at the National Bureau of Standards.

References

[*1*] Simms, D. L. and Hinkley, P. L., "Protective Clothing Against Flame and Heat," Fire Research Special Report No. 3, Joint Fire Research Organization, Department of Scientific and Industrial Research and Fire Officers Committee, Her Majesty's Stationary Office, London, England, 1960.

[*2*] Abbott, N. J. and Schulman, S., "Protection From Fire: Non-flammable Fabrics and Coatings," *Journal of Coated Fabrics*, Vol. 6, 1976, pp. 48-64.

[*3*] Brewster, E. P. and Barker, R. L., "A Summary of Research on Heat Resistant Fabrics for Protective Clothing," *American Industrial Hygiene Association Journal*, Vol. 44, 1983, pp. 123-130.

[*4*] Brewster, E. P. and Barker, R. L., "Bibliography of Published Information on Heat Resistant Fabrics for Protective Clothing," *Journal of Consumer Product Flammability*, Vol. 9, 1982, pp. 11-19.

[*5*] Stoll, A. M. and Chianta, M., "Method and Rating System for Evaluation of Thermal Protection," *Aerospace Medicine*, Vol. 4, 1969, pp. 1232-1238.

[*6*] Quintiere, J., "Radiative Characteristics of Fire Fighter's Coat Fabrics," *Fire Technology*, Vol. 20, 1974, pp. 153-161.

[*7*] Davies, J. M., McQue, B., and Hoover, T. B., "Heat Transferred by Decomposition Products from Cotton Fabrics Exposed to Intense Thermal Radiation," *Textile Research Journal*, Vol. 35, 1965, pp. 757-769.

[*8*] Baitinger, W. F., "Product Engineering of Safety Apparel Fabrics: Insulation Characteristics of Fire-Retardant Cotton," *Textile Research Journal*, Vol. 49, 1979, pp. 221-225.

[*9*] Schoppe, M. M., Welsford, J. M., and Abbott, N. J., "Resistance of Navy Shipboard Work Clothing Materials to Extreme Heat," Technical Report No. 148, Navy Clothing and Textile Research Facility, Natick, MA, 1982.

[*10*] Krasny, J. F., Allen, P. J., and Maldonado, A., "Burn Injury Potential of Navy Shipboard Work Clothing," Technical Report No. 146, Navy Clothing and Textile Research Facility, Natick, MA, 1983.

[*11*] Braun, E., Cobb, D., Cobble, V. B., Krasny, J. F., and Peacock, R. D., "Measurement of the Protective Value of Apparel Fabrics in a Fire Environment," *Journal of Consumer Product Flammability*, Vol. 7, 1980, pp. 15-25.

[*12*] Perkins, R. M., "Insulative Values of Single-Layer Fabrics for Thermal Protective Clothing," *Textile Research Journal*, Vol. 49, 1979, pp. 202-212.

[*13*] Perkins, R. M., Krasny, J. F., and Braun, E., "Insulative Values of Double Layers of Fabrics Exposed to Radiative Heat," *Proceedings*, 13th Annual Meeting, Information Council on Fabric Flammability, Atlanta, GA, 1979, pp. 88-96.

[*14*] Perkins, R. M., Krasny, J. F., Braun, E., and Peacock, R. D., "An Evaluation of Fabrics for Thermal Protective Clothing," *Proceedings*, 12th Annual Meeting, Information Council on Fabric Flammability, New York, 1978, pp. 212-230.

[15] Ross, J. H., "Thermal Conductivity of Fabrics as Related to Skin Burn Damage," *Journal of Applied Polymer Science: Applied Polymer Symposium*, Vol. 31, 1977, pp. 293-312.
[16] Maggio, R. C., "A Molded Skin Simulant Material with Thermal and Optical Constants Approximating Those of Human Skin," NS 081-001, Naval Material Laboratory, Naval Shipyard, NY, 1956.
[17] Stoll, A. M. and Chianta, M. A., "Heat Transfer Through Fabrics as Related to Thermal Injuries," *New York Academy of Sciences, Transactions*, Vol. 33, 1971, pp. 649-670.
[18] Derksen, W. L., Monahan, T. I., and Delhery, G. P., "The Temperature Associated with Radiant Energy Skin Burns" in *Temperature—Its Measurement and Control in Science and Industry*, Vol. 3, Part 3, Reinhold, New York, 1961, p. 171.
[19] Krasny, J. F., Singleton, R. W., and Pettengill, J., "Performance Evaluation of Fabrics Used in Fire Fighter's Turnout Coats," *Fire Technology*, Vol. 18, 1982, pp. 309-318.

Philip S. Jaynes[1]

A Method for Testing Fabrics with Molten Metals

REFERENCE: Jaynes, P. S., **"A Method for Testing Fabrics with Molten Metals,"** *Performance of Protective Clothing, ASTM STP 900*, R. L. Barker and G. C. Coletta, Eds., American Society for Testing and Materials, Philadelphia, 1986, pp. 475–486.

ABSTRACT: This paper describes a method for testing fabrics with molten metals. The test evaluates the fabric's ability to protect people who are accidentally splashed by metals such as molten iron, aluminum, or brass.

Some test results are discussed to illustrate the application of the test method. The test procedure is substantially the same as a test method for evaluating heat transfer through materials upon contact with molten substances, now being developed by ASTM Subcommittee F23.80 on Molten Substances, a subcommittee of ASTM Committee F-23 on Protective Clothing.

KEY WORDS: primary protection, secondary protection, protective equipment, calorimeters, molten metal splash test, molten substance splash test, safety, protective clothing, burns, burn prevention, testing, aluminized coats, time to blister, safety equipment, Stoll curve

In the metals business, it sometimes happens that a person is splashed by molten metal. Naturally, therefore, we want to find out the type of coat that will give him the best protection. We can put two different coats on two people working together, to test the coats. Perhaps there is an eruption of molten steel; both people are injured, one somewhat worse than the other. What are the possible explanations? It may be that one was closer to the eruption than the other, and we still do not know which coat was better.

Thus, it became necessary to set up a laboratory condition in which these conditions were simulated, yet controlled. It was necessary to be sure that any differences would be due to the coats, not to something else. With the help of many people at Inland Steel Co. and elsewhere, the author was able to develop a test method which meets these conditions. This same basic method is

[1]Senior safety engineer, Safety and Plant Protection, Inland Steel Co., East Chicago, IN 46312; currently safety consultant, Phil Jaynes & Associates, Valparaiso, IN 46383.

being adapted into a national standard by ASTM Committee F 23 on Protective Clothing, which has sponsored this volume.

Test Method

The basic procedure is to take some molten metal, pour it over the protective fabric being tested, and measure how much heat comes through the fabric. However, it must be done *very consistently* every time.

To conduct the tests, the author melted steel in a small furnace which held about a teacupful of metal (Fig. 1). The steel was weighed before it was put into the furnace and was poured directly on the test material from the furnace. The figure shows the equipment used. The metal was melted in the furnace; when it was ready, the arm that held the test cloth, was swung up close to the furnace. This avoided preheating of the fabric from the radiant heat and also prevented other unwanted variations.

A technique was also used in which the metal is melted in a larger furnace and then transferred into a pouring crucible. In this arrangement, the amount of metal poured is controlled by the size of the pouring crucible.

In both techniques, the temperature of the test metal is measured just before it is poured on the fabric. When the metal temperature drops to that specified, it is poured on the test fabric.

Complications

There were many factors, other than the fabrics tested, that could have affected the results (Table 1)—at least 21 that the author knows of. Controlling the temperature proved to be more critical than expected. In one of the preliminary tests made in developing the test procedures, two tests were made on the same kind of cloth. In one test the metal was poured at 1650°C (3000°F),[2] in the other at 1538°C (2800°F). It was quite surprising to find that a much worse burn had occurred at 1538°C (2800°F) than at 1650°C (3000°F). On examination of the test material, the explanation was discovered. With the higher-temperature pour, the molten metal had hit the material and run off very quickly. But with the lower-temperature material, some of the metal had solidified on the fabric. As it continued to cool slowly, the heat from the red hot metal continued to burn through the material; what had happened, in effect, was that heat transfer had occurred for 30 or 40 s with the lower-temperature material instead of for about 1/2 s as had occurred when the metal was poured at 1650°C (3000°F).

Since metals of different chemistry have different melting temperatures, the chemistry of the metal was quite critical. So it was necessary not only to control the temperature but also to make sure that the chemistry of the metal remained constant throughout the tests.

[2]All the measurements in this paper were originally in English units. They have been converted to metric units, in accordance with ASTM practice.

FIG. 1—*Test apparatus with a small induction furnace.*

Measuring the Results

There is a fairly reasonable assumption in tests of this sort that can be misleading. That assumption is that the worse the damage to the garment protecting the man, the worse the damage to the wearer. In some cases, this is not true. Some fabrics resist destruction by heat very well but at the same time are good conductors of heat. Although the fabric is not damaged, the heat is transferred through it to the human flesh underneath it, which becomes burned. On the other hand, the fabric may be destroyed but still block the heat in the process. One fabric actually takes up heat when it burns instead of

TABLE 1—*Factors affecting splash test results.*

Substance poured
Chemical composition
Temperature at pour
Freezing temperature of metal
Viscosity of metal
Calorimeter
Thickness
Material
Initial temperature
Number of thermocouples
Placement of thermocouples
Composition of thermocouples
Surface color
Sensor board
Angle from horizontal
Spacing of calorimeters
Pouring ladle
Height of fall of metal
Angle of ladle when stopped
Diameter of ladle
Depth of ladle
Shape of ladle
Rate of rotation of ladle
Ladle temperature
Time of pour
Tested fabric
Fiber composition
Weight of fabric
Weave of fabric
Density of fabric
Fabric coatings
Moisture content

giving it off; thus, in the act of being burned, the fabric actually cools the heat source. What we are most concerned about is what happens to the man underneath the garment, and not to the garment itself.

A researcher for the U.S. Navy, Alice Stoll, developed techniques for measuring the amount of heat required to provide various kinds of burns to human beings. What she found was that it was not so much the temperature that made the difference as the total amount of heat transferred.

For example, if you take a lighted candle flame and pass your hand through it fairly quickly, you will not be burned, although the flame is 800°C (1500°F). However, if you were to plunge your hand into boiling water, even for an instant, your hand would be badly burned, even though that water is only 100°C (212°F). Much more heat energy is transferred from the boiling water in a short period of time than is transferred to your hand from the burning gases of the flame. Water holds more heat and is capable of conducting heat to your hand faster than the flame of a candle.

Stoll also developed a device known as a calorimeter for measuring this heat transfer, which was used in these tests. It was mounted on the test board shown in Fig. 1, behind the test material. The loose wires, which can be seen in the figure, are running from the calorimeter to the recording instrument. Stoll also developed data by which this heat transfer information could be translated into a burn on a human being.

In addition to the actual amount of heat transfer, the amount of time during which that heat was being transferred is a factor in causing a human burn (Fig. 2). The main point of reference was the time at which the blister was first developed, represented by the dotted blister line on the graph. The length of time involved is indicated across the bottom of the chart under the heading, "Time from Initial Contact." This time continues even though the initial flow of molten metal may have lasted for only a part of a second. What happens is that the heat from the metal is transferred to the outer layer of fabrics, then to

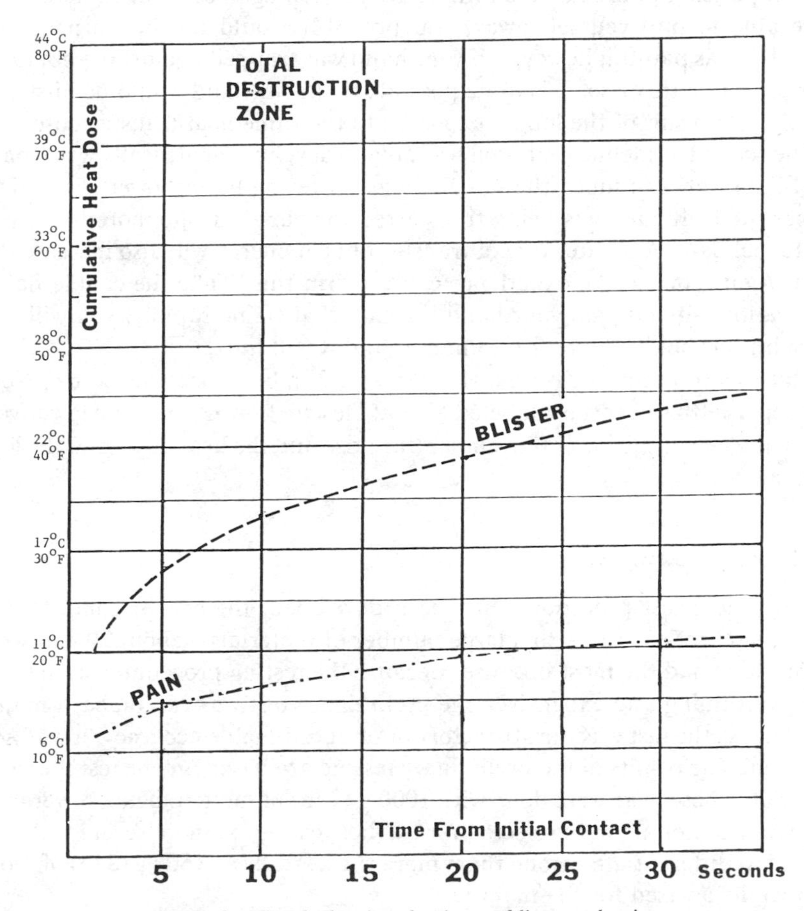

FIG. 2—*Graph showing the time to blister and pain.*

the under layer of fabrics, and then it is gradually absorbed by the human skin or test instrument.

The data on the left side are the cumulative heat energy dose. The calorimeter was set up in such a way that the more heat energy that was transferred, the higher the temperature of the recording instrument became. The instrument, or calorimeter, which was used is so designed that its heat rise is very close to what the heat rise would be for human flesh.

The combinations of heat rise and elapsed time required to produce a small blister are shown as a dashed line on the graphs. The point at which the blister line begins is after 1 s of exposure to heat; this would have required a 11-degree Celsius (20-degree Fahrenheit) temperature rise in the human flesh. At that point a small blister would have begun to form by the heat transfer. However, if it had taken 5 s for the human flesh to be raised 11 degrees Celcius (20 degrees Fahrenheit), this would be well below the blister line and would signify that no blister was formed. This may be surprising, but common experience confirms this finding. If you fell against a hot radiator and were able to pull yourself away, you probably would not be burned even though it was painful; however, if your hand was trapped against the hot radiator for a minute or so, it is quite possible that your hand would be blistered simply as a result of the longer exposure to the same heat transfer dose.

The second guideline that Stoll was able to develop scientifically was a pain line. Pain works in much the same way as blistering but at lower levels. The longer the heat energy is being transferred to your skin, the more likely you are to feel pain. A greater rate of transfer of heat energy will also increase the pain. Again, one's own experience can confirm this. Take the candle flame illustration—if you pass your hand through that flame rapidly, you will feel no pain; but hold it there for a minute, and it will hurt.

Thus, as you move up on the chart in Fig. 2, a burn becomes worse. Note the upper lefthand corner, labeled "Total Destruction Zone." The lower you are on the chart, the lower the temperature rise and the heat dose, and the less damage done.

Preliminary Testing

Once these test procedures had been developed, the author conducted a preliminary set of tests with a large number of materials—about 20—to seek those which had the most promise. Because the testing procedures were both time-consuming and expensive, one preliminary test was run on each material. Unless the test was unsatisfactory or of questionable accuracy, it was not repeated. The results of the preliminary testing are based on one test for each material. These tests were done with 1000 g (2 lb) of molten metal. Several of the better materials were selected for further testing. A more severe test was needed to differentiate among them more successfully so 1500 g (3 lb) of molten metal was used for the main tests.

Test Results

The results of the main tests are shown in Fig. 3. The top line on the chart is for 494 g/m^2 wool (nominally 32-oz wool). The fact that it is the highest on the chart for the materials tested, shows that this wool had the highest heat transfer and the least protection of the materials in this test.

The line labeled "E" indicates the results of the aluminized asbestos cloth. Like heavy wool, asbestos has been used for protective clothing in the metal industry to some degree for many years. Asbestos provides considerable protection and has saved many people from serious injury. However, asbestos is a relatively good conductor of heat, and it conducted heat through it to cause injury to the wearer even though the material itself was hardly damaged in the test.

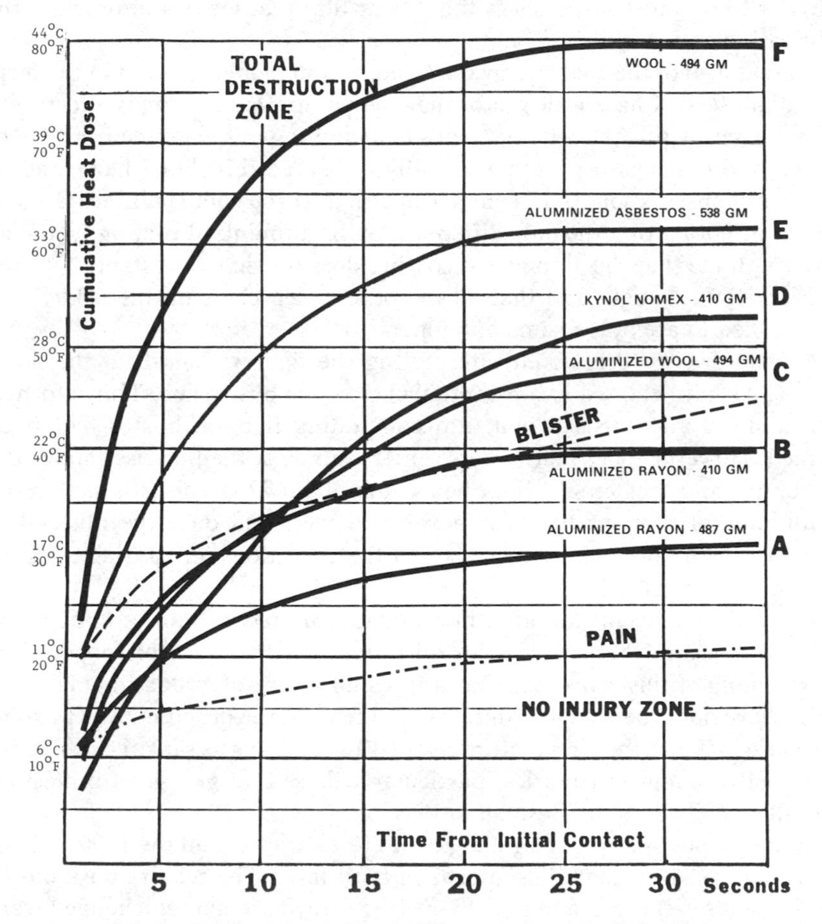

FIG. 3—*Graph of the results of 1500-g molten metal splash tests.*

The line labeled "D" on the chart represents the results for another fabric. Line D crosses the blister line at about 12 s, indicating that in just 12 s from the initial splash enough heat was transferred to cause a blister. At about 25 s, the line is far enough above the blister line that it probably caused a third-degree burn.

The metals industry began to use aluminized fabrics a number of years ago to cut down on the radiant heat present when molten metal is being poured. In the course of running these tests, the author observed that the aluminized coating also provided a considerable degree of protection: the aluminized coating acted as a lubricant, which caused the molten metal to slither off the protective fabric very quickly. When metal sticks to a fabric, the red hot metal continues to burn many times longer than if it does not stick, transferring energy through the fabric for a much longer time and causing a much worse burn. In most processes, the aluminum coating on a fabric is bonded in a Mylar film, and I suspect it is this Mylar film that melts and provides the lubricating action.

A great deal of the heat energy that comes from molten metal is in the form of radiant heat. I have little doubt that the aluminized coating is effective in reflecting away this radiant heat, thus contributing to the protective value of aluminized coatings. I have not actually measured this, but I have made a number of observations that seem to confirm it. If you compare Lines C and F in Fig. 3, noting that the only difference is the aluminized coating, you may agree with me that the aluminized coating does provide a substantial degree of protection. I might add that all the other researchers in this field that I have talked to are of the same opinion.

A few words further about interpreting the figures: Line B at the point where 30 s have elapsed is substantially below the blister line. This information alone would mislead you into concluding that no blister had been formed. However, note that at between 15 and 20 s, the line rises above the blister line and then crosses back below it at about 22 s. This tells us that the conditions for forming a blister have been met, and once the blister has formed, it will not go away even though the conditions for causing a blister may cease to exist.

Two useful concepts are used in evaluating protective fabrics. One is the time of exposure that will lead to blistering, discussed in this paper. The shortcoming of this concept is that it gives no means of evaluating materials which have not produced a blister. For this reason, the proposed ASTM standard also includes the temperature rise at 30 s. This lets us rank the protective ability of different test fabrics, particularly those that give good protection and allow no burn under test conditions.

In the graphs showing the 1000-g (2-lb) tests (Fig. 4) and the 1500-g (3-lb) tests (Fig. 3), three materials appear in both tests. The relative positions of the 410-g (16-oz) rayon and the 487-g (19-oz) rayon remain unchanged from those two conditions. However, the line for the 591-g (32-oz) wool has changed positions radically.

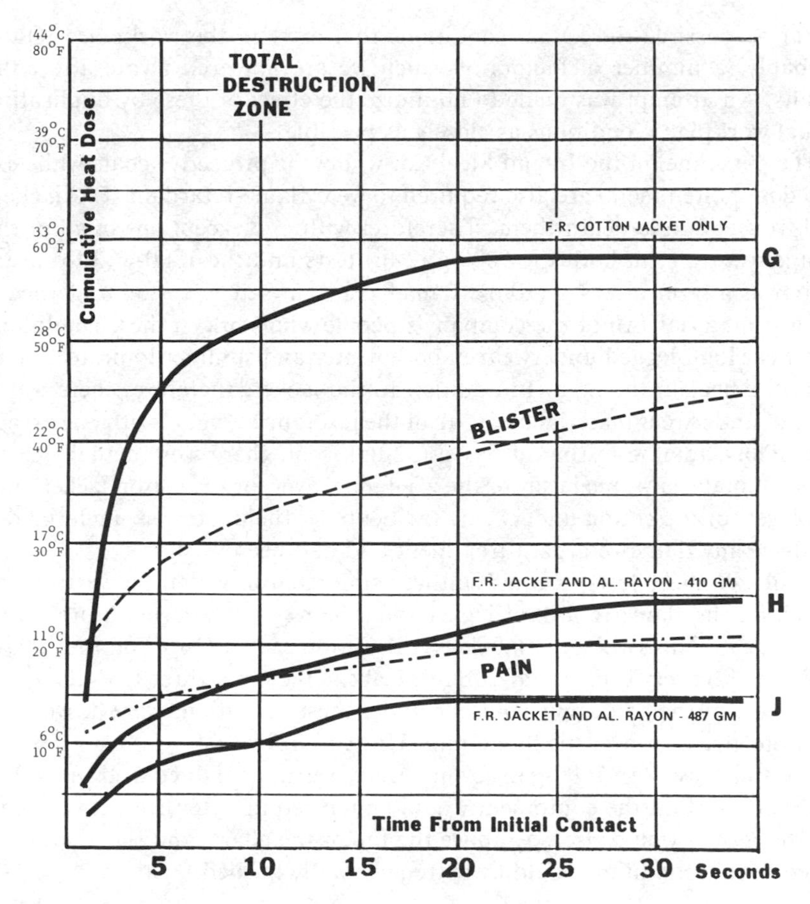

FIG. 4—*Graph of the results of 1000-g molten metal splash tests.*

The wool material used in the 1000-g (2-lb) test came through very much intact. Some brown scorch marks appeared on its surface, but no substantial damage to the material occurred. The molten metal had hit the wool and slithered off quite quickly. On the other hand, the wool used in the 1500-g (3-lb) test was quite badly damaged. At the point of impact, it was completely charred, and a small hole had appeared in the wool. In fact, the molten metal had pocketed in the wool, holding a small amount of hot metal at that point for the minute or so it took to lose its red color.

Apparently, the wool held up very well up to a point and then failed catastrophically. Perhaps this is the reason wool has been observed to give good protection; apparently, as long as the amount of molten metal is not too great, it does a good job, but it breaks down rapidly under the more severe conditions.

In conducting these tests, the author felt that it was important to simulate as far as possible the actual conditions that exist in the workplace. Quite probably, a number of factors of which we are not even aware affect the results. An attempt was made to minimize the effect of these by duplicating actual workplace conditions as closely as possible.

The personnel of the Inland Steel Co. who wear protective coats while exposed to molten metal are also required to wear flame-retardant (FR) jackets and trousers underneath them. Therefore, with one exception, in all of the company's tests, in both the 1000-g (2-lb) tests and the 1500-g (3-lb) tests, there was a layer of 284-g (10-oz) flame-retardant sateen cotton underneath the test material. All of the company's people who work in these conditions also wear long-legged underwear in both winter and summer to protect them from the heat in the area. In addition to the jacket, therefore, there was a layer of underwear material under all of the test fabrics, again with one exception. Thus, for the testing of 591-g (32-oz) wool, there were actually three layers of material—one layer for the wool, one layer for the cotton jacket, and one layer for the cotton underwear. Incidentally, the underwear material did not have any flame-retardant treatment and had been washed.

In one of these tests, a condition was simulated in which the wearer was struck on the flame-retardant jacket when he was not wearing a protective coat over it. This is Line G, labeled "FR Cotton Jacket Only" on the 1000-g (2-lb) test. Its very substantial difference above the other three lines indicates how badly burned a person would be under these conditions were he wearing no protective coat over his flame-retardant cotton jacket.

A similar test with 1500-g (3-lb) materials was not conducted, frankly, because of fear that the equipment would be burned up. However, it is safe to say the results would rise way above the top of the chart, and the wearer exposed to this condition would be extremely badly burned.

Testing with Different Metals

Those of us involved in this work have run many tests using different metals. The Aluminum Association and the Safety Equipment Manufacturers Association conducted a series of metal splash tests using aluminum. The FR-7 cotton fabrics, which work well for iron and steel, did poorly with aluminum. This led to development of a new fabric, FR-8, which protects well for aluminum but poorly for iron and steel. Similarly, aluminized rayon showed better primary protection against steel than wool, while wool tested better than the aluminized rayon in protecting against aluminum.

Application of Test Results

As mentioned earlier, at least 21 variables that can affect test results have been identified. Some of these are the temperature of the molten substance involved, the force, the angle of impact, the concentration of the splash, the quantity of metal, and the time of the splash event. A proposed ASTM molten metal splash test method now being developed is designed to control all these variables. Because of this, the results presented here may not apply to all operations.[3] The author and those who have helped him have made an honest effort to be as careful as possible in making these tests. The results are the best information we have; how they apply to other situations is something that others must evaluate.

Relating Laboratory Results to the Real World

How these test results compare with what happens in an actual accident, was demonstrated by an accident that occurred. In this accident, an eruption of molten metal splashed one of our employees, who was working at a continuous billet caster. This man was wearing flame-retardant cotton trousers and a wool jacket. He was not wearing an outer coat at all. The molten steel hit him at the point where the jacket overlapped his trousers and also hit him below that point, where he was protected by his trousers only. At the point where he was protected by two layers of material, one wool and one cotton, he received no burns although much of the wool jacket was destroyed. However, where he was protected only by the cotton clothing, he received a second-degree and some third-degree burns to his thighs. I was able to cut a piece from the back of the jacket and run a laboratory test on it. The test instruments indicated that with a 1000-g (2-lb) metal pour, the wearer would receive no burns where protected by both the wool and the flame-retardant cotton underneath. Where there was only the cotton layer, he would receive severe second-degree burns, or worse. Inasmuch as this was what had actually happened, I felt that the 1000-g (2-lb) pour test corresponded to the moderately severe metal explosion that had actually occurred at the continuous billet caster.

One other fact about this accident is worthy of note. There was a second man involved, who was standing much closer to the explosion than the man who was injured. However, he was wearing one of the aluminized rayon coats which were being used as a result of our research. Obviously, he had received a much larger molten steel splash, but he received no burns. There is no accident report on him because he did not need to come to the clinic for treatment.

The injured man also received one other burn which I have not yet mentioned—a small second-degree burn to his scrotum. It happened that the bottom of his wool jacket ended at exactly this point. I mention this in particular because we safety men sometimes have a great deal of difficulty in persuading people to wear the protective clothing they need. If the man's jacket had been

[3]Because of the variables involved neither the author nor the Inland Steel Co. can be responsible for any application that is made of these results.

a little shorter, he would have been burned in that part of his anatomy which most men value most highly. Many workers will be better motivated to wear protective coats by the thought of serious burns to their sexual organs than by any other means of persuasion. It very nearly happened to this foreman, and it certainly could happen to anyone who is not properly protected.

Conclusions

As a result of the author's testing program, Inland Steel Co. adopted the 487-g (19-oz) aluminized rayon coats as standard protection for all personnel working with molten metal. A considerable educational campaign was conducted, and appropriate orders were issued to this effect.

Because the coats are quite hot, there was a great deal of resistance to the program, both from the people who wore the coats and from supervisors.

Within the first few months of the program, three employees sustained heavy splashes of molten steel while wearing the new protective coats. One of these men received no burns, one received very minor spot burns, and the third man received a disabling injury. He sustained rather bad burns to both his hands and was hospitalized for some time. However, none of the three men received any burns of any degree to the parts of their bodies that were protected by the aluminized rayon coats. They all felt the coats had saved them from some very serious injuries. As the word of these incidents spread around the plant, people began to recognize the value of the coats and began to accept them in spite of the considerable discomfort they involved.

The author recognizes that there is still a good deal of room for improvement in these protective garments, although they have been a valuable addition to the company's safety program. He hopes this information may be helpful to others responsible for the protection of people.

James H. Veghte[1]

Functional Integration of Fire Fighters' Protective Clothing

REFERENCE: Veghte, J. H., **"Functional Integration of Fire Fighters' Protective Clothing,"** *Performance of Protective Clothing, ASTM STP 900*, R. L. Barker and G. C. Coletta, Eds., American Society for Testing and Materials, Philadelphia, 1986, pp. 487–496.

ABSTRACT: System integration of fire fighters' protective clothing requires the harmonious design of the coat, pants, helmet, boots, gloves, and breathing apparatus. Similarly, all clothing layers require a functional integration of characteristics such as resistance to cuts and punctures, mobility, waterproofness, fit, and durability. Thermal protective performance (TPP) rating analyses of several hundred protective clothing combinations have also shown the importance of thermally matching the vapor barrier and thermal liner under the outer protective shell. Inappropriate vapor barrier or thermal liner materials will melt under a Nomex III (95% Nomex I/5% Kevlar) or polybenzimidazole (PBI) outer shell if the thermal load is sufficient. TPP instrumentation permits moisture to be added to ensemble materials in order to assess the thermal merits of clothing combinations under operationally relevant fire fighting conditions. TPP data are presented to illustrate the protective merits of thermally integrating the clothing materials in structural fire fighters' clothing.

KEY WORDS: protective clothing, thermal, thermal protection, fire fighters' clothing

The recent advent of synthetic high-temperature-resistant fibers and blended materials has set the stage for the next step in the development of fire fighters' protective clothing: the functional design of fire fighters' protective ensembles.

Some materials used in the past in fire fighters' clothing are listed in Fig. 1. In the 1940s, world events triggered research in the development of new synthetic materials, as there were extreme shortages in natural fibers. Government support in this area of research continued in many countries of the world into the 1960s, when the National Aeronautics and Space Administration (NASA) began supporting the development of fire-resistant materials as a result of the tragic loss of three astronauts by fire. Nomex and polybenzimi-

[1]President, Biotherm, Inc., Beavercreek, Oh 45432.

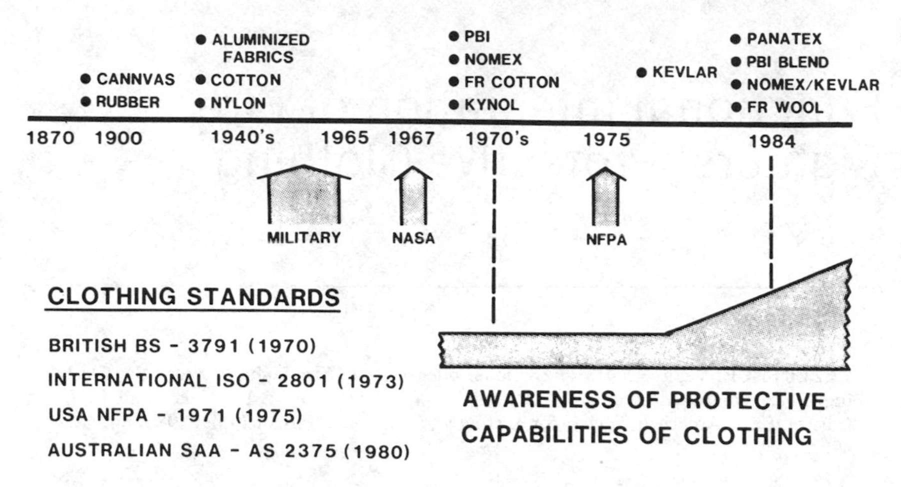

FIG. 1—*Historical development of fire fighters' protective materials and standards.*

dazole (PBI) were two synthetic fibers developed in this period. Since then, improvements in the high-temperature performance of these fibers and other materials have continued.

Another major contribution to the development of improved protective clothing in the United States has been the advent of the National Fire Protection Association (NFPA) standards. With considerable input from the National Bureau of Standards and fire service personnel across the country, a minimal performance standard was developed for structural fire fighters' clothing. The awareness of fire service people of the protective merits of their clothing began to increase as mandatory state or federal regulations dictated compliance. Fire service safety or protective equipment officers were appointed in many major cities to become knowledgeable about protective clothing.

In another area, a recent tabulation of fire fighter injuries was made by the International Association of Fire Fighters (IAFF). Lacerations, sprains, and inhalation of noxious fumes accounted for 72% of the 39 900 injuries reported in 1982, while burns, cold, and heat exhaustion injuries accounted for 11% [*1*]. Figure 2 shows the relationship between specific body regions and burn injury based on these data.

A recent survey of protective clothing used in nine major fire departments across the United States provides the clothing designer with user information [*2*]. A general summary of their comments follows:

1. The shoulder, elbow, and knee regions of the outer shell are excessive wear areas and should be reinforced with additional material.

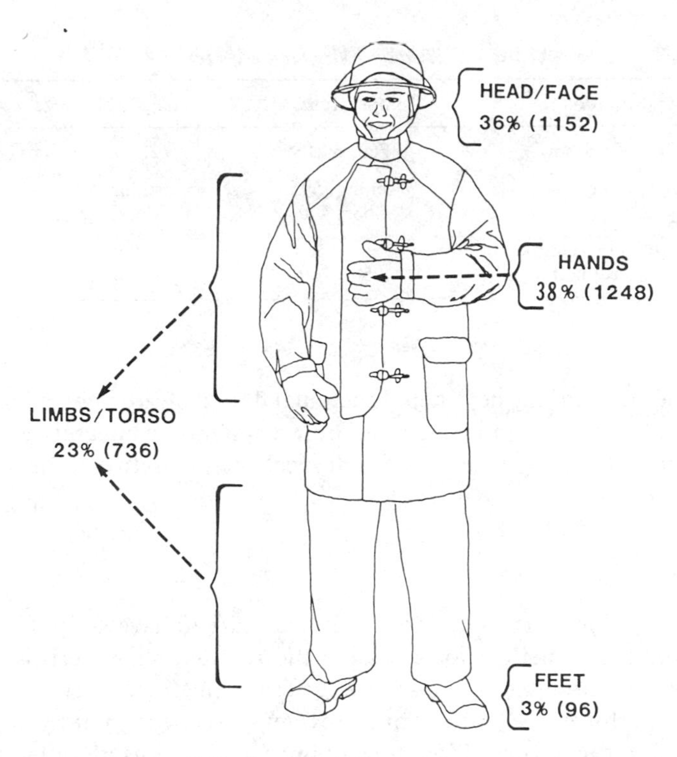

FIG. 2—*Burn injuries for 1982 of various body areas (percentage and total number of injuries)* [1].

2. Sweating or water leakage into clothing and icing are serious problems associated with fighting fires.
3. Coats pockets should be positioned so that they are accessible to the fire fighter even when the self-contained breathing apparatus (SCBA) torso strap is fastened.
4. One-third or more of the fire fighters feel hot or very hot while wearing their protective clothing in the summer. Ways of reducing this physiological stress on the fire fighter should be explored.
5. Gloves continue to be a major problem, as hands are a fire fighter's most important and most exposed body region.

Finally, before any design changes can be made, basic information is required to understand the level of thermal loads to which fire fighters are exposed. Table 1 [*3,4*] provides the ambient temperatures and energy flux associated with several conditions encountered by structural fire fighters. Flame contact and flashover conditions are unusual circumstances, and the typical conditions encountered by fire fighters during working fires are radiant thermal loads. Thus, the sophistication of higher-temperature-resistant materials and increased awareness among professional fire fighters and manufacturers

TABLE 1—*Structural fire temperatures* [*3,4*].

Situation	Temperature, °C	Energy, $cal/cm^2/s$
Flame contact	760 to 815	2.0 to 3.0
Flashover	ceiling: 595 floor: 400	ceiling: 1.1 floor: 0.5
Typical structural fire radiant level	95 to 200	0.2 to 0.6

has set the stage for the next step: functional design of fire fighters' protective ensemble. Such a design includes not only a systems or integrative approach concept but also design features which meet the protective needs of the fire fighter.

Methods

To address the thermal problem, it is necessary to have a rapid, accurate, and reproducible method for assessing the thermal characteristics for any given clothing ensemble and relating its effect on the fire fighter. The thermal protective performance (TPP) rating provides a precise laboratory method for evaluating new materials, different clothing combinations, the effect of moisture in clothing, and the effect of different radiant/convective heat loads on fabrics [*5,6*]. The multilayered clothing ensemble is exposed to a calibrated radiant/convective thermal flux, and a copper disk containing thermocouples records the heat transmitted through the material. This temperature curve is then compared with a second curve which relates time to second-degree burns. Thus, the TPP rating is a direct measure of the thermal protective capabilities of the clothing worn by a fire fighter.

Results and Discussion

Table 2 lists representative TPP values for some materials used in structural and proximity clothing. The first two ensembles show the effect of different vapor barriers on the overall thermal performance. Each assembly was exposed to two different sources of heat, a convective/radiant load (50%/50%) and a 100% radiant load. As these data show, the materials respond differently to actual flame impingement on the material than to a pure radiant load normally encountered by structural fire fighters. In structural fires, radiant exposures of 0.6 $cal/cm^2/s$ are encountered 90% of the time at all working fires [*3,7*]. This average radiant load is far less than the 2.0 $cal/cm^2/s$ loads used in this laboratory test. The last two materials in Table 2 show the dramatic effect on radiant heat protection of aluminizing the outer shell material.

TABLE 2—*TPP rating for protective clothing.*

Ensembles	Total Weight, g/m^2	Total Thickness, mm	Thermal Load, %, 2.0 cal/cm^2/s	TPP Rating
Nomex III shell Neoprene polycotton vapor barrier Nomex quilt liner	766	5.4	50/50 (convective radiant)	38.6
Nomex III shell Neoprene polycotton vapor barrier Nomex quilt liner	899	5.6	100 (radiant)	42.5
Nomex III shell Gore-Tex vapor barrier Nomex quilt liner	719	5.4	50/50	46.5
Nomex III shell Gore-Tex vapor barrier Nomex quilt liner	726	5.8	100	44.7
Aluminized Nomex I shell	302	0.4	100	67.7
Aluminized Kevlar shell	346	0.5	100	78.0

Table 3 compares combinations of materials for various clothing applications. Coating the inside of the outer shell with neoprene resulted in the highest TPP rating of these shell material combinations tested. The shell material embrittled with flame impingement, as does any Nomex III material. The next combination is a two-layer ensemble with the vapor barrier combined with a thermal insulative layer. The TPP rating for this combination is surprisingly high considering its weight and thickness. The next two ensembles compare two types of Gore-Tex moisture barriers. The Gore-Tex on Nomex ripstop used Nomex I, which shrinks and breaks open in comparison with E-89 in which the Gore-Tex is deposited on a 85/15 blend of Nomex/Kevlar. E-89 does not shrink enough to break open under these thermal loads. This difference in substrate is reflected in the higher TPP values. The next ensemble compares a neoprene vapor barrier to that of Gore-Tex. The use of Gore-Tex results in a higher TPP rating. In the following ensemble comparison, 1 and 2 g of water were added to the liner to simulate the amount of sweat measured in thermal liners during a field study in Seminole, Florida [*8*]. The TPP values were decreased dramatically. The increase in rating with 2 g of water compared with 1 g may illustrate the number of increased calories absorbed by the increased moisture in the clothing. Moisture in the liner material also prevented some thermal damage to the liner in comparison with the damage to a dry liner. No dye discoloration of the Nomex facecloth was observed in tests with the moist liners.

The next series of thermal exposures show differences in special design features. For example, the two collar ensembles' TPP ratings show the difference

TABLE 3—*TPP ratings for fire fighters' protective ensembles at 2.0 cal/cm²/s 50% radiant/50% convective load.*

Ensemble Description[a]	Area of Application	Weight, g/m²	Thickness, mm[b]	Time to Alarm, s		Time to Pain, s		Time to Blister, s	TPP Rating[c]
OS neoprene on 7.5-oz Nomex III	coat or	407	0.5						
VB neoprene on Nomex ripstop	pants body	258	0.3						
TL Nomex quilt		373	4.5						
Total		1038	5.7	6.2	+	27.5	=	33.7	67.4
	coat or								
OS 7.5-oz Nomex III	pants body	244	0.6						
VB neoprene on TL Nomex	OS yoke	...	...						
needle punch	design	654	2.8						
Total		899	3.4	6.6	+	17.3	=	23.9	47.9
OS 7.5-oz Nomex III	coat or	244	0.5						
VB neoprene on polycotton	pants body	288	0.3						
TL Nomex quilt		336	4.1						
Total		868	5.4	5.5	+	13.8	=	19.3	38.6
OS 7.5-oz Nomex III	coat or	241	0.6						
VB Gore-Tex on Nomex ripstop	pants body	126	0.2						
TL Nomex quilt		353	4.2						
Total		719	5.4	7.1	+	16.2	=	23.3	46.5
OS 7.5-oz Nomex III	coat or	251	0.6						
VB Gore-Tex on E-89 Nomex	pants body	119	0.6						
TL Nomex quilt		359	4.8						
Total		729	6.8	6.3	+	20.9	=	27.2	54.3
OS 7.5-oz Nomex III	coat or	241	0.6						
VB neoprene on Nomex ripstop	pants body	251	0.2						
TL Nomex quilt		342	4.2						
Total		834	5.4	4.8	+	16.1	=	20.9	41.8

OS 7.5-oz Nomex III	coat or	261	0.6						
VB neoprene on Nomex ripstop	pants body	258	0.3						
TL Nomex quilt	ensemble	315	4.0						
Other: 1 g water added	after sweating								
Total		834	4.9	2.2	+	9.3	=	11.5	23.0
OS 7.5-oz Nomex III	coat or	254	0.6						
VB neoprene on Nomex ripstop	pants body	251	0.3						
TL Nomex quilt	ensemble	353	4.9						
Other: 2 g water added	after sweating								
Total		858	6.3	3.0	+	13.3	=	16.3	32.6
OS 7.5-oz Nomex III		244	0.6						
VB neoprene on polycotton	collar	302	0.3						
TL not applicable	design	...	...						
Other: FR corduroy		363	0.7						
Total		909	1.7	9.0	+	5.4	=	14.4	28.9
OS 7.5-oz Nomex III		244	0.6						
VB neoprene + TL Nomex	collar	...	...						
needle punch	design	685	2.7						
Other: FR corduroy		363	0.8						
Total		1292	4.3	7.1	+	21.2	=	28.3	56.6
OS 7.5-oz Nomex III		244	0.5						
VB Neoprene on polycotton	storm flap	288	0.3						
TL not applicable	design	...	...						
Other: 7.5-oz Nomex 3		244	0.5						
Total		776	1.5	4.9	+	8.3	=	13.2	26.3
OS 7.5-oz Nomex III	facing or	241	0.6						
VB neoprene + TL Nomex	storm flap	...	...						
needle punch	design	624	2.8						
Other: 7.5-oz Nomex		241	0.6						
Total		1105	4.1	7.2	+	19.4	=	26.6	53.2

TABLE 3—*Continued.*

Ensemble Description[a]	Area of Application	Weight, g/m²	Thickness, mm[b]	Time to Alarm, s		Time to Pain, s		Time to Blister, s	TPP Rating[c]
OS 7.5-oz Nomex III	fly	241	0.6						
VB neoprene on polycotton	OS design	298	0.3						
TL — none									
Total		539	0.9	2.8	+	7.0	=	9.8	19.6
OS 7.5-oz Nomex III	shoulder	244	0.6						
Other: 7.5-oz Nomex III	and fly	244	0.6						
	OS design								
Total		488	1.2	3.3	+	5.2	=	8.5	17.0
Outer leather	elbow,	1200	1.7						
OS 7.5-oz Nomex III	knee,	244	0.6						
VB not applicable	shoulder								
TL not applicable	OS design								
Total		1444	2.2	2.1	+	9.1	=	11.2	22.3

[a]OS = outer shell, VB = vapor barrier; TL = thermal liner; 1 oz = 28 g.
[b]Total thickness may exceed the sum of parts because of air entrapment between layers.
[c]Each TPP rating is an average of three separate tests. A minimum TPP rating of 35.0 is being considered as a standard for coats or pants.

between neoprene on polycotton and the heavier, thicker neoprene on Nomex when they are sandwiched between Nomex III and flame-resistant (FR) corduroy material. Similarly, the next two storm flap constructions show the enhanced value of using the heavier, thicker neoprene Nomex needle punch. The last three special clothing layers show the protective merits of adding two layers of Nomex or using leather over Nomex to protect excessive wear areas.

Additional information presented in Table 3 shows the time to pain and the time between pain and second-degree burns, designated the alarm time. The pain and alarm times vary because of different constructions entrapping more or less air. These differences are seen in the different temperature response curves of the thermal sensor placed on the back of the clothing layers. In some clothing ensembles, the temperature can increase after the heat source is removed and result in further burn injury. Thus, empirical data demonstrate the performance characteristics of materials independent of weight, thickness, or density and provide the designer with the following information:

(*a*) where the weight should be reduced to meet a minimal TPP rating requirement of 35, which can enhance mobility and perhaps reduce sweating and fatigue;

(*b*) the effect of moisture in the clothing in drastically altering its protective capabilities, as seen in the decreased TPP rating; and

(*c*) the selection of the appropriate clothing materials to ensure thermal protection in specials areas such as the collar and fly and in excessive wear areas of the shoulders, elbows, and knees.

Conclusions

These data begin to allow the designer to assess the relative thermal merits of various clothing ensemble combinations and functionally tailor protective clothing. Such a design will increase the capability of the fire fighter to perform his work tasks without jeopardizing his thermal protection.

Acknowledgments

The thermal protective performance testing by Biotherm, Inc., was funded by Lion Uniform.

References

[1] *1982 Annual Death and Injury Survey*; International Association of Fire Fighters, Washington, DC, 1983.

[2] Veghte, J. H., "Protective Clothing Survey Results," *Fire Service Today*, Vol. 50, No. 8, 1983, pp. 16–19.

[3] Gempel, F. and Burgess, W. A., "Thermal Environment during Structural Fire Fighting," NFPCA Grant No. 76010, National Fire Prevention and Control Administration, Washington, DC, 1977.

[4] McCaffrey, B. J., Quintiere, J. G., and Harkleroad, M. F., "Estimating Room Temperatures and Likelihood of Flashover Using Fire Data Correlations," *Fire Technology*, Vol. 17, 1981, pp. 98-119.
[5] Behnke, W. P., "Thermal Protective Performance Test for Clothing," *Fire Technology*, Vol. 13, 1977, pp. 6-12.
[6] Veghte, J. H., "Rating Protective Clothing with TPP," *Firehouse*, Vol. 7, No. 11, 1982, pp. 18-19.
[7] Holcombe, V. B., "The Evaluation of Protective Clothing," *Fire Safety Journal*, Vol. 4, 1981, pp. 91-101.
[8] Veghte, J. H., "Testing Water Barriers in Protective Clothing," *Firehouse*, Vol. 9, No. 9, 1984, pp. 36-38.

Norman F. Audet[1] *and Kenneth J. Spindola*[1]

U.S. Navy Protective Clothing Program

REFERENCE: Audet, N. F. and Spindola, K. J., **"U.S. Navy Protective Clothing Program,"** *Performance of Protective Clothing, ASTM STP 900*, R. L. Barker and G. C. Coletta, Eds., American Society for Testing and Materials, Philadelphia, 1986, pp. 497-512.

ABSTRACT: The Navy Clothing and Textile Research Facility is involved in a major effort to develop fire retardant-heat protective clothing for Navy shipboard personnel encompassing both work and special protective clothing.

To determine the degree of protection afforded by the various articles of clothing being developed, both laboratory and field testing have been employed. Laboratory material tests have included vertical flammability resistance tests and protection time determinations for flame impingement exposure. Field tests have involved exposure of garments to total flame envelopment utilizing manikins equipped with paper tape temperature sensors to estimate the extent of burn injury sustained.

The usefulness of laboratory tests such as flame impingement tests to predict material/clothing performance under more realistic fire threat conditions is limited because of differences in exposure conditions between laboratory tests and full-scale fire tests. Laboratory tests do, however, provide an extremely useful screening tool to limit the number of candidate systems to be evaluated under field conditions.

This paper covers the nature of some of the clothing systems being developed, the laboratory and field evaluation procedures employed, and some of the laboratory and field test data accumulated on the materials and clothing items investigated.

KEY WORDS: fire retardant materials, protective clothing, fire resistance tests, fire protection tests

As part of the U.S. Navy's Battle Dress and Passive Fire Protection Programs, the Navy Clothing and Textile Research Facility (NCTRF) is responsible for the protective clothing development efforts. In these programs, fire protection is a major concern because exposure of personnel to fires can occur aboard ship both under battle conditions and during normal operations.

Developments completed to date in these programs include a lightweight fire retardant (FR) coverall, which is worn by personnel working in engineer-

[1]Technical director and textile technologist, respectively, U.S. Navy Clothing and Textile Research Facility, Natick, MA 01760.

ing spaces aboard ship and other hot-humid fire hazardous areas, and an FR utility uniform for wear by all shipboard enlisted personnel.

The material chosen for the coverall was a 153-g/m^2 (4.5-oz/yd^2), 95/5 Nomex/Kevlar blend fabic (Nomex 3), and the materials chosen for the utility uniform were a fire-retardant-treated (FRT), 100% cotton, 187-g/m^2 (5.5-oz/yd^2) chambray fabric for the shirt and an FRT, 100% cotton, 407-g/m^2 (12-oz/yd^2) denim fabric for the trousers. Both of these uniforms are pictured in Fig. 1.

Other FR articles of clothing currently under development are utility uniforms for officers and chief petty officers, cold weather clothing, an antiexposure suit, a damage control suit, antiflash gear, and chemical warfare protective clothing.

This paper focuses on the fire protection tests employed for the final selection of the materials to be used in the FR utility uniform for enlisted personnel. Information on the methods employed and results obtained is included, and the differences noted between laboratory and full-scale fire tests are discussed.

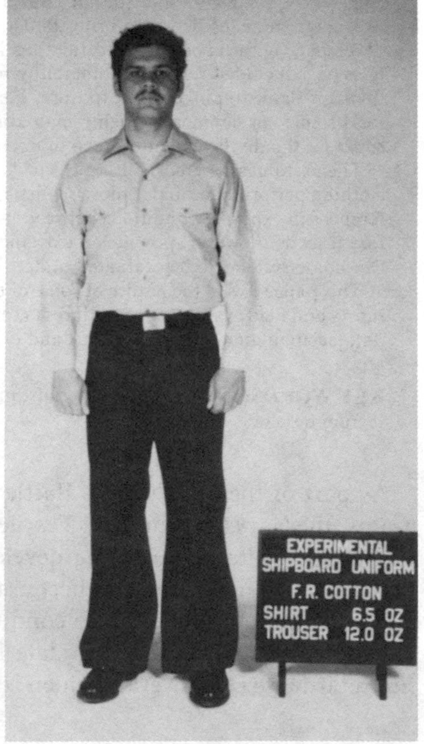

FIG. 1—*Flame retardant uniform selections.*

Materials

The principal uniform types and materials utilized in the final development tests of the FR enlisted utility uniform are shown in Table 1. Note that the shirting fabric finally selected [187 g/m^2 (5.5 oz/yd^2)] was lighter than the fabric tested [221 g/m^2 (6.5 oz/yd^2)] in order to improve comfort. The test uniforms were similar to those shown in Fig. 1.

These uniforms were selected as a result of an initial ship evaluation of several variations of materials and uniform types. A 153-g/m^2 (4.5-oz/yd^2) 95/5 Nomex/Kevlar blend coverall was selected by shipboard personnel as the best one-piece uniform, and a 221-g/m^2 (6.5-oz/yd^2) 100% FRT cotton chambray shirt and 407-g/m^2 (12-oz/yd^2) 100% FRT cotton denim trousers were selected as the best two-piece uniform.

In the final development tests, these uniforms were directly compared with each other for fire protection performance as well as other functional characteristics (comfort, durability, launderability, and so on) and overall user preference. This paper, however, will present only the fire protection data on these two uniform types.

Procedure

Fire Resistance

The fire resistance of the final candidate materials was evaluated new and after 25 and 50 simulated shipboard launderings using Navy Wash Formula II (Appendix I).

The vertical flammability test detailed in Method 5903 of Federal Test Method Standard for Textile Test Methods No. 191A was employed. Characteristics such as afterflame and afterglow times and char length were determined. This method judges the ability of a material to self-extinguish after removal of the flame source and the degree of material degradation caused by the flame exposure.

TABLE 1—*Fire retardant materials and uniforms evaluated.*

Uniform Type	Component	Material	Construction	Weight, g/m^2
One piece		95/5 Nomex/Kevlar	plain weave	153
Two piece	shirt	100% FRT[a] cotton	chambray	221
	trousers	100% FRT[a] cotton	denim	407

[a]FRT—tetrakis(hydroxymethyl)phosphonium chloride with gaseous ammonia cure (THPC-NH_3).

Flame Impingement

These tests were conducted by Albany International Research Corp. [*1*] under contract to the Navy using a device and methods developed by Stoll and Chianta [*2*]. Figure 2 depicts the apparatus and instrumentation employed, and Fig. 3 depicts schematically the exposure of the sample material to the burner flame.

In these tests, the materials and material assemblies were exposed to a 2.2-$cal/cm^2/s$ flame impingement flux for 3 s. The flux level was calibrated with a water-cooled heat flux transducer (Hy-Cal Engineering Model No. C-1300-A-15-072). A skin simulant sensor [*1*] was used to measure the temperature rise behind the material and material assemblies evaluated. The sensor was in direct contact with the material and material assemblies.

The time required to cause a second-degree burn injury (TBI) was estimated from curves developed by Stoll and Chianta. Three tests were conducted on each material and material assembly evaluated, and the results were averaged.

Fire Pit Tests

These tests were conducted to determine the thermal protection afforded by the candidate uniforms under full-scale fire conditions. The tests were conducted at the Naval Air Development Center.

In these tests, the uniforms were exposed to JP-4 fuel fires for 2 s (Navy protection requirement). The maximum allowable burn injury was 20% of body surface area (combined second and third-degree level burns). The 20% body surface area burn criterion employed relates to potential recovery from these injuries. Burns (second and third degree) involving more than 20% of the body surface area endanger life, and those involving more than 30% of the body surface area are usually fatal if adequate medical treatment is not provided within a short time [*3*].

The fire pit facility consisted of a 9-m-long by 6-m-wide metal tank having a depth of 0.3 m. A variable-speed rotating crane with cantilevered boom, equipped with a frame to carry a manikin dressed with test clothing, was used to propel the manikin across the length of the tank. A concrete block wall separates the tank from the crane control section and also isolates the test manikin from the fire before and after exposure to the fire. Attached to the crane manikin carry frame was a Hy-Cal Engineering No. C-1301-C-15-072 water-cooled heat flux transducer, which was used to determine the total heat energy to which the test clothing was exposed. Figure 4 shows a dressed test manikin before entering the flames, in the flames, and exiting the flames.

Fiberglass manikins coated with a fire-resistant paint were employed to simulate the configuration of the human body. Paper tape temperature sensors affixed to leather patches (Fig. 5) were mounted on the surface of the

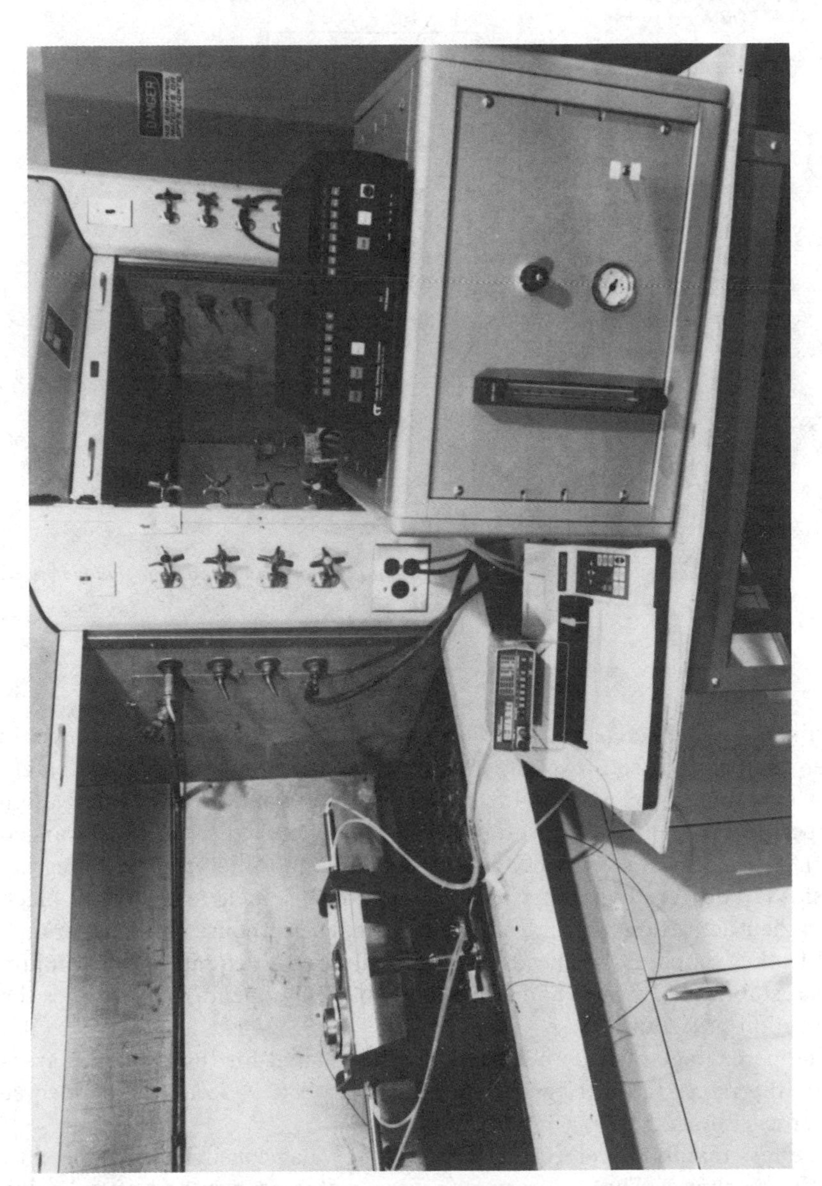

FIG. 2—*Flame impingement test apparatus.*

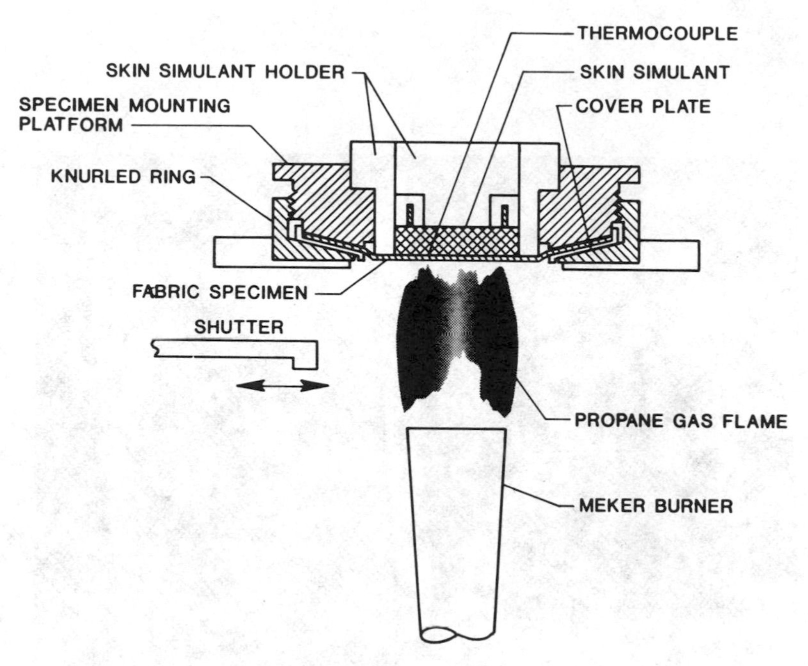

FIG. 3—*Cross-sectional view of the specimen mounting assembly and skin simulant in its holder.*

manikins at 20 discrete body sites in the torso, leg, and arm areas. A set of seven temperature-sensitive paper tapes was affixed to each leather patch. Each tape was 0.8 cm wide by 4.8 cm long and was stamped with its activation temperature value. The activation temperature of each set of tapes ranged from 99 to 138°C (220 to 280°F) with incremental differences of 5 degrees Celsius (10 degrees Fahrenheit). The tapes change shade from gray to black when their activation temperature is reached. The response of the paper tape and leather patch assemblies had been calibrated to equate to burn injury levels established by Stoll and Chianta (Table 2). Appendix II describes the calibration procedure.

The percentage of body burn area was estimated for those surface areas where the paper tapes triggered at the 138°C (280°F) level (second-degree blister level burn).

The test manikins were dressed with underwear consisting of a tee-shirt and boxer shorts. The test garments were evaluated new and after having been subjected to 25 simulated shipboard launderings using Navy Wash Formula II (Appendix I). The number of tests conducted on each of the test garments in both the new and laundered state are shown in Table 3.

FIG. 4—*Fire pit facility.*

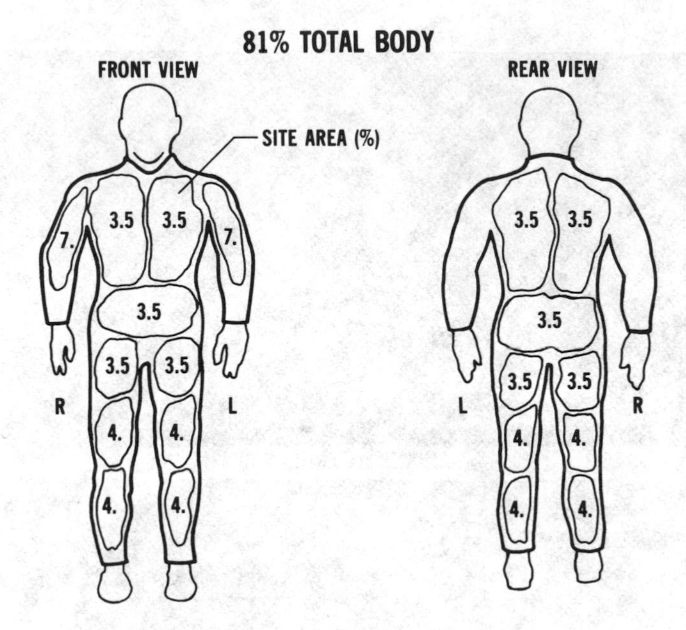

FIG. 5—*Manikin paper tape temperature sensor sites.*

TABLE 2—*Relationship of burn level to paper tape activation temperature.*[a]

Degree of Burn Injury	Paper Tape Activation Temperature, °C (°F)
Pain	116 (240)
Survival	127 (260)
Blister	138 (280)

[a]Body burn area was calculated for a tape activation temperature of 138°C (280°F).

TABLE 3—*Number of test garments for each type of uniform evaluated in a new and laundered condition.*

Test Garment	Condition	Number Tested
95/5 Nomex/Kevlar coverall	new	10
	laundered	3
FRT cotton chambray shirt and denim trousers	new	10
	laundered	5

Results

Fire Resistance

Table 4 shows the vertical flammability results for the Nomex/Kevlar and FRT cotton materials when new and after 25 and 50 simulated shipboard launderings. All of the materials showed acceptable fire resistance in these tests. This performance did not change appreciably from the new condition and after 50 launderings.

Flame Impingement

Table 5 shows the flame impingement results adapted from Ref *1* for 3-s exposures to a 2.2-cal/cm^2/s heat flux for the Nomex/Kevlar and cotton materials alone and in assembly with cotton underwear. Since data were not available on the 407-g/m^2 (12-oz/yd^2) FRT cotton denim material, data for a nontreated, 100% cotton 339-g/m^2 (10-oz/yd^2) material was substituted.

The results obtained were not surprising; they showed that, in general, the heavier materials provided better protection, as depicted by lower temperature increases in the skin simulant sensor and higher estimated TBIs with increasing material and material assembly weights. Both cotton fabrics were heavier than the Nomex/Kevlar blend fabrics and showed a higher TBI than the Nomex/Kevlar blend fabrics when compared alone or in assembly with cotton underwear fabric.

Linear regression analysis showed excellent correlation (0.93) between temperature change and material and material assembly weights at 3 s and for

TABLE 4—*Vertical flammability performance of fabrics before and after simulated shipboard laundering.*[a]

Material		Condition								
		New			Launderings					
					25			50		
Type	Weight, g/m^2	After Flame, s	After Glow, s	Char Length, cm	After Flame, s	After Glow, s	Char Length, cm	After Flame, s	After Glow, s	Char Length, cm
95/5% Nomex/Kevlar	153	0	18	7.1	0	13	7.6	0	16	8.4
100% FRT cotton	221	0	2	8.9	0	1	7.6	0	1	7.4
	407	0	1	7.1	0	1	7.4	0	1	6.4

[a]Navy Wash Formula II (Appendix I).

TABLE 5—*Temperature rise in skin simulant during flame impingement (heat flux—2.2 cal/cm²/s).*

Material		Assembly Weight, g/m²	Temperature Rise, °C		TBI, s
Outer	Underwear		3 s	Maximum	
95/5% Nomex/Kevlar		156[a]	29.1	35.5	1.7
	cotton	265[b]	20.4	29.5	2.9
100% FRT cotton		234	21.1	26.6	2.7
	cotton	343	12.3	19.6	5.9
100% cotton		350	14.3	18.0	4.8
	cotton	458	12.0	16.6	6.0

[a]Weight of outer material.
[b]Weight of total material assembly.

the maximum temperature rise measured (Fig. 6). The two temperature versus material and material assembly weight curves were nearly parallel (12% difference in slopes), indicating that the rate of afterheating resulting from heat storage in the materials did not generally cause significantly greater elevations in sensor temperatures because of differences in material and mate-

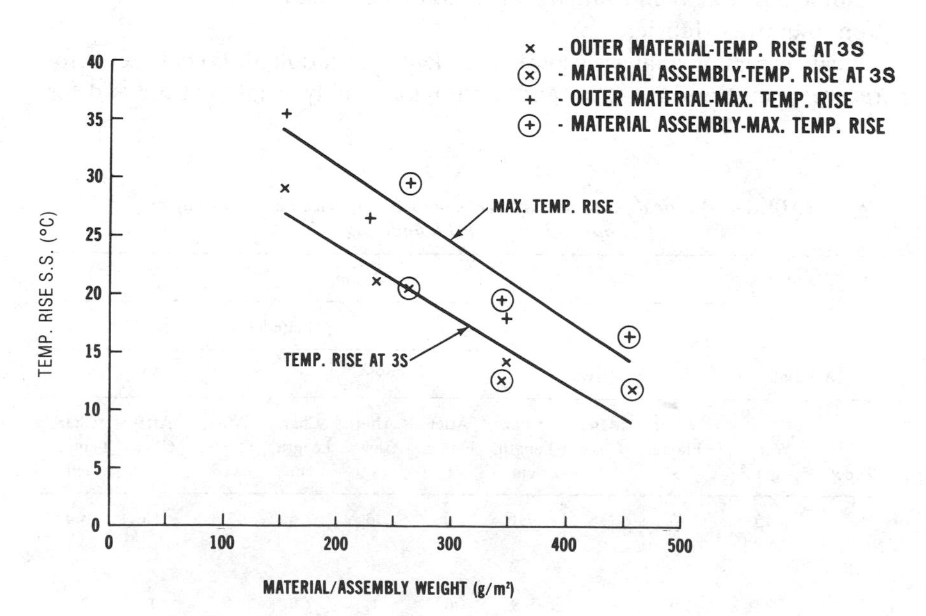

FIG. 6—*Temperature rise in skin-simulant (SS) sensor versus the outer material and material assembly weights.*

rial assembly temperatures, weight, and specific heats. However, the assemblies showed greater changes in temperature resulting from afterheating than the singular materials (Table 5).

The benefit of layering materials, not considering afterheating, was most noticeable when the temperature rise at 3 s for the heavier outershell cotton material (350 g/m^2) was compared with the lighter-weight cotton outershell fabric combined with the cotton underwear (343 g/m^2), producing approximately the same total weight. In this case, there was more than a 1-s increase in estimated TBI for the assembly based on the temperature rise data at 3 s. In comparing the maximum temperature rise for these same materials, layering shows no particular benefit.

Fire Pit

Table 6 shows the individual total heat and the percentage of body area burn injury data for the Nomex/Kevlar coverall and FRT cotton chambray/denim two-piece uniform before and after 25 launderings, as well as the average and standard deviation results. As can be seen, the results were quite

TABLE 6—*Individual total heat—estimated blister level burn injury data from 2-s JP-4 fuel fire tests.*

	Coverall 95/5 Nomex/Kevlar		Chambray/Denim 100% FRT Cotton	
Garment Condition	Total Heat, cal/cm^2	Body Burn, %	Total Heat, cal/cm^3	Body Burn, %
New	6.1	16	8.3	7
	5.9	16	6.7	7
	6.4	27	6.6	0
	3.5	16	6.4	0
	8.1	42	5.7	7
	6.6	8	6.0	7
	9.5	37	5.1	0
	7.5	42	7.2	0
	6.4	31	3.9	0
	11.5	53	8.2	19
Average	7.1 ± 2.2	29 ± 15	6.4 ± 1.3	5 ± 6
25 shipboard launderings	3.5	16	5.2	2
	6.9	22	6.6	4
	6.8	25	6.8	5
			4.4	0
			5.2	2
Average	5.7 ± 1.9	21 ± 5	5.6 ± 1	3 ± 2

variable from test to test, particularly the data on percentage of body area burn injury.

The 20% body area burn criterion was not exceeded in any test with the cotton uniform, whereas the average value with the Nomex/Kevlar coverall exceeded this criterion with individual burn area estimates as high as 53%.

Because of the variability in results, linear regression analysis was performed on the data, and the curves are shown in Fig. 7. As can be seen, the extent of body area burn injury for the Nomex/Kevlar coverall was significantly greater than for the cotton uniform at any specific total heat level whether in a new condition or laundered.

Extrapolation of the curves for the new condition to the 0% body area burn ordinate would indicate that no burn would be sustained with the Nomex/Kevlar coverall at a total heat of 2 cal/cm^2 and 4.5 cal/cm^2 for the cotton two-piece uniform.

Considering the design configuration differences between the two uniforms—the Nomex/Kevlar coverall has one common weight throughout the uniform whereas the two-piece cotton uniform has different weights for the shirt and trousers and a combination of these two weights where the materials overlap in the lower torso area—it was estimated that the average weight of the cotton uniform was 363 g/m^2 (10.7 oz/yd^2) compared with an average

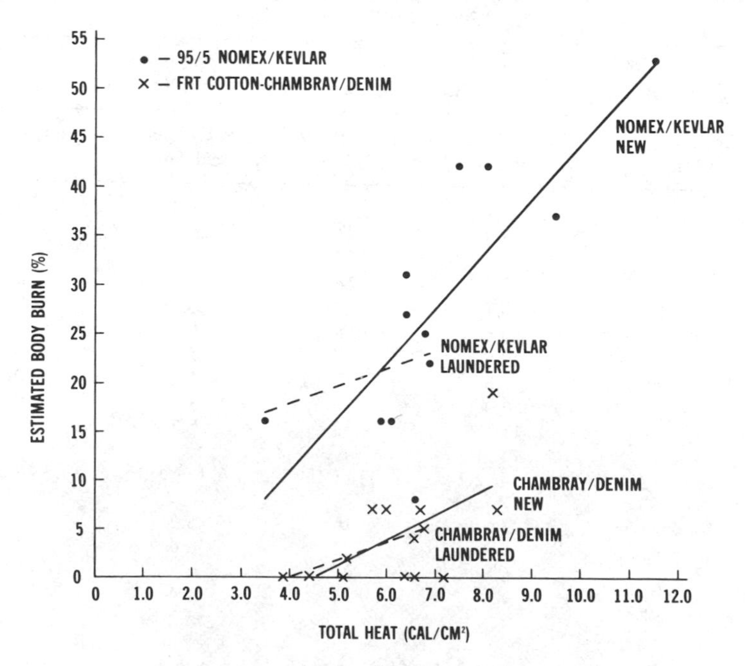

FIG. 7—*Percentage of body burn versus total heat from new garments and laundered garments.*

weight for the Nomex/Kevlar coverall of 156 g/m^2 (4.5 oz/yd^2). When the weight ratio between the cotton and Nomex/Kevlar uniform (2.4) is compared with the 0% body area burn total heat ratio between the cotton and Nomex/Kevlar uniforms (2.25), it can be seen that the amount of protection relates quite closely to weight effects.

Figure 8 shows the incidence of burns recorded at the various body areas monitored. There were small differences between the Nomex/Kevlar uniform and the cotton two-piece uniform in those areas covered by underwear (upper and lower torso) and the arm areas where the weights of the uniforms are not substantially different. The major area where substantial differences in incidence of burn occurred was in the leg area. These differences in the leg area were believed to be related to several factors:

1. *Uniform design*—The cotton denim trousers have a flared leg construction and provide a more substantial air gap between the trousers and the leg below the knee than the Nomex/Kevlar coverall, which had a straight leg construction. It is believed that some of this advantage was maintained, even under dynamic test conditions, because of the additional stiffness of the cotton fabric versus the Nomex/Kevlar fabric and the better dimensional stability of the cotton fabric versus the Nomex/Kevlar fabric in high-temperature exposures.
2. *Trouser weight*—The cotton denim trousers were 2.7 times heavier than the Nomex/Kevlar material.

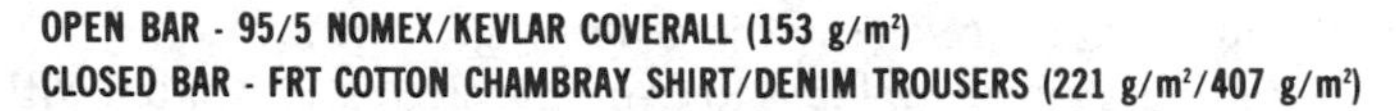

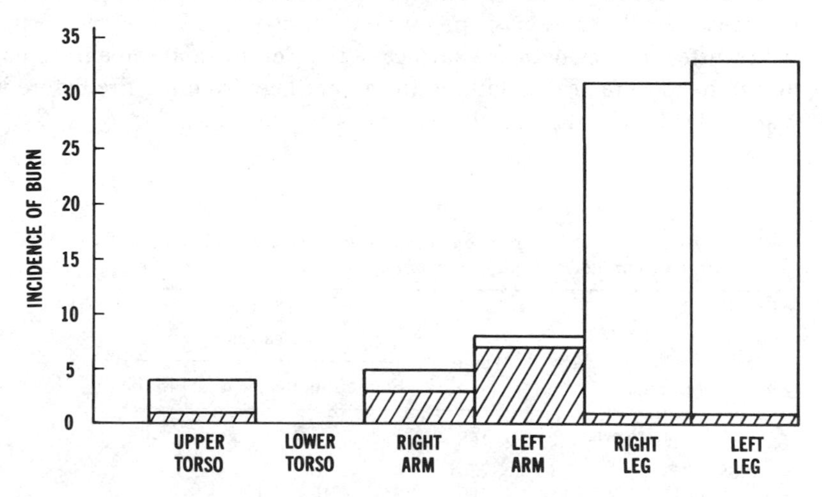

FIG. 8—*Incidence of burns for different body sections.*

3. *Thermal shrinkage*—The cotton denim trousers remained dimensionally stable during the fire exposures whereas the Nomex/Kevlar material showed significant thermal shrinkage. The thermal shrinkage of the Nomex/Kevlar material eliminated any air gaps between the garment and the manikin in some locations. For instance, the outline of the paper tape leather patch temperature sensors was clearly visible below the coverall material in the thigh area of the legs. The shrinkage also caused a reduction in the trouser length, which in some tests exposed the bottom of the legs to direct fire exposure.

Discussion

Fire resistance tests (vertical flammability) for fire retardant materials provided little insight into the thermal protection these materials provide to wearers. Laboratory tests similar to the flame impingement device described herein provide a better estimate of the efficacy of potential materials to provide adequate burn protection but tend to overestimate the total heat protection compared with fire pit results.

Table 7 estimates total heat protection based on the flame impingement and fire pit test results using the total heat values computed for the singular fabrics from the flame impingement test results and the 0% body area burn threshold total heat values from the fire pit tests. As can be seen, the total heat protection estimated from the fire pit results are substantially lower than the flame impingement estimates. The difference relates in part to the fact that the Stoll-Chianta TBI estimate method does not consider heating effects after cessation of exposure in the flame impingement tests.

Other differences may be related to dynamic effects associated with the fire pit tests caused by the motion of the manikin through the flames increasing the heat transfer coefficient at the surface of the clothing and some preexposure and postexposure of the clothing to radiant heat from the fire before it enters and after it exits the fire pit.

TABLE 7—*Total heat protection estimated for Nomex/Kevlar and FRT cotton fabrics from flame impingement and fire pit results.*

Material	Weight, g/m^2	Total heat, cal/cm^2	
		Flame Impingement	Fire Pit
95/5 Nomex/Kevlar	156	3.7	2.0
100% FRT cotton	234	5.9	4.5
	350	10.6	

In both tests, protection estimates appear to be strongly related to the weight of the materials, but this relationship differs between the tests because of systematic differences between the two tests such as material sensor configuration differences and the dynamics of the two methods.

Conclusions

Laboratory tests that judge the thermal protection offered by various materials such as the flame impingement tests discussed herein provide a useful tool in estimating thermal protection differences between materials. However, in converting skin-simulant sensor temperature changes to protection times using the Stoll-Chianta method, protection times are overestimated because afterheating effects are not accounted for. Changes in specimen weights and temperature rise data in the skin-simulant sensor correlated well for the materials studied.

Full-scale fire pit tests appear to better estimate protection differences between materials and uniform configurations for flame envelopment conditions of the type described herein, since this process more closely mimics the potential fire hazard faced by Navy personnel. As in the flame impingement tests, material weight appeared to be the most significant factor in relating protection differences between the material/uniform types.

Acknowledgment

The authors wish to thank Albany International Research for their efforts in developing the flame impingement data used herein, Dr. George Kidd of the Naval Air Development Center for conducting the fire pit tests, Richard Wojtaszek of the Navy Clothing and Textile Research Facility for his major contribution to the development of the Navy's FR utility uniform for enlisted personnel, and the Naval Sea Systems Command, which sponsored the program.

APPENDIX I

Navy Formula II—Hot Formulas Without Bleach [60°C (140°F)]

Classification: Cotton, synthetic blend colored—khaki, dungaree, etc.

P-D-245-C detergent
Hard/soft water—Type I
Seawater—Type II[a] 100-lb load basis

Step	Notes	Operation	Cycle Time, min	Water Temperature, °F[b]	Water Level, in.[c]	Supplies, 100-lb Basis[d]
1	[e]	break/suds	10	140	4	8 oz detergent 16 oz alkali 2 oz nonionic
2		drain	1			
3		flush/suds	6	140	4	
4		drain	1			
5		spin	1			
6		rinse	3	140	4	
7		drain	1			
8		rinse	3	140	4	
9		drain	1			
10	[f,g]	sour	4	120	4	2-oz sour blue 12-oz instant starch
11		drain	1			
12		final spin	4			

[a]For seawater washing: Use seawater in Steps 1 and 3. Use Type II detergent. Use fresh water in Steps 6, 8, and 10.

[b]Degrees Celsius = (degrees Fahrenheit − 32) ÷ 1.8.

[c]4 in. = 10.2 cm.

[d]1 lb = 0.45 kg; 1 oz = 28.35 g.

[e]Add nonionic while water is being added.

[f]Bacteriostats are added in this operation, if required.

[g]Add starch and run for 10 min in the manual mode when starch is required.

APPENDIX II

Paper Tape-Leather Patch Calibration Procedure

To calibrate the paper tape-leather patch assemblies against the Stoll-Chianta burn injury curves, a quartz lamp radiant heat tester was employed. A water-cooled heat flux transducer (Medtherm Corp. Model 64-2-20) was initially used and placed 1.25 cm behind a single layer of fabric, and the radiant heat load to the fabric was increased in discrete increments for exposures of 2 s until the average heat flux measurements behind the test fabric were equivalent to the Stoll-Chianta pain, survival, and blister levels for this same time duration. The heat flux transducer was then replaced by the paper tape-leather patch assemblies and, employing the same radiant heat loads and 2-s exposures used to determine the pain to blister levels with the transducer, the highest tape-activation temperature for each of these conditions was recorded.

References

[1] Schoppes, M., *Resistance of Navy Shipboard Work Clothing Materials to Extreme Heat*, Contract No. N00140-81-C-BA83, Albany International Research Corp., Dedham, MA, October 1982.

[2] Stoll, A. and Chianta, M., *A Method and Rating System for Evaluation of Thermal Protection*, Naval Air Systems Command, Washington, DC, December 1968.

[3] Sirois, J., *Standard First Aid Training Course*, NAVEDTRA 10081-N, Naval Education and Training, Naval Air Station, Pensacola, FL, 1982.

Heat Stress, Fit Testing, and Other Performance Requirements for Protective Clothing

Richard R. Gonzalez,[1] *John R. Breckenridge,*[1] *Clement A. Levell,*[1] *Margaret A. Kolka,*[1] *and Kent B. Pandolf*[1]

Efficacy of Heat Exchange by Use of a Wettable Cover over Chemical Protective Garments

REFERENCE: Gonzalez, R. R., Breckenridge, J. R., Levell, C. A., Kolka, M. A., and Pandolf, K. B., **"Efficacy of Heat Exchange by Use of a Wettable Cover over Chemical Protective Garments,"** *Performance of Protective Clothing, ASTM STP 900,* R. L. Barker and G. C. Coletta, Eds., American Society for Testing and Materials, Philadelphia, 1986, pp. 515–534.

ABSTRACT: A continuing effort is being made to reduce the dangers of heat exhaustion for persons working in heat with impermeable chemical protective garments (CPGs). One approach that appears feasible is the use of a wettable cover surrounding the CPG. The efficacy of evaporation of the heat extracted is a function of the evaporative heat transfer coefficient, h_e (in watts per square metre per torr), and the gradient, ΔP_{sk}, of the saturation vapor of pressures (in torrs) between the wettable cover surface, P_c, and the ambient air, P_w. The h_e value is dependent on wind movement [based on a convective heat transfer coefficient, (in watts per square metre kelvin)], the appropriate Lewis numbers for air, L_a (2.2 K/torr), and the layers of wettable covers, L_{lay}. Copper manikin and experimental studies showed that (1) when ΔP_{sk} is reduced, as in severe hot and humid conditions, evaporative heat removal is minimal; (2) model predictions for CPG gave supplementary cooling (increased skin heat loss) ranging from 40 W at 35°C, $P_w = 29.5$ torr, and low air movement, to almost 200 W for a hot and dry environment of 50°C, 18 torr, with a 5-m/s wind. The predicted water requirements to maintain the cover at 100% wetness under these conditions ranged from 3 to 32 g/min, respectively.

KEY WORDS: wettable cover, impermeable garment, evaporative heat loss, cooling efficiency, clothing insulation, clothing temperature, heat stress prediction, protective clothing

Several methods have been tried for relief of heat stress for workers wearing chemical protective clothing [*1–7*]. Among these methods are (1) increasing

[1]Chief, Biophysics Branch, biophysics consultant, research physical scientist, research physiologist, and director, respectively, Military Ergonomics Division, U.S. Army Research Institute of Environmental Medicine, Natick, MA 01760-5007.

the permeability of the garments to water vapor transfer [5,6], (2) delivering filtered ambient air into impermeable suits, and (3) using other auxiliary cooling methods [6, 7]. All of these methods require special mechanisms that become intrusive to the wearer. One approach, which is less well defined, is the use of a wettable cover over an impermeable garment [8]. The theory is that skin heat loss can be increased from such a wet cover over a garment. Heat loss comes through four avenues: external air movement, V; the consequent increases in the convective heat transfer coefficient, h_c; the ambient air vapor content; and layering of the wetted cover and insulative properties between the cover and the skin of the wearer. The latter property establishes the efficacy of skin cooling in relation to the total possible amount of water capable of being evaporated [1].

This paper describes a biophysical approach, which allows prediction of the cooling benefits from a wetted cover on an impermeable ensemble. The factors considered are air temperature, dew point, wind speed, and ensemble heat transfer characteristics. Cases for validation of the model are presented with wet cover tests on an electrically heated manikin and with computer predictions that provide for increases in body heat loss with multiple ensembles.

Theoretical Analysis

Dry Heat Exchange and Cooling Efficiency

In developing the heat exchange properties in this paper the authors are primarily interested in clothing systems with an impermeable layer underneath a wettable cover that allows quantifiable evaporation. Other clothing layers, dry or wet, or nude skin may be found under the impermeable barrier. Figure 1 is a schematic of the exchange properties.

In the case in which the subject is not sweating and the layers are kept dry, a cooling efficiency may be described from Burton's [9] original analogy as

$$F_{cl} = \frac{I_a}{f_{cl} I_{\text{tot}}} \tag{1}$$

where I_a is the boundary air layer for nude skin which, expressed in clo units[2] [10], is given by

$$I_a = \frac{6.46}{h_r + h_c} \tag{2}$$

[2]A clo unit of insulation is equal to 0.155 K · m²/W; 1 clo is roughly the insulation of a winter suit with a vest at 58 W/m² (1 metabolic unit) in still air, and 21°C temperature.

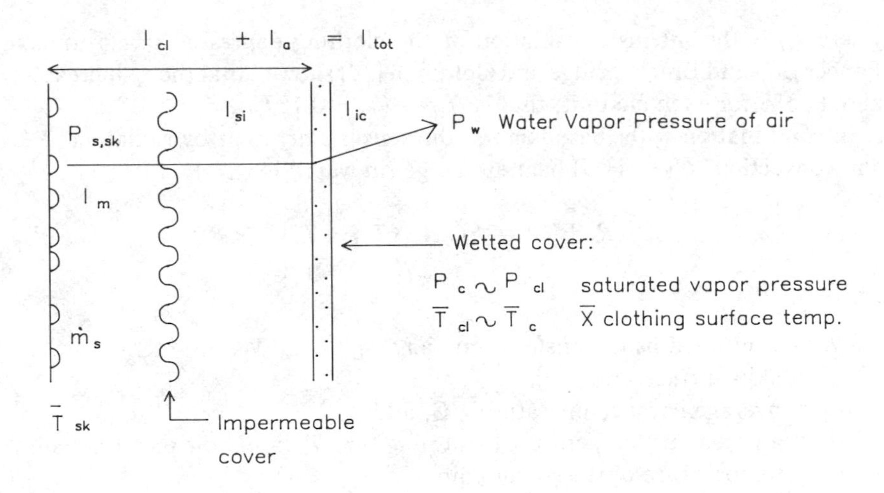

Sensible Insensible Solar

Skin Heat Loss or Gain = (R+C) + H_e + H_{sl}

$$H_{sk} = h'(T_a - \overline{T}_{cl}) + w \cdot h_e'(P_c - P_w)$$

I_m, h', h_e' effective transfer coefficients

T_a, $\overline{T}_{cl}$, P_w environmental parameters

w, $\overline{T}_{sk}$, $P_{s,sk}$, I_{tot}, $\dot{m}_s$ biophysical & physiological parameters

FIG. 1—*Complete heat balance of wetted cover. The symbols are defined in the text.*

where 6.46 converts kelvin square metres per watt to clo units. The radiation coefficient, h_r, is a function of surface and air temperatures, the percentage of the ensemble surface area present in radiant exchange, and the emissivity of the surface site. The coefficient for convection, h_c, is a function primarily of effective air movement, V, over the garment layer. In Eq 1 (Fig. 1), I_{tot} includes the combined insulation of clothing as well as its boundary layer, and f_{cl} is the ratio of clothing surface to skin surface area [*11–13*]. This factor modifies the boundary layer insulation as the clothing surface area increases, and thereby expands the surface area open for heat transfer [*12*]. Another way of expressing this factor is

$$I_{\text{tot}} = \frac{I_{cl} + I_a}{f_{cl}} \tag{3}$$

where I_{cl} is the intrinsic insulation of the clothing expressed in clo units. Fanger [11] and Breckenridge and Goldman [14] showed that the f_{cl} increases about 15% for each clo unit; that is, $f_{cl} = (1 + 0.15I_{cl})$.

In combination with these factors, the sensible dry heat by radiation, R, and convection, $C(R + C)$ heat exchange (in watts) is

$$(R + C) = hf_{cl}A_{sk}(\bar{T}_c - T_a) \tag{4}$$

where

h = combined heat transfer coefficient, $h_r + h_c$, $W/(m^2 \cdot K)$,
A_{sk} = skin surface area, m^2,
$\bar{T}_c$ = average cover temperature, °C, and
T_a = ambient air temperature (assuming that $T_a = \bar{T}_r$, the mean radiant temperature of the surroundings, °C).

Evaporative Heat Loss from a Wetted Cover

Evaporation rate, H_e (in watts), may be described by

$$H_e = A_{sk}f_{cl}wh_e(\bar{P}_c - P_w) \tag{5}$$

where

w = wetted fraction of the clothing surface, %,
h_e = evaporative heat transfer coefficient, $L_a \cdot h_c$, where L_a is the Lewis number (2.2 K/torr) [12],
$\bar{P}_c$ = average water vapor pressure at the cover surface, torr, and
P_w = ambient water vapor pressure, torr.

Typically, the combination $A_{sk}f_{cl}$ in Eq 5 includes the total surface area (Fig. 1), and $(A_{sk}f_{cl}w)$ includes only the wetted cover area.

In field situations, the gradient, $\bar{P}_c - P_w$, is not easily determined since $\bar{P}_c$ is obtained from the cover temperature, T_c, which is not always known. However, by accounting for all heat gains or losses at the clothing surface, T_c may be equated to an effective clothing surface temperature, $\bar{T}_{cl}$. This heat balance is described by

$$H_{sk} + (R + C) + H_e + H_{sl} = 0 \tag{6}$$

where H_{sk} is the total heat gain from the skin (in watts). H_{sl} refers to the amount of solar heat absorbed at the clothing surface. This component will not be considered in the present paper but was considered in a previous technical report [4]. However, appropriate equations to determine solar heat load as a contribution from direct, diffuse, and terrain-reflected sunlight as a function of solar angle and posture have been developed [11,14]. In our analysis, H_{sl} is independent of $\bar{T}_{cl}$ and will be constant when $T_a = \bar{T}_r$, as in the treatment here.

Assuming that the skin surface and inner clothing layers are dry, Eq 6 will quantify heat gains and losses for wet and dry cover conditions. The cooling efficiency (as in Eq 1) may be derived by determining H_{sk} with a wet cover minus H_{sk} for a completely dry cover. In our case for no solar load and no H_e, $H_{sk} = (R + C)$.

Determination of $\bar{T}_{cl}$

In Eq 6, all the heat gains are considered positive. When the temperature and vapor pressure gradients in Eqs 4 and 5 are rewritten as $(T_a - \bar{T}_c)$ and $(P_w - \bar{P}_c)$, respectively, dry heat loss and evaporative heat loss have a *negative sign.* Additionally, assuming that $\bar{T}_c \approx T_{cl}$ and $P_c = \bar{P}_{cl}$, the saturated vapor pressure at $\bar{T}_{cl}$,[3] Eqs 4 and 5 become

$$(R + C) = A_{sk} f_{cl} h (T_a - \bar{T}_{cl}) \tag{4a}$$

$$= A_{sk} \frac{T_a - \bar{T}_{cl}}{\dfrac{I_a}{f_{cl}}} \tag{4b}$$

and

$$H_e = A_{sk} f_{cl} w h_e (P_w - \bar{P}_{cl}) \tag{5c}$$

These simplifying assumptions introduce an error in the cover vapor pressure, P_c, and hence in H_e, since the two temperatures will rarely be equal. Under most conditions the wetted cover temperature will average less than $\bar{T}_{cl}$, because the dry surface areas will not be cooled. Using $\bar{P}_{cl}$ in equations, in place of P_c in Eq 5, will then result in an overestimate of evaporative cooling (H_e will assume too high a negative value). However, this overestimate will not be as great as it appears. Equation 6 shows that, if H_e is overestimated, a corresponding increase in $[(R + C) + H_{sk} + H_{sl}]$ is required to balance the equation. With constant solar heating, H_{sl}, this means that $\bar{T}_{cl}$ must be lowered to effect the required increase; this reduction in $\bar{T}_{cl}$ will tend to bring it into line with the actual wet cover temperature, thereby reducing any overestimate in evaporative heat loss.

[3] $\bar{P}_{cl}$ may be estimated by the Antoine equation (Eq 12), $P_{cl} = \exp[18.6686 - 4030.143/(\bar{T}_{cl} + 235)]$ torr.

With a clothing system containing an impermeable layer, the total transfer of heat from a sweat-wetted skin to the clothing surface involves both sensible and evaporative flows up to the impermeable barrier, but only sensible flow between it and the clothing surface. Breckenridge [*15*] has shown that evaporative heat transfer can occur between the skin and vapor barrier if the two are at different temperatures, even without a net water loss. The explanation involves a cyclic process in which heat is absorbed from the skin by sweat evaporation, transferred across the inner insulating layers along with the vapor, and then released when the vapor condenses on the impermeable barrier, which is at a lower temperature than the skin; the condensate subsequently wicks back toward the skin to sustain the cycle. Experimental evidence [*15*] suggests that the rate of evaporative heat transfer in this process is proportional to the vapor pressure gradient between the skin and impermeable layer, which is the difference between saturated vapor pressures at the skin and impermeable layer temperatures, respectively. Beyond the impermeable layer, the flow, H_{sk}, to the clothing surface is all sensible heat. Assuming no heat storage at the vapor barrier, this sensible heat flow must equal the sum of the sensible and evaporative flows from the skin to the vapor barrier. Thus, the temperature, T_i, of the inner face of the impermeable layer may be determined by writing the H_{sk} equations for the two zones (see Fig. 1).

For heat transfer from the skin to the impermeable layer

$$H_{sk} = \left(\frac{6.46 A_{sk}}{I_{si}}\right)[(\bar{T}_{sk} - T_i) + L_a i_m (P_{s,sk} - P_i)] \tag{7}$$

where

I_{si} = insulation between skin and impermeable layer, clo,
$\bar{T}_{sk}$ = mean skin temperature, °C,
T_i = temperature of the inner face of the impermeable layer, °C,
$P_{s,sk}$ = saturated vapor pressure at $\bar{T}_{sk}$, torr,
P_i = saturation vapor pressure at T_i, torr, and
i_m = coefficient of moisture permeability [*16*] of the skin to the impermeable layer (ND).

The effective evaporative heat transfer coefficient is $(L_a \cdot i_m / I_{si})$, with i_m the factor describing the resistance of fabrics to evaporative transfer (varying from 0 = no transfer to 1.0 = maximal) [*16*].

Heat flow from the impermeable layer to the clothing surface involves sensible heat flow, in which

$$H_{sk} = \frac{6.46 A_{sk}}{I_{ic}(T_i - \bar{T}_{cl})} \tag{8}$$

where I_{ic} = insulation (in clo units) from the inner face of the impermeable layer to the clothing surface.

The two equations may be solved for T_i (in degrees Celsius) to determine the inner face temperature of the impermeable layer

$$T_i = \frac{\bar{T}_{sk} + L_a \cdot i_m(P_{s,sk} - P_i) + (I_{si}/I_{ic})T_{cl}}{1 + \frac{I_{si}}{I_{ic}}} \tag{9}$$

The dependency of the saturation vapor pressure of the impermeable layer on T_i is not easily handled as a simple solution of the outer clothing temperature. Since only a limited range of values is needed for accurate evaluation of P_i, T_i may be suitably defined as a linear function in terms of P_i. This method was used before by Gonzalez and Gagge [*17*] in defining a humid operative temperature for a "linear dew point" temperature, in which

$$T_{\text{dew}} = \frac{25.3 + P_w}{1.92} \tag{10}$$

substituting T_i and P_i for T_{dew} and P_w in Eq 9 and solving for P_i

$$P_i = 1.92T_i - 25.3 \tag{10a}$$

This would be accurate (to ± 0.5 torr) in measuring P_i from 25 to 40°C [*17,18*], the T_i range in which impermeable evaporative heat transfer to the skin is of major importance. When substituting Eq 10*a* in Eq 9 and determining H_{sk} from the impermeable layer to the clothing surface, Eq 8 becomes

$$H_{sk} = \frac{A'[(\bar{T}_{sk} - \bar{T}_{cl} + L_a \cdot i_m(P_{s,sk} + 25.3) - 4.2i_m\bar{T}_{cl})]}{B} \tag{11}$$

where

$A' = 6.46A_{sk}/I_{ic}$ and
$B = 1 + 4.2i_m + I_{si}/I_{ic}$.

Final Solution of $\bar{T}_{cl}$

The equation for predicting $\bar{T}_{cl}$ may now be solved by substituting Eqs 11, 4*a*, and 5*a* into the complete heat balance equation (without H_{sl}) for a wet cover. This equation (in degrees Celsius) is

$$\bar{T}_{cl} = \frac{hT_a f_{cl} + L_a h_c w f_{cl} P_w - L_a h_c w \bar{P}_{cl} f_{cl} + A'\left(\frac{\bar{T}_{sk} + L_a i_m (P_{s,sk} + 25.3)}{B}\right)}{C} \tag{12}$$

where

P_{cl} is evaluated by the Antoine equation [*12*],

$$A' = \frac{6.46}{I_{ic}},$$

$$B = 1 + 4.2i_m + \frac{I_{si}}{I_{ic}}, \text{ and}$$

$$C = hf_{cl} + \frac{6.46(1 + 4.2i_m)}{I_{ic} \cdot B}$$

After $\bar{T}_{cl}$ is solved by iteration of Eq 12, separate avenues of exchange may be evaluated. In summary, these are

$$R + C = A_{sk} f_{cl} h (T_a - \bar{T}_{cl})$$

$$H_e = L_a A_{sk} f_{cl} w h_c (P_w - \bar{P}_{cl})$$

$$H_{sk} = -[(R + C) + H_e]$$

For a totally dry skin condition, H_{sk} can be determined using

$$H_{sk} = \frac{6.46 A_{sk}}{I_{tot}} (\bar{T}_{sk} - T_{cl}) \tag{13}$$

where I_{tot} refers to the intrinsic clothing (in clo units) of the entire system involving the sum of the insulations, I_{si} and I_{ic}, and the temperature gradients, $(\bar{T}_{sk} - T_i)$ and $(T_i - \bar{T}_{cl})$. For the dry case, $(P_{s,sk} - P_i) = 0$, and i_m in Eq 12 is effectively zero. The cooling benefit for a wet cover is derived from the difference in skin heat loss, H_{sk}, with the cover wet and dry, respectively.

Procedure

Equipment, Instrumentation, and Coefficients

The heat loss data for validating the preceding mathematical skin-cooling model were obtained using a standard electrically heated life-size copper manikin covered with a form-fitting cotton "skin" [*4*]. This skin could be left dry to represent a nonsweating condition or could be completely wetted by spraying with water to simulate a sweating man with a 100% sweat-wetted skin surface. The manikin was dressed in standard tropical combat fatigues, a loose-fitting two-piece impermeable garment with a hood that covered all areas except the feet, hands, and face, and then a wettable cotton layer, also two-piece. The ensemble also included cushion sole socks, combat boots, and ankle-height rubber bootees on the feet; flexible rubber gloves over the hands and wrists; and a standard Army chemical protective mask over the face. The impermeable garment was elasticized at the ankles and wrists and around the facial opening to provide a tight seal; it was worn outside the tops of the combat boots and the wrist gauntlets on the gloves, and it covered the edges of the mask. The wettable outer cover provided similar coverage except that the bottoms of the trousers extended just to the top of the combat boots, well above the impermeable layer elastic seals. The trouser legs, sleeves, and neck opening on the cover were generously cut and were thus not in close contact with the impermeable layer, so that air could freely circulate between it and the cover during exercise.

We conducted experiments in a controlled temperature–humidity chamber with an air movement of approximately 0.3 m/s. For most of the experiments, the air temperature was closely controlled at about 27 ± 0.1°C, and the relative humidity was near 50%. A few experiments were run under other conditions—27°C and 20% relative humidity, and 34°C with both 25% and 90% relative humidity—to provide a wider range for testing the prediction equations and the estimation accuracy of some of the ensemble parameters, such as f_{cl} and w, which could not be measured.

During the experiments, the manikin was positioned on a Sauter platform scale, which provided a printout at 30-s intervals. These data provided a record of water evaporation during the wet cover experiments; the average rate of loss, which was obtained using a least-squares fit on the data and converted to an equivalent evaporative heat loss rate, permitted a comparison with the evaporative heat loss rate, H_e, predicted by the mathematical model. A large pan was positioned between the manikin and scale to reduce the likelihood that, in the event the cover was excessively wetted, drippings falling to the floor of the chamber would be recorded as an evaporative water loss.

The procedures used in operating the manikin were standard for this device [*14*]. Heating was accomplished using a proportional-power temperature controller, coupled to a series of thermistor sensors in the manikin, which held an average surface temperature during a given experiment constant within 0.1 degrees Celsius at about 32°C. With this close control, it could be assumed that the power input equaled the manikin heat loss, H_{sk}. Power was continuously recorded during a run, utilizing a thermal wattmeter connected to a single-pen strip-chart recorder. The average manikin surface temperature, $\bar{T}_{sk}$, was calculated by equally weighting recorded outputs from 21 copper-constantan thermocouples uniformly distributed, by area, over the copper shell. The air temperature, T_a, was also recorded at approximately 70-s intervals, using multiple paralleled thermocouples located at various heights around the manikin.

The linear radiation coefficient, h_r, was assumed to be constant at a value of 4.6 W/(m$^2\cdot$K). This assumption will not result in serious error since the practical range of air temperatures requiring a wetted cover would be limited to about 20°. Considering all the factors, h_r would not be expected to vary by more than 20%; the maximum inaccuracy in h_r can be limited to about 10% by choosing this value of 4.6 W/(m$^2\cdot$K), which applies for an average of cover and air temperatures of about 35°C [*12*].

The convection coefficient, h_c, is affected primarily by the effective air movement over the clothing surface. Several relationships between h_c and air movement have been published. A recent relationship developed by Mitchell and Whiller [*13*] for standing men appears to agree closely with results obtained locally using a heated copper manikin and was used to express h_c in this paper

$$h_c = 8.3\,V^{0.6} \qquad \text{W/(m}^2\cdot\text{K)} \tag{14}$$

where V is the combined air movement from the subject motion and the ambient wind, in metres per second.

The boundary layer insulation, I_a, based on $I_a = 6.46/(h_r + h_c)$, was 0.75 clo for air motion of 0.3 m/s.

The wetted fraction, w, of the clothing surface was estimated at 0.8 (80%) for the ensemble used in this study, since the cover did not extend over the essentially impermeable boots, gloves, or mask. Deductions of 11, 6, and 3%, respectively, were made for these dry surface areas.

Data Collection

Manikin Runs

Experiments were not carried out until the manikin had been fully equilibrated. For the dry skin runs, at least 3 h was allowed for this equilibration. The experiments lasted a minimum of 10 min.

For those runs in which the "skin" on the manikin was wetted, the clothing items were either removed or opened sufficiently to allow the skin to be thoroughly wetted with a spray bottle. During the wetting, the manikin heaters remained energized so that the average surface temperature was maintained near its control level. The clothing was then replaced and the system allowed to reequilibrate for at least 30 min before recordings of power, temperature, and so on, were made, that is, until the power level and the various temperatures had become practically constant. Similar rules were followed for wet cover runs. Water additions were made at intervals to keep the cover wet; no experiment was begun until the system had completely restabilized.

Three, and sometimes four, experiments were run on a single day except where problems in obtaining a desired chamber environment delayed the equilibration process. Runs were made at approximately hourly intervals. Prior to each day's runs, the manikin was "redressed" by opening and dropping the ensemble and removing the gloves, then replacing the items, readjusting the cover, and so on. This process was employed to prevent the layers from gradually settling against the manikin with time and also to introduce variability in dressing as a factor in the study.

Additional thermocouples were attached to the inside of the impermeable layer and on the wettable cover surface, respectively, during the preliminary experiments conducted to establish the permeability index, i_m, of the clothing inside the vapor barrier, the skin-to-barrier insulating value, I_{si}, the barrier-to-cover surface insulation, I_{ic}, the combined intrinsic insulation, I_{cl}, and the total insulation including the surface air layer, I_{tot}. Each of these insulating values were calculated in clo units using the general equation

$$I_x = \frac{6.46 A_{sk}(T_1 - T_2)}{H_{sk}}$$

where $(T_1 - T_2)$ was the applicable temperature gradient in degrees Celsius, that is, $(\bar{T}_{sk} - T_i)$ for I_{si}, $(T_i - T_{cl})$ for I_{ic}, $(\bar{T}_{sk} - \bar{T}_{cl})$, and $(\bar{T}_{sk} - T_a)$ for I_{tot}. During these insulation measurements, both the manikin skin and the cover were left dry.

The permeability index, i_m, for the layers beneath the impermeable vapor barrier was obtained by operating the manikin with its skin wet but with the cover dry. The applicable equation for calculating i_m was Eq 7.

Model Runs

Similar procedures were used in the experiments for testing the prediction model except that no thermocouples were used on the impermeable layer. However, measurements were made of the wettable cover temperature at four sites (the left breast, the middle of the back, the front of the left thigh, and the right calf) to provide an average that could be compared with the value of $\bar{T}_{cl}$ calculated using the heat balance equation. These four thermocouples were sewn to the cover surface rather than cemented or taped, to produce the minimal effect on evaporation from the cover. Experiments were run with the manikin skin both dry and wet to provide as complete a check as possible on the model, even though in the practical situation the wearer's skin would be expected to be wet at any activity level in the heat with such an impermeable garment.

Results and Discussion

Ensemble Properties

The average value for the intrinsic insulation, I_{cl}, from seven experiments at 27°C and 50% relative humidity with a dry skin layer and dry cover on the manikin was 0.29 K · m²/W, or 1.9 clo. The total insulation, including the surface air layer, was 0.38 K · m²/W, or 2.46 clo. The insulation of the surface air layer, or I_a/f_{cl}, was estimated from the difference between I_{cl} and I_t, as 0.087 K · m²/W, or 0.56 clo. Assuming that $I_a = 6.46/(h_r + h_c)$, with values $h_r = 4.6$ W/(m² · K) and $h_c = 4$ W/(m² · K), f_{cl} had a value of 1.32.

Values of the other parameters for the ensemble were obtained from separate experiments under similar environmental conditions with the manikin controlled to produce a mean skin temperature, $\bar{T}_{sk}$, of about 33°C. The indicated values obtained by temperature measurements within the clothing were

I_{si} (skin to impermeable layer insulation) = 1.20 clo
I_{ic} (impermeable layer to clothing surface insulation) = 0.70 clo
i_m (permeability index of layers beneath impermeable barrier) = 0.43

Heat Exchange Model Predictions

Table 1 gives the experimental data for the wet-cover results and a comparison of the predicted and observed clothing temperature, $\bar{T}_{cl}$, cover evaporative cooling loss, H_e, skin heat loss, H_{sk}, and net cooling benefits at the body surface. The latter condition results from evaporation at the outer cover. Of a total of 15 separate runs, the last 3 were conducted with the manikin skin completely wetted. The dry skin runs are grouped according to the air relative humidity condition. Runs 1 through 8 were averaged, ±1 SD, since individual runs were all very close to 50% relative humidity. The model predictions and predicted $\bar{T}_{cl}$ for these runs show little variability, as indicated by the small standard deviation. The experimental results were more variable, but the predicted and observed values are satisfactory. In the 50% relative humidity series, the predicted cover evaporative loss (Column 8) and that derived from scale weight loss (column 4) differed by only 10 W on the average ($P > 0.05$, not significantly different). Likewise, the differences between measured and predicted skin heat loss (Columns 2 and 9) were very small for all runs (a maximum of 12 W). These results indicate that the prediction equation is valid for the nonsweating, inactive condition. Fow low humidity conditions, the predicted differences were as much as 9.6 W (Run 10).

Table 1 also shows the comparison between the H_{sk} predicted (Column 9) and that measured (Column 2) for the wet skin experiments (Runs 13 to 15). This agreement may be fortuitous since only a limited number of measurements were taken to evaluate I_{si} and I_{ic}, and i_m of the fatigues under the

TABLE 1—*Experimental and predicted results and cooling benefit with wetted cover.*

Condition	Direct Measurements							Predicted Results					
				Cover									
	T_a, °C	$\bar{T}_{sk}$, °C	P_w, torr	Manikin Loss, W	Water Loss, g/min	Evaporation Cooling, W	Clothing Surface Temperature, $\bar{T}_{cl}$, °C	$\bar{T}_{cl}$ Prediction, °C	$R + C$, W	H_e, W	Skin Heat Loss, H_{sk}, W	Dry Cover, H_{sk}, W	Net Cooling at Outer Cover, W
Columns[a]			(1)[a]	(2)	(3)	(4)	(5)	(6)	(7)	(8)	(9)	(10)	(11)
Dry skin													
50% RH, Runs 1 through 8													
$\bar{X}$	26.4	33.1	13.3	62.1	3.52	144	24.2	23.1	70.0	134	63.8	32.6	31.2
SD	0.2	0.5	0.1	4.3	0.18	7.5	0.8	0.2	0.3	2.4	2.2	1.6	0.5
20% RH, $\bar{X}$ Run 9	26.0	33.4	5.0	78.6	5.5	224	22.6	19.9	127	213	85.8	36.2	49.6
Run 10	34.4	34.3	10.0	46.1	6.4	262	28.4	25.4	187	242	55.7	−0.5	56.2
90% RH, $\bar{X}$ Run 11	26.5	33.7	23.2	48.6	4.0	67.0	26.8	26.5	1.3	47.5	46.3	34.9	11.4
Run 12	34.5	34.8	37.2	9.7	3.7	61.2	33.2	33.5	20.9	28.8	7.9	1.2	6.8
Wet Skin (100%)													
50% RH, Runs 13 through 15, $\bar{X}$	26.6	33.3	13.0	99.3	10.3	172	26.0	24.0	54.5	154	99.2	47.1	52.2

[a]The column numbers in parentheses indicate the following information related to those columns:

1. Continuous wet bulb and dry bulb.
2. Corresponds to skin heat loss.
3. From the Sauter balance measurement.
4. From Column 3, · λ · 60 W · h/g.
5. From direct thermocouple measurements.
6. Predicted clothing surface from Eq 12.
7. Radiation plus convection.
8. Evaporative heat loss of cover.
9. Column 7 minus Column 8.
10. Total heat gain from the skin.
11. Net cooling benefit (Column 9 minus Column 10).

impermeable layer. From temperature data collected on the dry system—that is, with both the manikin skin and the wettable cover dry—the ratio I_{si}/I_{ic} was determined from the equation

$$\frac{I_{si}}{I_{ic}} = \frac{\bar{T}_{sk} - T_i}{T_i - \bar{T}_{cl}}$$

where T_{sk}, T_i, and T_{cl} were *local* skin, impermeable layer, and clothing surface temperatures measured over a given manikin skin site, that is, with all three thermocouples over the same segment of the copper surface. The ratio I_{si}/I_{ic} determined by averaging values for four segments (on the back, chest, stomach, and thigh) was 1.8; I_{si} and I_{ic} were, accordingly, 1.2 and 0.7 clo, respectively (since the ensemble intrinsic insulation, I_{cl}, was equal to 1.9 clo).

The permeability index, i_m, for the fatigues, calculated from data obtained in wet skin, dry cover runs, and using 1.2 clo for I_{si}, was 0.43. These values of insulation and i_m for the fatigues seem reasonable based on the values for fatigues alone, namely, a total insulation, including surface air layer, of 1.37 clo and an i_m of 0.43. In the ensemble, the impermeable layer overlying the fatigues would reduce the insulation of their surface air layer to some extent, but probably by no more than 30%, or 0.2 clo, since the impermeable layer was rather stiff and loose fitting. The i_m value for the fatigues would not be expected to change when the impermeable layer was added. Unfortunately, the predicted H_{sk} is rather sensitive to the value of I_{si} chosen. Using a value of 1.37 clo for I_{si}, and a corresponding value for i_m of 0.51 (higher than normally measured at low air movement), results in a prediction for H_{sk} of 110 W for Run 13, or 11.4 W higher than when an I_{si} of 1.20 clo was used. Any inaccuracy in measuring T_{sk}, T_i, or T_{cl} when determining I_{si}/I_{ic} obviously has an important effect on the predicted skin heat loss, H_{sk}. The generally close agreement between the predicted and measured skin heat losses in Runs 13 to 15 in Table 1 indicates that the various heat exchanges with a wet skin are probably correctly handled on the model and that realistic predictions of skin cooling can be made if the ensemble parameters are correct.

Cooling Benefits from Cover Evaporation Losses

The predicted increases in skin heat loss caused by evaporation from the cover are given in Column 11 of Table 1. Each value was obtained by subtracting the calculated manikin heat loss, H_{sk}, with a dry cover from that with wet cover. For the dry skin runs, these results may also be obtained by multiplying the predicted H_e values by Burton's efficiency factor, in Eq 1, that is

$$\text{Cooling benefit} = H_e\left(\frac{I_a}{f_{cl}I_{\text{tot}}}\right)$$

For I_a equal to 0.75 clo at 0.3 m/s air movement (from Eq 9) and I_{tot} equal to 2.46 clo, the cooling efficiency, $I_a/(f_{cl}I_{tot})$, is 0.234, or 23.4%.

Burton's equation is not applicable to the wet skin runs because, even with the vapor barrier, there is evaporative heat transfer from the skin to the impermeable layer. This evaporative transfer in effect reduces the thermal insulation of the inner layers and, by extension, the effective value of I_{tot}. Cooling efficiency and the benefits of cover evaporation are, therefore, increased.

In Runs 13 to 15, the average efficiency, that is, the benefit derived by H_e, is 33.9%, or a 45% increase over the dry skin value. The true wet skin efficiency, unlike that for dry skin, is not constant since ($P_{sk} - P_i$) is not a linear function of the temperature gradient, $\bar{T}_{sk} - T_i$. As ($\bar{T}_{sk} - T_i$) increases, sensible heat transfer from the skin to the impermeable layer changes linearly, but ($P_{sk} - P_i$) and the evaporative transfer increase by a larger percentage, thus lowering the effective insulation between the skin and impermeable layer. The highest cooling efficiencies may therefore be expected when the $\bar{T}_{sk} - T_i$ gradient is large, as in cool environments or when the cover temperature is greatly depressed below ambient by rapid evaporation.

Model Predictions

The predicted values of the minimal water requirements (in grams per minute) to keep the cover wet for a 1.8-m^2 man wearing the experimental ensemble in various combinations of air temperature, relative humidity, and two wind speeds are shown in Fig. 2. The values of skin net heat dissipation with a wet cover are given in Fig. 3. These predictions are based on the 0.3-m/s values for I_{si}, I_{ic}, and permeability index, i_m, obtained in the chamber experiments and do not include any effects of solar radiation. A mean skin temperature, $\bar{T}_{sk}$, of 37°C, which would be typical for a stressed man in an impermeable ensemble, has been assumed in making the predictions. Both parameters have been plotted on a psychrometric format with ambient vapor pressure on the y axis and $T_a(=\bar{T}_r)$ on the x axis. The effects of still air ($V =$ 0.3 m/s) are depicted by the stipled lines for each intersection of the percent relative humidity lines. The effect of wind speed ($V = 5$ m/s) is shown by the solid vertical line for T_a from 30 to 50°C.

Figure 2 shows clearly that water requirements for a given wind speed decrease as the ambient water vapor increases at each given ambient dry bulb temperature. On the other hand, water requirements necessarily increase, and the requirements are higher and rise more rapidly with the wind, when relative humidity is low but dry heat stress is increased. A 10-degree Celsius change in air temperature has a change about equivalent to a 10-torr change in water vapor pressure at high wind speed. By comparing the relative cooling benefits (not shown in the figure) and water requirements, a prediction is that cooling efficiency of the water evaporated from the cover would fall off as the wind speed increases. Such a reduced efficiency would be predicted by the

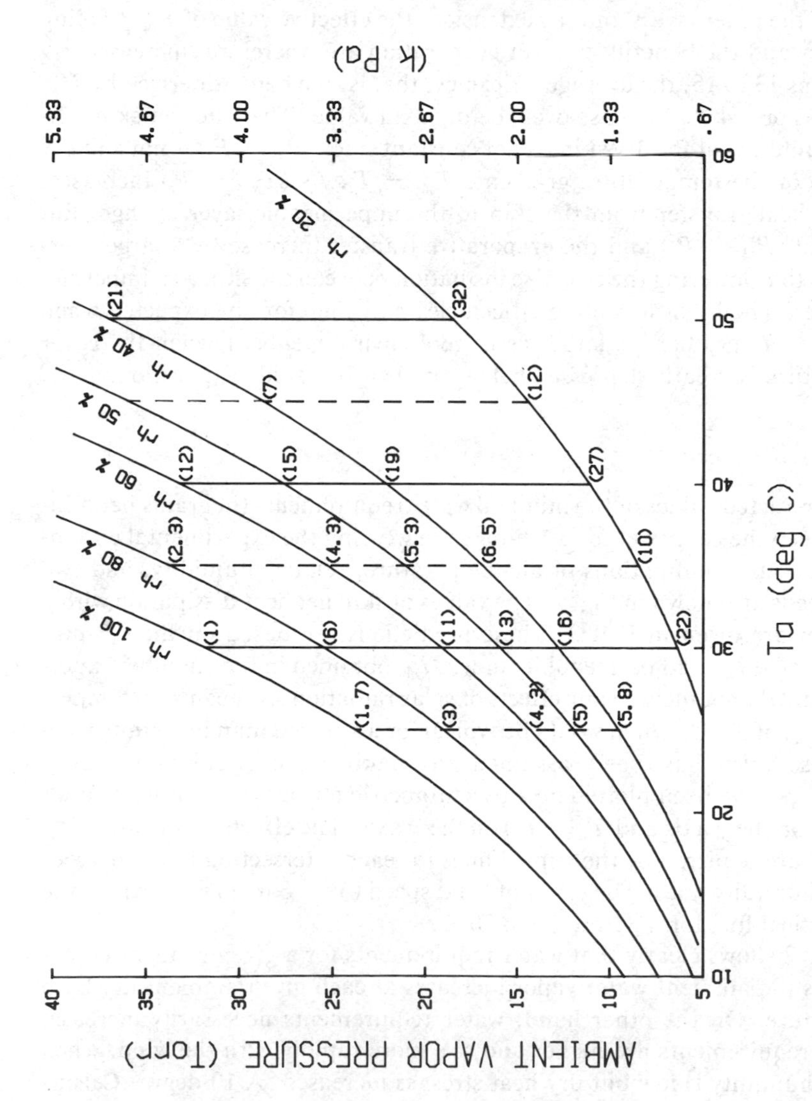

FIG. 2—*Minimal water requirements (in grams per minute) to maintain a wet cover for a standard man (A_s = 1.8 m^2). Numerals (N) crossing the relative humidity line indicate the grams per minute for still air (stipled vertical line) and for 5-m/s air movement.*

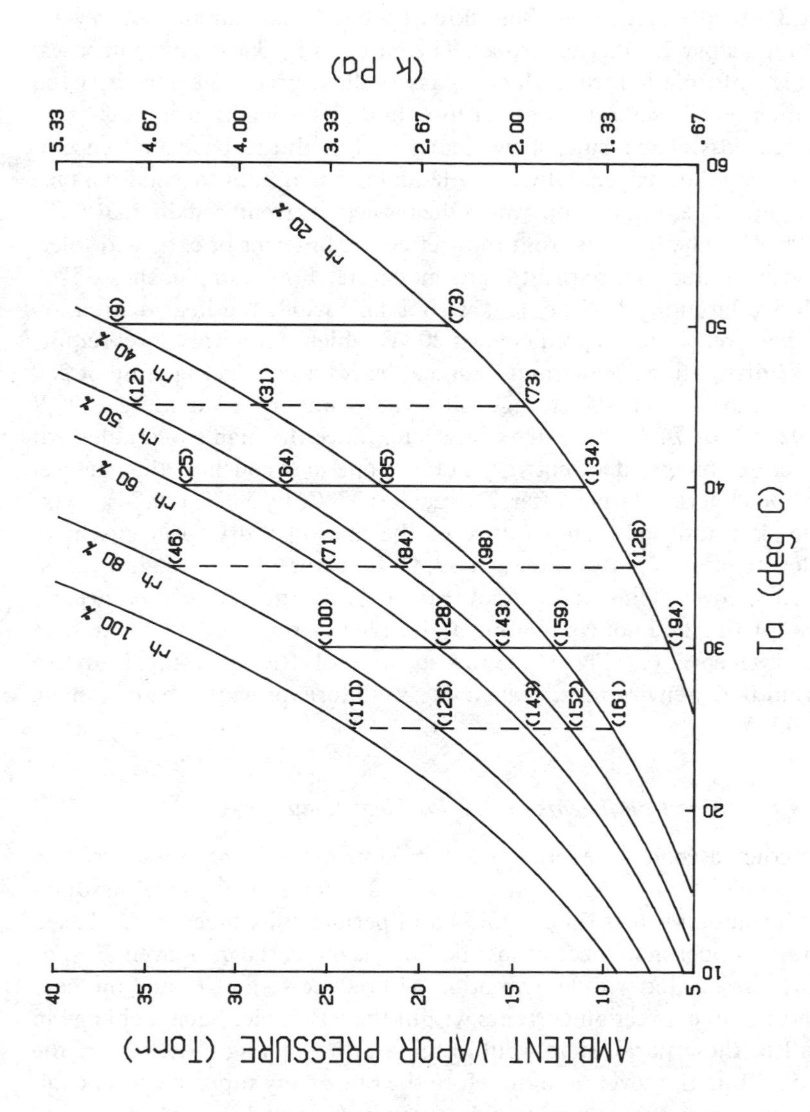

FIG. 3—*Net skin heat loss for still air and 5-m/s wind movement on a psychrometric format. The symbols are the same as in Fig. 2.*

Burton equation (Eq 1), since I_a rapidly decreases with wind albeit total insulation decreases more slowly.

Net Skin Heat Loss with a Wet Cover

Figure 3 provides prediction of net skin heat loss for still air and 5 m/s wind movement on a psychrometric format. The data in Fig. 3 serve two purposes: (1) they give information from which to assess the degree of balance between heat production and subsequent heat loss, and (2) they permit inspection of the skin heat loss. The results show clearly that, with this particular ensemble, heat dissipation is generally not adequate to maintain thermal balance during extended periods of moderate to heavy activity (required dissipation of 300 to 400 W). The benefits from the wet cover, in terms of extended tolerance time in hotter environments, are moderate. For example, in a 35°C, 50% relative humidity environment with 1 m/s wind, the heat dissipation without the cover wetted is predicted as 20 W, which, for a work level requiring 300-W dissipation for thermal balance, leaves a heat storage rate of 280 W. On the other hand, the storage rate with a wet cover would be 208 W (300 − 92 W), or 74% of the dry cover rate. Since the man's tolerable heat storage before collapse does not vary greatly, one may conclude that the wet cover extends tolerance time, after $\bar{T}_{sk}$ reaches 37°C, by 35% (1/0.74). Typically, the skin and body temperature of the man in a dry cover ensemble would elevate more rapidly and reach 37°C $\bar{T}_{sk}$ before that of the individual with a wetted cover. Thus, the wet cover could easily mean the difference between completing and not completing a given heat stress task. Wet skin heat loss from a wet cover (see Fig. 3) shows a substantial efficacy in still air up to a 5-m/s wind in dry environments when $P_w \leq 15$ torr, provided the metabolic load $\leq$300 W.

Effects of Pumping Coefficients and Solar Heat Load

The predicted cooling benefits of a wet cover are minimal values for the ensemble, which will be exceeded during body motion; the model assumes constant insulating values (in clo units) and permeability indexes, i_m, based on the static copper man measurements, but the increased air movements, or "pumping" associated with body motion will reduce I_{si} and I_{ic} and increase i_m, by setting up convection currents within the ensemble. Such a change in one or all of these parameters would increase the cooling efficiency of the evaporation from the cover, and therefore the rate of any supplementary cooling. Calculations simply assuming a decrease in I_{si} from 1.2 to 0.8 clo (which is quite a reduction owing to the effect of body motion) show that, at 30°C and 50% humidity, supplementary cooling for the 1 to 5-m/s range increases by 10 to 11 W, producing a net skin dissipation increase of from 17 to 20 W; the latter is higher since dry cover dissipation is increased by reducing the ensemble insulation. At 45°C, 20% humidity with the same I_{si} reduction, supplementary cooling increases by 17 to 22 W, but the net cooling increases by only 12 to 16 W (less because the dry cover heat gain at the skin is higher). Additional benefits would accrue from an increase in i_m. Further work is obviously required to define the effects of wind on the three parameters and to describe any changes with pumping during body motion, in terms of an effective wind velocity, V_{eff}, by conducting physiological studies on human subjects [*19*]. Such determinations have been made for four standard military ensembles, but not for an impermeable-type system.

The predictions in this report have not included any effects of solar radiation absorbed at the surface of the ensemble. This source of heating would be handled by the term H_{sl} in the heat balance (Eq 6). H_{sl} is independent of T_{sl}, but effects on $\bar{T}_{cl}$ would change all the other factors in Eq 6, namely, $(R + C)$, H_e, and H_{sk}. The effect on H_e plays an important role in reducing the additional heat load at the skin surface resulting from solar radiation, commonly called the solar heat load, but this treatment is beyond the scope of this paper and is treated in other reports [*4,12,14*].

Conclusions

In concluding this discussion of the benefits of a wet cover over an impermeable clothing system, the authors emphasize that the predictions which have been made are necessarily approximations because of uncertainties in some of the factors in the prediction equations. Those associated with wind and body motion have been discussed. Others which have not been defined exactly are the vapor pressure–temperature relationship at the impermeable layer and the radiant heat transfer coefficient, h_r, which has been assumed to be linear and constant.

Finally, the effects of a wet cover in reducing heat stress on the person would depend on the availability of an adequate supply of water. The maximal indicated water requirement (32 g/min at 50°C, 20% relative humidity, high wind) (see Fig. 2) is about half the amount that can be held in a standard U.S. Army helmet and about two thirds that which can be held by a helmet liner. If careful wetting of the cover is performed, such an amount of water might be applied without dripping. If the water is simply poured over the head, even a full helmet may not prove sufficient to wet the cover completely, and large amounts of water will drip or be splashed onto the ground. It is clear that some method of uniformly applying water, such as a hand pump spray system, will produce more satisfactory results without wasting water. The amount of water applied should probably be automatically titrated so that enough water is provided to last at least 1 h at each wetting. One good application of wettable covers for $M \leq 300$ W/m^2 may be under conditioning in which there is an abundance of water, such as in ships at sea, during fire fighting operations, or among pesticide workers for whom toxic fumes require wearing of impermeable, wettable garments.

Acknowledgments

The authors are grateful for the efforts of Edna R. Safran in the processing of this paper and to William L. Holden for computer drafting of the figures. Portions of this work have been documented in a U.S. Army Research Institute of Environmental Medicine Technical Report [*4*].

Disclaimer

The views, opinions, and findings contained in this report are those of the author(s) and should not be construed as an official Department of the Army position, policy, or decision unless so designated by other official documentation.

References

[*1*] Goldman, R. F. and Breckenridge, J. R., "Current Approaches to Resolving the Physiological Heat Stress Problems Imposed by Chemical Protective Clothing Systems," *Army Science Conference Proceedings*, Vol. 1, 1976, p. 447.

[*2*] Minard, D., *Military Medicine*, Vol. 126, 1961, p. 261.

[*3*] Martin, H. deV., and Goldman, R. F., *Ergonomics*, Vol. 15, 1972, p. 337.

[*4*] Breckenridge, J. R., "Use of a Wetted Cover to Reduce Heat Stress in Impermeable Clothing," U.S. Army Research Institute & Environmental Medicine Report T7/80, Natick, MA, 1980.

[*5*] Fonseca, G. F., "Effectiveness of Four Water-Cooled Undergarments and a Water-Cooled Cap in Reducing Heat Stress," U.S. Army Research Institute & Environmental Medicine Report T23/76, Natick, MA, 1975.

[*6*] Shapiro, Y., Pandolf, K. B., Sawka, M. N., Toner, M. N., Winsmann, F. R., and Goldman, R. F., *Aviation and Space Environmental Medicine*, Vol. 53, 1982, pp. 785–789.

[*7*] Toner, M. M., Drolet, L. L., Levell, C. A., Levine, L., Stroschein, L. A., Sawka, M. N., and Pandolf, K. B., "Comparison of Air Shower and Vest Auxiliary Cooling During Simulated Tank Operations in the Heat," U.S. Army Research Institute & Environmental Medicine Report T2/83, Natick, MA, 1983.

[*8*] Darling, R. C., Johnson, R. E., Moreira, M., and Forbes, W. H., "Physiological Tests of Impermeable Suits," Harvard Fatigue Laboratory Report No. 89, Oct. 1943.

[*9*] Burton, A. C. and Edholm, O. G., *Man in a Cold Environment*, Hafner, New York, 1969.

[*10*] Gagge, A. P., Burton, A. C., and Bazett, H. C., *Science*, Vol. 94, 1941, p. 428.

[*11*] Fanger, P. O., *Thermal Comfort*, McGraw-Hill, New York, 1972, p. 33.

[*12*] Gagge, A. P. and Nishi, Y., *Handbook of Physiology*, 1975, Chapter 5, pp. 69–92.

[*13*] Mitchell, D. and Whiller, A., "Cooling Power of Underground Environments," *Mine Ventilation Journal*, Vol. 17, 1972, pp. 140–151.

[*14*] Breckenridge, J. R. and Goldman, R. F., *Journal of Applied Physiology*, Vol. 31, 1971, p. 659.

[*15*] Breckenridge, J. R., *Textile Research Journal*, Vol. 37, 1967, p. 809.

[*16*] Woodcock, A. H., *Textile Research Journal*, Vol. 32, 1962, p. 628.

[*17*] Gonzalez, R. R. and Gagge, A. P., *A.S.H.R.A.E. Transactions*, Vol. 79, 1973, pp. 88–96.

[*18*] Nishi, Y., Gonzalez, R. R., and Gagge, A. P., *A.S.H.R.A.E. Transactions*, Vol. 81, 1975, p. 183.

[*19*] Givoni, B. and Goldman, R. F., *Journal of Applied Physiology*, Vol. 32, 1972, p. 812.

John Davies[1]

A Dynamically Insulated Heat-Protective Clothing Assembly

REFERENCE: Davies, J., **"A Dynamically Insulated Heat-Protective Clothing Assembly,"** *Performance of Protective Clothing, ASTM STP 900*, R. L. Barker and G. C. Coletta, Eds., American Society for Testing and Materials, Philadelphia, 1986, pp. 535-555.

ABSTRACT: This paper describes the design development and evaluation of a thermally comfortable pressure suit and associated equipment, using the vortex cooling effect for use during access penetrations into the United Kingdom-designed, advanced gas-cooled nuclear reactor system (AGR). Also discussed is the method of assessment of clothing materials. The system design of a heat balance equation using the Ranque Hilsch cooling effect produced by a vortex cooling tube, used in conjunction with a ventile clothing assembly, and the operation of a system to achieve adequate protection for persons working for extended periods in abnormal environments are demonstrated.

KEY WORDS: protective clothing, dynamically insulated suit, vortex cooling tube

Regardless of how efficiently designed any machine or process is, invariably some form of protective clothing or equipment will be required to insulate the human workers from hazardous environmental factors. Much progress has been made in the last few decades in protecting the human in industry from a great variety of hazards. The expanding nuclear energy program around the world has presented further challenges in protecting persons during routine servicing and maintenance work and under emergency conditions.

This paper describes the design, development, and evaluation of a thermally comfortable pressure suit and associated equipment using the vortex cooling effect, for use during maintenance access penetrations into the United Kingdom-designed advanced gas-cooled nuclear reactor system (AGR). Men working inside the concrete pressure vessel of the AGR require protection from heat (up to 60°C) and noise (up to 130 dB) in addition to radiological hazards.

[1]Senior engineer, Central Electricity Generating Board, London, United Kingdom.

A normal physical working environment includes the following conditions:

(*a*) Temperature—23°C (relative humidity 45%);
(*b*) Air—at one bar, uncontaminated and dust free;
(*c*) Acceleration—1 g steady;
(*d*) Noise—less than 80 dB;
(*e*) daylight; and
(*f*) circadian rhythm in phase with the local time.

The advent of AGR presented an unusual set of environmental working conditions. The working environment under controlled conditions for planned maintenance access in the pressure vessel was set as follows:

(*a*) temperature—60°C (metal contact temperatures of 70 to 80°C);
(*b*) air—contaminated, requiring a supply of respirable air;
(*c*) noise—up to 132 dB;
(*d*) darkness for initial entry, with limited artificial lighting to be installed; very difficult access and egress; cramped conditions generally; and
(*e*) around-the-clock shift work.

When human beings are required to perform physical work in hostile environments, a number of alternatives are open to ameliorate the situation. First, attempts may be made to reduce the risk by (*a*) using acceptable engineering control methods, (*b*) reducing the work load, or (*c*) achieving some compromise between these two possibilities. Second, a work system may be devised whereby exposure durations are limited to those for which the risk is kept to a relative minimum. However, there will be circumstances in which, for one reason or another, these alternatives will be uneconomical, impractical, or inadequate, or an emergency condition may arise as a result of a plant or process failure overriding the engineering control.

The environmental conditions in the AGR for initial entries requires work to be carried out wearing protective clothing. The features influencing the design and selection of personal protective systems for a given environment can be summarized as follows:

(*a*) environmental conditions,
(*b*) activity expected,
(*c*) duration and intensity of exposure, and
(*d*) protection efficiency afforded by the systems available.

A basic problem with protective clothing is that of achieving and maintaining a microclimate around the wearer in which all physiological processes can function without undue strain.

Physiological strain results when the heat balance of the body cannot be maintained and heat accumulates within the body. The heat balance of the body is the balance between the body's metabolic heat and the heat exchanges occurring between the body surface and its environment.

If it is impossible or impractical to reduce the heat to an acceptable level for the work required, then two alternatives are possible: the duration of exposure can be limited according to the working conditions, or special ventilated protective clothing can be provided. In some extreme cases both alternatives may have to be applied. Air-ventilated protective clothing can be classified into two distinctive systems.

In one system, the cool air flows parallel to the body surfaces and escapes at specific points, such as at the ankles, wrists, or neck, or impermeable materials are used and ventilation occurs at filter points designed into the garment—this system is called axial ventilation.

In the second system, the air passes through a fabric in the opposite direction to the flow of body heat; this system is defined as dynamic insulation. By passing through the fabric, the ventilating air can pick up and carry away from the wearer four or five times as much heat as it can when directed to specific ventile positions. This system also has an advantage insofar as the ventilating air is also used to dissipate heat flowing in from the elevated ambient environment as well as to cool the body.

Air hoses are also vital to the success of any ventilated clothing assembly that is going to be used for deep penetrations into a hot working environment. A rise in air temperature in the hose can defeat the correct functioning of the clothing system. A solution to this problem is to cool the air just before it enters the ventilated suit. A compact device is essential.

For the purpose of entry into an AGR pressure vessel, a hot environmental work suit based on the principle of dynamic insulation, used in conjunction with a vortex tube cooler, was developed. Davies [*1*] has reported on this device using the principle of the Ranque Hilsch cooling effect, together with an associated air-hose assembly.

Hot Environmental Protective Suit

The clothing assembly, shown in Figs. 1 and 2, consists of three pieces: a helmet assembly, an outer garment, and an inner-garment. The helmet assembly is complete with a molded polycarbonate visor, a liner of the same material as the inner suit, and an outer covering of the same material as the outer suit, giving a permeable surface area of 0.214 m^2. This helmet assembly is fastened by three slide fastener closures to a one-piece outer garment constructed of close-woven cotton fabric, with physical properties as shown in Table 1. The garment has a front entry closed with a slide fastener; the sleeve and trouser ends are elasticated to prevent the escape of air. The surface area of the garment is approximately 2.12 m^2 and designed to fit over, without constricting or crushing, an inner garment (Fig. 2) manufactured from open-cell nitrogen-blown polyurethane foam, with a density of 3.8 to 5 kg/m^2, covered on both sides with lightweight nylon fabric; its physical properties are shown in Table 1. Inside this garment is fitted a removable air distribution

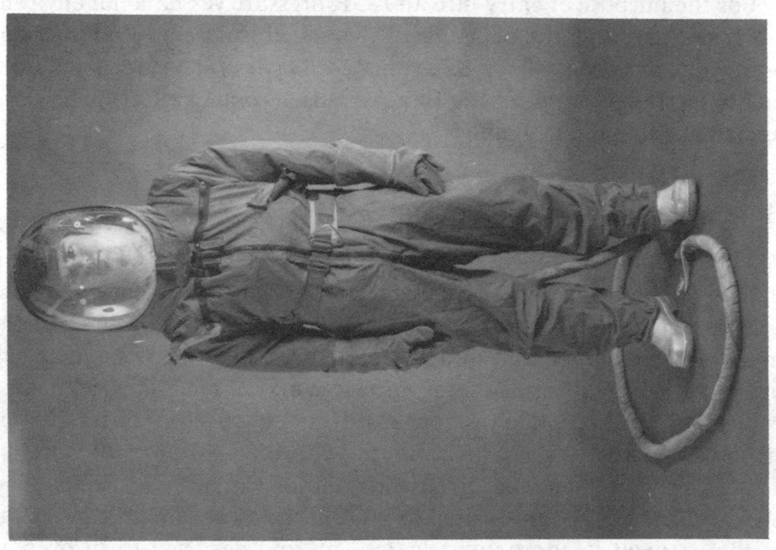

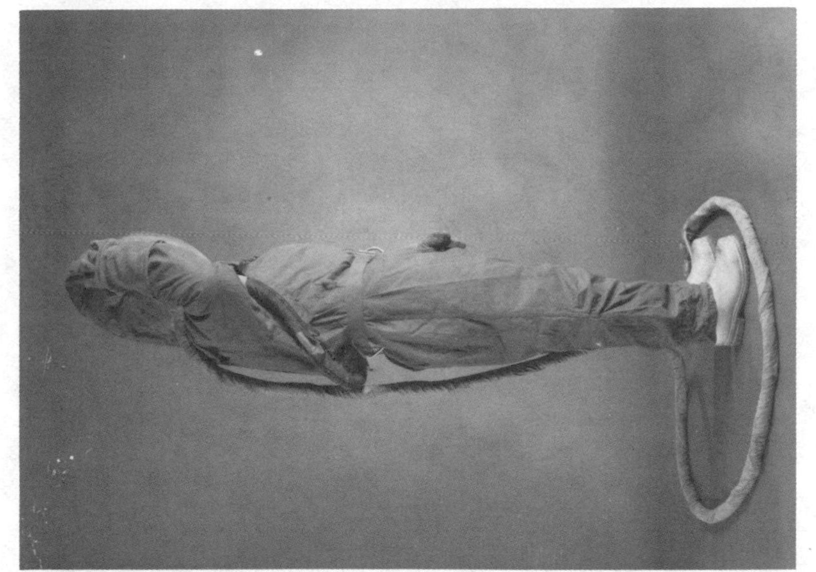

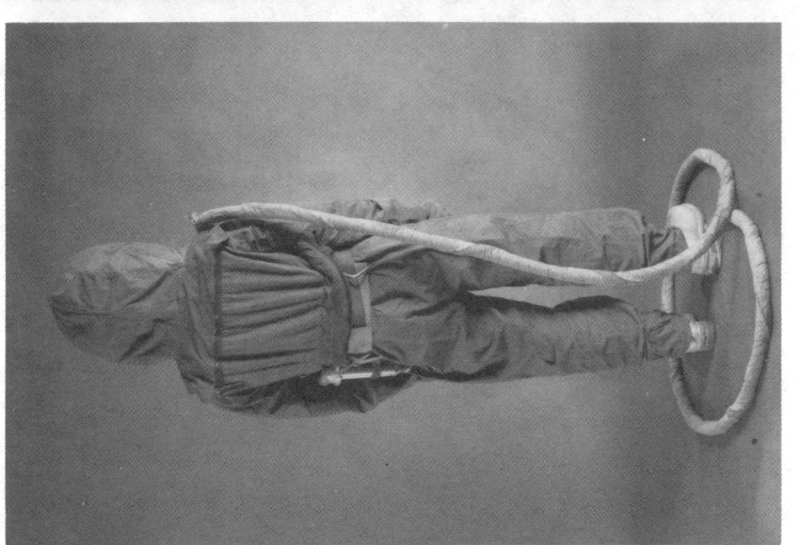

FIG. 1—*Hot environmental clothing assembly (complete with vortex tube).*

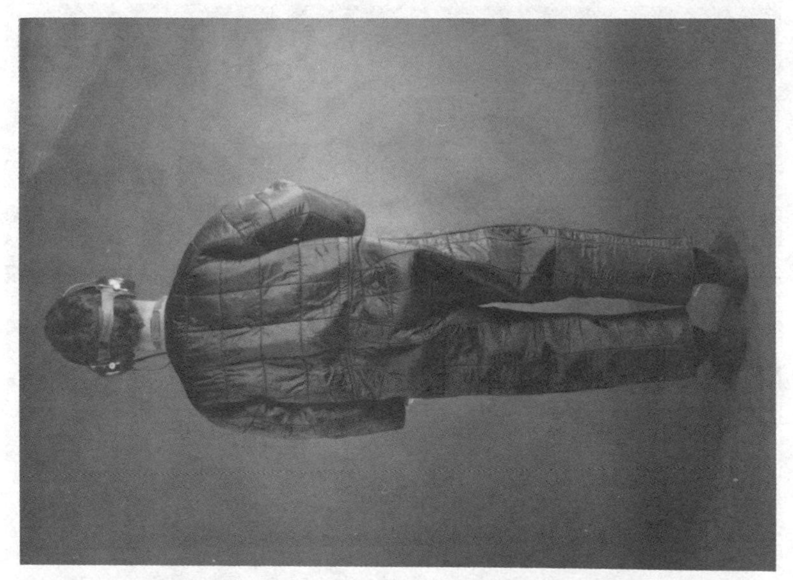

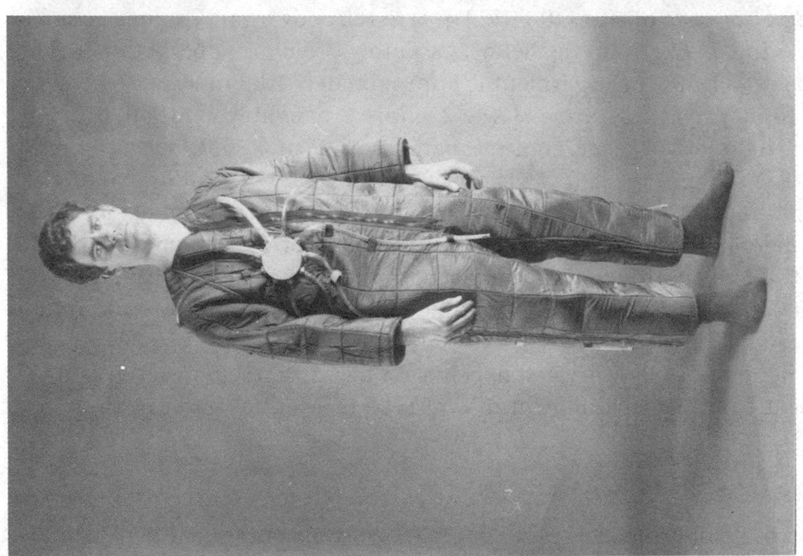

FIG. 2—*Hot environmental inner suit (complete with cooling distribution system).*

TABLE 1—*Physical properties of materials used for a hot environment work suit.*

Physical Properties	Fabric Type		
	Cotton L28	Nylon No. 556	Polyurethane Foam
Weave	oxford	plain	...
Weight, g/m^2	340	60	2.5
Breaking strength, Kp			
Warp	80	20	...
Weft	70	20	...
Threads/cm			
Warp	64	16	...
Weft	25	16	...
Air permeability, cm^3/cm^2/s	0.2	52.0	20.0

cooling system, shown in Fig. 3. Seven outlets from the distribution manifold feed an air supply to the body, arms, and legs through 6.5-mm-bore plastic tubing, the air passing into the suit through small vent holes pitched from the sealed ends of the tube. Two 10-mm-bore plastic supply tubes lead air from the manifold to the air distribution system in the helmet for cooling and breathing. At an air flow of 0.012 kg/s, the air distribution in the suit is designed to give 50% (0.006 kg/s) to the arms, legs, body back, and groin (Cooling Tubes 1 to 6), 10% (0.0012 kg/s) to the body front (Cooling Tube 7), and 40% (0.0049 kg/s) to the helmet assembly (Cooling Tubes 8 to 12). The surface area of this inner garment is approximately 1.83 m^2.

The total permeable area of the complete assembly is approximately 2.33 m^2, its function being both mechanical protection and thermal insulation. Air is fed into the suit for two primary purposes—breathing and cooling—and since a common supply is used, the quality of the air must meet the requirements of respirable air. The basic purpose of the system design is to ensure that the air brought into contact with the body facilitates heat removal from the body by increasing the sensible heat of the air and by evaporating and sweeping away sweat.

The metabolic heat generated depends on a wearer's activity level, and radiant heat from the environment may also be transmitted through the suit.

Dynamic Insulation Characteristics

The experiments reported here concern the testing of the clothing assembly to determine if dynamic insulation was established and to compare the experimental data obtained with that predicted by the equation derived by Crockford and Goudge [*2*], in which the thermal conductance was found to be proportional to the logarithm of the air flow through the fabric of the garment. The method used to evaluate the thermal conductance of the material assem-

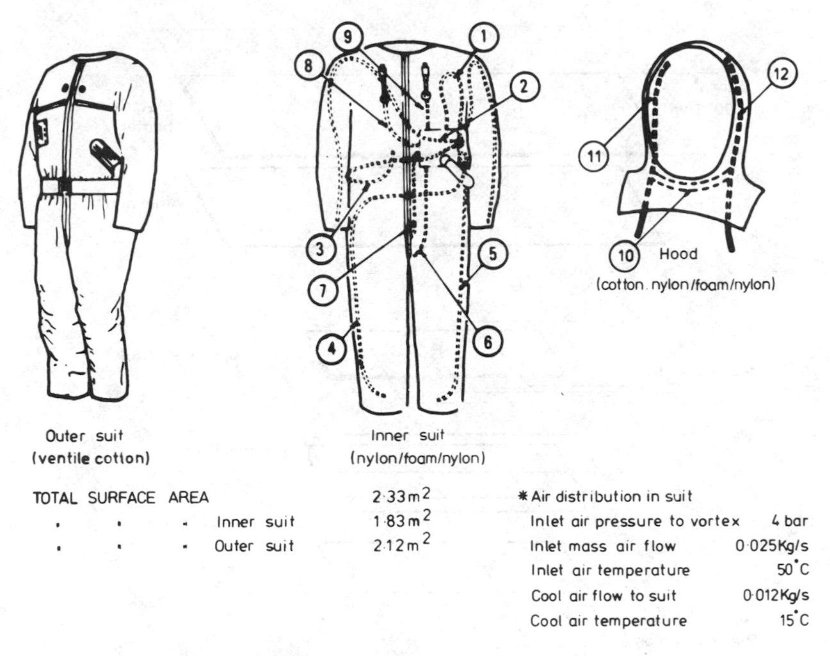

Cooling tube details

Tube reference		Length	No. of	Pitch	Air * distribution
No	Position	mm	holes	mm	Kg/s
1	Left arm	1350	10	114	
2	Right arm	1575	10	117	
3	Back body	1125	10	100	0·0059
4	Right leg	1550	14	117	
5	Left leg	960	18	61	
6	Grain	495	8	100	
7	Front body	300	10	38	0·0012
8	Hood connection	266	0	–	
9	Hood connection	170	0	–	
10	Hood front	300	5	50	
11	Hood right	350	5	50	0·0049
12	Hood left	350	5	50	

FIG. 3—*Air distribution system for the prototype hot environmental work suit.*

bly and the results presented in the tables are those reported by Crockford and Karim [*3*], and the apparatus used is shown in Fig. 4. This allows air to be fed through material assemblies placed on a metal grid of known surface area. The heater placed above the assembly acts as a radiant heat source, and temperature changes can be measured with thermocouples and heat flow with heat flow disks.

The results of a series of tests using this apparatus are shown in Table 2. The heater voltage was set at a value that brought the outer surface of the

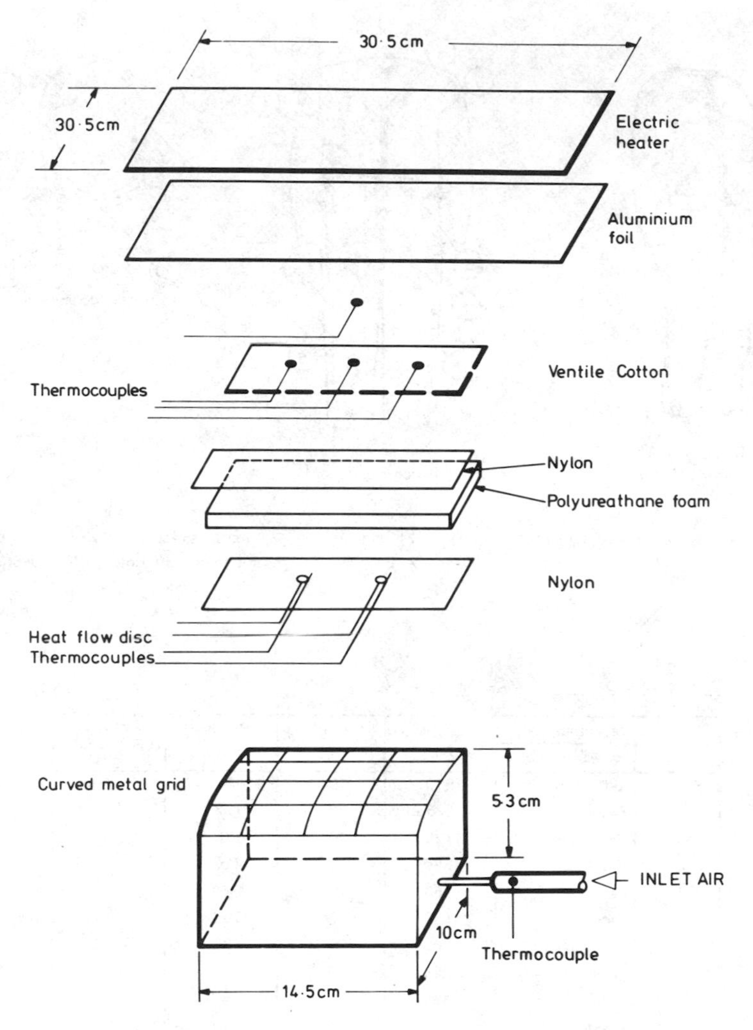

FIG. 4—*Exploded view of the apparatus used for determining the thermal conductances of multilayer fabrics.*

material assembly to 60 and 112°C. Air was supplied at 20°C. The temperatures were measured by 40-standard wire gage (swg) copper/constantan thermocouples threaded into the inner and outer surfaces, two on each surface, and the heat flow was measured by two heat flow disks, 5 mm in diameter, fastened to the outer surface of the inner fabric layer of the assembly. Measurements were taken at a number of air flow rates, from 0 to 20 L/min, as the flow was increased from zero and as it returned to zero from the maximum. The readings obtained at each flow rate were averaged. It was noted

TABLE 2—*Conductance of garment materials at 60 and 112°C initial outer surface temperatures.*

Air Flow		Inner Temperature, °C	Outer Temperature, °C	Temperature Gradient, Δt, °C	Heat Flow, W/m^2	Conductance, $kcal/m^2/min/°C$
L/min	m/min					
0	0	40.5	62.5	22.0	116.5	0.077
2	0.175	36.5	62.0	25.5	105.5	0.060
5	0.416	31.0	57.5	26.6	74	0.0406
10	0.833	27.5	48.5	21.0	30	0.0209
15	1.249	26.0	42.0	16.0	16	0.0144
20	1.666	25.0	38.5	13.5	9.6	0.0108
0	0	64.5	112.0	47.5	278	0.0850
2	0.175	57.0	110.0	53.0	253	0.0695
5	0.416	47.0	106.0	59.0	203	0.0502
10	0.833	37.5	86.5	49.0	94	0.0280
15	1.249	33.0	76.0	43.0	51	0.0171
20	1.666	31.5	66.0	34.5	33	0.0141

that the heat flow disks had some effects on the air flow through the fabric lying above them; however, the thickness of the assembly used in this study, 6.5 mm, is such that the disks do not appear to influence the results seriously. If, however, the air stream quickly reformed behind the disk and any elevation of temperature above the disk was reduced by conduction to cooler surrounding material, the heat sink in this apparatus—the box itself—is small, thus enabling the system to respond quickly to changes in air flow.

The conductance of the assembly increased as expected with the increase in the radiant load, but at both heat loads there was an unexpected change in the conductance curve at high rates of air flow. The increase of 10% in the static thermal conductance between 60 and 112°C is suggested to be due to the increase in thermal conductivity of air, which would account for about 8.6%, and to increased radiant heat transfer as the temperature gradient across the assembly is increased from 22 to 48°C. A predicted dynamic insulation curve for the material assembly is shown in Fig. 5, calculated from the following formula given by Crockford and Goudge.

$$\log y = \frac{-(\log t + 0.2785)}{1.0755} - (0.0453 + 0.719t)x$$

where

y = the conductance, $kcal/m^2/min/°C$,
t = the thickness of material assembly, mm, and
x = air flow, m/min.

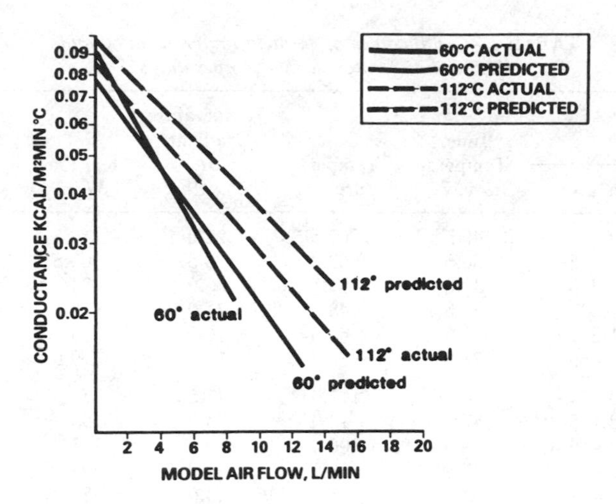

FIG. 5—*Relationship between the conductance of the material assembly and the predicted values at 60 and 112°C.*

Thermal Conductance—Complete Assembly

The thermal conductance of the complete assembly was determined by applying a standard intensity of radiant heat to the outside of the suit when worn by a manikin in a hot room and by measuring the temperature on the inside and outside of the suit with thermocouples and the heat flow through the materials with heat flow disks. Four thermocouples were placed in close proximity to the heat flow disks on the inside of the inner garment. The outer surface temperature was measured at the chest, abdomen, right and left lumbar region, and center back. The assembly was tested at five air flows of 0.13, 0.26, 0.40, 0.54, and 0.70 m^3/min for 10 min each; the sequence was repeated five times, and the values shown in Table 3 are the mean of the thermocouple readings at each site. The curves plotted, shown in Fig. 6, are the conductance values for the assembly.

The outer garment was removed, and the inner garment was tested alone. The averaged results are shown in Table 3 and Fig. 6. With the outer garment removed because of the very low permeability of the inner suit, the thermal conductance of the assembly increased substantially due to the loss of the insulating layer of air trapped between the inner and outer suit. At the higher rates of air flow, it was noted that the conductivity of the inner suit approached that of the complete assembly, presumably through its high initial conductance and the increase in the efficiency with which the ventilating air is used in dynamic insulation.

The objective of these investigations was to test for the establishment of dynamic insulation in a clothing assembly by determining its thermal conductance at a number of different air flows. The air flow/conductance curve (Fig. 6) indicates that dynamic insulation was established. This observation,

TABLE 3—*Averaged results of five tests for conductance of the complete suit assembly and the inner suit only at a radiant temperature of 60°C.*

	Air Flow		Air Temperature, °C	Outer Surface Temperature, °C	Inner Surface Temperature, °C	Temperature Gradient, Δt, °C	Heat Flow, W/m^2	Conductance, $kcal/m^2/min/°C$
	m^3/min	m/min						
Complete suit assembly	0.130	0.065	41.50	57.25	43.50	13.75	50	0.0525
	0.263	0.131	37.50	56.75	41.50	15.25	54	0.0514
	0.409	0.204	36.50	57.00	39.00	18.00	53	0.0448
	0.542	0.271	30.50	56.00	36.50	19.50	54	0.0402
	0.700	0.350	30.00	54.00	34.25	20.25	49	0.0351
Inner suit only	0.130	0.065	36.00	57.25	45.50	11.75	63	0.0782
	0.263	0.131	29.00	56.25	41.25	15.00	69	0.0668
	0.409	0.204	26.50	56.75	39.25	17.50	67	0.0560
	0.542	0.271	24.50	56.50	38.00	18.50	62	0.0479
	0.700	0.350	23.00	54.00	35.00	19.00	57	0.0440

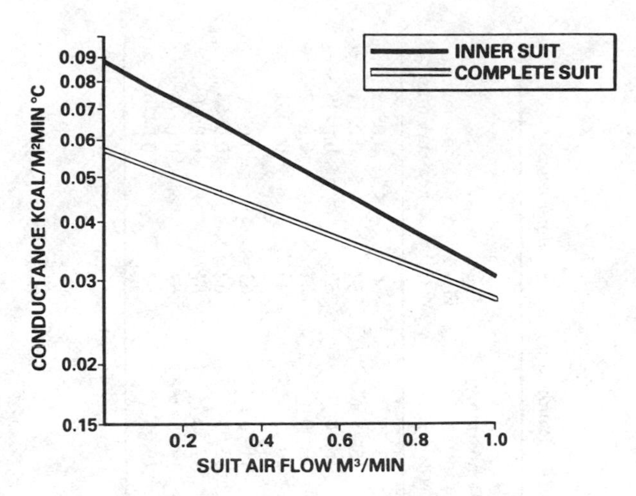

FIG. 6—*Relationship between the air flow and thermal conductance for the complete suit assembly and for the inner suit only.*

however, indicates that dynamic insulated suits may be effective at reducing heat flows through the outer suit fabric only if they are used beyond the air flows at which dynamic insulation is counterbalanced by an increase in temperature gradient across the whole garment. The air flow required to achieve this depends on the garment, but with the assembly tested, no significant drop in heat entry occurred up to the maximum air flow used. It must be noted, however, that this result will alter if the materials of the outer garment are changed. The differences between the performance of the material assembly, as predicted by the preceding equation, and the actual performance could be due to the action of the radiant heat load influencing its conductance, and the experiment with a surface temperature of 112°C and zero air flow was sketched to enable a direct comparison with the equation to be made.

The lower static and dynamic conductance values, compared with those predicted, indicate that the material assembly used for the inner garment is more suited for dynamic insulation than the material assemblies used in the experiments from which the formula was derived. Textile assemblies will therefore vary.

Ranque Hilsch Cooling Effect

Most systems developed to achieve microenvironmental air conditioning for men at work have required the use of special clothing—protective suits into which air is delivered from an outside source. The delivery of cool respirable air can pose some practical engineering problems. For instance, the flow of cool air through a length of hose can result in reheating of the air from its exposure to a radiant heat source. Such was the case in the access penetrations into the AGR in which a 70-m length of hose is used.

To reduce the effect of heat pickup in the hose, a vortex tube cooler (Fig. 7) is worn as part of the clothing assembly and is carried on the belt, as shown in Fig. 1. Air entering the cooler is forced through nozzles at near-sonic velocity into a vortex generator which causes the air to separate into hot and cold fractions. Valves (Item 12 in Fig. 7) located at the bottom of the tubes control the amount and temperature interval, in degrees Kelvin, of the cold air entering the suit, as shown in Fig. 8.

Heat Balance Equation

Consideration of the heat transfer within a microclimate can be shown by the following heat balance equation

$$H_e + H_m = H_a + H_s$$

where

H_e = heat transmitted through the suit from the environment,
H_m = metabolic heat,
H_a = sensible heat pickup by the cooling air, and
H_s = heat lost through sweating.

To undertake a work load in a known environment, the first step in designing a system is to assess the quantity and temperature of the air to be supplied to the microclimate. To meet these conditions, the potential cooling capacity of the air, H_a, can be based on the cooling air flow, its specific heat, and the temperature difference and the air and skin temperature.

The heat loss by sweating, H_s, can be limited by either of two considerations: (*a*) the limiting sweat rate, 1.4×10^{-4} kg/s, or (*b*) moisture pickup of air flow through the clothing assembly. Three factors influence this, namely, the air flow, the inlet temperature, and humidity. For a range of inlet air flows to a clothing assembly, the capacity of the inlet air to remove heat plotted against the inlet temperature is shown in Fig. 9.

The sensible heat component, H_a, can be calculated from the following formula

$$H_a = F(T_s - T_i)10^3 \text{ W}$$

where

F = air flow, kg/s,
T_s = 33°C, and
T_i = inlet temperature.

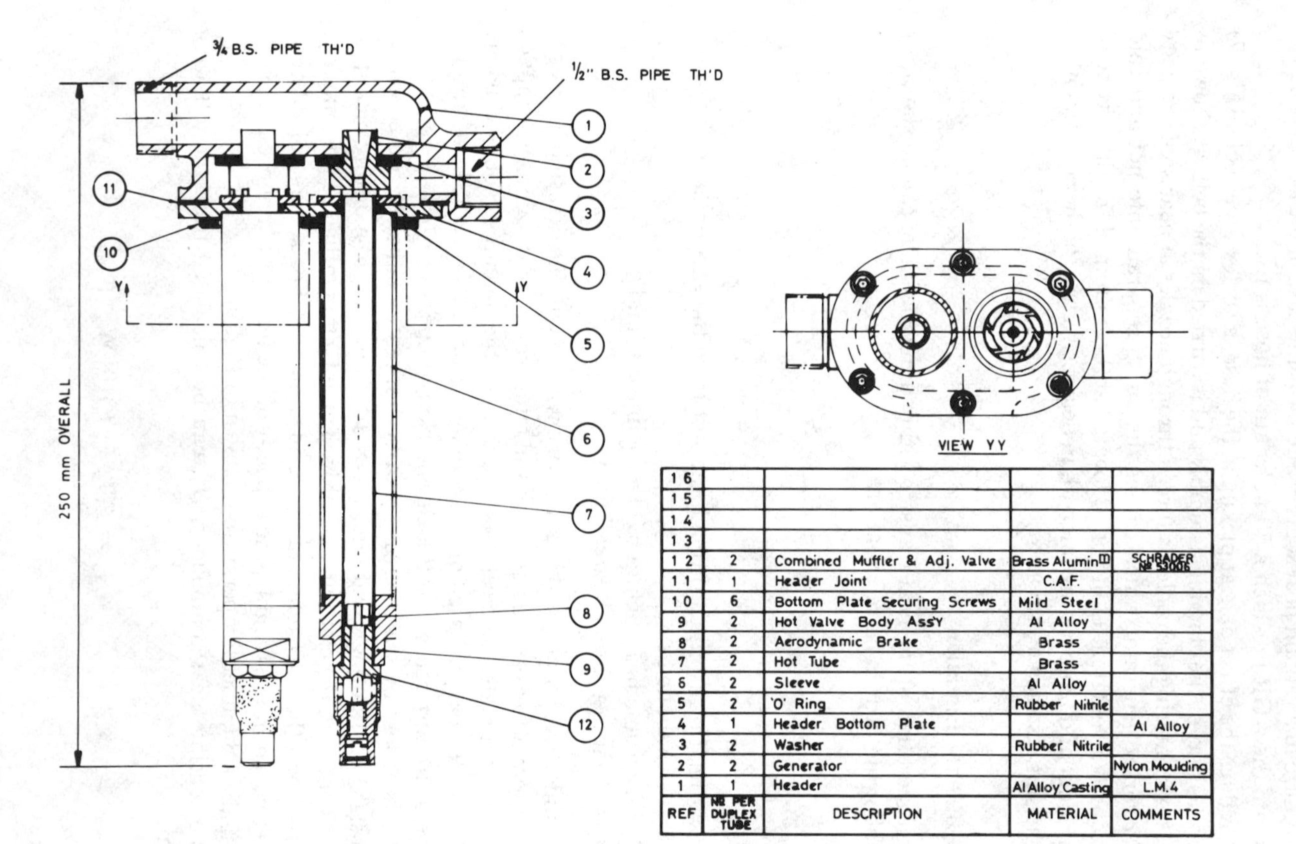

REF	Nº PER DUPLEX TUBE	DESCRIPTION	MATERIAL	COMMENTS
16				
15				
14				
13				
12	2	Combined Muffler & Adj. Valve	Brass Alumin[1]	SCHRADER Nº 53006
11	1	Header Joint	C.A.F.	
10	6	Bottom Plate Securing Screws	Mild Steel	
9	2	Hot Valve Body Ass'Y	Al Alloy	
8	2	Aerodynamic Brake	Brass	
7	2	Hot Tube	Brass	
6	2	Sleeve	Al Alloy	
5	2	'O' Ring	Rubber Nitrile	
4	1	Header Bottom Plate		Al Alloy
3	2	Washer	Rubber Nitrile	
2	2	Generator		Nylon Moulding
1	1	Header	Al Alloy Casting	L.M.4

FIG. 7—*Duplex vortex tube.*

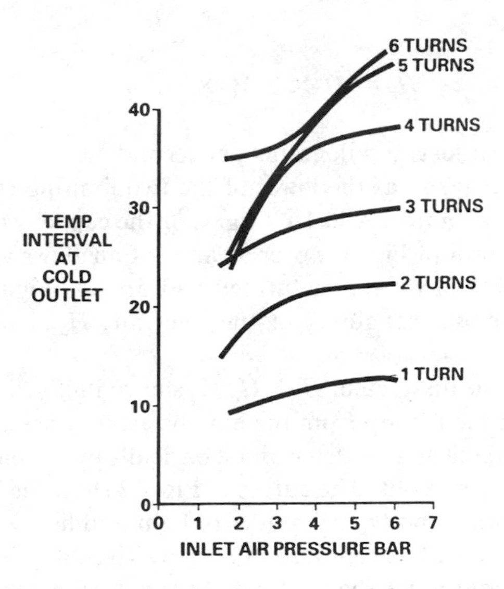

FIG. 8—*Relationship between the inlet air pressure and the temperature interval, in degrees Kelvin, with varying hot valve openings.*

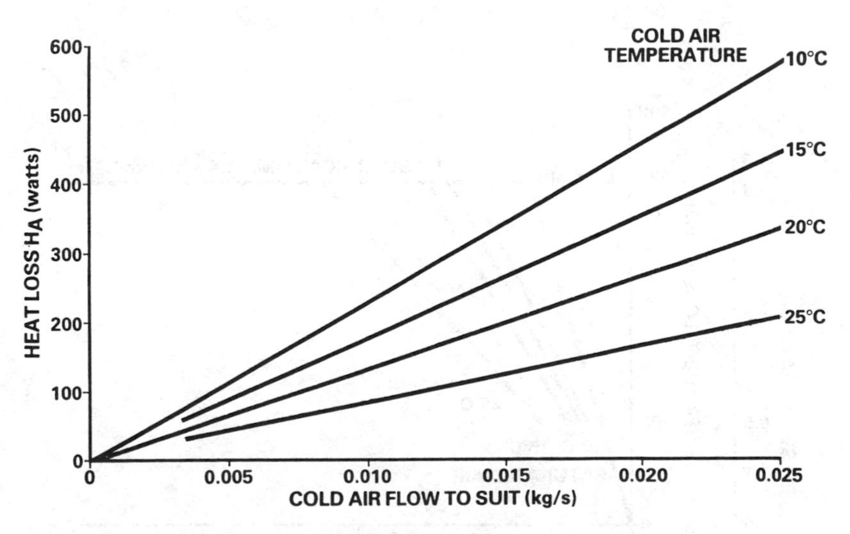

FIG. 9—*Sensible heat loss.*

The heat lost through sweating, H_s, can be calculated from the following formula

$$H_s = Q \times 2.41 \times 10^6 \text{ W}$$

where Q is the sweat loss, in kilograms per second.

This sweat loss is taken as the lower of the two limiting criteria—the (assumed) physiological limit, 1.4×10^{-4} kg/s, or the capacity of the air to take up moisture, as shown in Fig. 10, over a range of air flows at inlet temperatures of 10, 15, 20, and 25°C of saturated and dry air. Figure 11 shows the combined heat removal capability of the inlet air, $H_a + H_s$, under these conditions.

The total heat to be dissipated, $H + H_m$, is shown in Fig. 12, and from this it can be seen that the pickup from the environment varies directly with the temperature. Metabolic heat is determined basically by the nature and extent of the work being carried out. The curves in Fig. 12 show the heat dissipation appropriate for resting and typical grades of light and heavy work (the mean rate) and of heavy work (continuous). If, after a given application, the environmental temperature and the postulated work load are known, then the heat removed can be determined from Fig. 12. This heat rate can be referred to Fig. 11, and these curves can be used to determine the maximum inlet air temperature for a given inlet air flow.

It is important to note that these curves represent the limiting condition: for a given inlet temperature, the flow given is the *minimum*, or alternatively, for a given flow the inlet temperature shown is a *maximum*.

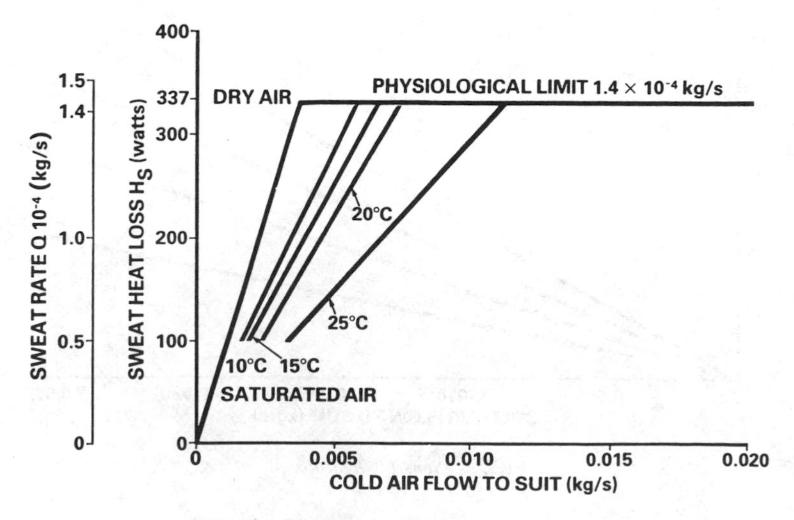

FIG. 10—*Moisture uptake and sweat heat loss.*

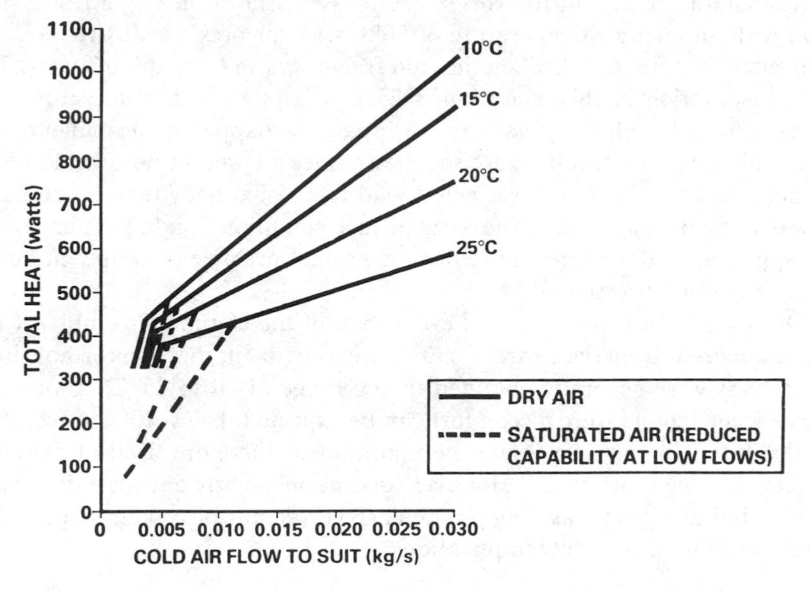

FIG. 11—*Total heat removal capability of air*, $H_a + H_s$.

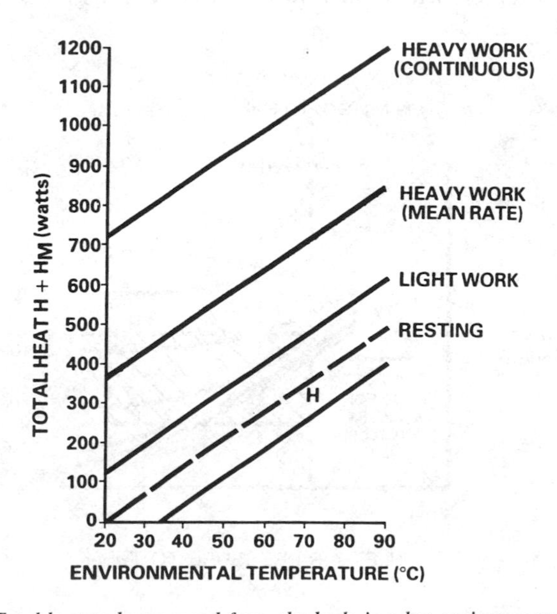

FIG. 12—*Total heat to be removed from the body in a hot environmental work suit.*

The characteristics of the vortex cooler used, shown in Fig. 13, are such that with an inlet air temperature of 50°C with air pressure at five bars, air will enter the suit at 15°C when the cold fraction valve is set at four turns. The air consumption in this condition is 25 L/s (at standard temperature and pressure). Although many wearers are apparently happy with a smaller quantity of air, caution must be exercised in use since a lower air flow would result in poor cooling. This, in turn, could lead to a rising body temperature and loss of critical facility. Since the onset of such conditions might not necessarily be apparent to the wearer himself, it is normal practice to supply more air than is absolutely essential.

During practical physiological evaluation of the clothing assembly, it became apparent from the wearers' subjective assessment that comfortable inlet conditions were generally obtained in the range of 10 to 25°C. Above this range some temperature discomfort can be expected; below 10°C discomfort is also felt. The inlet air should be maintained, therefore, if at all possible within this "comfort zone." However, operation slightly outside this zone is considered acceptable as long as the correct amount of cooling is provided and should have no safety implications.

Conclusion

The preceding analysis is based on the performance of small-scale material evaluations and the complete clothing assembly evaluation in a climate laboratory. It has attempted to show how estimates can be made of the perfor-

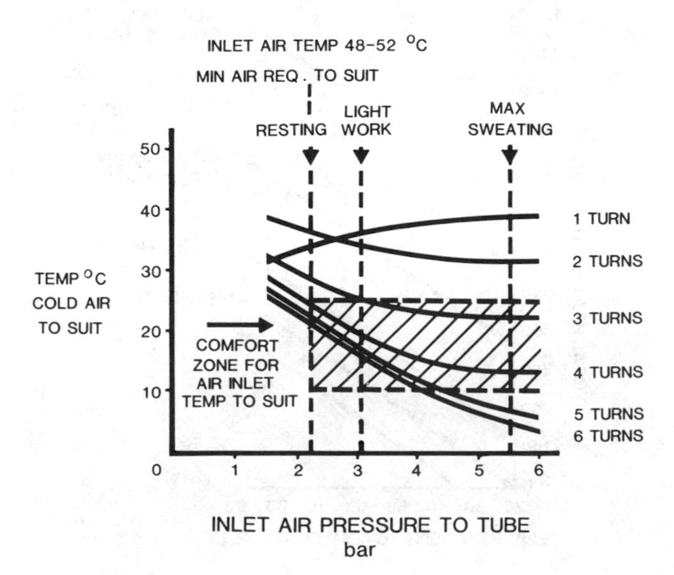

FIG. 13—*Duplex vortex tube performance characteristics with an inlet air temperature of 48 to 52°C.*

mance requirements of a dynamically insulated clothing assembly under various conditions of cooling. Many variables can affect the performance of the clothing assembly, but if its usage can be fairly clearly envisaged, the following list of essential requirements that must be met under any conditions of use is suggested:

1. The wearer should be fit and have been medically examined and pronounced fit for the work to be undertaken.
2. The clothing assembly should be worn in an atmosphere in which the air is substantially at atmosphere pressure.
3. The radiant heat to which the suit is subjected should not be significantly above what might be expected from the ambient environmental temperature.
4. The assembly should be worn over underclothing only.
5. The assembly should be worn with suitable hand and foot protection.
6. The maximum period for the assembly to be worn should be limited to 4 h.

Thus, the basic principles of working in an abnormal thermal environment have been set out and solutions to the problems have been obtained by the design of a protective clothing assembly to be used in conjunction with a vortex tube cooler, which gives a satisfactory heat balance and control in a simple, practical way.

References

[1] Davies, J., "The Vortex Tube," Report Ref. SER/ENG/08/77, Central Electricity Generating Board, London, U.K., 1977.
[2] Crockford, G. W. and Goudge, J., *Proceedings of the Royal Society of Medicine*, Vol. 63, 1970, p. 1012.
[3] Crockford, G. W. and Karim, M., *Annals of Occupational Hygiene*, Vol. 17, 1974, p. 111.

John T. McConville[1]

Anthropometric Fit Testing and Evaluation

REFERENCE: McConville, J. T., **"Anthropometric Fit Testing and Evaluation,"** *Performance of Protective Clothing, ASTM STP 900*, R. L. Barker and G. C. Coletta, Eds., American Society for Testing and Materials, Philadelphia, 1986, pp. 556–568.

ABSTRACT: A major challenge facing designers and manufacturers of protective clothing and equipment is the establishment of a sizing system that will accommodate the body size variation of the user population as well as provide a good fit for individual users. This has been variously achieved by means ranging from hit-or-miss trial-and-error methods to the careful application of anthropometric design data. Ultimately, the success of the design and the sizing system can be established only by an anthropometric fit test in which garments or equipment are worn or used by subjects representing the size range of the user population and carefully evaluated for fit and function. Described here are guidelines for conducting such a fit test.

KEY WORDS: anthropometry, fit tests, sizing, design, body size variability, protective clothing

Each item of specialized clothing or personal-protective equipment is designed to meet a specific need and must fit and function within specified limits if it is to be considered a successful solution to a given problem. Each is the end product of an intensive effort by a number of researchers, designers, and manufacturers continuously engaged in a series of decisions over trade-offs involving function, comfort, cost, appearance, and safety, to name but a few.

One means—probably the only means—of ensuring the success of these efforts is the timely use of a hands-on fit test in which garments or equipment are tried by individuals representing the body size and physical capabilities of the user population, and fully evaluated for fit and function.

Ideally, some form of fit test should be employed at various critical stages of design and sizing. Unfortunately, this does not usually occur. Often, a fit test is called for only *after* problems have developed in the fit or function of the manufactured item. Most often, the fit test is the next-to-last step in the

[1]President, Anthropology Research Project, Inc., Yellow Springs, OH 45387.

development of the item and is conducted prior to its production for purposes of validating the sizing program and establishing the tariff—that is, a listing of how many items of each size should be manufactured for the accommodation of the user population.

While human factor evaluation procedures have been exhaustively cataloged in various publications, particularly for military users, procedures for conducting anthropometric fit tests are never included in these documents.

It is my intention here to describe such fit testing and to suggest that its use early in the design cycle will prevent countless problems later. Most of our experience in this field has been with military clothing and equipment, and examples cited here will concern protective clothing designed for military uses. It should be noted emphatically, however, that the principles are the same for items used by other populations.

The anthropometric fit test can be separated into three distinct phases:

(*a*) pretest preparation,
(*b*) testing and evaluation, and
(*c*) analysis and reporting.

Pretest Preparation

Selection of the fit-test site and the subjects to be used should be made with the end users of the test item in mind. If the item to be tested is a one-piece coverall designed for use in a variety of occupations, the choice of subjects and sites will be a wide one. If the item is designed for a specialized population, however, the test sample must be an operational group knowledgeable about the item to be tested and able to wear the item under the conditions for which it is designed. For example, a random test subject wearing a 1.5-kg-plus flight helmet can rarely be objective about comfort or preference, whereas an aircrew member who may have worn such an item for several thousand flight hours will take a more useful view of the matter. He or she can rapidly determine if the item is better or worse than previously used similar items. More important, a flier will be able to determine if the test item will integrate with the other items of the personal protective clothing and equipment normally worn and whether it will allow him or her to function effectively.

The size of the test sample depends on various factors, not the least of which is the availability of appropriate subjects. While there is no specific formula for arriving at a proper number for a given case, several guidelines can be articulated. In general, fewer than 20 subjects will produce a biased sample unlikely to represent accurately the ultimate range of body sizes in the user population. In any case, there should be at least 3 to 5 subjects for each size of the test item.

The investigator should aim for a sample group covering 90 to 98% of the body size variability for the anthropometric variables of interest. There are two categories of such anthropometric variables. Of prime importance are the key variables on which the sizing of the test item is based. These dimensions may range from hand circumference and hand length for gloves, to stature and weight for a one-piece overgarment.

An expanded series of body size dimensions that may interact with the test item is also measured on each subject. These additional measurements are usually taken as comparative data for use in determining whether the sample represents the larger user population. Thus, in testing the MC-1 oxygen mask, lip length and total face length were the two key dimensions measured to obtain the indicated size. In addition, several facial measurements, such as face breadth and nose length, were taken for use as comparative data. Height, weight, and age are always routinely recorded for the same purpose.

Having selected the list of variables that will be measured on each subject, the investigator should prepare a percentile table showing the 1st through the 99th percentile values for each of the variables in the overall population or in selected segments of it. Such data are available from a number of large-scale anthropometric surveys of military and civilian populations which have been conducted in the last few decades. They are stored in a data bank that is maintained by the author's company and periodically updated as new data become available. A table of percentile values is used for field checking the measurement data as they are obtained and for ensuring that the test sample represents a comprehensive range of relevant body sizes.

The investigator should also secure, in advance of the test, a bivariate frequency table for the key sizing variables. A bivariate table shows how individuals from a given population are distributed over the entire size range of any two selected measurements. Each subject, as he is measured, is "entered" in the appropriate spot on the table, thus providing the investigator, at a glance, with a graphic indication of the range of subjects being tested.

The bivariate table in Fig. 1 was used in a fit test of a nine-size British chemical defense coverall sized on the basis of stature and chest circumference. The 36 subjects used in the test are represented by dots on the bivariate table and were judged to represent adequately the sizing range of the user population.

The test questionnaire represents the skeleton on which the study of the fit test will depend. Often there will be one for the investigators to list basic information (for example, name, age, and work experience where relevant), record measurements and sizes, and enter the results of any special tests (Fig. 2). A second questionnaire, filled out by the subject, seeks largely subjective information (such as the quality of fit, comfort, and ability to perform needed functions). It is essential to plan the questionnaires with care; they should be comprehensive in scope and clear in their intent, and they should have room for possible additions to be made in the field. Information can be discarded later if it is not needed, but it can seldom be retrieved if it was not observed or recorded at the time of the test.

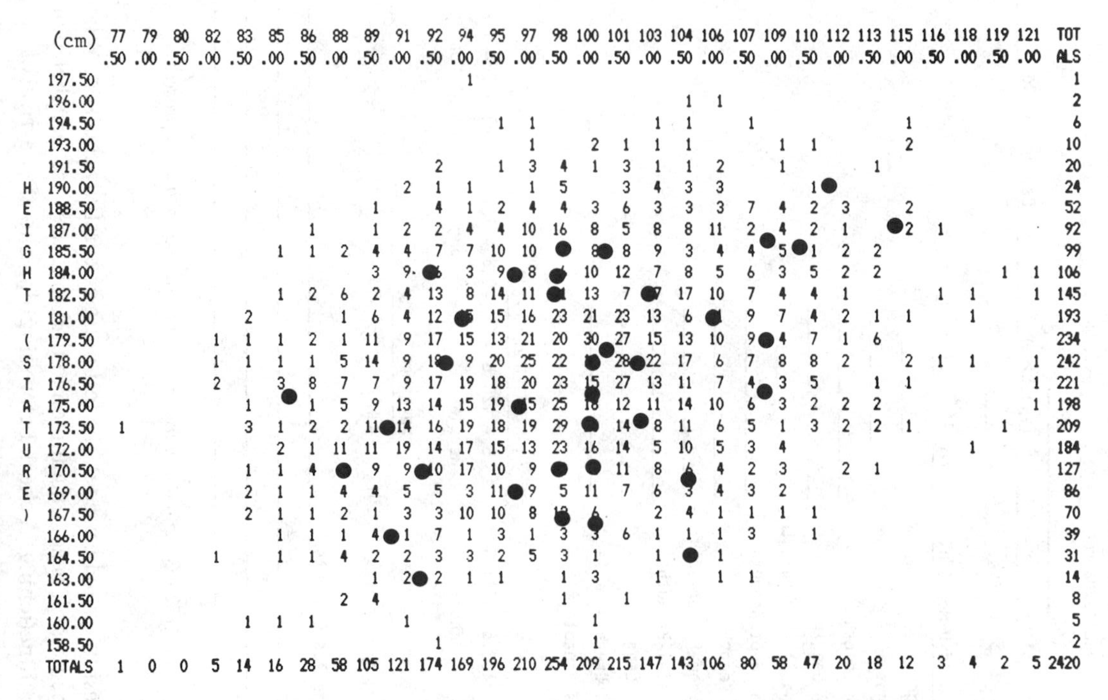

CHEST CIRCUMF"ENCE

	(cm)	77.50	79.00	80.50	82.00	83.50	85.00	86.50	88.00	89.50	91.00	92.50	94.00	95.50	97.00	98.50	100.00	101.50	103.00	104.50	106.00	107.50	109.00	110.50	112.00	113.50	115.00	116.50	118.00	119.50	121.00	TOTALS
	197.50												1																			1
	196.00																			1	1											2
	194.50													1	1				1	1		1					1					6
	193.00														1		2	1	1	1			1	1			2					10
	191.50											2		1	3	4	1	3	1	1	2		1			1						20
H	190.00										2	1	1		1	5		3	4	3	3			1								24
E	188.50									1		4	1	2	4	4	3	6	3	3	3	7	4	2	3		2					52
I	187.00							1		1	2	2	4	4	10	16	8	5	8	8	11	2	4	2	1		2	1				92
G	185.50						1	1	2	4	4	7	7	10	10	[illegible]	8	8	9	3	4	4	5	1	2	2						99
H	184.00									3	9	[illegible]	3	9	8	[illegible]	10	12	7	8	5	6	3	5	2	2				1	1	106
T	182.50						1	2	6	2	4	13	8	14	11	[illegible]	13	7	[illegible]	17	10	7	4	4	1			1	1		1	145
	181.00					2			1	6	4	12	[illegible]	15	16	23	21	23	13	6	[illegible]	9	7	4	2	1	1		1			193
(	179.50				1	1	1	2	1	11	9	17	15	13	21	20	30	27	15	13	10	9	4	7	1	6						234
S	178.00				1	1	1	1	5	14	9	18	9	20	25	22	[illegible]	28	22	17	6	7	8	8	2		2	1	1		1	242
T	176.50				2		3	8	7	7	9	17	19	18	20	23	15	27	13	11	7	4	3	5		1	1				1	221
A	175.00					1		1	5	9	13	14	15	19	[illegible]	25	18	12	11	14	10	6	3	2	2	2					1	198
T	173.50	1				3	1	2	2	11	14	16	19	18	19	29	[illegible]	14	8	11	6	5	1	3	2	2	1			1		209
U	172.00						2	1	11	11	19	14	17	15	13	23	16	14	5	10	5	3	4						1			184
R	170.50					1	1	4	[illegible]	9	9	10	17	10	9	[illegible]	[illegible]	11	8	6	4	2	3		2	1						127
E	169.00					2	1	1	4	4	5	5	3	11	9	5	11	7	6	3	4	3	2									86
)	167.50					2	1	1	2	1	3	3	10	10	8	[illegible]	6		2	4	1	1	1	1								70
	166.00						1	1	1	4	1	7	1	3	1	3	3	6	1	1	1	3		1								39
	164.50				1		1	1	4	2	2	3	3	2	5	3	1		1	[illegible]	1											31
	163.00									1	2	2	1	1		1	3		1		1	1										14
	161.50								2	4						1		1														8
	160.00					1	1	1			1						1															5
	158.50											1					1															2
	TOTALS	1	0	0	5	14	16	28	58	105	121	174	169	196	210	254	209	215	147	143	106	80	58	47	20	18	12	3	4	2	5	2420

FIG. 1—*A bivariate frequency table used in the fit test of a British chemical defense coverall.*

SIZING/DESIGN EVALUATION
MBU-12/P ORAL NASAL OXYGEN MASK

NAME______________________ Age _____ Subject No. _______

AeroRating______________________ Career Flight Hours__________

Operational Squadron____________ Command____________

Current A/C_____________________

Helmet Size_____________ Type Liner________ Bayonet Style________

Current Size Mask ______________

Measured Dimensions	mm	%ile
Height	_______	_______
Weight	_______	_______
Face Length	_______	_______
Nose Length	_______	_______
Subnasale-Menton Lgth	_______	_______
Nasal Root Breadth	_______	_______
Lip Length	_______	_______
Bizygomatic Breadth	_______	_______
Bigonial Breadth	_______	_______
Head Length	_______	_______
Head Breadth	_______	_______
Nose Breadth	_______	_______

INDICATED MBU-12/P

102.5 - 111.5	111.5 - 120.5	120.5 - 129.5	129.5 - 138.5
Short	Regular	Long	X-Long

PRESSURE CHECK

Safety 41M 43M 45M >45M

Subject's Comments____

____

____

____

FIG. 2—*Questionnaire used in the fit test and evaluation process.*

The test items to be used in the test must be carefully checked. When using prototype items it is not uncommon to find that sizes are mislabeled or equipment malfunctioning. A check before shipping and after arrival of the gear at the test site is mandatory.

Data Collection Phase

A team of two investigators is recommended for the conduct of the test. One briefly interviews each subject for the basic background information required on the form and explains the nature of the test if a prior briefing has not taken place. While it is often tedious to repeat this explanation over and over again, it is essential to secure the full cooperation of the subject and it serves, also, to put him or her at ease. One investigator measures the subject for the variables of interest. The other acts as recorder and, referring to previously prepared percentile tables, watches for apparent anomalies in proportions as, for instance, head and face length percentiles widely at variance with each other or a face length in the 90th percentile with a chin-to-base-of-nose length in the 20th percentile. Such an anomaly may be an error in the measuring but may also reflect the actual configuration of the subject's face. Disproportionality is carefully noted since it may affect the fit of the item or later explain a particular observation or complaint—undue pressure or oxygen leakage at some point on a protective mask, for example.

In addition to continuous scanning of the measurement data, the recorder also locates each subject on the previously prepared bivariate table. By this means, investigators can keep track of the distribution of the test subjects as the test proceeds and do some last-minute recruiting of subjects if necessary to ensure that a reasonably representative sample of the user population will be tested.

Following the measurement session, an investigator helps the subject don and adjust the test item. When a satisfactory fit has been achieved, the subject is exercised in any one of a variety of means selected to test the fit and function of the garment in actual use. Flight garments and equipment may be tested in a simulator. In the case of oxygen masks, the subject is usually put through tests on a console designed to test leakage. For garments designed for ground crew, a standard familiar task may be used. For a full-face respirator, the fit may be tested against a challenge aerosol and a protection factor determined.

In all cases, both investigators watch carefully for any noteworthy details. These include decrements in the ability to perform necessary functions or interferences with mobility, as well as stress lines on the garment itself during the exercises.

In the case of a Canadian chemical defense overgarment, for example, the indicated sizes looked too big in the shoulders and arms when first tried on by some subjects (Fig. 3). When squatting, these subjects encountered significant pulling along the back of the garment and were unable to raise their arms above shoulder level (Fig. 4). Contributing to the stress in this garment was the short underarm seam created by the dropped armhole of the overly large sleeve and the shoulder seam's location out on the arm rather than on the shoulder. Fit-testing earlier in the design/production cycle would have turned up this problem before the garment was manufactured.

Often the test item is designed as part of an assemblage. The investigators must study the nature of the integration with other garments or pieces of equipment. Are garments designed to be worn over or under other garments too restrictive? Do they cause undue bunching of fabric in the secondary gar-

FIG. 3—*Canadian chemical defense overgarment.*

ment? Can the subject wear glasses with it? Figure 5 illustrates a case in which the fit test revealed that the edge rolls of a test helmet butted up against the flier's goggles, forcing the goggles away from the face. This was significant because the helmet was designed as a jump helmet for pararescue personnel, and ill-fitting goggles are likely to be blown off the head by wind blast.

It may be possible to effect a fix during the test itself. In the fit test of a two-piece chemical defense overgarment to be worn by Air Force personnel, and already in use by Army personnel, a gap in the waist-back between the overblouse and trousers occurred when the subject bent over or extended his arms (Fig. 6). Since such a breach is unacceptable in a chemical defense garment, this defect was cause for rejection of the item. Three snap fasteners were affixed to the garment, and subsequent testing showed that the fasteners prevented the separation of the blouse and trousers (Fig. 7). Although this solu-

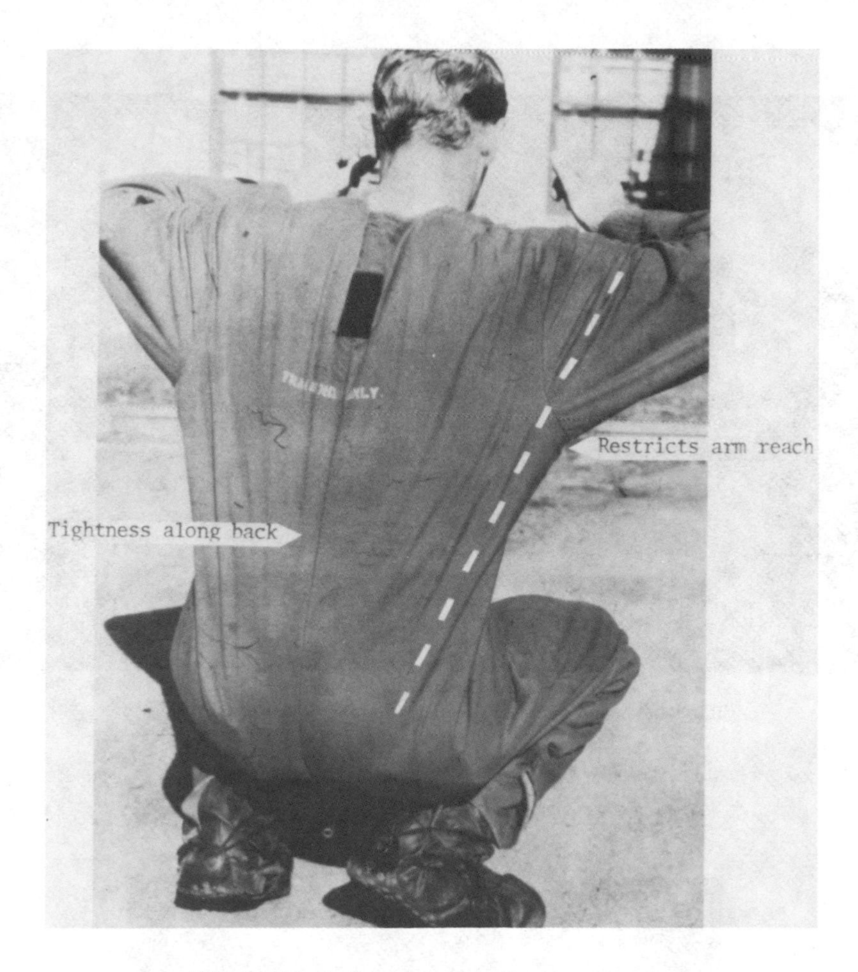

FIG. 4—*Restrictions of the Canadian overgarment.*

tion worked, such quick fixes are usually only interim answers to problems which are far better dealt with in the original design of the item.

The waist circumferential adjustment straps on trousers were also found to be insufficient during this fit test. A 2.54-cm (1-in.) increment added to each side of the strap during the course of the test proved to be sufficient to provide proper waist fit for all the subjects tested. Figure 8 shows the strap before and after modification.

While a camera is not essential to the conduct of a fit test, it is highly recommended. As demonstrated here, photographs of stress lines and defects are excellent means of documenting and illustrating findings. A photograph is often clearer and more convincing than several paragraphs of detailed description.

The final step in the testing procedure is debriefing. After the subject has

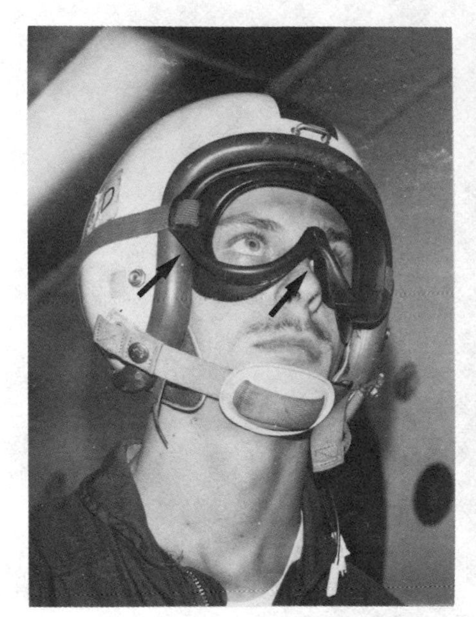

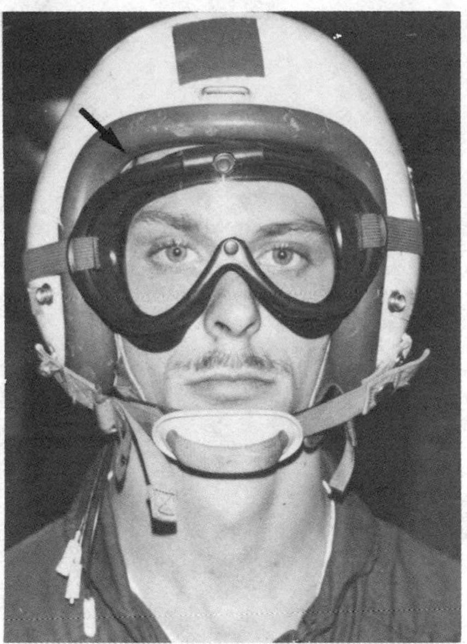

FIG. 5—*Poor integration between test helmet assembly and goggles.*

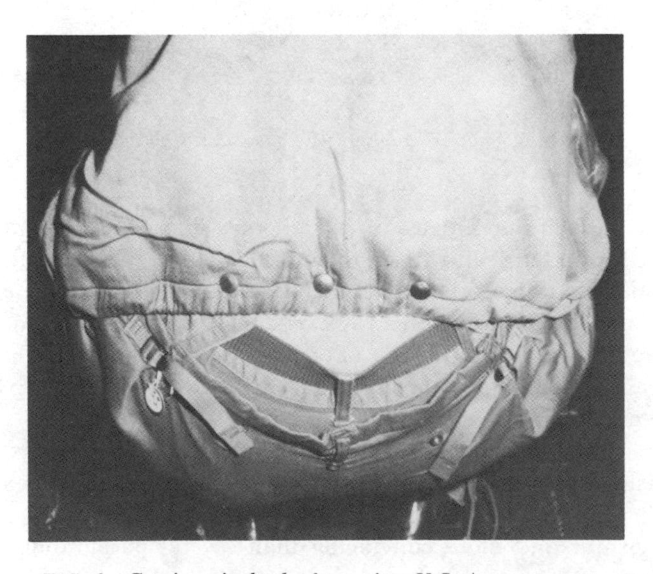

FIG. 6—*Gap in waist-back of two-piece U.S. Army garment.*

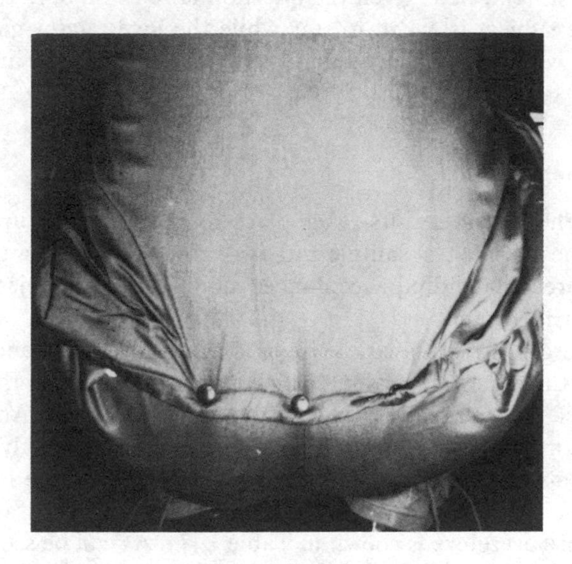

FIG. 7—*Waist-band closure accomplished with three snap fasteners to the garment in Fig. 6.*

FIG. 8—*Adjustment straps for the trousers in Fig. 6 before and after modification.*

removed the test item, he is given the questionnaire to fill out. When that is completed, the subject is asked to stay while the investigator goes over the filled-out form with the subject, to make sure all the responses are clear and to the point.

Analysis and Reporting

Analysis of the fit test results takes place in two steps: documenting the representativeness of the test sample and analyzing the results of the test with a view toward recommending acceptance, modification, or rejection of the item under consideration.

Use of the percentile and bivariate frequency tables to track and adjust the test sample as the fit test goes along will already have ensured that the subjects represent a reasonably full range of the user population. An additional means of assessing the representativeness of the sample is to analyze the measured dimensions of the sample and compare them with those found in the user population. Such a comparison for a sample used to test an eight-size chemical defense overglove is shown in Table 1 [*1*]. As can be seen, the sample closely reflects the larger population with regard to hand dimensions.

In assessing the acceptability of the item, investigators should summarize the material garnered from the questionnaires, presenting results, wherever possible, in quantitative terms—How many, or what percentage of subjects rated the fit "excellent," "good," or "poor"? How many preferred the item to a previously used version? How many subjects were not well fitted in their indicated sizes? These results are then studied and significant trends are reported. It is important, for example, to note whether certain difficulties crop up across the whole gamut of sizes or seem concentrated in one sector of the size range. It was found, for example, that the poor integration between the

TABLE 1—*Anthropometric comparison of fit test sample with 1967 U.S. Air Force population for hand size dimensions [1].*

	Fit Test Sample (n = 33)		USAF Flying Population (n = 2420)	
Variable[a]	Mean	Standard Deviation	Mean	Standard Deviation
Height (reported), in.	70.47	2.78	69.58	2.37
Weight (reported), lb	172.80	26.10	173.06	19.65
Hand length, cm	19.17	0.80	19.11	0.82
Palm length, cm	11.11	0.55	10.83	0.54
Hand breadth, cm	8.91	0.38	8.90	0.41
Hand Thickness,[b] cm	3.01	0.25	2.97	0.12
Hand circumference, cm	21.31	0.82	21.55	0.94
Wrist circumference, cm	16.97	1.09	17.59	0.92

[a]1 in. = 25.4 mm; 1 lb = 0.45 kg.
[b]From USAF flying population—1950.

pararescue helmet and the goggles described earlier occurred only with the smaller helmets.

Although the major concern of the test revolves around fit, function, and integration, the subjects' comments regarding comfort and preference are also of concern. If an item is good in terms of fit and function but repeatedly causes a pressure point that threatens to become intolerable after long-term wear, the item is unacceptable until that problem can be resolved. It is here that it becomes important to have experienced test subjects who are familiar with the item or class of item under consideration.

An important element of the fit test report is establishment or verification of the procurement tariff—that is, how many items of each size should be manufactured for the accommodation of the user population. The fit test will indicate what body sizes are fitted by what size garments or end items. This, in conjunction with the knowledge of the body size distribution of the population, provides the basis for the preparation of the procurement tariff. Table 2 [*2*] shows a tariff for a twelve-size high-altitude glove.

This tariff was based on the number of subjects included in each size category of the hand circumference/hand length bivariate table expressed as a percentage of the total sample. The fit test is used to verify the tariff by demonstrating that test subjects are fitted in their indicated sizes. Tariffs can be adjusted at this time as necessary.

Summary

Investigators can, and should, recommend modifications that would improve the fit and function of the test item. Ultimately, of course, responsibility for the further disposition of the item lies with the program manager or manufacturer. It is the central purpose of the fit test report to provide the wherewithal to arrive at these decisions.

In conclusion, may we again stress the importance and cost-effectiveness of including the anthropometric fit test early in the research and development cycle. All too often, the test is scheduled so late that the sizing and design of

TABLE 2—*Suggested procurement tariff for high-altitude gloves per thousand pairs [2].*

Size	Units per 1000	Size	Units per 1000
A	55	G	74
B	62	H	167
C	34	I	84
D	140	J	24
E	208	K	43
F	80	L	29

the end item has become fixed, and when a problem surfaces everyone involved in its manufacture is looking for a "quick fix" to prevent its rejection. Such last-minute jury rigging seldom, if ever, holds up over the life cycle of the item and often results in costly and time-consuming resizing or redesign of the item being developed.

References

[1] Alexander, M. and McConville, J. T. in *Chemical Warfare Equipment and Procedures*, R. D. Guidry and S. R. Schick, Eds., Special Project Final Report, TAC Project 74E-057T, U.S. Air Force Tactical Air Warfare Center, Eglin Air Force Base, FL, July 1977, pp. 2-B-2–2-B-10.

[2] Barter, J. T. and Alexander, M., "A Sizing System for High Altitude Gloves," WADC TR 56-599, Wright Air Development Center, Wright-Patterson Air Force Base, OH, Dec. 1956.

Kathleen M. Robinette[1]

Anthropometric Methods for Improving Protection

REFERENCE: Robinette, K. M., **"Anthropometric Methods for Improving Protection,"** *Performance of Protective Clothing, ASTM STP 900*, R. L. Barker and G. C. Coletta, Eds., American Society for Testing and Materials, Philadelphia, 1986, pp. 569–580.

ABSTRACT: Many jobs involve contact with dangerous materials or hazardous environments requiring workers to wear various types of protective clothing and equipment. The fit of these items can directly affect both the worker's performance and the degree of protection provided and is, therefore, of paramount concern. Often, a stringent conformity to the body is required to afford the necessary physiologic protection and minimize degradation of performance. Traditional sizing practices are proving inadequate for protective wear in many instances; however, anthropometric analysis can be a very effective means of resolving the sizing and design problems. An anthropometric sizing analysis is based on the concept of dividing the population into subgroups of individuals who are similar for one or two relevant body size dimensions, then analyzing the remaining anthropometric data for each subgroup to arrive at appropriate dimensional design values. In this paper, the development and use of anthropometric programs for personal protective clothing and equipment is discussed.

KEY WORDS: anthropometry, protective equipment, clothing, sizing, body size, protective clothing

Worker safety concerns almost all industries, and as a result, tens of thousands of American workers wear some type of protective clothing or equipment. If it does not fit properly, such protective gear is not only hazardous but can hamper the worker's performance. Yet sizing problems for protective equipment and clothing are more serious than those for traditional clothing. First of all, many protective items do not have the sizing history that traditional clothing has, so educated guesses are not quite as educated. Neither the items themselves, such as respirators, nor the materials they are made of—for example, flame retardant fabrics or rubberlike materials—have been around very long. Second, manufacturers of protective items often try to fit everyone into a limited set of sizes, sometimes only three—Small, Medium, and Large.

[1]Research physical anthropologist, Workload and Ergonomics Branch, Human Engineering Division, U.S. Air Force, Wright-Patterson Air Force Base, OH 45433.

Yet, to protect effectively and function well, the items must often fit better than street clothing. This report discusses some of the problems that exist with current sizing practices and presents a method for sizing that can improve the situation.

Discussion

In the development of sizing criteria for their items, many manufacturers start with a Size Medium. This size can actually be many different things. Some manufacturers use average values for a series of dimensions relevant to the item. Some select a person, often referred to as a "Greek god" or a "Cinderella" or a "perfect 10," to represent Medium. This person may be someone who "appears average," who meets some "average" criteria for one or two dimensions, or who simply looks good. Then, to create the other sizes, these patterns or forms are scaled up and down. This may seem like a reasonable approach but, in reality, people do not come as scale models of each other. In fact, Daniels [*1*] clearly demonstrated that no one is average in every body dimension. People come in a variety of shapes that combine some average with some smaller and larger dimensions. The following figures (Figs. 1 and 2), illustrate in a simplified manner why this can be a problem.

Diagrams *A* through *F* in Fig. 1 constitute a sample of three-dimensional blocks intended to represent different types of people. Taking 20 units as average, all the blocks are average in one respect but not in others. Diagram *g* is a block that is average in all three dimensions and represents the mythical average person. Diagrams *h* and *i* are scaled versions of *g*. In short, blocks *a* through *f* represent people, and *g* through *i* represent sizing forms. If a cover were made to fit Block *g*, it would not be large enough for one dimension to fit any of the *a* through *f* blocks. A scaled-up version of the Medium block, represented by Block *i*, would cover Blocks *a* through *f*, but it would be too large in two of the three dimensions. This Large size would be too large by 20 units in one direction and by 10 units in another for a total of 30 units of offset.

The same sort of problems occur with the "Greek god" method of sizing, as is shown in Fig. 2. Block *g* in this figure represents Size Medium and will fit only blocks like the Greek god, Block *b*. Block *h* represents a scaled-down medium, and Block *i* a scaled-up medium.

Sizes are scaled in many ways. Common to them all is the fact that if one dimension is made smaller they are all made smaller, and conversely if one is made larger they are all made larger. In this paper two different methods are employed: proportional scaling, in which the proportion between dimensions remains the same, and constant scaling, which is scaling by increasing or decreasing the dimensions by equal (constant) amounts. In Fig. 1, these two methods produced equivalent results because Size Medium is a cube. In Fig. 2 Size Small is scaled down using the former method and Size Large using the latter. Both methods are used because this appears to provide the best possi-

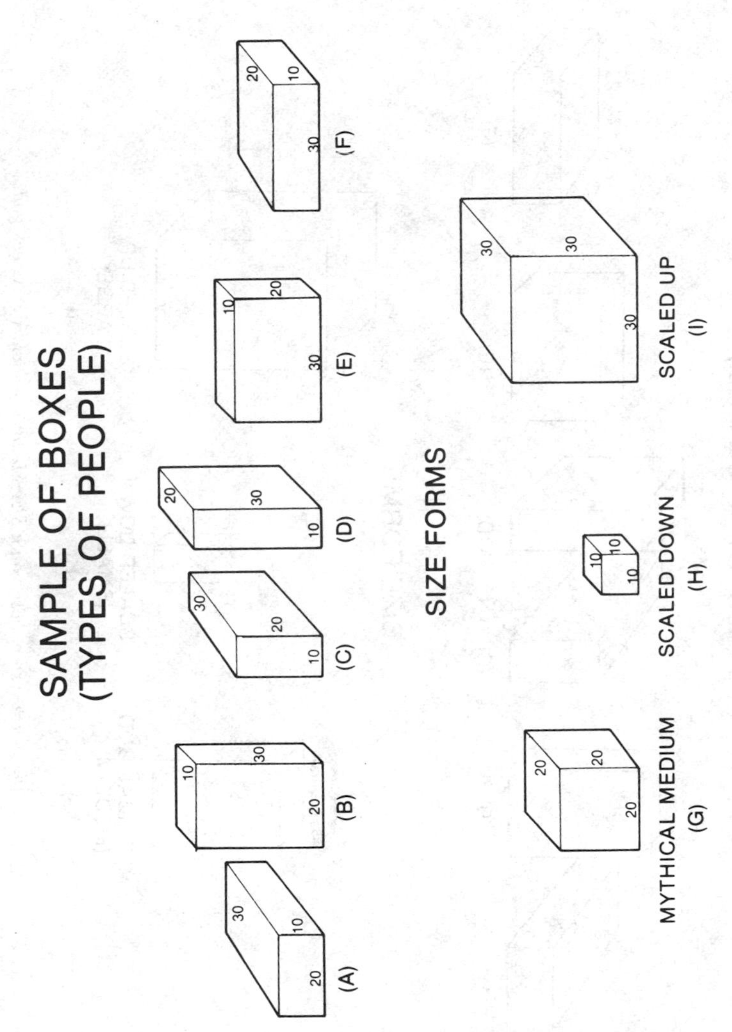

FIG. 1—*A three-size system based on a Size Medium, which is a composite of average values.*

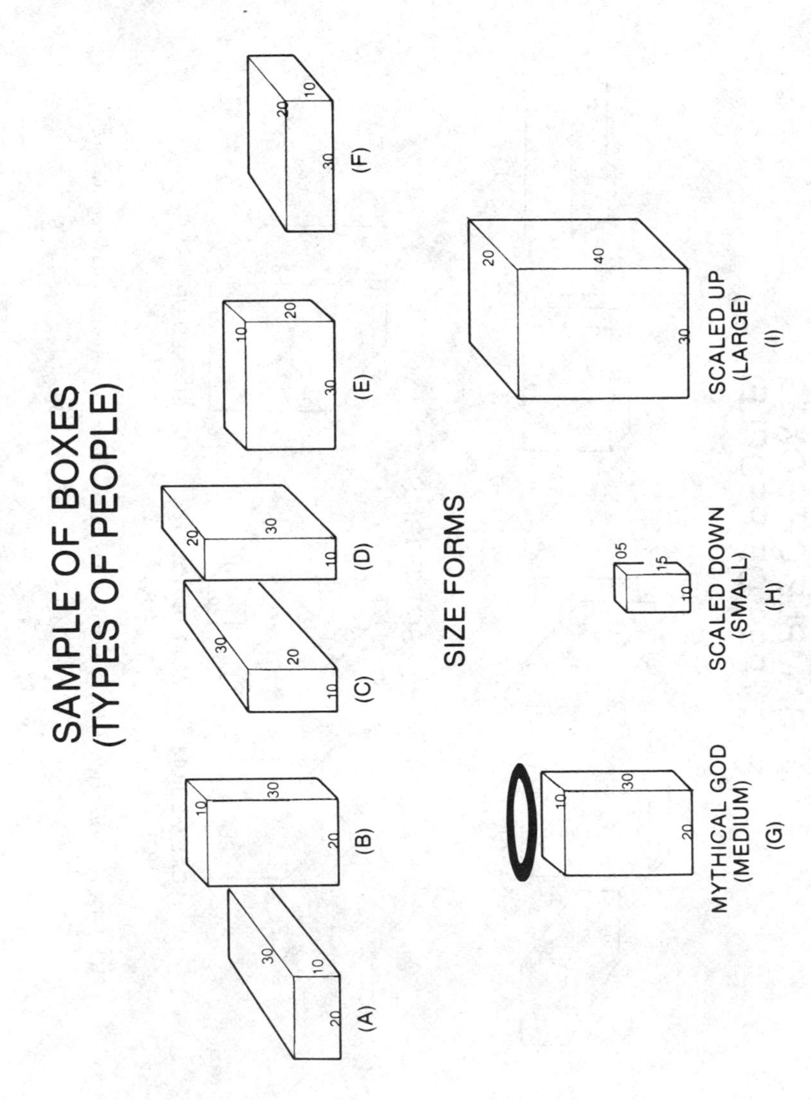

FIG. 2—*A three-size system based on a Size Medium, which is a "Greek god."*

ble fit using scaling methods in this case. As can be seen, Size Small will not fit any of the *a* through *f* blocks. Size Large, the scaled-up Greek god, will cover *e* and *f* blocks, but with a total of 30 units of gap in one place or another, and will not fit *a* or *c* blocks because it is not deep enough. (Recall that Block *i* represents a size for an item, and Blocks *a* through *f* represent people. Turning Block *i* to cover Blocks *a* or *c* would be analagous to putting on a pair of coveralls sideways.) Therefore, this three-size system covers only four of the six blocks.

As stated previously, different manufacturers use different Size Mediums; therefore, if a suitable size cannot be found from one manufacturer, it may be possible to find one from another. For traditional clothing, people also have the option of altering their clothes; this has aided the effectiveness of the preceding type of sizing strategy for this type of clothing. With protective equipment, however, adjustment by the home sewer is not feasible. First, not many home sewers know how to work with butyl or charcoal-impregnated materials, nor would they know how to alter respirators, boots, gloves, or goggles. Furthermore, equipment and clothing designed for protective purposes have to be more wear resistant than normal clothing. A seam torn because of home alteration can be a serious matter. So the alteration option should not be considered a remedy.

Unlike designers in civilian industries, designers of military equipment have been working with sizing from an anthropometric approach, on and off, since World War II. In other words, we have been investigating and applying methods for summarizing body size into usable sizes rather than redoing sizing schemes on a trial-and-error basis. Many mistakes have been made as fallacious assumptions are uncovered but, overall, we have made progress and now feel we have an effective technique for developing sizing systems for equipment and clothing.

This technique, which we call an anthropometric sizing system, could be used by the industrial sector as well, where in fact the need is probably greater, since the variety of body types in industry is more diverse than is found in the military services.

Basically, the technique involves four steps:

1. Select one or two dimensions which are key to the end item.
2. Divide a sample of the user population into subgroups of people who are similar for these key dimensions.
3. Summarize the variability in all other relevant dimensions within each of the subgroups.
4. Select design values for all the dimensions to accommodate the variance for each subgroup.

Many dimensions are usually needed to design an item. "Key" dimensions—those on which the sizing of the item is based—are dimensions that collectively have a strong relationship with most of the other dimensions important to the item. It is usually necessary to select at least two—one to control a vertical variance in people, in other words, heights and lengths, and one to control a horizontal variance, in other words, breadths, circumferences, and depths. Two commonly used key dimensions are waist circumference and pants' inseam for trousers. Stature and weight are a good key dimension pair for coveralls—stature to control vertical variation, and weight to control the horizontal. For respirators and oxygen masks, face length and face breadth might be a good pair.

Dividing a sample of the user population into subgroups, for Step 2, is analogous to defining the size categories in terms of the key dimensions. Once the size categories are defined, the subgroups of people who fall within each are treated as separate samples. The samples are described using summary statistics. Among those statistics are variance estimates which describe the range of variability for each dimension. Using these as guidelines, design values are selected to accommodate the people within each category.

Figure 3 shows a bivariate frequency plot and an example of an eight-size system developed for garments, such as coveralls, for U.S. Army women [*2*]. In this system, weight was selected as the horizontal control dimension and stature as the vertical dimension. The scale for weight runs along the left side of the graph, from 38.1 to 99.8 kg (84 to 220).[2] The scale for stature runs across the top, from 147.3 to 184.2 cm (58 to 72.5 in.). The numbers in each little cell represent the number of women at that particular combination of weight and stature. Each of the eight large blocks represents a size category. The one heavily outlined, Size X-Large Regular, includes women who are 77.1 to 88.4 kg (170 to 195 lb) and 161.3 to 171.5 cm (63.5 to 67.5 in.).

As stated earlier, for each category, summary statistics are computed. A sample of the summary statistics for the Size X-Large Regular from this sizing system is shown in Table 1. The range for stature and weight that defines the category is shown at the top of this table. Some of the dimensions considered to be important for this sort of garment are shown in the first column at the left. The "midsize value" is similar to a mean, although we compute it from a regression midpoint. A variance estimate [which is called a size standard deviation (SZ-SD) in this example] is shown, and the Range to Be Accommodated column lists an estimated range of 5th to 95th percentile values within a size for each dimension.

The boxed values, called "Recommended Values," are regression estimates from the largest stature and weight values in the category. These values are often used as design values because they are large enough to cover most people within a size and they are additive. For some dimensions, alternatives to these are used, such as a 99th percentile value, when a particular dimension needs to be exaggerated. These must be used sparingly, however, because percentile values are not additive and can result in an adverse effect

[2]Original measurements were in English units.

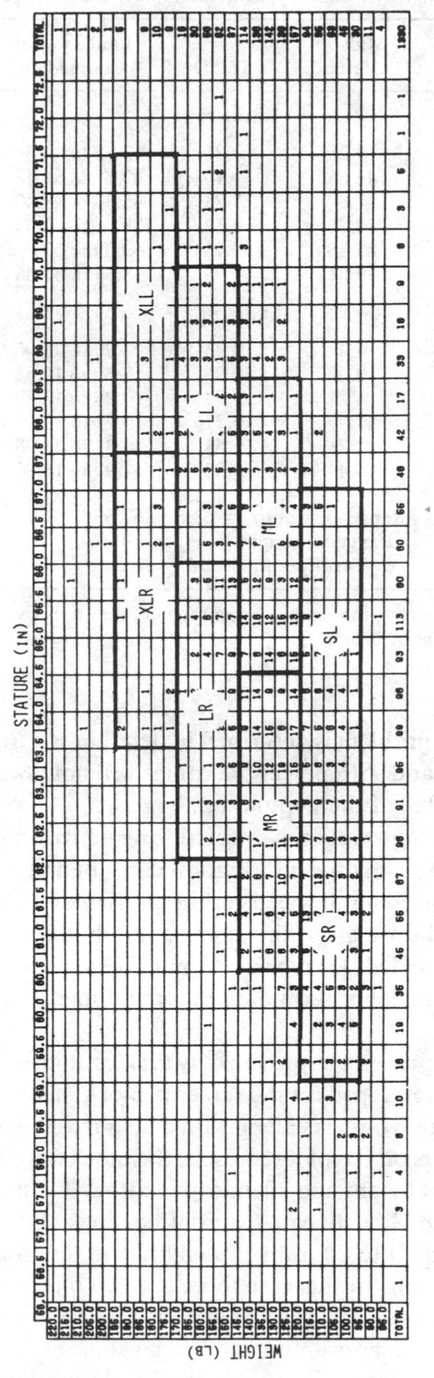

FIG. 3—*A bivariate frequency plot of stature and weight values with an eight-size system displayed.*

TABLE 1—*Eight-size system for the total body, Size X-large regular.*

Variable Name[a]	Midsize Value	SZ-SD	Range to Be Accommodated	Recommended
		HEIGHTS		
Ankle height	4.25	0.36	3.65 to 4.84	4.39
Axilla height	49.66	1.10	47.85 to 51.47	51.32
Bust point height	47.37	1.23	45.34 to 49.40	49.96
Crotch height	30.50	1.12	28.65 to 32.34	31.62
Stature	65.50	1.15	63.59 to 67.41	67.50
Waist height	40.61	1.24	38.56 to 42.66	42.01
		LENGTHS		
Acromion-axilla	4.48	0.35	3.90 to 5.06	4.64
Axilla to waist	9.23	0.96	7.66 to 10.81	9.45
Neck to bust point	11.24	0.70	10.08 to 12.39	11.53
Sleeve inseam	17.99	0.77	16.72 to 19.27	18.57
Sleeve outseam	21.71	0.83	20.34 to 23.08	22.43
Waist back	16.49	0.92	14.97 to 18.00	16.92

Range for stature = 63.5 to 67.5 in.
Range for weight = 170 to 195 lb
Tariff per thousand = 16.5

[a]Units are inches. To convert from inches to centimeters, multiply by 2.54. To convert from pounds to kilograms, divide by 2.205.

somewhere else. This problem is described in detail in Ref *3*. Contrary to popular belief, using 5th and 95th percentile values will not always fulfill the goal of accommodating 90% of the population.

A separate set of design values is selected for each size category for all the important dimensions, using the information described previously. The selection of the values is based on the needs of the particular design problem. The design values selected for a given size represent only the people contained in that size and are not scaled down or up from other sizes. One way to use the system is to create a model or form based on the dimensions from each category.

To illustrate the improvement provided by this technique, an anthropometric sizing system was developed for the same six blocks shown in Figs. 1 and 2. Shown in Fig. 4 are the six blocks along with a height and width bivariate plot of them. The large dots on the plot represent the blocks; for example, Block *a* is the dot at 10 units of height and 20 units of width. Also shown are three size categories, X, Y, and Z, delineated by dotted lines. Category X contains blocks less than or equal to 20 units in height, and less than or equal to 20 units in width, and so on. In this case there are two blocks in each category, and they are circled for clarity. The key dimension design values are the largest for each category. Therefore, for Size Y the values are 20 and 30 units for height and width, respectively. Since only two blocks exist per category, there

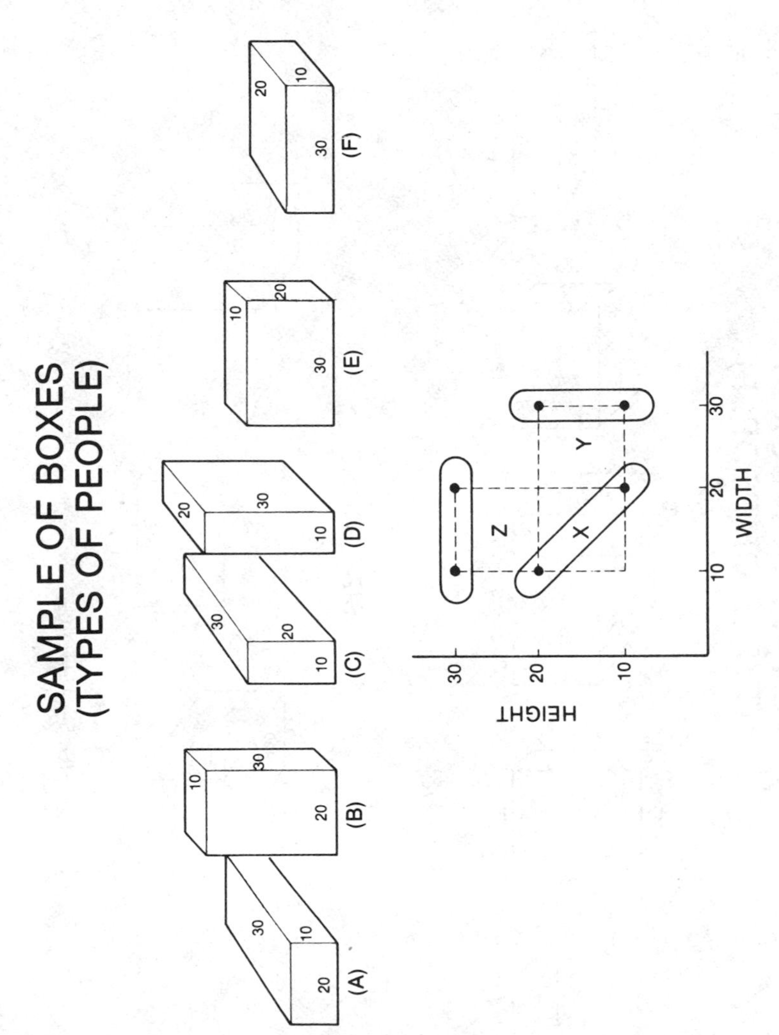

FIG. 4—*A bivariate plot of height and width with a three-size system displayed.*

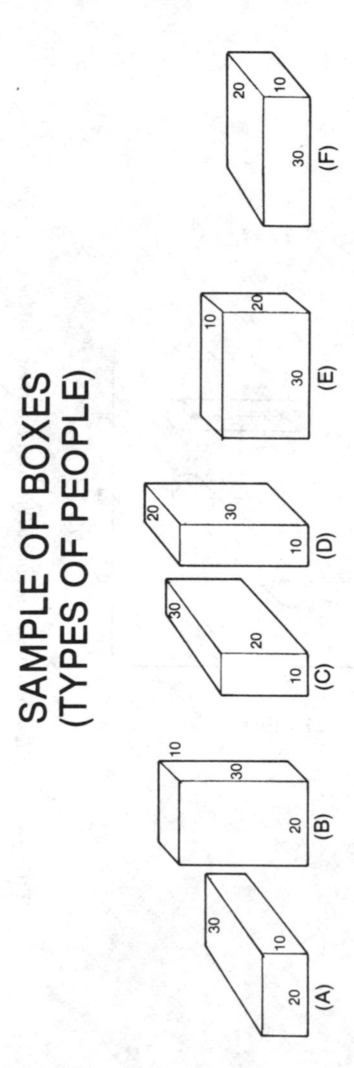

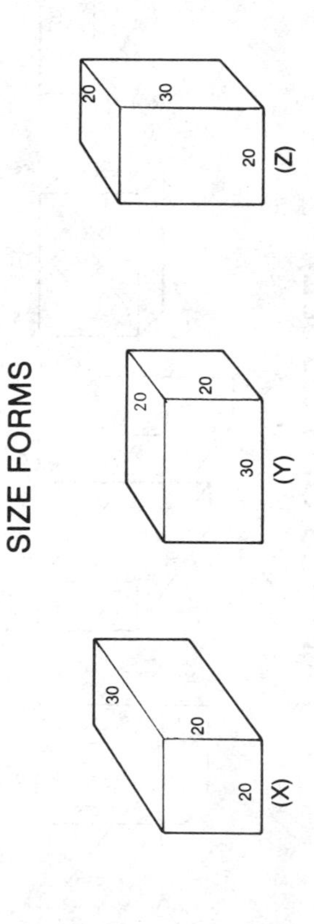

FIG. 5—*A three-size ''anthropometric'' sizing system.*

are no 5th or 95th percentile values. The best variance estimate for the third dimension is simply the range. Regression values are also not reasonable to use in this case, because of the small sample size. Therefore, the best design values for the third dimension is simply the top of the range. The resulting forms for each size are shown in Fig. 5.

Size X fits Blocks *a* and *c*, Size Y fits Blocks *e* and *f*, and Size Z fits Blocks *b* and *d*, all within a total of ten units of gap. Using this system, with the same number of sizes, all of the blocks are covered, and they cover with less error than the scaled Medium or the scaled Greek god sizes, with the exception of the fit of the Greek god in the Greek god size. After all, you cannot beat a custom fit!

In addition to improving the quality of fit, an anthropometric sizing system has several other advantages. Both the manufacturer and the buyer will have a good idea about the number of people who will fit in each size. They can use this information to determine how many of each size to produce or purchase. Table 2 is a list (tariff) of the approximate number of women per 1000 who would wear each size in the eight-size program described earlier. This is equivalent to the approximate number of items to order per size.

Another advantage to this system is that the individual user can identify relatively easily which size to wear. Most people who will wear the item will be able to determine their sizes by measuring themselves rather than by trial and error. For example, in the previous eight-size system, they need know only their stature and weight values.

Conclusion

This paper demonstrates how an anthropometric system can result in items that fit more people better. This means that employees will be safer and more productive in the workplace. It also means that waste will be cut considerably.

TABLE 2—*Eight-size system for the total body, categories and tariff.*[a]

Size	Stature, in.	Weight, lbs	Tariff per Thousand, No.
Small Regular	59.0 to 63.0	95.0 to 120.0	152
Small Long	63.0 to 67.0	95.0 to 120.0	101
Medium Regular	60.5 to 64.4	120.0 to 145.0	303
Medium Long	64.5 to 68.5	120.0 to 145.0	218
Large Regular	62.0 to 66.0	145.0 to 170.0	107
Large Long	66.0 to 70.0	145.0 to 170.0	93
X-Large Regular	63.5 to 67.5	170.0 to 195.0	17
X-Large Long	67.5 to 71.5	170.0 to 195.0	9

[a]To convert from inches to centimeters, multiply by 2.54. To convert from pounds to kilograms, divide by 2.205.

First of all, if items are more realistically proportioned, huge sizes will not be necessary just to make room for single larger body proportions. Second, sizes that do not fit anyone are eliminated. Third, this system makes it easier to determine what sizes people wear, and how many of each size to make and order.

References

[1] Daniels, G. S., "The 'Average Man'?," TN WCRD 53-7, Aerospace Medical Research Laboratory, Aerospace Medical Division, Air Force Systems Command, Wright-Patterson Air Force Base, OH, 1952.

[2] Robinette, K. M., Churchill, T., and McConville, J. T., "Anthropometric Sizing Systems for Army Women's Field Clothing," Technical Report Natick/TR-81/026, U.S. Army, Natick Research and Development Laboratories, Natick, MA, 1981.

[3] Robinette, K. M., and McConville, J. T., "Alternatives to Percentile Models" in *1981 SAE Transactions*, Society of Automotive Engineers, Warrendale, PA, 1982, pp. 938–946.

Claire C. Gordon[1]

Anthropometric Sizing and Fit Testing of a Single Battledress Uniform for U.S. Army Men and Women

REFERENCE: Gordon, C. C., "**Anthropometric Sizing and Fit Testing of a Single Battledress Uniform for U.S. Army Men and Women,**" *Performance of Protective Clothing, ASTM STP 900*, R. L. Barker and G. C. Coletta, Eds., American Society for Testing and Materials, Philadelphia, 1986, pp. 581-592.

ABSTRACT: Women in the U.S. Army currently wear field uniforms that were designed and sized for men. Downsizing the male patterns to fit women has accentuated the fitting problems caused by sexually dimorphic body proportions. Thus, female body dimensions must be considered in the design and sizing of Army field clothing. The integrated sizing system for men and women presented here was derived mathematically from the U.S. Army's anthropometric data base. Key sizing dimensions were established by selecting those variable pairs which minimize within-size variations of other body dimensions important to garment design. Design values for additional body dimensions critical to garment fit were calculated from regression equations that used the key sizing variables to predict the dimensions of the largest individual who will wear each size. Appropriate ease was added to these nude body dimensions to arrive at an empirically defined grading system. Fitting trials will test the hypothesis that this anthropometric sizing program can provide an adequate fit for both Army men and women.

KEY WORDS: anthropometry, clothing sizing, fit testing, protective clothing

With increasing numbers of women entering the U.S. Army and assuming a wider variety of roles than ever before, there is clearly a need to provide field clothing, personal protective equipment, and work spaces that accommodate the body sizes and, more important, the body proportions of both men and women. The majority of U.S. Army personnel currently wear a two-piece work uniform, the battledress uniform (BDU), originally designed and sized for men, which was later scaled down to smaller and shorter sizes for women

[1]Research anthropologist, U.S. Army Natick Research and Development Center, Natick, MA 01760.

to use (see Fig. 1). These down-sized male shirts and trousers retain their inherently male proportions and therefore do not fit women well in those body dimensions that are most sexually dimorphic. When an Army woman is issued a shirt that is appropriate for her chest size, it does not fit properly over the hips; when a shirt covers the woman's hips, the shoulders are often too big. Similarly, when the trousers are issued to Army women, sizes that accommodate their hips have waist and inseam dimensions that are too large.

A major logistical burden and a substantially increased cost to the taxpayer would be associated with the introduction of two separate field uniform systems for men and women. Thus, anthropometric sizing methods have been applied to derive a single sizing system that will accommodate both men and women. This paper reports on progress to date in the development and testing of two integrated sizing systems for the U.S. Army's battledress uniform—one for the shirt and one for the trousers.

Derivation of the Sizing Systems

As in any anthropometric sizing study, the first step was locating an appropriate data base from which to derive the sizing systems and design values. Body dimensions from the U.S. Army's two most recent surveys of Army personnel were used in the derivation of the shirt and trouser sizing systems—the 1966 survey of 6682 Army men [*1*] and the 1977 survey of 1330 Army women [*2*].

The second step was the selection of key sizing dimensions for the shirt and trousers. Key sizing dimensions should partition the user population into sizing subgroups whose members are similar to one another in the body dimensions most important to garment fit. In choosing one or two dimensions for sizing a garment, one thus seeks to minimize within-size variation of the user population. The best key sizing dimensions to choose are those that result in the smallest average ranges for other critical body dimensions.

Using the average ranges to be accommodated as a measure of within-size variation, three candidate pairs of key dimensions were compared for the integrated sizing system of the shirt: (1) chest circumference and stature, (2) chest circumference at armscye and stature, and (3) shoulder circumference and stature. The integrated ranges to be accommodated for each system were contrasted with male-only ranges for the standard chest circumference/stature sizing system in order to estimate the "cost" in fit of integrated sizing.

As can be seen in Table 1, all three pairs of key dimensions produced size groupings of the male and female user population that were relatively homogeneous for linear dimensions. This indicates that stature is a good key dimension for controlling linear variation in the upper body of integrated user populations. The three pairs of key dimensions differed, however, in within-size variation of circumferential dimensions. The traditional key dimensions for U.S. Army work shirts, chest circumference and stature, produced

FIG. 1—*The U.S. Army battledress uniform.*

TABLE 1—*Average ranges to be accommodated for critical dimensions in shirt sizing systems.*[a]

Critical Dimension	Male Standard, Stature/Chest Circumference	Sex-Integrated Proposed		
		Stature/Chest Circumference	Stature/Chest Circumference at Armscye	Stature/Shoulder Circumference
		LENGTHS		
Shoulder-elbow	1.8	2.0	2.0	2.0
Shoulder height	4.2	4.2	4.2	4.2
Sleeve inseam	2.6	2.7	2.7	2.6
Waist back	4.1	4.3	4.2	4.1
Interarmscye	3.4	3.5	3.5	3.9
Shoulder	2.5	2.5	2.5	2.5
		CIRCUMFERENCES		
Armscye	3.2	4.6	4.0	3.6
Chest	2.9	2.9	2.9	7.8
Hip	5.2	8.3	9.4	10.7
Neck	2.2	3.5	3.2	2.8
Shoulder	4.8	7.7	6.1	3.0
Waist	6.9	8.2	7.0	8.2
Wrist	1.0	1.5	1.4	1.3

[a]All values are expressed in inches; 1 in. = 25.4 mm.

within-size ranges for such sexually dimorphic dimensions as shoulder and hip circumference, which were 160% of those present in the standard system. This indicates that grouping men and women of similar chest sizes does not guarantee that other upper body circumferences will be similar. Although the chest circumference at armscye key dimension slightly reduced within-size variation for many circumferential dimensions, the within-size variation of shoulder circumference was still 127% of that present in the standard sizing system. However, with shoulder circumference as a key dimension, sex-integrated fit at the shoulder greatly improved, and chest and hip circumferences were the only dimensions adversely affected. Since shoulder circumference is more critical to the fit of this style of work shirt than chest and hip circumference, shoulder circumference and stature were selected as key dimensions in the single sizing system of this item.

Similar methods were employed in the selection of key sizing dimensions for the trousers (see Table 2). When the traditional sizing system—crotch height and waist circumference—is applied to an integrated user population, it is apparent that within-size variation for hip and thigh circumference is much too large. When hip circumference is substituted for waist circumference as a key dimension, hip and thigh variations are well controlled, but the average range to be accommodated in waist circumference more than doubles. This situation is preferable to that of larger hip and thigh ranges, how-

TABLE 2—*Average ranges to be accommodated for critical dimensions in trouser sizing systems.*[a]

Critical Dimension	Male Standard, Crotch Height/ Waist Circumference	Sex-Integrated Proposed	
		Crotch Height/ Waist Circumference	Crotch Height/ Hip Circumference
	Heights		
Calf	3.0	3.0	3.0
Crotch	2.9	2.9	2.9
Knee	3.2	3.2	3.3
	Circumferences		
Calf	2.8	3.0	3.2
Hip	4.9	9.0	2.9
Thigh	4.4	7.0	3.8
Waist	3.8	3.8	11.6

[a]All values are expressed in inches; 1 in. = 25.4 mm.

ever, because waist circumferences can be altered by adjustment straps, whereas hip and thigh circumferences cannot be adjusted easily. Hip circumference and crotch height were thus chosen as key dimensions for the single-sized trousers.

In order to contrast potential key dimensions, a 20-size system was arbitrarily used for computing the within-size variation because this closely approximates the 22-size system present in the standard BDU. However, answering the question, "How many sizes?," is always an important step in the derivation of an anthropometric sizing system. Obviously, the more sizes there are, the less within-size variation there will be, and the better the garment will fit. But one rapidly reaches a limit beyond which increasing the number of sizes does not improve garment fit enough to justify the logistical complications and additional taxpayer's burden that ensue. Again, the magnitude of within-size variation can be used to compare different sizing systems and identify the optimum number of sizes required. Integrated sizing systems with 15, 20, and 25 size categories were compared for use with the U.S. Army's work shirt and trousers. A 20-size system significantly reduced the within-size variation from that of a 15-size system for both shirt and trousers. However, the 25-size systems did not significantly reduce within-size variation in comparison with the 20-size systems [*3*]. Thus, a 20-size system was chosen for both the shirt and the trousers.

Figures 2 and 3 present the overall structure of the anthropometric sizing systems derived for the U.S. Army's single sizing program. Tables 3 and 4 present the relative frequencies of men and women expected to wear each size garment based on data from the 1966 and 1977 Army surveys. The shirt sizing program provides a theoretical coverage of 98% of the males and 96% of the

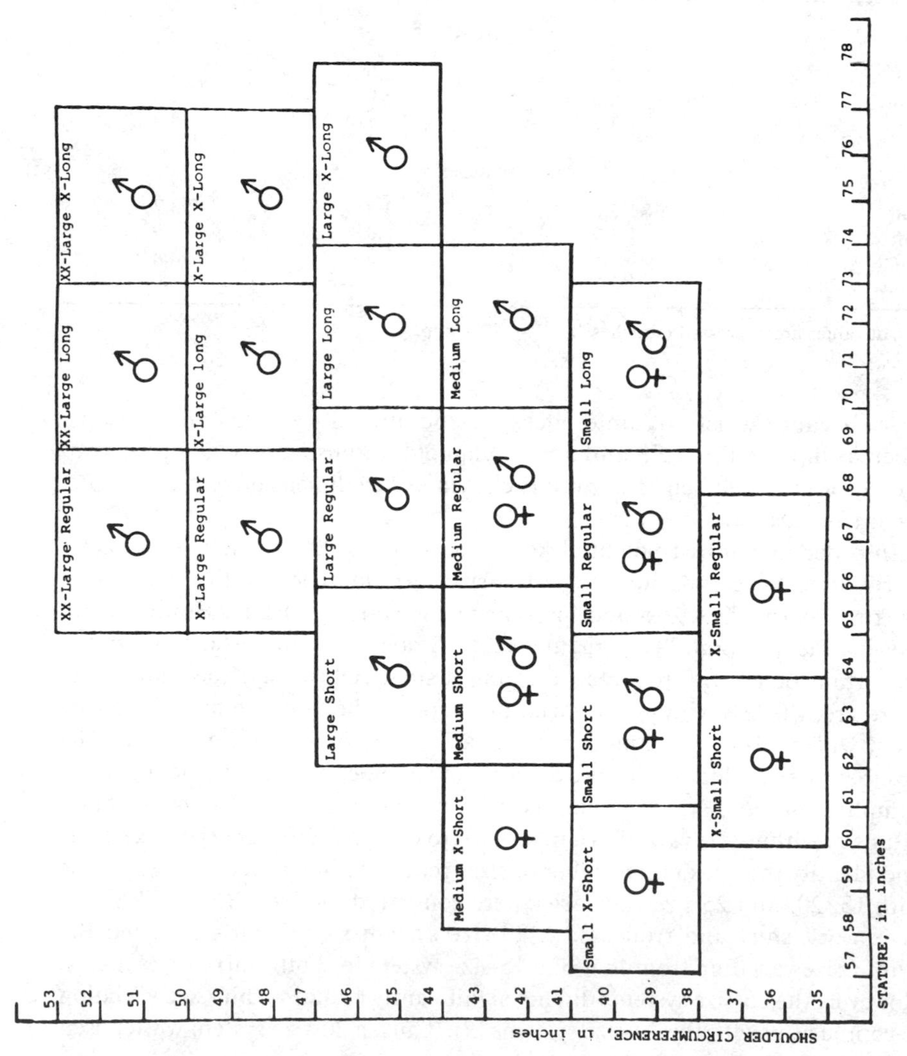

FIG. 2—*A 20-size system for the battledress uniform shirt (1 in. = 25.4 mm).*

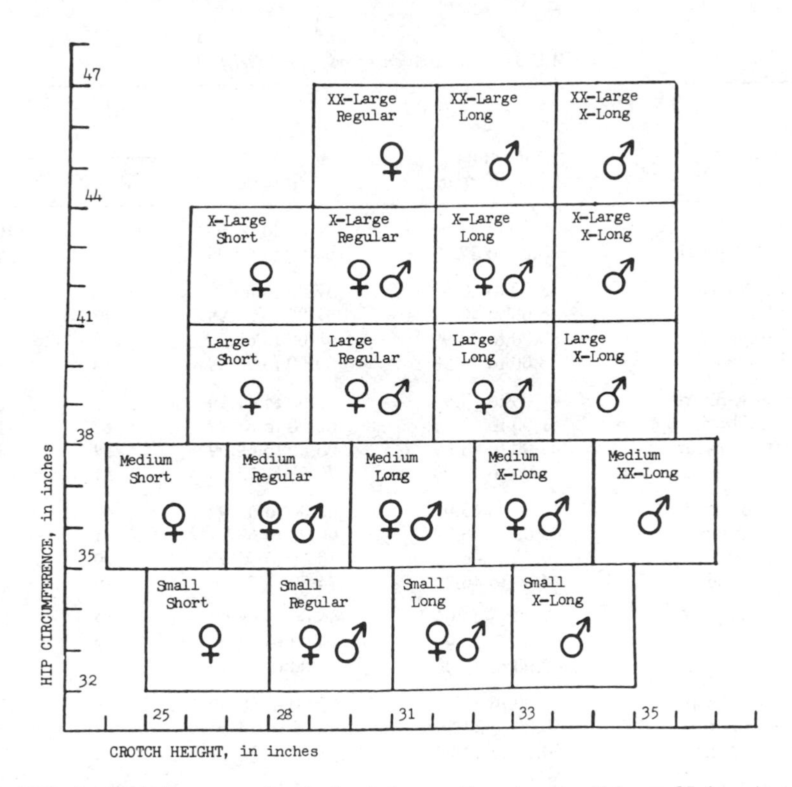

FIG. 3—*A 20-size system for the battledress uniform trousers (1 in. = 25.4 mm).*

females with 20 sizes; the trouser sizing program provides theoretical coverage of 98% of each sex with 20 sizes.

As can be seen in Table 3, the representation of each sex varies considerably among the size categories. In fact, the single sizing system for the shirt is composed of three separate groups of sizes: four of the smallest, shortest sizes will be worn primarily by women, eleven of the largest, longest sizes will be worn primarily by men, and five average sizes will be worn by both men and women. These three shirt size groupings are identified by gender in Fig. 2. A similar, though less pronounced, situation exists for the trouser sizing system presented in Table 4 and Fig. 3. There are five women's size categories, six men's size categories, and nine integrated size categories. These three size groupings have important consequences for determining garment shapes and will be referred to further in the next section.

Construction and Grading of Master Patterns

After determining the structure of the single sizing systems, anthropometric design values were generated for each size category of the shirt and of the

TABLE 3—*Shirt size categories and tariff.*[a,b]

Size	Shoulder Circumference	Stature	Tariff per Thousand, No.	
			Males	Females
X-Small Short	35.00 to 37.99	60.00 to 63.99	0	146
X-Small Regular	35.00 to 37.99	64.00 to 67.99	0	80
Small X-Short	38.00 to 40.99	57.00 to 60.99	0	45
Small Short	38.00 to 40.99	61.00 to 64.99	12	281
Small Regular	38.00 to 40.99	65.00 to 68.99	38	190
Small Long	38.00 to 40.99	69.00 to 72.99	14	15
Medium X-Short	41.00 to 43.99	58.00 to 61.99	0	22
Medium Short	41.00 to 43.99	62.00 to 65.99	62	131
Medium Regular	41.00 to 43.99	66.00 to 69.99	222	70
Medium Long	41.00 to 43.99	70.00 to 73.99	81	7
Large Short	44.00 to 46.99	62.00 to 65.99	47	6
Large Regular	44.00 to 46.99	66.00 to 69.99	231	7
Large Long	44.00 to 46.99	70.00 to 73.99	129	0
Large X-Long	44.00 to 46.99	74.00 to 77.99	10	0
X-Large Regular	47.00 to 49.99	65.00 to 68.99	47	0
X-Large Long	47.00 to 49.99	69.00 to 72.99	72	0
X-Large X-Long	47.00 to 49.99	73.00 to 76.99	13	0
XX-Large Regular	50.00 to 52.99	65.00 to 68.99	6	0
XX-Large Long	50.00 to 52.99	69.00 to 72.99	12	0
XX-Large X-Long	50.00 to 52.99	73.00 to 76.99	4	0
Total			1000	1000

[a]All values except tariffs are expressed in inches; 1 in. = 25.4 mm.
[b]Theoretical coverage of the male population, 98.3%; theoretical coverage of the female population, 95.6%.

trousers. Anthropometric design values are the relevant nude body dimensions of the largest individual who will be expected to wear each particular size and serve as guidelines for the clothing designer and grader. To create an anthropometric model for each size that is proportionally correct, multiple regression equations were used to predict the nude body dimensions of each individual whose key dimensions were the same as the upper limits of each size category. Unlike percentiles, which are not additive, design values calculated from multiple regression equations can be combined without distorting normal body proportions.

Table 5 presents the anthropometric design values for several sizes of shirt. In the Small X-Short size, design values were set by regression equations calculated from female body dimension data because this size will be worn almost exclusively by women. Similarly, because the X-Large Long size will be worn almost exclusively by men, those design values were set by regression equations derived from male data. The Medium Regular size will be worn by

TABLE 4—*Trouser size categories and tariff.*[a,b]

Size	Hip Circumference	Crotch Height	Tariff per Thousand, No.	
			Males	Females
Small Short	32.00 to 34.99	25.00 to 27.99	0	26
Small Regular	32.00 to 34.99	28.00 to 30.99	32	79
Small Long	32.00 to 34.99	31.00 to 33.99	115	19
Small X-Long	32.00 to 34.99	34.00 to 36.99	40	0
Medium Short	35.00 to 37.99	24.00 to 26.99	0	15
Medium Regular	35.00 to 37.99	27.00 to 29.99	19	240
Medium Long	35.00 to 37.99	30.00 to 32.99	220	189
Medium X-Long	35.00 to 37.99	33.00 to 35.99	231	18
Medium XX-Long	35.00 to 37.99	36.00 to 38.99	26	0
Large Short	38.00 to 40.99	26.00 to 28.99	0	69
Large Regular	38.00 to 40.99	29.00 to 31.99	61	205
Large Long	38.00 to 40.99	32.00 to 34.99	148	68
Large X-Long	38.00 to 40.99	35.00 to 37.99	44	0
X-Large Short	41.00 to 43.99	26.00 to 28.99	0	14
X-Large Regular	41.00 to 43.99	29.00 to 31.99	13	38
X-Large Long	41.00 to 43.99	32.00 to 34.99	33	15
X-Large X-Long	41.00 to 43.99	35.00 to 37.99	11	0
XX-Large Regular	44.00 to 46.99	29.00 to 31.99	0	5
XX-Large Long	44.00 to 46.99	32.00 to 34.99	5	0
XX-Large X-Long	44.00 to 46.99	35.00 to 37.99	2	0
Total			1000	1000

[a]All values except tariffs are expressed in inches; 1 in. = 25.4 mm.
[b]Theoretical coverage of the male population, 98.3%; theoretical coverage of the female population, 98.5%.

both males and females. Females set the chest, waist, and hip dimensions because, given equal shoulder circumferences and stature, women are larger for these dimensions than men. All the other design values are set by male-derived regressions because, given equal stature and shoulder circumferences, men are larger than women for these dimensions. Design values such as those in Table 5 were then used by clothing designers for master pattern construction and grading.

In a traditional (single-gender) sizing system, one would ordinarily request a single master pattern, usually a size in the center of the bivariate distribution. The shape of this master pattern would then be carried throughout all sizes by using standard pattern grading techniques. With an integrated sizing system, however, the center of the bivariate distribution is an integrated size that is essentially a compromise in shape between men and women. To grade this unusual shape up into the male-only sizes and down into the female-only sizes would unnecessarily compromise the fit and appearance of the shirts

TABLE 5—*Anthropometric design values for three sizes of shirt.*[a,b]

Dimensions	Small X-Short	Medium Regular	X-Large Long
Key dimensions			
Shoulder circumference	41	44	50
Stature	61	70	73
Critical dimensions			
Shoulder elbow length	12.6[f]	14.8[m]	15.5[m]
Shoulder height	49.8[f]	57.7[m]	60.5[m]
Sleeve inseam length	16.8[f]	19.5[m]	20.4[m]
Waist back length	15.4[f]	17.9[m]	19.0[m]
Interarmscye	15.2[f]	15.2[m]	17.2[m]
Shoulder length	5.8[f]	6.4[m]	7.0[m]
Armscye circumference	15.1[f]	17.4[m]	19.5[m]
Chest circumference	36.3[f]	38.9[f]	41.8[m]
Hip circumference	38.1[f]	42.2[f]	41.5[m]
Neck circumference	12.9[f]	14.6[m]	15.9[m]
Waist circumference	29.5[f]	32.0[f]	36.8[m]
Wrist circumference	5.8[f]	6.7[m]	7.2[m]

[a]All values are in inches; 1 in. = 25.4 mm.
[b]Key to abbreviations:
[f] = derived from female regressions.
[m] = derived from male regressions.

and trousers in these essentially gender-specific regions of the single-sizing systems. For this reason, three master patterns were requested instead of the usual one: a female master pattern in size Small X-Short, a male master pattern in size X-Large Long, and an integrated master pattern in size Medium Regular. This approach was taken in both the shirt and the trouser sizing systems.

Anthropometric design values for each master size were then submitted to clothing designers for the construction of master patterns. As mentioned previously, anthropometric design values are the nude body dimensions of the largest individual expected to wear each size. Thus, some amount of ease must be added to the anthropometric values during pattern construction for personal comfort and freedom of movement. Quantification of the ease needed for each critical anthropometric dimension is not as simple as it sounds, since the specification of nude versus ease values for every critical garment dimension is not a part of the industry's pattern drafting tradition. As a working hypothesis, exactly the same amount of ease for each critical dimension was added as is present in the standard battledress uniform sizing system. These values were derived empirically by predicting the anthropometric dimensions of the largest male expected to wear each standard BDU size and subtracting them from the analogous finished shirt and trouser dimensions.

After master pattern construction, computerized pattern grading was applied to each pattern to produce the full range of sizes in the shirt and trouser sizing systems. Grading increments for adjacent sizes were obtained by subtracting the garment design values of the smaller or shorter size from the larger or longer size for every critical dimension specified.

Fit Testing of Shirts and Trousers

A fit test of garments made over these patterns is planned for early 1985 at one of the U.S. Army's training posts where large numbers of test participants may be conveniently obtained. The goal of the fit test will be to quantify the fit characteristics of the garments so that statistical methods may be applied to the test data that will result in systematically derived, quantitative recommendations for pattern modification. Essential to this process will be the acquisition of anthropometric and tailoring data on test participants whose body dimensions represent the full range of body sizes and shapes present in the user population.

The fitting trials will proceed in three phases: (1) collection of body measurements on the test participants, (2) collection of qualitative and quantitative static fit data on the test participant wearing his or her best-fitting size, and (3) collection of qualitative and quantitative fit data with the test participant performing a predetermined series of body movements to simulate dynamic fit characteristics of the garments. Dimensional data on the test garments themselves will be obtained prior to the actual test as part of a quality control examination. Administration of the fit test will involve close cooperation between Natick Research and Development Center's anthropologists, psychologists, and clothing designers.

With data in hand on predicted and actual garment sizes, wearers' body dimensions, and quantitative fit characteristics of the best-fitting sizes, patterns and pattern grading may be modified as necessary to improve the fit of the integrated garments. When pattern modification is complete, a more extensive wear test of the integrated sizing garments is planned. In the wear test, Army men and women will be issued their best-fitting sizes in both the standard and the integrated battledress uniforms. The standard and integrated shirts and trousers will be worn alternately during field exercises. Quantitative evaluations of fit and user feedback will then be used to determine whether or not the sex-integrated sizing systems are an improvement over the standard shirt and trouser sizing systems.

Conclusions

This application of statistically derived design values to master pattern construction and grading is the most extensive of its kind to date. Through close interaction between clothing designers and anthropologists, anthropometric dimensions and landmarks that are analogous to garment dimensions have been identified for use in pattern drafting and grading. Perhaps more important, many anthropometric dimensions previously provided to designers have been discarded because they are not analogous to garment dimensions, and new ones have been used in their place. Early results in the single-sizing program have also pointed out the need to quantify the ease required for each critical body dimension and how this ease may vary with body size. Finally, in implementing the sizing programs described here, we have begun to quantify systematically what heretofore has been a jealously guarded art: the dimensional interaction between human body size and shape and garment size and shape. While these methods may never replace the skills and experience of an expert clothing designer for custom fit, there is little doubt that they have broad application in the ready-to-wear clothing industry, both military and civilian.

Acknowledgments

The original formulation of a single sizing program for the U.S. Army's battledress uniform was done by Anthropology Research Project Inc., Yellow Springs, Ohio, under contract to the U.S. Army. The late Robert Manson White served as contract monitor for the integrated sizing research and was also the Natick Research and Development Center's anthropologist during the 1966 and 1977 anthropometric surveys that form the data base used in this project. Rosemary Lomba served as project officer for the integrated battledress uniform and contributed significantly to the close working relationships between anthropologists and clothing designers so necessary to this project. William Amico and Barbara Quinn, of the Clothing and Uniform Division at Natick, contributed many ideas on the special problems of grading an anthropometric sizing system. Dr. Carolyn Bensel, chief of the Human Factors Group, provided scientific and technical guidance in this project, as well as critical comments on this manuscript. Critical comments and valuable editorial suggestions for improving this manuscript were also offered by Charles Williams, chief of the Life Support Systems Division; Leonard Flores, chief of the Armor Section; and Edna Albert, technical publications editor of the Technical Library Division. Figures 2 and 3 were drawn by Linda Ng of New York University.

References

[*1*] White, R. M. and Churchill, E., "The Body Size of Soldiers: U.S. Army Anthropometry—1966," Technical Report 72-51-CE (AD 743 465), U.S. Army Natick Laboratories, Natick, MA, 1971.

[*2*] Churchill, E., Churchill, T., McConville, J. T., and White, R. M., "Anthropometry of Women of the U.S. Army—1977; Report No. 2, The Basic Univariate Statistics," Technical Report NATICK/TR-77/024 (AD A044 806), U.S. Army Natick Research and Development Command, Natick, MA, 1977.

[*3*] McConville, J. T., Robinette, K. M., and White, R. M., "An Investigation of Integrated Sizing for U.S. Army Men and Women," Technical Report NATICK/TR-81/033 (ADA 109 496), U.S. Army Natick Research and Development Laboratories, Natick, MA, 1981.

Norman Wilson[1]

Incendiary Behavior of Electrostatic Spark Discharges from Human Beings

REFERENCE: Wilson, N., "**Incendiary Behavior of Electrostatic Spark Discharges from Human Beings,**" *Performance of Protective Clothing, ASTM STP 900*, R. L. Barker and G. C. Coletta, Eds., American Society for Testing and Materials, Philadelphia, 1986, pp. 593-603.

ABSTRACT: An examination has been made of the critical voltages on the body for the ignition of the most easily ignitable mixtures of methane and air and of hydrogen and air by electrostatic spark discharges from the body to earthed electrodes of optimum dimensions.

At normal temperatures, the critical body voltages for the ignition of methane and hydrogen are 5.5 and 1.3 kV, respectively. The effects on the critical body voltage of increasing the temperature of the gas and the electrode independently to up to 80°C indicate that the primary factor controlling the ignition of methane is the temperature of the electrode, which in the present work is the cathode. An increase in the temperature of the electrode generally causes a fall in the critical body voltage. With the electrode at 80°C, the lowest voltage on the body for an ignition is 4.8 kV.

An increase in the gas and electrode temperatures has little effect on the critical body voltage at which hydrogen is ignited, the lowest value remaining at 1.3 kV.

The author suggests that the importance of the electrode rather than the gas temperature on the critical body voltage, particularly with methane, arises from the effects of electron emission at the electrode on the incendiary behavior of the gas, but this needs to be confirmed.

KEY WORDS: electrostatic sparks, critical body voltage, ignition hazard, methane, hydrogen, minimum ignition energy, temperature effect, electron emission, corona discharge, protective clothing

Electrostatic charges are produced almost continuously on the human body during movement, primarily because of the contact with and separation from the external surfaces of the outer clothing. The electric field from the charge on the clothing is directed mainly toward the body, which can be regarded as

[1]Principal research officer, Apparel and Domestic Textiles, Shirley Institute, Manchester, England M20 8RX.

an electrical conductor, and instantly causes the body to become charged by induction. The effect of the charge can be to raise the voltage on the body to a high value. Charges are also produced at the interfaces between the sublayers of clothing and between the skin and the underwear. However, as such charges occur in equal quantities of opposite polarity and in close proximity to one another, they have little or no effect on the body voltage.

With the body in a charged condition and insulated from earth, it is possible that on approaching an earthed or large conductor sparks may pass directly from the body or from the clothing to the conductor. In the presence of a flammable atmosphere, such sparks can be of sufficient energy to ignite the atmosphere, causing a fire or an explosion. Of the two types of spark, that from the body is likely to be the more hazardous because a large proportion of the available energy can be released at once.

Recognizing that the principal hazard is that of sparks from the body, a number of investigators [*1–6*] have examined the static propensity of garments on the body by charging the outer garment by rubbing it with another material and measuring the voltage induced on the body. The criterion for safety often used is that the energy of the charge induced on the body should not exceed the minimum ignition energy (MIE) of the relevant flammable atmosphere. However, in some tests [*3*] in which coal gas was ignited by a spark from the body, it was found that the energy of the charge on the body at which an ignition occurred was many times greater than the MIE of the gas. It was suggested that one factor contributing to the discrepancy was an instantaneous loss of heat to the metal electrodes (quenching) between which the spark was passed. This effect of quenching occurs when, for a given gas mixture, the distance between the electrodes is of a critically low value. Further work [*7*] with stoichiometric mixtures of coal gas and air and natural gas and air, in which sparks were passed between the finger and an earthed steel electrode, also gave large discrepancies between the critical energy of the charge on the body for an ignition and the corresponding MIEs of the gas mixtures. It was noted that another factor, not previously considered, that could account for a major part of the discrepancy is the effect of the electrical resistance of the body on the distribution of the energy during a discharge. This would be expected to restrict the mobility of the charge (current) and to absorb part of the available charge energy as heat, the rest going to the spark gap.

Experiments with spark discharges from the body to electrodes of various diameters [*8,9*] have shown that for natural gas (methane) the critical parameter that determines whether or not an ignition occurs is the voltage on the body. This critical voltage was found to fall with a decrease in the size of the electrode to which the spark was passed, but with thin electrodes it increased again because of corona discharging [*8*]. It was concluded [*9*] that the discrepancy between the energy of the charge on the body at which an ignition occurred and the MIE of the gas is due largely to the quenching effect of the electrodes. However, from studies of the charge-time profiles of sparks from the body [8,10], the following has been shown: Firstly, the spark discharge consists of a number of discrete sparks in which the first is the largest and alone is likely to cause an ignition. Second, from measurements of the energy released in the spark gap, it has been shown that a sizable proportion of the charge energy released by the body in the first spark is absorbed by the body resistance as heat. The author concluded [10] that the need for a greater charge energy on the body than the MIE of the gas mixture to cause an ignition arises primarily from the fragmentation of the discharge into discrete sparks and the absorption of part of the energy by the body resistance as heat.

Thus, the single important factor to have emerged from the various researches into the ignition of flammable atmospheres by sparks from the body to a given electrode in normal ambient conditions is that the incendivity of the spark depends on the body voltage rather than on the energy of the charge stored on the body, providing, of course, that the latter is greater than the MIE of the atmosphere.

In view of the wide range of ambient conditions in which the vapors from liquid fuels and flammable gases are handled in hospital, industrial, and military situations, it is of considerable importance to establish the effects of gas and electrode temperature on the critical body voltages required for the ignition of flammable atmospheres.

Purpose of the Work

The objective of the present work was to determine the critical voltages on the body required for the ignition of selected flammable gases by spark discharges from the body to earthed steel electrodes, with gas and electrode temperatures of up to 80°C. Combinations of gas and electrode temperatures were examined in order to separate out any individual effects of such changes on the ignition behavior of the gases.

Two gases have been investigated: one is the most easily ignitable mixture of methane and air (8.5% concentration volume/volume), and the other is a stoichiometric mixture of hydrogen and air (29% concentration volume/volume). The MIE of the former is practically the same as that of natural gas, and is in the same region as those of saturated hydrocarbon vapors such as gasoline. Hydrogen gas was included mainly because of its much higher sensitivity to ignition compared with methane.

Ignition Apparatus

A general view of the ignition equipment is shown in Fig. 1. There are three flowmeters and their control valves—A, B, and C in the figure—calibrated for air, hydrogen, and methane, respectively. The selected gas and air are dried by passing them through separate columns of calcium chloride, D in

FIG. 1—*Ignition equipment.*

Fig. 1, and are brought together through concentric tubes at the entrance to a heat exchanger, E, which is mounted on top of the ignition chamber, F. Blow back of flame into the heat exchanger and beyond is prevented by means of metal gauzes placed at both ends of the heat exchanger and at the top of the ignition chamber.

The heat exchanger comprises a brass tube wrapped with a 120-W electrothermal coil. The gas-air mixture passes through the tube, which contains brass infill to ensure adequate mixing. The ignition chamber, F, comprises a box of polycarbonate sheet with a value of about 1.2 L. The gas mixture flows downward through the box and out into a fume cupboard inside which the whole piece of equipment is mounted. Baffles of polycarbonate fixed to the base of the ignition chamber constrict the outlet in order to avoid changes in the concentration of the gas mixture by the diffusion of air into the box.

The gas temperature is controlled by a platinum temperature probe, G, mounted inside the ignition chamber and connected to a Jumo digital display controller, H. This controller is set manually to the required gas temperature and controls to within ± 1 degree Celsius.

The earthed electrode to which the spark is passed is screwed into the electrode mounting, I, which has an annular cavity through which hot or cold ethylene glycol is circulated from a reservoir, J, through a refrigeration unit (not shown in Fig. 1). The temperature of the glycol is regulated by a Thermomix circulator at the reservoir. The temperature of the electrode is displayed digitally by means of a Comark electronic thermometer, K. The thermocouple is inserted into a hole 2 mm in diameter along the axis of the electrode mounting and to within a distance of 2 mm from the tip of the electrode.

When an ignition occurs it is important, for safety, that the combustion of the gas be suppressed as quickly as possible. For this purpose a 'Fireye' flame detector is used which operates through an ultraviolet (UV) sensor, L, which is mounted just below and to one side of the ignition chamber. At the moment the gas is ignited, the detector closes the solenoid valve, M, and stops the flow of gas. The latter is restarted by pressing the button, N, which causes the solenoid valve to open.

A high-voltage direct-current (D-C) supply with a variable output of from 0 to 15 kV, and with an inbuilt safety resistor of 20 M Ω, is used for charging the body to a known potential.

Experimental Procedure

Ignition Tests

The ignition tests were done by the operator standing with the footwear removed on a platform of Perspex, which isolates the body from earth. The procedure was to set the high-voltage supply to a known voltage, as indicated on a static voltmeter. The methane gas and air were turned on and the flow rates adjusted to give the required mixture. The operator raised the potential on the body to the required voltage by touching the high-voltage terminal with the forefinger and then brought the finger near the steel electrode inside the ignition chamber, pausing for a few seconds to ensure that the gas-air mixture was unperturbed before passing the spark. One hundred attempts were made to ignite the gas in this manner, and when no ignition occurred the voltage was increased in steps until one or more ignitions were produced. Preliminary tests with earthed electrodes of different diameters showed that the lowest critical body voltage was achieved when the spark was passed between the finger and the 0.5-mm-diameter electrode. Attempts to ignite the gas by using a 0.3-mm-diameter electrode were unsuccessful owing to corona discharging.

Preliminary ignition tests with hydrogen gas showed that the lowest body voltage at which the gas was ignited was achieved by using pointed electrodes of 0.1-mm diameter, one mounted inside the ignition chamber and the other attached to a brass rod held in the hand.

Ignition tests with each of the gases were done with the earthed electrodes at 0, 20, 40, 60, and 80°C and with the gases at 25, 40, 60, and 80°C.

Minimum Ignition Energies of the Gas Mixtures

The MIEs of the gas-air mixtures were determined by discharging sparks of known energy from a cylindrical air capacitor of 6.3-pF capacitance through the gas mixtures. A low value of capacitance was deliberately chosen to ensure that the required energy can be stored on the capacitor at a potential which is high enough to avoid quenching effects at the electrodes. In Fig. 2 Capacitor A rests on a laboratory jack in a position just below the earthed electrode inside the ignition chamber and was charged by means of the high-voltage supply through a high-voltage decoupling resistor of 160 GΩ (B in Fig. 2). By this means sparks were passed to the earthed electrode at regular intervals of a few seconds. The voltage on the capacitor when a discharge occurred was measured directly by means of a recorder and a static voltmeter (not shown)

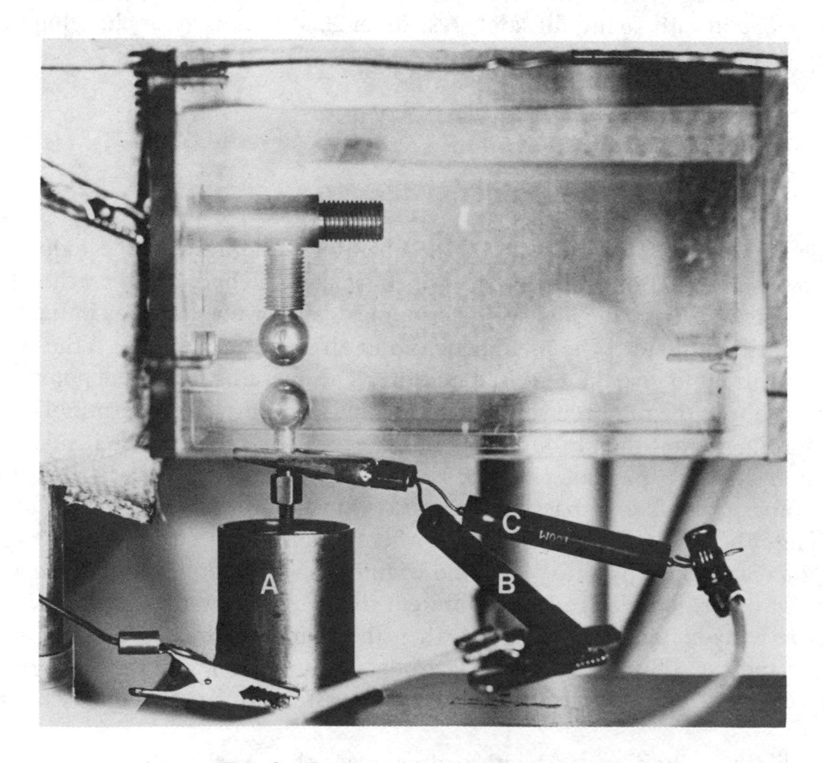

FIG. 2—*Equipment for determining MIE.*

connected to the capacitor by a decoupling resistor of 100 MΩ. The 12-mm-diameter electrodes, shown in Fig. 2, were used with the methane-air mixture and 1-mm-diameter electrodes with the hydrogen-air mixture.

For the purpose of determining the energy of the charge stored on the body when an ignition occurs, it is necessary to know the electrical capacitance of the operator when he stands on the Perspex platform. This was done by charging the operator to a known voltage (3.0 kV) and sharing the charge on the body with a standard capacitor of 1.0-μF capacitance, which is large compared with the capacitance of the body. By measuring the voltage across the standard capacitor, one can calculate the initial charge on the body. The capacitance of the body is given by the ratio of this charge to the value of the initial voltage on the body.

Results and Discussion

The values of the critical body voltage at which the methane-air gas mixture was ignited by sparks from the finger to the 0.5-mm in diameter electrode are given in Table 1.

The accuracy of the individual values is to ±0.1 kV. By referring to Table 1, one can see that in normal temperature conditions (gas at 25°C and electrode at 20°C) the minimum voltage on the body for an ignition of the most sensitive mixture of methane and air is 5.5 kV. The effect on the critical body voltage of increasing the gas temperature from 25°C up to 80°C at a given electrode temperature is hardly significant for electrode temperatures of 0, 20, 40, and 60°C, but shows a decrease when the electrode is at 80°C. The behavior of the critical body voltages at the lower electrode temperatures is, at first sight, surprising since an increase in gas temperature would be expected to cause a fall in the MIE of the gas, resulting in a decrease in the critical body voltage. However, a number of factors must be considered in assessing the effect of gas temperature on the critical body voltage:

(*a*) *Thermal effect on the MIE of the gas*. The MIE of the gas should fall, providing the rise in gas temperature is enough to significantly augment the

TABLE 1—*Critical body voltage for the ignition of methane.*

Electrode Temperature, °C	Electrode Diameter, mm	Critical Body Voltage, kV, at Various Gas Temperatures			
		25°C	40°C	60°C	80°C
0	0.5	5.7	5.9	5.8	5.8
20	0.5	5.5	5.7	5.8	5.7
40	0.5	5.4	5.7	5.3	5.3
60	0.5	5.5	5.4	5.2	5.5
80	0.5	5.5	5.0	4.8	5.0

heat produced by the spark. This, in turn, should lead to a fall in the critical body voltage.

(*b*) *Effect on gas concentration*. The concentration of the most easily ignitable mixture of methane and air is 8.5% volume/volume, and any deviation to the rich or lean side of this mixture will cause an increase in its MIE. In the present work, the gas mixture is heated at constant pressure (atomspheric), and the volumes of the methane and air increase in the same proportion (Charles' law). The volume concentration of methane in the mixture, therefore, remains unchanged, and on this account the MIE of the mixture and the critical body voltage for an ignition of the mixture also remain unchanged.

(*c*) *Effect on the diffusion rates of the gases*. The most easily ignitable mixture of methane and air is slightly to the lean side of the stoichiometric mixture. This is caused by the difference in the diffusion rates of methane and of the oxygen in the air. Methane, being lighter than oxygen, diffuses more rapidly into the combustion zone than does oxygen, thereby increasing the concentration of the burning mixture to slightly above the stoichoimetric value. This effect is compensated by using a leaner gas mixture.

The effect of increasing the gas temperature at constant pressure is to cause a reduction in the densities of the methane and oxygen in the mixture. As the rates of diffusion of the gases are inversely proportional to the square root of their densities (Graham's law), the relative rates of diffusion of the gases into the combustion zone remain unchanged. It follows that no change should occur in the MIE of the mixture or in the critical body voltage for an ignition.

(*d*) *Effect on gas density*. A reduction in the densities of the component gases with an increase in temperature at constant pressure is equivalent to a reduction in the pressure of the gases at constant volume. Work on a stoichiometric mixture of methane and air [*11*] has shown that a reduction in gas pressure causes an increase in the MIE of the mixture and in the critical quenching distance between the electrodes. Both of these effects are expected to cause an increase in the critical body voltage at which the gas is ignited.

Thus, of these four factors, only Factors *a* and *d*, individually, should affect the critical body voltage, and they are in opposition to each other. By referring to Table 1, one can see clearly that the largest effect on critical body voltage arises from an increase in the temperature of the electrode. In all cases, particularly at gas temperatures above 25°C, the general trend is for a decrease in the critical body voltage which with an increase in the temperature of the electrode.

A similar effect was observed for the measured MIEs of the gas mixture, as is shown in Table 2. An increase in gas temperature of from 25 to 80°C has little or no effect on the MIE of the gas mixture at a given electrode temperature. These results imply that the counteracting effects of Factors *a* and *d* are either small individually or tend to cancel each other. However, a rise in the temperature of the electrode to which the spark is passed of from 20 to 80°C, produces a significant fall in the MIE of the gas. Because the electrode in

TABLE 2—*Minimum ignition energy of methane.*

Electrode Temperature, °C	Minimum Ignition Energy, mJ, at Various Gas Temperatures	
	25°C	80°C
20	0.30	0.31
80	0.26	0.26

these tests is the cathode, it is possible that electron emission from the electrode will play a part in the discharge mechanism, providing the field, or potential gradient, at the surface of the electrode is large enough. Heating the electrode to 80°C might cause an electron emission similar, at a lower potential gradient, to that with the electrode at 20°C, resulting in a lower minimum ignition energy.

The fall in MIE of the gas mixture at 25°C with an increase in the electrode temperature (see Table 2) is not accompanied by a fall in the critical body voltage at the same gas temperature, for reasons which are not understood.

The electrical capacitance of the operator doing the tests is 92 pF, and the corresponding energy of the charge on the body at 4.8 kV, the lowest critical body voltage observed in these tests (Table 1), is 1.06 mJ. This is approximately four times the MIE of the gas mixture (0.26 mJ), and, as was discussed earlier, this situation arises from the fact that only a small proportion of the energy reaches the spark gap, the rest being lost through spark fragmentation and losses to the body resistance as heat [*8,10*].

The critical body voltages at which the hydrogen-air mixture was ignited by sparks from the pointed hand-held electrode to a similar electrode inside the ignition chamber are shown in Table 3.

By referring to Table 3 one can see that for normal temperature conditions, that is, the gas at 25°C and the electrode at 20°C, the critical body voltage for the ignition of the gas mixture is 1.3 kV. In general, the effect of increasing the gas temperature at a given electrode temperature or of increasing the elec-

TABLE 3—*Critical body voltage for the ignition of hydrogen.*

Electrode Temperature, °C	Electrode Diameter, mm	Critical Body Voltage, kV, at Various Gas Temperatures			
		25°C	40°C	60°C	80°C
0	0.1	1.4	1.4	1.4	1.6
20	0.1	1.3	1.4	1.4	1.4
40	0.1	1.3	1.4	1.4	1.5
60	0.1	1.3	1.4	1.4	1.4
80	0.1	1.4	1.3	1.3	1.3

trode temperature at a given gas temperature on the critical body voltage is small. Similarly, the values of the MIEs of the gas mixture determined in normal temperature conditions (0.0223 mJ) and with the gas and electrode at 80°C (0.0237 mJ) indicate a relatively small effect of gas and electrode temperature on the ignition behavior of hydrogen compared with methane.

The same factors that determine the incendiary behavior of methane should also apply to hydrogen, except that with the latter the diffusion rates of hydrogen and oxygen into the combustion zone are slow compared with the rate of flame propagation in the mixture. As a result of this, the most easily ignitable mixture of hydrogen and air is practically the same as the stoichiometric mixture. It follows from the critical body voltage and MIE results that the counteracting Factors *a* and *d* discussed earlier are either small or tend to cancel each other. Further, any effects of electron emission at the electrode on the incendiary behavior of the gas must also be small, if not absent, possibly because of a low potential gradient at the electrode.

The energy of the charge on the body of the operator at 1.3 kV, the lowest critical body voltage at which the hydrogen-air mixture was ignited, is 0.078 mJ. This is approximately 3.5 times the MIE of the gas mixture, which indicates that, as with methane, a significantly higher charge energy than the MIE of hydrogen is required on the body to ignite hydrogen by a spark from the body.

Conclusions

Ignition tests with the most easily ignitable mixture of methane and air by sparks from the body to an electrode of optimum diameter have shown that, with the gas and electrode at normal temperatures, the critical voltage on the body for an ignition is about 5.5 kV. Similar tests with the gas and electrode temperatures set independently to values of up to 80°C show that, except with the electrode at 80°C, gas temperature has practically no effect on the critical body voltage.

The factor that primarily controls the critical body voltage is the temperature of the electrode to which the spark is passed when the latter is the cathode. An increase in the temperature of the electrode generally causes a fall in the critical body voltage. With the electrode at 80°C, the lowest voltage on the body at which ignitions occur is 4.8 kV, and the corresponding stored charge energy is about four times the MIE of the gas mixture.

The MIE of the methane-air mixture is also practically independent of gas temperature up to 80°C but falls significantly with an increase in electrode temperature of from 20 to 80°C.

The importance of electrode rather than gas temperature on the incendiary behavior of methane is thought to be due to electron emission from the electrode.

Similar ignition tests with a stoichiometric mixture of hydrogen and air by sparks from the body to an electrode of optimum diameter show that the effects of increasing the gas or the electrode temperature, or both, on the critical body voltage are generally small. The lowest voltage on the body to ignite the mixture is about 1.3 kV, and the corresponding energy of the charge on the body is about 3.5 times the MIE of the gas mixture.

The effect of electrode temperature, particularly on the incendiary behavior of methane is of considerable fundamental and practical importance in view of the variety of metal surfaces to which sparks may be passed in practice. Further work on the present lines with electrodes of different metals and alloys, and including tests at subzero temperatures, would provide further valuable information on the hazardous aspects of static charges on the body and on the physical factors governing the incendivity of spark discharges from human beings.

Acknowledgments

The author acknowledges with pleasure the help of T. Haigh of the Shirley Institute in constructing the ignition apparatus and in doing some of the experimental work. My thanks are also due to the U.K. Ministry of Defence for sponsoring a larger program of work, of which the present paper forms a part.

References

[1] Guest, P. G., Sikora, V. W., and Lewis, B., "Static Electricity in Hospital Operating Suites," Bulletin 520, U.S. Department of the Interior, Bureau of Mines, Washington, DC, 1953.

[2] Potter, A. E., Jr. and Baker, B. R., "Static Electricity in the Apollo Spacecraft," NASA Technical Note TN-5579, Manned Spacecraft Center, Houston, TX, 1969.

[3] Henry, P. S. H., "Risks of Ignition due to Static on Outer Clothing," *Static Electrification*, Institute of Physics Conference Series, No. 11, 1971, p. 212.

[4] Bajinskis, G. and Lott, S. A., "Electrostatic Hazards from Insulated Operators," Report 521, Defence Standards Laboratories, Australian Defence Scientific Service, Maribyrnong, Victoria, Australia, 1972.

[5] Wilson, L. G. and Cavanagh, P., "Electrostatic Hazards due to Clothing," Report No. 665, Defence Research Establishment, Physics and Test Section, Ottawa, Canada, 1972.

[6] Orner, G. M., "Static Propensity of Navy Clothing," Technical Report No. 109, Navy Clothing and Textile Research Unit, Natick, MA, 1974.

[7] Wilson, N., "The Risks of Fire or Explosion due to Static Charges on Textile Clothing," *Journal of Electrostatics*, Vol. 4, No. 1, 1977, p. 67.

[8] Wilson, N., "The Nature and Incendiary Behaviour of Spark Discharges from the Body," *Static Electrification*, Institute of Physics Conference Series No. 48, 1979, p. 73.

[9] Tolson, P., "The Stored Energy Needed to Ignite Methane by Discharges from a Charged Person," *Journal of Electrostatics*, Vol. 8, Nos. 2 and 3, 1980, p. 289.

[10] Wilson, N., "The Ignition of Natural Gas by Spark Discharges from the Body," *Static Electrification*, Institute of Physics Conference Series No. 66, 1983, p. 21.

[11] Lewis, B. and von Elbe, G., *Combustion, Flames and Explosion of Gases*, Academic Press, London, 1961, p. 328.

Summary

Summary

The International Symposium on the Performance of Protective Clothing, on which this publication is based, may have been the most important forum on protective clothing in recent years. Its purpose was to bring together, for the first time, all those interested in protective clothing for occupational exposures, to present the findings of relevant research and, perhaps most important, to stimulate discussion and further development efforts.

The response to the symposium was overwhelming, not only in the numbers of participants and diversity of interests but also in the expertise of the authors. Further symposia are planned.

The papers in this volume are organized into sections by topic area. The sections are arranged to distinguish between chemical protection, thermal protection, and other related topics that are generally important to the performance of protective clothing.

Summaries of the papers follow.

Permeation Resistance of Chemical Protective Clothing Materials

Test Methodology

Davis, Feigley, and Dwiggins set the tone for technical discussions on permeation test methods by reviewing the performance of glove materials challenged by liquified coal. In carrying out their work, these authors compared performance measured by a test method using flame ionization detection with performance measured by a test method using radio-labeled, single-species detection. The data show that test methods that do not specifically quantify the compounds of toxicological significance can provide misleading information on the degree of protection provided. A method responding to all volatile components of a mixture may yield results that differ substantially from results of a method responding to a single substance only.

Perkins and Ridge studied the utility of infrared spectroscopy as the detection methodology in the ASTM Test for Resistance of Protective Clothing Materials to Permeation by Hazardous Liquid Chemicals (F 739-81). The authors concluded that infrared (IR) detection provides the ability to monitor a permeation test constantly without the need to take periodic, discrete samples. IR systems are relatively stable and do not require a full-time attendant. Breakthrough time is very easy to discern. Perkins and Ridge acknowledge that there are disadvantages, but the disadvantages may have workable solutions.

The third paper, written by *Spence*, addresses the difficult problem of directly comparing and characterizing the permeation resistance of different protective clothing materials. Spence cites this problem as particularly severe when attempting to compare generic materials containing varying amounts of raw materials and additives. This paper presents a solution: a standard set of permeation tests using nine chemicals selected on the basis of their solubility parameters. The rationale and usefulness of such a standardized test battery are discussed.

Recent technological advances have resulted in the use of absorptive fabrics in protecting against low levels of toxic vapors. In this final paper on test methodology, *Baars, Eagles, and Emond* describe a test apparatus and method designed for evaluating the performance of such absorptive fabrics. Carbon tetrachloride at a concentration of 5.0×10^{-6} g/cm^3 was used as the standard challenge in an apparatus based on a design given in the military specification for cloth, laminated, nylon tricot knit, polyurethane foam laminate for chemical protection (MIL-C-43858). Permeated carbon tetrachloride was detected by using flame ionization.

Performance Data

Henry utilized ASTM Test F 739-81 to generate data on the permeation resistance of glove or ensemble materials exposed to chlorine gas, liquid hydrogen cyanide, and 37% formaldehyde solution. In his paper, Henry presents his performance results, but also ties the final selection process into an evaluation of the resistance of materials to physical hazards.

Forsberg, Olsson, and Carlmark then present a model testing program for meeting performance requirements of gloves to be used against metal cutting fluids in workshops and in steel handling processes. Chromium and nickel particulates can be carried through some glove materials by cutting fluids. The author combines permeation resistance with data on puncture resistance and tensile strength.

In his paper, *Mellström* carries the use of permeation testing a step further. He describes a data base combining permeation resistance data with medical and industrial hygiene information to provide an up-to-date source for assessing risk of chemical exposure. Currently in an experimental stage, the data base is available for use on a microcomputer through the Swedish National Board of Occupational Safety and Health.

A more detailed analysis of the permeation phenomenon is provided by *Schlatter and Miller* in their paper on material thickness versus permeation resistance. The authors, through testing five industrial chemicals against different thicknesses of five unsupported glove materials, conclude that a change in thickness is more likely to affect breakthrough time than the steady-state permeation rate.

In the last paper in this section, *Winter* describes a system developed for archiving permeation resistance data in a computer data base. This system is intended to provide the safety and health professional with rapid access to performance data in graphical form. Further, the professional would be able to compare directly different clothing materials against a single chemical or different chemicals against a single clothing material.

Resistance to Pesticides—Field Performance and Cleaning Procedures

Field Performance

Worker exposure to agricultural chemicals is just now emerging as an area of critical concern to safety and health practitioners. Although exposures have always existed, the geographical diversity and nature of farm workers themselves have contributed to a slow recognition of the extent and seriousness of the problem.

In the first two articles of this section on pesticides and protective clothing, the authors review the current state of the use and integrity of protective clothing. *Nielsen and Moraski* take a historical perspective to help identify monitoring and research needs pertaining to the agricultural worker. Priorities for a protective program, under the guidance of the U.S. Environmental Protection Agency, are discussed. *Rucker, McGee, and Chordas* then explore attitudes and practices of California pesticide applicators. The three coauthors identify growers as the group at highest risk because of their use of more hazardous substances, combined with a superficial understanding of the need to take precautions.

Branson, Ayers, and Henry prepared a paper describing their laboratory evaluation of seven protective clothing fabrics. The effectiveness of the fabrics was determined by exposing them to carbon-14 labeled solutions of commercially used pesticides. Their results indicate that some currently worn fabrics do not provide as much protection as commonly believed.

Lloyd follows this work in the United States with a description of a far-reaching study conducted in the United Kingdom. The author reviews a three-step process involving field and laboratory data gathering aimed at the development of national and international standards of performance for protective clothing used in pesticide spraying.

In another paper on the effectiveness of protective clothing, *Laughlin, Easley, and Hill* explore the impact of various fabric parameters and pesticide characteristics on permeation and penetration. The fabric parameters investigated were fiber content and functional finishes, while the pesticide characteristics were concentration and formulation. The work evolved as an effort to address a growing concern about accidental spilling of concentrated or diluted pesticides onto protective garments used by home gardeners, commercial applicators, and field workers.

Finally in this section, *Hobbs, Oakland, and Hurwitz* look at the effects of barrier finishes on the pesticide penetration and comfort of woven and disposable nonwoven fabrics used by pesticide applicators. A laboratory test procedure was developed to evaluate the resistance of fabrics to oil-based and water-based pesticide carriers used as aerosol sprays.

Cleaning Procedures

Considerable research emphasis has been placed on decontamination of protective clothing after exposure to and contamination by pesticides. Both home and commercial laundering are currently in wide use, but with questionable effect.

Keaschall, Laughlin, and Gold evaluated the amount of residue remaining on or in three fabrics exposed to each of eleven pesticides. The fabrics were all identical cotton/polyester blends, but one was unfinished, the second was treated with a renewable consumer-applied fluorocarbon finish, and the third was treated with a commercially applied fluorocarbon finish. The fabrics were laundered using one of three treatments: a heavy-duty liquid detergent, a heavy-duty liquid detergent with a prewash spray, and a heavy-duty liquid detergent with an agriculturally marketed pretreatment. The resulting data show that residues after laundering vary considerably.

In a similar research program, *Leonas and DeJong* evaluated how functional finishes and different levels of laundering affect a fabric's ability to prevent or inhibit pesticide penetration. Three fabrics—cotton, cotton/polyester, and polyester—treated with selected functional finishes—durable-press, soil-release, and water-repellent finishes—were laundered 0, 10, 30, and 50 times. All the specimens were exposed to methyl parathion spray. Again, there was a wide range in measured performance.

Obendorf and Solbrig used electron microscopy to determine the relative distribution of malathion and methyl parathion on cotton/polyester unfinished and durable-press fabrics before and after laundering. In their paper, the authors conclude that laundering can remove up to 70% of each pesticide from the surfaces of the fibers, but has little effect on the pesticide accumulated within the lumen of cotton fibers. Results with the durable-press-finished fabric did not differ significantly from those for the unfinished fabric.

Risk Assessment of Chemical Exposure Hazards in Selecting Protective Clothing

Most chemical protective clothing demonstrates some level of permeation by toxic chemicals. A steady-state rate of release with the garment material acting as a toxic reservoir is possible after breakthrough of a chemical. The question therefore arises: What is an acceptable level of dermal exposure?

Many variables distinguish the level of risk to a worker. These include the toxicity of the chemical or chemicals, the absorption into the skin, and the rate and location of contact. Confounding the evaluation of an acceptable level of skin exposure to toxic chemicals are the pragmatic issues of combining costs with overall benefits, including reuse.

This section of the STP deals with the need to assess risks and costs concurrently.

Mansdorf, in the lead paper in this section, suggests the need for a series of recommended dermal exposure levels and a field-validated method for quantitatively measuring exposures inside protective garments, including gloves. If developed, these suggestions would directly lead to a correlation between the steady-state permeation rate after breakthrough and acceptable dermal exposure levels.

As a means of measuring the absorption of organic solvents through the skin, *Boman and Wahlberg* describe an animal model intended to replace human experimentation. In their paper, the authors demonstrate that the animal model is also applicable to testing the protective effect of glove materials.

Schwope has prepared a paper that provides background information for those with little or no knowledge of chemical permeation through the skin. He presents an overview of the structure of the skin, methods for assessing skin permeation, and examples of published data.

Coletta and Spence then continue in the theme of risk assessment by describing five key elements important to the success of a chemical protective clothing management program. These elements form the basis for an ongoing risk/benefit analysis intended to ensure maximum employee protection within constraints imposed by operating and business considerations. The paper concludes with a case study.

A paper by *Forsberg* groups protective clothing materials into several different classes or categories on the basis of permeation data. Categories are proposed, based on three intervals of breakthrough time and five levels of steady-state permeation rate. The author also proposes applying a labeling and classification system to provide ready information for the risk assessment process.

In the last paper on risk assessment, *Berardinelli and Roder* develop the concept of quantitatively testing protective clothing in the field. The two authors present a scenario for evaluating clothing materials' resistance to degradation, permeation, and penetration by liquid chemicals.

Testing the Chemical Resistance of Seams, Closures, and Fully Encapsulated Suits

Laboratory evaluation of the chemical resistance of protective clothing materials provides base-level data on expected performance. However, the integrity of complete garments and their accessories before and after use is just as important as that of the base materials.

The first paper in this set, by *Berardinelli and Cottingham*, presents an evaluation of penetration through bulky seams and closures using the ASTM Test for Resistance of Protective Clothing Materials to Penetration by Liquids (F 903-84). Eight different seams and closures were tested against several liquids. Specimens without seams or closures served as controls.

Garland, Goldstein, and Cary describe the use of a test booth for evaluating fully encapsulated chemical suits in a simulated work environment. Included in the test protocol are three exercises simulating typical work activities. A number of suit designs and concepts, including limited use of disposal clothing, have been tested. The results have led to modifications in design to improve performance.

Moore writes about his thesis that one set of use and maintenance guidelines may be developed to aid users of fully encapsulated suits. This may be true regardless of the fact that six or seven manufacturers currently draw upon three or four base materials for their individual designs and configurations. Moore notes that his data show that the differences mostly occur in closure systems, dump valves, glove-to-sleeve joints, and pressure-sealing zippers, rather than in the bodies of the suits themselves.

The "age-old" issue of protective clothing decontamination for reuse after chemical exposure is addressed by *Ashley* in this section's final paper. The author looks at two key issues. The first is a decontamination methodology to remove polychlorinated biphenyls (PCBs) from fire-fighters' protective clothing. The second is a technique for minimizing damage to protective clothing while subjected to the decontamination process. Removal efficiencies up to greater than 99% for PCBs have been achieved using a Freon solvent technique.

Laboratory Measurement of Thermal Protective Performance

A paper by *Hoschke, Holcombe, and Plante* reviews the merits and deficiencies of some existing and proposed test methods for evaluating fabrics intended for protection against radiant heat and flame. Standard test methods for thermal protection of clothing and materials, including British Standard 3791, International Organization for Standardization (ISO) Standard 6942, and the ASTM Test for Thermal Protective Performance of Materials for Clothing by Open-Flame Method (D 4108-82) are considered, as well as the radiant heat test method proposed in Federal Aviation Administration Report FAA-RD-75-176. These test methods are discussed on the bases of the heat source, the design of the heat sensor, and the method of specimen mounting. The authors suggest modifications to improve the testing apparatus and procedures.

In another paper, *Holcombe and Hoschke* describe the role of the protective garment during intense heat exposure and discuss how current laboratory test methods relate to typical heat hazards. They express the concern that current standards pay too little attention to the nature of the working environment. According to the authors, consideration should be given to development of a test to simulate intense radiant heat exposure, which, they argue, better simulates flame exposure incidents than convective heat tests. They believe that in setting performance requirements for protective clothing, it is essential that the relative risks be properly balanced and that undue emphasis not be placed on aspects that are not warranted according to injury statistics. They propose that more attention be devoted to the need for materials that reduce the risk of metabolic heat stress in hot working environments.

Schoppee, Welsford, and Abbott discuss factors affecting the thermal performance of protective clothing. The authors believe that the effect of heat exposure on the strength of fabrics is an important consideration in their protective ability. They show that strength retention during short-term exposure to high-intensity radiant and convective heat depends on the temperature reached at a given instant during exposure. It is independent of the mechanism of heat absorption. Their experiments predict that high-temperature materials such as polybenzimidazole (PBI) and Nomex/Kevlar can provide a few extra seconds of protection to the extreme heat of a large fire. This paper covers problems associated with the use of an instrumental skin-simulant device to determine the rate of conductive heat transfer through fabrics of various kinds. It concludes that ranking fabrics by the maximum temperature achieved in the skin simulant is the only reliable way of using data from this device.

Shalev and Barker describe the use of novel experimental techniques to explain the characteristic responses of PBI, aramid, and flame-resistant cotton fabrics in laboratory tests of thermal protective performance. They show that thermophysical properties change extensively during exposure to intense heat (2.0 $cal/cm^2/s$) in a thermal protective performance (TPP) test. The polymer-to-air ratio emerges as an important variable predicting TPP in flame exposures. The mechanism by which thermal insulation is maintained depends on the fiber or flame-retardant (FR) finish. The authors speculate that the ablative effect of moisture loss, along with retention of the air volume fraction, contributes to improving the thermal protective performance of PBI fabrics in a convective exposure. According to this study, the ability of a fabric to maintain surface fibers impacts significantly in flame heat transfer. The findings presented here demonstrate why vapor-phase active flame retardants, which suppress air gap ignition when treated cotton fabrics are tested, perform better in a TPP exposure than cotton fabrics treated with solid-phase flame retardants.

Schleimann-Jensen and Forsberg describe a new test method for evaluating protective materials in exposures to infrared (IR) radiation. Their test equipment uses an IR thermometer which, without contacting the test specimen, computes the surface temperature by measuring radiation transmitted through the fabric. The authors report that this method is useful as a means of characterizing the emissivity and reflectivity properties of protective materials in radiant heat exposure. They suggest that the method is suited for use as a screening test for materials intended for high-intensity radiation exposures.

Evaluating Materials for Thermal Protective Clothing

Bouchillon summarizes the physical, thermal, and fire protection characteristics of fabrics prepared from Celanese PBI fibers blended with other fibers (for example, FR rayon, aramids). Data are presented on the properties of PBI including flame resistance, thermal stability, chemical resistance, and comfort. Bouchillon describes applications for PBI in industrial protective apparel. He discusses the properties of PBI-containing fabrics measured in laboratory tests intended for predicting performance in high-temperature gloves, foundry and fire proximity equipment, and fire-fighters' protective clothing.

Benisek, Edmondson, Mehta, and Phillips discuss the performance of FR wool and other heat-resistant fabrics in tests against flame and exposure to molten metal. The types of fibers and flame-retardant compounds are shown to play an important role in the transfer of dangerous heat through clothing fabrics. The authors find that fabric construction parameters are important in protection against flames and molten metal. They conclude that moisture plays a large role in thermal protective performance. In multilayer clothing assemblies moisture in outer fabrics increases protection against flame, while the accumulation of moisture (body perspiration) in the inner layer decreases protection against flame. The authors discuss the results of interlaboratory evaluations of two draft ISO test methods for molten metal splash.

The paper by *Baitinger and Konopasek* investigates the thermal protective performance of FR-treated cotton fabrics. The thermal insulative performance of single-layer and multiple-layer fabric assemblies is compared in purely radiant and convective heat exposures ranging from 0.30 to 2.0 cal/cm^2/s in intensity. The authors find that, at 1.0 and 2.0 cal/cm^2/s heat flux levels, fabric weight and thickness are directly related to thermal protection. Air spacing in multiple-layer clothing systems is found to be a significant contributor to insulation.

Mischutin and Brown describe cotton fabrics treated with Caliban flame-retardant finish. The Caliban finish is described as an aqueous dispersion of decabromodiphenyloxide and antimony trioxide. They discuss a new hydrophylic butyl acrylic binder system designed to improve the water vapor and air permeability of Caliban-treated cotton fabrics, without loss of softness or durability. They present data on the physical properties, flame resistance, and performance of Caliban-treated cotton fabrics in a molten aluminum pour test.

Dixit reviews the development of high-temperature gloves made with Zetex, a trademark name for a highly texturized fiberglass fabric. He compares

Zetex fabrics and asbestos fabrics in a laboratory test intended to measure the insulation that gloves provide in handling hot metal pipes. Dixit reports that Zetex gloves, although lighter in weight, provide better insulation than gloves made with asbestos. He compares the abrasion resistance of texturized glass fabric and asbestos fabrics, including performance in a splash test against molten iron.

Clothing Systems for Industrial and Fire-Fighting Applications

Krasny reviews principles of clothing protection in fire situations and gives examples of test results on materials used in heat protective garments. He discusses turnout coats for structural fire fighters and work garments to be worn in areas where accidental fire may occur. He gives examples of measurements of heat protective properties. The materials covered are single layers of fabrics appropriate for work uniforms, the same type of fabric combined with four popular underwear fabrics, and typical fire-fighters' turnout coat assemblies, consisting of a shell fabric, vapor barrier, and thermal barrier.

Jaynes describes a method developed at the Inland Steel Co. to evaluate the resistance of protective clothing materials to molten metal splash. He discusses factors affecting molten metal splash testing and compares the performance of various heat-resistant fabrics in these tests. Jaynes describes, from the viewpoint of a safety engineer, the relationship between laboratory measurements of molten metal splash resistance and protection in an industrial environment.

Veghte discusses factors to be considered in the design of fire-fighters' protective clothing. He reviews the development of heat resistant materials and standards for thermally protective clothing. Veghte reports on a recent tabulation of fire-fighters' injuries and shows the relationship between specific body regions and burn injury. He summarizes data on the temperatures and heat flux levels associated with conditions encountered in fighting structural fires. He presents the results of TPP tests made on a number of fabric ensembles used in fire-fighters' protective equipment. These data are presented to illustrate the merits of matching the thermal responses of the vapor barrier and thermal liner under the outer protective shell. Veghte cites examples illustrating the effect that moisture in the thermal liner component has on the protective insulation of fabric ensembles.

Audet and Spindola describe the U.S. Navy's effort to develop fire-retardant/heat-protective clothing for shipboard personnel. They discuss the use of laboratory material tests including vertical flammability tests and protection time determinations for both flame impingement and radiant heat exposures. They describe field testing involving exposure of garments on instrumental manikins to estimate the extent of burn injury sustained or protection times provided by various clothing ensembles. They find that the usefulness of laboratory tests for predicting clothing performance in actual fire situations is somewhat limited because of the variability of the exposure encountered in a large-scale fuel fire. However, they conclude that laboratory tests are extremely useful as screening tools to limit the number of candidate systems that need to be evaluated under field conditions.

Heat Stress, Fit Testing, and Other Performance Requirements for Protective Clothing

A paper by *Gonzalez, Breckenridge, Levell, Kolka, and Pandolf* addresses the need to reduce the danger of heat exhaustion for persons who must wear impermeable chemical protective garments in hot environments. They suggest that a feasible approach is the use of a wettable cover over the impermeable garment. The theory is that an increase in skin heat loss can occur from such a wet cover over a garment. The authors describe a biophysical approach which allows prediction of the cooling benefits from a wetted cover over an impermeable ensemble.

Davies describes the development and evaluation of a specialized clothing system designed to provide a thermally comfortable environment for workers during access to an advanced gas-cooled nuclear reactor system. The system uses a vortex cooling effect to maintain a comfortable environment inside a pressure suit. The author concludes that this approach provides adequate protection for persons working for extended periods in abnormal environments.

A systematic method for garment fit testing and evaluation is covered in a paper by *McConville*. McConville believes that a major challenge facing designers and manufacturers of protective clothing and equipment is the establishment of a sizing system to accommodate the body size variation of the user population as well as to provide a good fit for individuals. He advocates replacing trial-and-error methods with careful application of anthropometric design data. He provides guidelines for conducting such a fit test. Examples cited in this paper concern protective clothing for military uses. However, the author suggests that the principles could be applied for the sizing of protective apparel for other wearer populations.

Robinette also reports problems with current garment sizing practices and presents a method for sizing intended to improve the situation. Her paper demonstrates how an anthropometric system can result in garments that fit people better. She discusses several benefits to be gained by using a systematic sizing approach for protective clothing. Robinette cites advantages gained in workers' safety and productivity. She lists the benefits associated with realistically proportioned clothing and better information concerning the demand for particular garment sizes.

Gordon demonstrates the application of an anthropometric approach to garment sizing and fit testing. Her paper reports on the development of integrated sizing systems for the U.S. Army's battledress uniforms. She discusses the problems associated with the introduction of two separate field uniform systems for men and women. She shows how anthropometric sizing methods are applied to derive a single sizing system for both men's and women's uniforms.

The danger of fire or explosions caused by electrostatic spark discharge from the human body or from clothing is discussed in a paper by *Wilson*. The purpose of this research was to determine the critical voltages required for the ignition of flammable gases by spark discharges from the body. The author describes experiments with spark discharge from the body to metal electrodes of various diameters and temperatures. He concludes that the temperature of the metal surface to which the spark is passed is the main factor controlling the body voltage required to ignite mixtures of methane and air.

Concluding Remarks

The research discussed in this volume contributes to the body of authoritative literature on protective clothing. The editors hope that these findings will focus additional efforts in this important area, an area that has impact on the safety and health of hundreds of thousands of workers throughout the world.

Roger L. Barker

North Carolina State University School of Textiles, Raleigh, NC 27650; symposium chairman and editor.

Gerard C. Coletta

Risk Control Services, Inc., Tiburon, CA 94920; symposium chairman and editor.

Indexes

Author Index

Subject Index

B

F

G

H

I

K

L

M

T

U

V

W

Z